Plant Responses and Tolerance to Metal/Metalloid Toxicity

Plant Responses and Tolerance to Metal/Metalloid Toxicity

Special Issue Editors

Mirza Hasanuzzaman
Masayuki Fujita

MDPI • Basel • Beijing • Wuhan • Barcelona • Belgrade

Special Issue Editors
Mirza Hasanuzzaman
Department of Agronomy,
Faculty of Agriculture,
Sher-e-Bangla
Agricultural University
Bangladesh

Masayuki Fujita
Faculty of Agriculture,
Kagawa University
Japan

Editorial Office
MDPI
St. Alban-Anlage 66
4052 Basel, Switzerland

This is a reprint of articles from the Special Issue published online in the open access journal *Plants* (ISSN 2223-7747) from 2019 to 2020 (available at: https://www.mdpi.com/journal/plants/special_issues/metal_toxicity).

For citation purposes, cite each article independently as indicated on the article page online and as indicated below:

LastName, A.A.; LastName, B.B.; LastName, C.C. Article Title. *Journal Name* **Year**, *Article Number*, Page Range.

ISBN 978-3-03936-198-4 (Hbk)
ISBN 978-3-03936-199-1 (PDF)

Contents

About the Special Issue Editors

Mirza Hasanuzzaman is a professor of Agronomy at Sher-e-Bangla Agricultural University, Dhaka, Bangladesh. He received his PhD in Plant Stress Physiology and Antioxidant Metabolism from the United Graduate School of Agricultural Sciences, Ehime University, Japan, with a scholarship from the Japanese Government (MEXT). Later, he completed his postdoctoral research at the Center of Molecular Biosciences (COMB), University of the Ryukyus, Okinawa, Japan, with a Japan Society for the Promotion of Science (JSPS) postdoctoral fellowship. Subsequently, he became an adjunct senior researcher at the University of Tasmania, with the Australian Government's Endeavour Research Fellowship. Prof. Hasanuzzaman has been devoting himself to research in the field of Crop Science, particularly focusing on Environmental Stress Physiology, since 2004. Prof. Hasanuzzaman has published over 100 articles in peer-reviewed journals and books. He has edited 13 books and written 35 book chapters on the important aspects of plant physiology, plant stress responses, and environmental problems in relation to plant species. These books were published by internationally renowned publishers. Prof. Hasanuzzaman is a research supervisor for undergraduate and graduate students and has supervised 20 MS students so far. He is an editor and reviewer for more than 50 peer-reviewed international journals and the recipient of the Publons Global Peer Review Award 2017, 2018 and 2019. He is an acting editor of Plants—MDPI. Prof. Hasanuzzaman is an active member of about 40 professional societies and is Publication Secretary of the Bangladesh Society of Agronomy. He has been honored by different authorities due to his outstanding performance in different fields like research and education. He received the World Academy of Science (TWAS) Young Scientist Award 2014. He has attended national and international conferences in different countries and has presented 25 papers and posters at these conferences (in the USA, the UK, Germany, Australia, Japan, Austria, Sweden, Russia, etc.)

Masayuki Fujita is a professor at the Laboratory of Plant Stress Responses, Faculty of Agriculture, Kagawa University, Kagawa, Japan. He received his BSc in Chemistry from Shizuoka University, Shizuoka, and MAgr and PhD in Plant Biochemistry from Nagoya University, Nagoya, Japan. His research interests include physiological, biochemical, and molecular biological responses based on secondary metabolism in plants under various abiotic and biotic stresses; phytoalexin, cytochrome P450, glutathione S-transferase, and phytochelatin; and redox reactions and antioxidants. In the past decade, his works have been focused on oxidative stress and antioxidant defense in plants under environmental stress. His group investigates the role of different exogenous protectants in enhancing antioxidant defense and methylglyoxal detoxification systems in plants. He has supervised 4 MS students and 13 PhD students. He has about 150 publications in journals and books, and has edited 10 books.

Preface to "Plant Responses and Tolerance to Metal/Metalloid Toxicity"

In the industrial era, toxic metal stress is one of the major concerns for the environment. Fast industrialization and the growing population coincide, resulting in the generation and dissemination of huge amounts of toxic metals in the environment. Toxic metals and metalloids result in growth reduction, altered physiology, and metabolisms, and diminishing cellular integrity. Toxic metals also alter redox homeostasis by influencing reactive oxygen species generation. They also interfere with nutrient and water uptake. As a result, both crop yield and quality are hampered.

Considering these effects, plant biologists, breeders, and agronomists are working hard to find approaches to enhancing plants' tolerance to metal/metalloid toxicity. Remarkable progress has been made in understanding metal chelation, antioxidant defense, and phytoremediation. A number of biotechnological tools and molecular approaches also contribute to such developments.

We are very pleased to present this Special Issue, which aims to exhibit research on plant responses and tolerance to various metals/metalloids. In this Special Issue, 15 research articles and 4 reviews have been published by expert researchers in the field of plant physiology, environmental sciences, soil science, agricultural sciences, plant biotechnology, plant molecular biology and other related areas. The editors have made a special effort to bring together research on both the morphological and molecular responses of plants to metal/metalloid stress, as well as different mechanisms to make plants more tolerant to these adversities. These articles are different but complementary perspectives on plant–metal/metalloid interaction.

I would like to give special thanks to the authors for their outstanding and timely work and for producing such high-quality articles. We are highly thankful to the reviewers for their rigorous evaluation of manuscripts and their effort in improving the articles. Thanks to Shuang Zhao and other Assistant Editors of Plants for their prompt handling of the manuscript.

Mirza Hasanuzzaman, Masayuki Fujita
Special Issue Editors

plants

MDPI

Article

Modulation of Cadmium Tolerance in Rice: Insight into Vanillic Acid-Induced Upregulation of Antioxidant Defense and Glyoxalase Systems

M.H.M. Borhannuddin Bhuyan [1,2], Khursheda Parvin [1,3], Sayed Mohammad Mohsin [1,4], Jubayer Al Mahmud [5], Mirza Hasanuzzaman [6,*] and Masayuki Fujita [1,*]

1 Laboratory of Plant Stress Response, Department of Applied Biological Sciences, Faculty of Agriculture, Kagawa University, Miki-Cho, Kita-gun, Kagawa 761-0795, Japan; razon_sau@yahoo.com (M.H.M.B.B.); hirasau@gmail.com (K.P.); mohsinsau.ac@gmail.com (S.M.M.)
2 Citrus Research Station, Bangladesh Agricultural Research Institute, Jaintapur, Sylhet 3156, Bangladesh
3 Department of Horticulture, Sher-e-Bangla Agricultural University, Dhaka 1207, Bangladesh
4 Department of Plant Pathology, Sher-e-Bangla Agricultural University, Dhaka 1207, Bangladesh
5 Department of Agroforestry and Environmental Science, Sher-e-Bangla Agricultural University, Dhaka 1207, Bangladesh; jamahmud_bd@yahoo.com
6 Department of Agronomy, Sher-e-Bangla Agricultural University, Dhaka 1207, Bangladesh
* Correspondence: mhzsauag@yahoo.com (M.H.); fujita@ag.kagawa-u.ac.jp (M.F.)

Received: 26 December 2019; Accepted: 1 February 2020; Published: 4 February 2020

Abstract: Cadmium (Cd) is a toxic heavy metal that enters the human food chain from the soil via plants. Increased Cd uptake and translocation in plants alters metabolism andreduces crop production. Maintaining crop yield therefore requires both soil remediation andenhanced plant tolerance to Cd. In this study, we investigated the effects of vanillic acid (VA) on Cd accumulation and Cd stress tolerance in rice (*Oryza sativa* L. cv. BRRI dhan54). Thirteen-day-old rice seedlings treated with $CdCl_2$ (1.0 and 2.0 mM) for 72 h showed reduced growth, biomass accumulation, and water and photosynthetic pigment contents, as well as increased signs of oxidative stress (elevated levels of malondialdehyde, hydrogen peroxide, methylglyoxal, and lipoxygenase) and downregulated antioxidant and glyoxalase systems. Cadmium-induced changes in leaf relative turgidity, photosynthetic pigment content, ascorbate pool size, and glutathione content were suppressed by VA under both mild and severe Cd toxicity stress. The supplementation of VA under Cd stress conditions also increased antioxidant and glyoxylase enzyme activity. Vanillic acid also increased phytochelatin content and the biological accumulation factor, biological accumulation co-efficient, and Cd translocation factor. Vanillic acid, therefore appears to enhance Cd stress tolerance by increasing metal chelation and sequestration, by upregulating antioxidant defense and glyoxalase systems, and by facilitating nutrient homeostasis.

Keywords: abiotic stress; antioxidant defense; metal toxicity; methylglyoxal; oxidative stress; organic acid

1. Introduction

The global human population is increasing rapidly, and feeding growing numbers of people has become a challenging task for farmers. Likewise, plant biologists face related challenges in their search for plant varieties that can produce sufficient food on farmlands that are increasingly prone to abiotic stresses that cause huge crop production losses. Among these abiotic stresses, metal/metalloid toxicity is one of the most common and most deleterious.

Metal/metalloid toxicities are largely a consequence of environmental pollution due to rapid industrialization. One of the most harmful industrial metal pollutants is cadmium (Cd), as it is highly toxic to virtually all life forms, including humans, animals, and plants, and it readily enters the trophic chain primarily via plants [1]. Cadmium is rapidly taken up by plants and accumulates in various tissues due to its very high mobility and hydrophilic nature [2]. Plants that accumulate Cd show stunted growth due to cadmium-induced toxicity, which can also induce chlorosis, epinasty, and disruptions in pollen germination and pollen tube growth. Cadmium stress can also cause alterations in chloroplast ultrastructure, thereby inhibiting photosynthesis and CO_2 fixation, as well as disturbing N and S metabolism [3].

Cadmium is a redox-inactive metal in nature, and thus it does not generate reactive oxygen species (ROS) directly by the Haber–Weiss reaction [4]. Instead, it alters the function of electron transport chains and disrupts antioxidant activities, thereby creating oxidative stress conditions in plants [5]. Cadmium also has a strong affinity for protein side chains and nitrogenous bases, and thus it can damage proteins and nucleic acids to further alter oxidative phosphorylation processes [5]. The generation of ROS in living organisms is an unavoidable consequence of aerobic metabolism. These Cd-induced oxidative stress reactions indirectly lead to overproduction of ROS (singlet oxygen, $^1O_2$; superoxide, $O_2^{\bullet-}$; hydrogen peroxide, H_2O_2; hydroxyl radical, OH) and initiate chain reactions that damage important plant biomolecules [6]. However, plants are sessile organisms and cannot move away from environmental sources of toxic pollutants such as Cd. Instead, they must rely on specially developed biological mechanisms that allow them to tolerate the presence of toxic metals. These mechanisms involve processes for avoidance, exclusion, excretion, binding, chelation, and compartmentalization of Cd [7].

Plants have limited strategies for avoidance of Cd toxicity, but their inherent capacity for dealing with oxidative stress is also not sufficient to cope with metal-induced oxidative damage. The plant's antioxidant defense system consists of antioxidants that are enzymatic (superoxide dismutase [SOD], catalase [CAT], ascorbate peroxidase [APX], monodehydroascorbate reductase [MDHAR], dehydroascorbate reductase [DHAR], glutathione reductase [GR], glutathione-*S*-transferase [GST], and glutathione peroxidase [GPX]) and nonenzymatic (ascorbic acid [AsA], glutathione [GSH], tocopherol, and phenolic compounds). These enzymatic and nonenzymatic antioxidants function to counteract ROS production and serve as cellular redox buffers [6]. Metal/metalloid toxicity, and especially Cd stress, also induces the production of a highly cytotoxic compound, methylglyoxal (MG), which causes structural damage to cellular components, destroys DNA, and creates mutations [8,9]. The damage from MG is diminished in plants through the action of the glyoxylase system, which consists of the glyoxalase I (Gly I) and glyoxalase II (Gly II) enzymes that detoxify MG [9].

Both enzymatic and nonenzymatic antioxidants work simultaneously to combat the deleterious effects of oxidative stress caused by exposure to abiotic factors such as Cd. However, another notable response to several abiotic stresses is a marked accumulation of phenolic compounds [10–14]. Some phenolics have the capability to scavenge ROS directly as a way to reduce oxidative stress [11]. In recent decades, plant biologists have uncovered numerous different phytoprotectants in the form of secondary metabolites, hormones, nutrients, and allelopathic compounds, which all offer advantages to plants under stress conditions [15]. Metabolic manipulation of these compounds in modern agriculture therefore has great potential to improve crop plant tolerance to abiotic stresses such as Cd pollution.

One crop plant that is particularly hampered by Cd toxicity is rice—the most important cereal crop that is consumed by the half of the world community as a staple food [4]. Rice has also been reported to show dramatic stress-related increases in phenolics, vanillic acid (VA) in particular, which increases sevenfold in rice seedlings subjected to flooding [11]. The aim of the present study was therefore to examine the effects of application of exogenous VA on rice seedlings under Cd stress, with a particular focus on the antioxidant defense and glyoxalase systems, as well as nutrient homeostasis during the early seedling stage. To the best of our knowledge, this study is the first to investigate improvements in Cd stress tolerance in rice by VA-induced modulation of ion homeostasis, antioxidant defenses, and the glyoxalase system.

2. Results

2.1. Vanillic Acid Prevented the Suppression of Growth and Biomass Accumulation under Cd Toxicity

Cadmium exposure resulted in a decrease in plant growth, visible as a reduction in plant height and root length as well as a limited relative elongation of the shoot (RSE) and root (RRE). Cadmium toxicity also reduced the overall seedling biomass (Figure 1).

Figure 1. Visual images of the morphological differences in rice (*Oryza sativa* L. cv. BRRI dhan54) seedlings grown under different treatments: VA (50 μM vanillic acid), Cd1 (1.0 mM $CdCl_2$), andCd2 (2.0 mM $CdCl_2$).

Compared to the control condition, Cdstress (1.0 and 2.0 mM) reduced plant height by 14% and 23% and root length by 28 and 33%, respectively. The RSE was reduced by 53% and 79% and the RRE by 87% and 93%, respectively, by 1.0 and 2.0 mM $CdCl_2$ (Table 1). A similar result was also observed for biomass accumulation. However, supplementation of the growth media with exogenous VA in the presence of Cd increased the shoot and root length, their relative elongation percentages, and biomass accumulation under both doses of Cd (Table 1; Figure 1).

Table 1. Effects of exogenous VA on plant height, root length, relative elongation of shoot and root, shoot and root fresh weight, and shoot and root dry weight of rice (*Oryza sativa* L. cv. BRRI dhan54) seedlings under Cd toxicity stress. C, VA, Cd1, and Cd2 indicate control, vanillic acid (50 μM), $CdCl_2$ 1.0 mM, and $CdCl_2$ 2.0 mM, respectively. RSE and RRE indicate relative elongation of shoot and relative elongation of root, respectively. Means (± SD) were calculated from three replications ($n = 3$) for each treatment. Values with different letters are significantly different at $p \leq 0.05$ applying Fisher's least significant difference (LSD) test.

Treatments	Plant Height (cm)	RSE (%)	Root Length (cm)	RRE (%)	Shoot FW (mg plant^{-1})	Root FW (mg plant^{-1})	Shoot DW (mg plant^{-1})	Root DW (mg plant^{-1})
C	15.5 ± 1.1 a	100.0 ± 0.0 b	7.1 ± 0.5 a	100.0 ± 0.0 b	66.4 ± 5.6 ab	36.7 ± 3.1 ab	10.8 ± 0.9 ab	6.6 ± 0.5 ab
VA	16.2 ± 1.3 a	113.5 ± 0.8 a	7.4 ± 0.4 a	106.8 ± 0.9 a	70.5 ± 5.3 a	38.2 ± 2.9 a	11.4 ± 0.8 a	6.8 ± 0.6 a
Cd1	13.4 ± 0.8 bc	46.1 ± 8.9 d	5.1 ± 0.3 bc	20.6 ± 6.2 e	54.5 ± 3.2 c	33.3 ± 1.9 b	8.8 ± 0.5 c	6.0 ± 0.4 b
Cd1 + VA	15.1 ± 0.2 ab	86.6 ± 6.7 c	5.6 ± 0.5 b	41.1 ± 0.4 d	61.3 ± 4.6 bc	36.1 ± 2.5 ab	9.9 ± 0.7 bc	6.5 ± 0.5 ab
Cd2	11.9 ± 1.1 c	12.1 ± 3.1 e	4.7 ± 0.4 c	6.1 ± 1.9 f	58.6 ± 4.5 bc	32.4 ± 2.7 b	9.5 ± 0.8 bc	5.8 ± 0.5 b
Cd2 + VA	15.1 ± 1.0 ab	88.0 ± 2.7 c	5.8 ± 0.5 b	45.9 ± 3.5 c	62.0 ± 3.5 bc	38.2 ± 2.1 a	10.1 ± 0.6 bc	6.8 ± 0.4 a

2.2. Vanillic Acid–induced Accumulation, Translocation, and Detoxification of Cd

Cadmium exposure induced a dose-dependent increase in the Cd content of the rice seedlings. Significant differences were noted in the Cd content of the shoots and roots, and Cd content was much higher in the root than in the shoot. Addition of VA to the growth medium further increased the Cd content in the shoot (Figure 2A) and the root (Figure 2B).

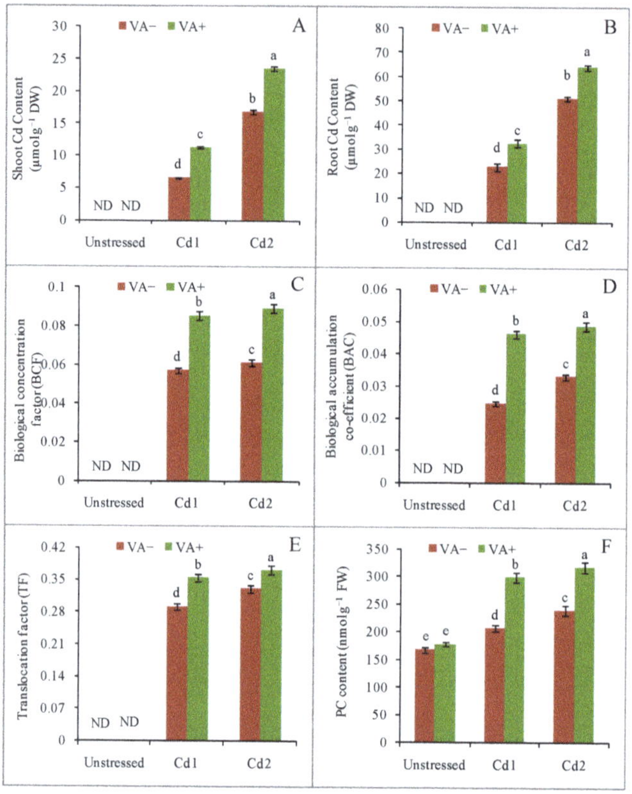

Figure 2. Effects of exogenous VA on Cd content of shoot (**A**), root (**B**), biological accumulation factor (BCF) (**C**), biological accumulation co-efficient (BAC) (**D**), translocation factor (TF) (**E**), and phytochelatin (PC) (**F**) contents of rice (*Oryza sativa* L. cv. BRRI dhan54) seedlings under Cd toxicity stress. Mean (±SD) was computed from three replications of each treatment. Bars with dissimilar letters are significantly different at $p \leq 0.05$ from Fisher's least significant difference (LSD) test.

The biological accumulation factor (BCF), biological accumulation co-efficient (BAC), and Cd translocation factor (TF) also showed dose-dependent increases in response to Cd, and these values were further increased by VA supplementation (Figure 2C–E).

Phytochelatin (PC) content also showed a dose-dependent increase in response to Cd stress. Compared to the control, the PC content increased by 23% and 47% in response to exposure to 1.0 mM and 2.0 mM $CdCl_2$, respectively. Vanillic acid supplementation further increased the PC content at both Cd doses (Figure 2F).

2.3. Vanillic Acid Prevented theLoss ofLeaf Relative Turgidity, Proline Accumulation, and Loss of Photosynthetic Pigment under CdToxicity

Cadmium stress altered the leaf turgidity (RT) of the rice seedlings. Compared to control leaves, leaf RT was reduced by 13% and 18% at mild and severe Cd stress, respectively. Treatment with VA in the presence of Cd suppressed the losses in RT, and thus RT values were higher by 9% and 14%, respectively, compared to the respective stress treatments (Figure 3A). Cadmium toxicity promoted osmolyte accumulation in the rice seedlings leaves, as indicated by a huge increase in proline (pro) content, but treatment with VA suppressed the Cd-induced accumulation of pro (Figure 3B).

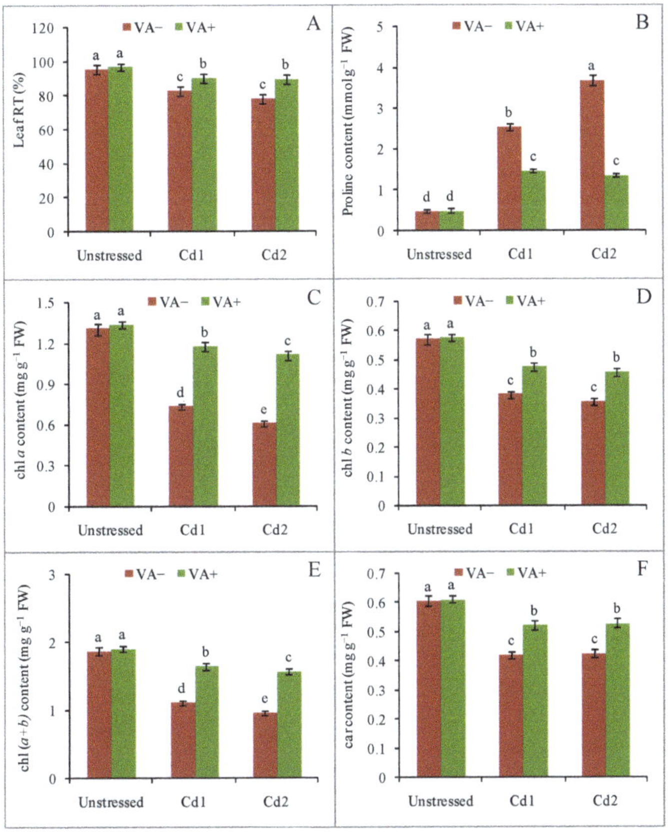

Figure 3. Effects of exogenous VA on leaf turgidity (RT) (%) (**A**), pro content (**B**), chlorophyll (chl) *a* (**C**), chl *b* (**D**), chl (*a* + *b*) (**E**), and carotenoid (car) (**F**) contents of rice (*Oryza sativa* L. cv. BRRI dhan54) seedlings under Cd toxicity stress. Mean (± SD) was computed from three replications of each treatment. Bars with dissimilar letters are significantly different at $p \leq 0.05$ from Fisher's least significant difference (LSD) test.

Cadmium toxicity stress also caused a loss of photosynthetic pigments and induced chlorotic symptoms (Figure 1). Compared to the control seedlings, both mild and severe Cd stress caused a sharp decrease in chlorophyll (chl) *a*, chl *b*, and carotenoids (car), whereas VA treatment in the presence of Cd suppressed these losses of pigments (Figure 3C–F).

2.4. Vanillic Acid Suppressed the Induction of Oxidative Stress Markers under Cd Toxicity

Malondialdehyde (MDA), which is produced as a byproduct of lipid peroxidation, was increased in the leaf tissue at both doses of Cd. Compared to the control seedlings, MDA content increased by 80% and 124% in response to mild (1.0 mM $CdCl_2$) and severe (2.0 mM $CdCl_2$) Cd stress, respectively. However, VA supplementation in the presence of Cd suppressed this increase in MDA content compared with the respective stressed seedlings (Figure 4A).

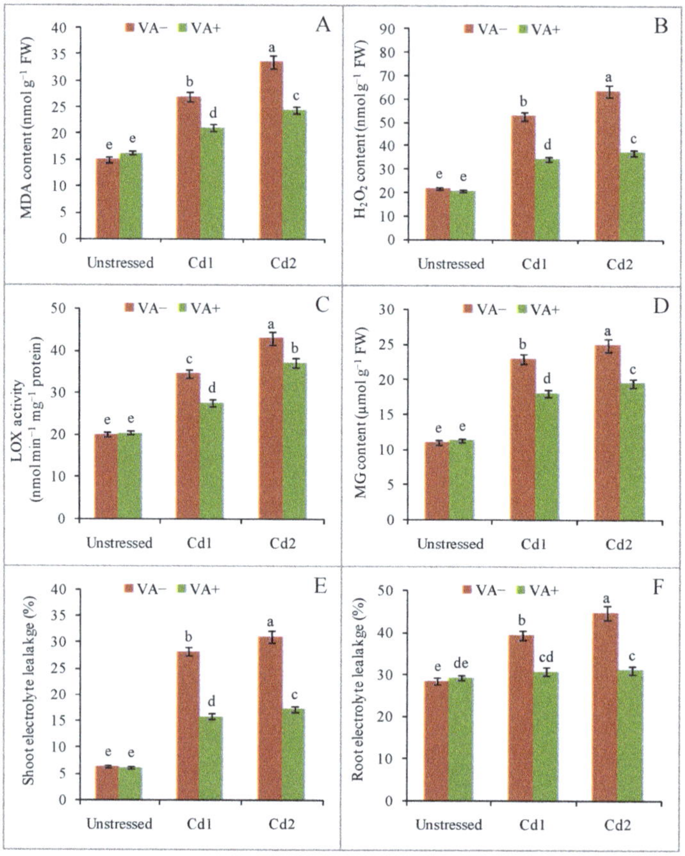

Figure 4. Effects of exogenous VA on malondialdehyde (MDA)(**A**) and H_2O_2 (**B**) content, lipoxygenase (LOX) activity (**C**), methylglyoxal (MG) content (**D**), and electrolyte leakage (EL) of shoot (**E**) and root (**F**) of rice (*Oryza sativa* L. cv. BRRI dhan54) seedlings under Cd stress. Mean (±SD) was computed from three replications of each treatment. Bars with dissimilar letters are significantly different at $p \leq 0.05$ from Fisher's least significant difference (LSD) test.

Similar to MDA, H_2O_2 content increased noticeably in rice leaf tissue upon Cd exposure (Figure4B). Compared with the control seedlings, H_2O_2 content increased by 143% and 193% in mild (1.0 mM $CdCl_2$) and severe (2.0 mM $CdCl_2$) Cd stress-exposed seedlings, respectively. However, VA supplementation in the presence of Cd suppressed this H_2O_2production by 21% and 27%, respectively, compared with the respective Cd-stressed seedlings (Figure 4B).

In line with MDA and H_2O_2 content and lipoxygenase LOX activity increased noticeably in rice leaf tissue upon Cd exposure (Figure 4C). In comparison with the control seedlings, LOX activity increased by 72% and 114% following mild (1.0 mM $CdCl_2$) and severe (2.0 mM $CdCl_2$) Cd stress, respectively. Vanillic acid supplementation in the presence of Cd suppressed this increase in LOX activity by 19% and 13%, respectively, compared with their respective stressed seedlings (Figure 4C). In a similar way, the toxic MG content was increased due to Cd toxicity in a dose-dependent way, which was further suppressed by VA supplementation in the presence of Cd under both doses of Cd toxicity stress (Figure 4D).

Lipid peroxidation caused electrolyte leakage (EL) from the leaf and root tissue. In comparison with control shoots, EL increased by 346% and 391% in the 1.0 mM $CdCl_2$ and 2.0 mM $CdCl_2$ treatments, respectively. Conversely, VA supplementation in the presence of Cd suppressed the EL by 43% and 44%, respectively, compared with their respective stressed seedlings (Figure 4E).

Similarly, the root EL increased by 38% and 57% in 1.0 mM $CdCl_2$ and 2.0 mM $CdCl_2$-exposed seedlings, respectively, whereas VA supplementation in the presence of Cd suppressed EL by 38% and 30%, respectively, compared with their respective cadmium-stressed seedlings (Figure 4F).

2.5. Vanillic Acid Suppressed Changes in Nonenzymatic Antioxidant Content due toCd Toxicity

Among the nonenzymatic antioxidants, AsA content decreased by 75% and 84% due to mild and severe Cd toxicity, respectively, in comparison with control seedlings (Figure 5A). By contrast, the dehydroascorbate (DHA)content increased by 106% and 131% under mild and severe Cd stress, respectively (Figure 5B). Therefore, the redox pair ratio of AsA/DHA was reduced by 87% and 93% due to mild and severe Cd stress, respectively (Figure 5C). Vanillic acid supplementation in the presence of Cd suppressed the AsA decrease and the DHA content increase, leading to higher AsA/DHA ratios under both mild and severe Cd toxicity compared to their respective stress treatments (Figure 5A–C).

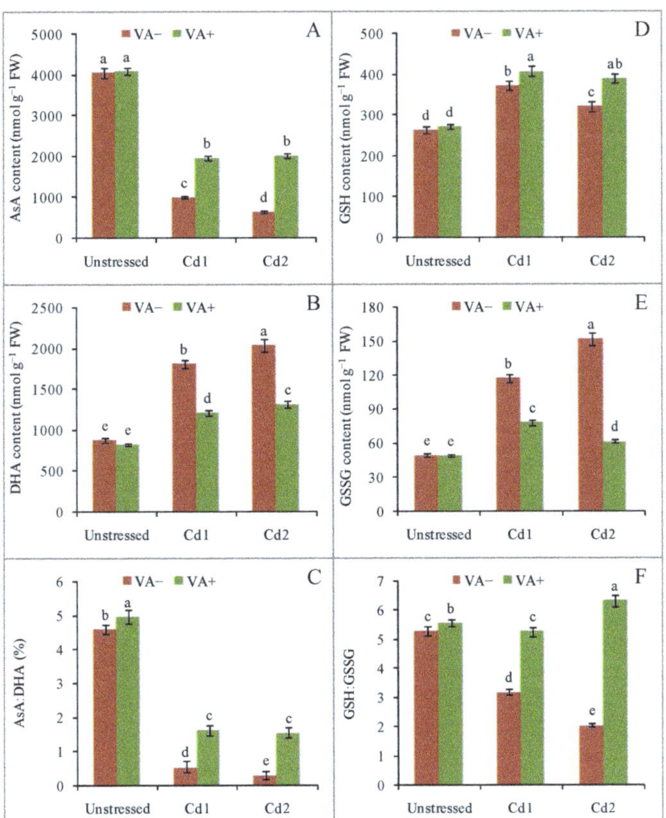

Figure 5. Effects of exogenous VA on ascorbic acid (AsA) (**A**) and dehydroascorbate (DHA)(**B**) contents, AsA/DHA ratio (**C**), reduced glutathione (GSH) (**D**) and oxidized glutathione (GSSG) (**E**) contents, and GSH/GSSG ratio (**F**) of rice (*Oryza sativa* L. cv. BRRI dhan54) seedlings under Cd stress. Mean (±SD) was computed from three replications of each treatment. Bars with dissimilar letters are significantly different at $p \leq 0.05$ from Fisher's least significant difference (LSD) test.

Although a sharp decrease in AsA content was observed due to Cd toxicity, both GSH (41% and 21%) and oxidized glutathione (GSSG) (136% and 206%) content increased under mild and severe Cd toxicity, respectively, in comparison with the control seedlings (Figure 5D,E). However, the redox couple ratio GSH/GSSG was reduced to 39% and 61%, respectively, under mild and severe Cd toxicity when compared to the control (Figure 5F). Addition of VA in the presence of Cd further increased the GSH content, but strongly suppressed the GSSG increase induced by Cd stress, resulting in an increase in the redox ratio for GSH/GSSG compared with their respective stress treatments (Figure 5D–F).

2.6. *Vanillic Acid Induced Antioxidant Enzyme Activity under Cd Toxicity*

Among the antioxidant enzymes, the activity of the potent H_2O_2scavenger APX was increased by 12% and 39% by mild and severe Cd toxicity, respectively, compared to the control. Vanillic acid application in the presence of Cd promoted a further increase in the APX activity, by 24% and 12%, compared to the respective stress treatments (Figure 6A). Similar to the effect on APX activity, MDHAR activity was increased due to Cd toxicity, but the effect was greater for mild Cd stress (38%) than for severe Cd stress (15%). Vanillic acid supplementation in the presence of Cd further increased MDHAR activity by 15% and 46% under mild and severe Cd stress, respectively, compared to their respective control seedlings (Figure 6B). In contrast to the effects on APX and MDHAR activity, DHAR activity was decreased under both mild (18%) and severe (33%) Cd stress, respectively, compared to control, whereas VA supplementation in the presence of Cd increased the DHAR activity by 35% and 90% under mild and severe Cd stress, respectively, compared to their respective Cd stress treatments (Figure 6C). The GR activity was also increased (7% and 23%) under mild and severe Cd toxicity, respectively. However, a substantial increase (26% and 8%) in GR activity was observed under mild and severe Cd toxicity, respectively, by VA in the presence of Cd compared to their respective stressed seedlings (Figure 6D).

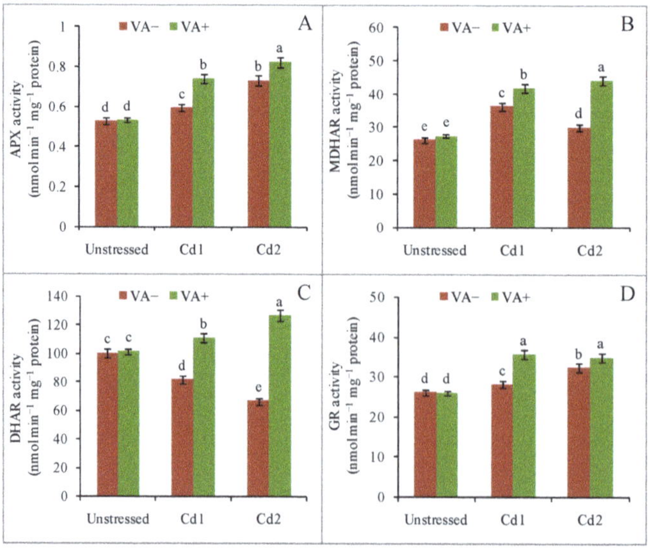

Figure 6. Effects of exogenous VA on activities of ascorbate peroxidase (APX) (**A**), monodehydroascorbate reductase (MDHAR) (**B**), dehydroascorbate reductase (DHAR) (**C**), and glutathione reductase (GR) (**D**) activity of rice (*Oryza sativa* L. cv. BRRI dhan54) seedlings under Cd toxicity stress. Mean (±SD) was computed from three replications of each treatment. Bars with dissimilar letters are significantly different at $p \leq 0.05$ from Fisher's least significant difference (LSD) test.

Mild and severe Cd toxicity increased the activities of the antioxidant enzymes SOD (9% and 13%, respectively) and GPX (35% and 38%, respectively) compared to control seedlings. Vanillic acid supplementation in the presence of Cd further increased SOD (9% and 14%) and GPX (12% and 22%) activity compared to their respective stress treatments (Figure 7A,C). By contrast, CAT activity decreased by 30% and 35%, due to mild and severe Cd stress, respectively, but VA supplementation in the presence of Cd suppressed the losses in CAT activity under both mild (33%) and severe (39%) Cd toxicity, respectively, compared to the respective stress treatments (Figure 7B). The activity of GST showed no changes under control and VA supplementation conditions, but GST activity was reduced by Cd stress, where 46% and 70% increases of GST activity was observed under both mild and severe Cd stresses, respectively, in the presence of VA (Figure 7D).

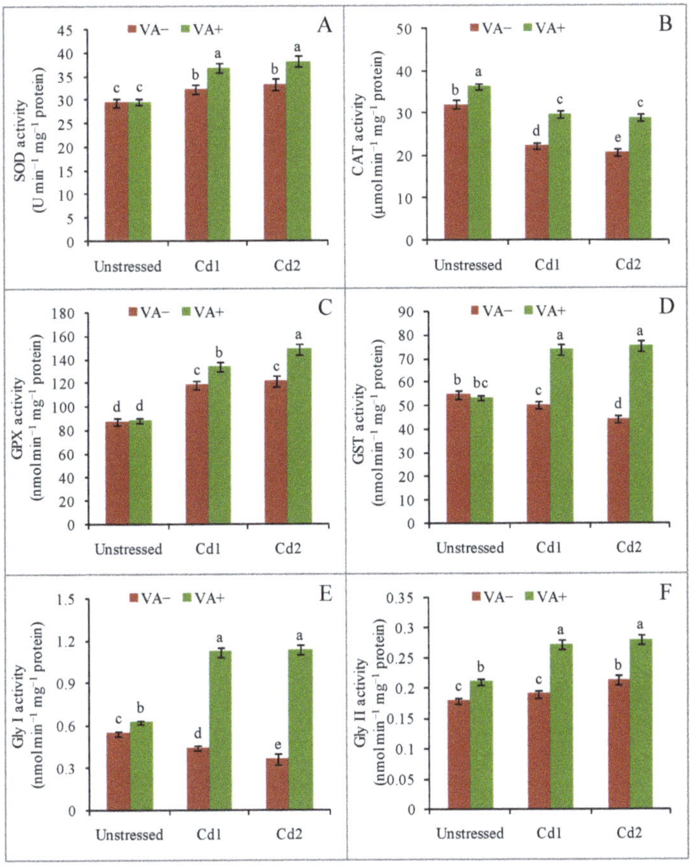

Figure 7. Effects of exogenous VA on activities of superoxide dismutase (SOD) (**A**), catalase (CAT) (**B**), glutathione peroxidase (GPX) (**C**),glutathione-S-transferase (GST) (**D**),glyoxalase I (Gly I) (**E**) and glyoxalase II (Gly II) (**F**) activity of rice (*Oryza sativa* L. cv. BRRI dhan54) seedlings under Cd toxicity stress. Mean (± SD) was computed from three replications of each treatment. Bars with dissimilar letters are significantly different at $p \leq 0.05$ from Fisher's least significant difference (LSD) test.

2.7. Vanillic Acid induced Glyoxalase Enzyme Activity under Cd Toxicity

Glyoxalase I activity was reduced by 19% and 34% under mild and severe Cd toxicity (Figure 7E), respectively, whereas Gly II activity increased by 6% and 18% (Figure 7F). Therefore, MG content increased under both mild and severe Cd toxicity, respectively (Figure 4D). Vanillic acid supplementation in the presence of Cd greatly increased Gly I activity by 154% and 213% following mild and severe Cd stress, respectively, and Gly II activity was also increased by 43% and 31% under both mild and severe Cd toxicity, respectively, compared to the respective stress treatments (Figure 7E,F).

2.8. Vanillic Acid Induced Mineral Homeostasis under Cd Toxicity

Shoot and root K content was decreased upon Cd toxicity stress, but the losses were suppressed by VA supplementation in the presence of Cd (Figure 8A,B). Similarly, shoot and root Ca content decreased under Cd stress, but VA supplementation in the presence of Cd suppressed both shoot and root Ca losses under both mild and severe Cd toxicity, with strong suppression observed in the shoot (Figure 8C,D). Shoot and root Mg content was reduced by Cd, but this effect was suppressed by VA supplementation, particularly in response to severe Cd stress (Figure 8E,F).

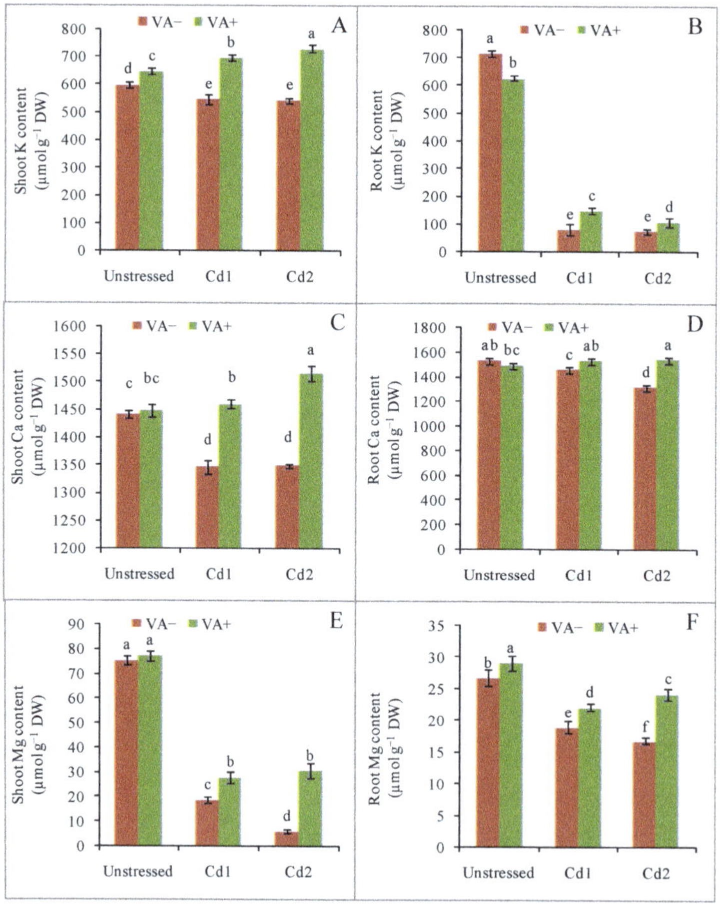

Figure 8. Effects of exogenous VA on the contents of shoot and root K, Ca, and Mg((**A**) shoot K; (**B**) root K; (**C**) shoot Ca; (**D**) root Ca; (**E**) shoot Mg; (**F**) root Mg) of rice seedlings under Cd toxicity stress. Mean (± SD) was computed from three replications of each treatment. Bars with dissimilar letters are significantly different at $p \leq 0.05$ from Fisher's least significant difference (LSD) test.

2.9. CorrelationAnalysis of the Different Vanillic Acid Responses

The correlation analyses showed that plant growth and physiological parameters were negatively correlated with oxidative stress markers (e.g., MDA and H_2O_2), whereas oxidative stress markers were negatively correlated with the components of the antioxidant defense system (Table S1a–c). Similarly, the glyoxalase system activity was negatively correlated with the MG content (Table S1c).

3. Discussion

Metal/metalloid contamination in the soil and environment has a negative effect on plants [16], as these compounds at high soil concentrations inhibit growth and disturb many biochemical processes, thereby leading to several negative consequences, particularly during the early stages of plant growth and development [17,18]. Cadmium is ubiquitous in many environments and highly toxic to all forms of life, and thus it causes significant yield losses due to its phytotoxicity [19]. By contrast, phenolic compounds, and especially low molecular weight organic acid (LMWOA) forms, are reported to alleviate the toxic effects of Cd stress [20]. As an LMWOA, VA has been shown to ameliorate the abiotic stress caused by drought and flooding [11,14]. Vanillic acid is also reported to increase plant growth and productivity, as well as upgrade the soil microbial community, in several test plant species [21–23]. However, a role for VA in ameliorating metal/metalloid toxicity, including Cd, has not been established. Therefore, the aim of the present study was to document the effects of exogenous VA on Cd toxicity responses in rice (*O. sativa* L. cv. BRRI dhan54) at the early seedling stage.

Cadmium causes several deleterious morphological effects at different stages of plant growth and development, including growth and biomass accumulation [3]. We observed a reduction in biomass under Cd stress in the present study. Inhibition of the supply of essential ions required for seedlingbiomass accumulation could be one of the reasons for the reduction in growth and biomass. Similar studies on growth parameters have also reported reduced growth and biomass accumulation in various plant species in response to Cd stress [24,25]. In the present study, the inclusion ofVA in the growth medium under Cd stress conditions prevented the suppression of biomass accumulation by Cd in rice seedlings. Previous reports have suggested that LMWOAs are capable of increasing growth and biomass under abiotic stresses, including Cd toxicity [24,25], indicating that they can alleviate the abiotic stress responses induced by metal/metalloids and allow greater tolerance to pollutants such as Cd.

Any agricultural strategy for increasing plant tolerance to Cd first requires an understanding of Cd accumulation as well as the Cd transport behavior inside plant organs. The recognition that LMWOAs are potent alleviators of metal/metalloid stress and that they can accumulate inside plant cells suggests that they have a dual role in regulating metal/metalloid stress in plants [24–26]. Similar to other previous findings, we also observed an increased accumulation of Cd in both shoots and roots of our rice seedlings, which showed a three- to fourfold greater increase in Cd content in the root compared with the shoot [24–26]. Interestingly, the VA treatment in the presence of Cd further increased the shoot and root Cd content. Previous studies have shown that application of LMWOAs in many plant species increased shoot and root metal/metalloid content, which might indicate metal scavenging as well as metal chelation and sequestration. Therefore, our findings are supported by previous findings of other researchers [24].

The visible symptoms of Cd stress in rice seedlings, including loss of biomass, were suppressed by the addition of VA despite the increased Cd accumulation in both roots and shoots. This suggested that even with high Cd accumulation, inclusion of VA in the presence of Cd in the growth medium alleviated Cd toxicity. Our findings also validated the results of other researchers [24]. Organic acids are renowned worldwide for their association with metal/metalloid stress tolerance in plants [24–26]. We also observed increases in the BCF and BAC values in response to VA in the cadmium-stressed rice seedlings. The reason for this might have been an increase in Cd solubility due to a decrease in pH by VA in the growth medium. Moreover, a VA-assisted release of strong ligands in the growth medium might assist in increased Cd uptake by the rice roots and subsequent translocation to the shoots [27]. Conversely, LMWOAs can chelate Cd to reduce the toxicity of free Cd in the growth medium [28]. Furthermore, the chelated intermediate might be translocated more efficiently through xylem [29]. Therefore, in our study, the observed effects of VA could indicate chelation of the free Cd in the growth medium, translocation to the growing shoots, and sequestration of the toxic Cd in cell vacuoles. In support of this statement, we also observed an increase in PC content under both doses of Cd, which further increased sharply by inclusion of VA in the growth medium, in agreement with previous findings [28]. Previous reports have suggested that PC, an oligomer synthesized in the plant cell in response to metal toxicity, acts as a binder for metals and transfers them for sequestration away from cell metabolic activities [30]. The activity of PC synthase in plant cells, as well as the formation of PCs, have been positively correlated with free Cd levels that can cause phytotoxicity symptoms [31]. The observed increase in PC content due to VA could further aid in ensuring plant tolerance and survivability, while also reducing oxidative stress, under Cd toxicity.

Plants exposed to metal/metalloid stress show a range of secondary stress symptoms, especially osmotic changes [32]. Therefore, plants also respond in various ways to mitigate water balance to cope with abiotic stresses. For example, osmotic adjustment and regulation of water content or water potential are some of the adaptation mechanisms used by plants to tolerate Cd exposure [33]. In our study, in response to Cd exposure, rice seedlings also showed lower water content under both mild and severe stress, in agreement with previous reports [24,33]. However, application of exogenous VA in the presence of Cd suppressed the changes in water balance seen in Cd-stressed seedlings in our study, hence suggested the role of LMWOAs in increasing water content in leaf tissues of cadmium-stressed plants. Besides maintaining water status, the biosynthesis of osmolytes is one of the vital strategies used by plants to mitigate the water balance changes under stress conditions. For example, biosynthesis and accumulation of proline, glycine betaine (GB), and trehalose (Tre) allows osmotic adjustment of Na^+ stress inside cells to maintain water balance [34]. We found an increased pro content in rice seedling leaf tissues under both mild and severe Cd stress, but this accumulation was suppressed by VA under both mild and severe Cd stress. The exogenous application of VA prevented the changes in water balance in cadmium-stressed rice seedlings so the plants could reduce the requirement for biosynthesis of pro.

Metal/metalloids hamper the growth and development of the plant but they also affect physiological processes. One of the most deleterious effects of metal/metalloid toxicity is the destruction of photosynthetic pigments. The biosynthesis of chl and its content decreases in many plant species in response to toxic metal/metalloid stress, including Cd stress [28,33]. Our results indicated a marked decrease in photosynthetic pigment content in terms of the chl (a + b) contentin rice seedling leaves under Cd stress, as well as a decrease in car content. These decreases in pigment content due to Cd stress might be linked with the inhibition of several enzymes, leading to a disruption of pigment biosynthesis. Some researchers have also hypothesized that peroxidative breakdown of photosynthetic pigments, as well as the lipids of the chloroplast membrane, occurs in response to abiotic stress due to increased generation of ROS [35]. However, this effect was suppressed when VA was included in the growth medium, in agreement with previous published findings [33]. Improved chl and car content might be associated with an increased sequestration of Cd or an increased biosynthesis and/or decrease in the destruction of pigment complexes [24,25,33].

Similar to other environmental stresses, Cd stress mediates an enhanced generation of ROS, including $O_2^{\bullet-}$, OH, and H_2O_2 [3]. These ROS are potentially capable of triggering membrane lipid peroxidation, as well as damaging amino acids, proteins, nucleotides, and nucleic acids. Membrane damage resulting from lipid peroxidation produces MDA, a major indicator of oxidative stress [9]. In our experiment, H_2O_2 and MDA content, as well as LOX activity, were markedly increased by Cd stress in a dose-dependent fashion. Simultaneously, the EL also increased. These responses might be attributed to a significant increase in ROS due to Cd, and they may also be responsible for the observed increases in the activity of the enzymes responsible for the degradation of lipid peroxides, which represent the attempt to control membrane damage and EL. Thus, our results are in line with those of previous researchers who observed similar responses to Cd-induced toxicities [24]. Higher H_2O_2, MDA, and EL levels also indicate that the antioxidant system defenses may be inadequate. However, the inclusion of VA in the growth medium suppressed the production of MDA and H_2O_2, while increasing LOX activity and ameliorating the EL observed in the Cd-stressed seedlings. These responses most likely reflect an enhanced activity of the ROS scavenging antioxidant defense system in the VA-treated seedlings. These observations are also in line with previous work [24,28] that reported are version of Cd-induced damage by LMWOAs.

Plants have evolved various mechanisms for protection against abiotic stress. A prime example is the antioxidant system, which consists of various enzymes and other non-enzyme components that are distributed throughout cell components (chloroplast, cytoplasm, apoplast, mitochondria, peroxisomes, and membranes) and allow dissipation of overproduced ROS. The end result is the ability to acclimate to unfavorable environments and to maintain growth and physiological functions. The content of these non-enzymatic components (AsA and GSH) and the potent activities of antioxidant enzymes (SOD, CAT, APX, MDHAR, DHAR, GR, GPX, and GST) have been documented to modulate growth under different stress conditions, including Cd stress [24,33,36].

The AsA–GSH cycle, or the 'Foyer–Halliwell–Asada' pathway, operates to scavenge H_2O_2 in plant cells [37]. AsA functions as the major antioxidant for reducing H_2O_2, OH, $O_2^{\bullet-}$, and $^1O_2$ levels. We found significant decreases in AsA content in the cadmium-stressed plants when compared with the control plants, which was mainly due to an increase in APX activity and decreased DHAR activity. The MDHAR activity increased under mild Cd stress, but it was not sufficient to repress the AsA content. A similar decrease in AsA content was also reported in other crops [24,25,27]. The upregulation of APX activity was also previously reported [25]. However, inclusion of VA in the presence of Cd increased the AsA content, as well as APX and DHAR activity, in cadmium-stressed rice seedling leaf tissue. This finding suggests that exogenous VA might play vital roles in AsA regeneration, in agreement with the findings of other researchers working on various other phytoprotectant chemicals [25,27,28,33].

Glutathione is another major component in the AsA–GSH cycle and is critical for AsA regeneration, xenobiotic detoxification, and metal/metalloid sequestration, as well as other mechanisms involved in stress tolerance [15]. Glutathione is linked with the AsA–GSH cycle for detoxification of H_2O_2, and it can also detoxify H_2O_2 via GPX/GST and xenobiotics through GST. Glutathione is also vital for MG detoxification via the Gly I and Gly II enzymes and has signaling properties [9]. In the present study, we found increased GSH content under both mild and severe Cd stress, together with increased GR and GPX activity. These changes might reflect aCd stress response involving the conversion of GSH to GSSGfor AsA recycling, whereas GPX activity is simultaneously increased to scavenge the overproduced H_2O_2, in agreement withother published findings [28]. The addition ofVA to the growth medium in the presence of Cd allowed further increases in the GSH content under both mild and severe Cd stress while also increasing the GR activity to enhance the recycling of GSSG to GSH. By contrast, the GST activity decreased in VA-treated cadmium-stressed seedlings, perhaps because the accumulation of $H_2O_2$was prevented by the enhanced antioxidant activity in the cells.

Superoxide dismutase activity was also increased by Cd stress, but the increase was suppressed byVA application in the presence of Cd. Reports have suggested that SOD is the plant's first-line defense for scavenging toxic $O_2^{\bullet-}$ radicals and converting them to H_2O_2. Therefore, the increased SOD activity in our study might be attributed toan increased $O_2^{\bullet-}$ content, which was suppressed by VA. Our findings are in line with those of other researchers [24], as many studies on plant responses to various stresses have shown significant alterations in CAT activity and have identified CAT as the most efficient H_2O_2 scavenging enzyme [38]. In the present study, the CAT activity decreased under Cd stress, in agreement with the findings of Praveen et al. [39]. However, adding VA in the presence of Cd suppressed the decrease in CAT activity, in accordance with the previous findings [25].

A VA-induced MG-mediated inhibition of glycation has been reported in animal cells [40], but no similar reports have been published for plant cells. In our study, Cd toxicity decreased the activity of Gly I but increased Gly II enzyme activity, thereby increasing the MG content in rice seedling leaf tissues. The inclusion of VA in the growth medium increased both the Gly I and Gly II enzyme activity, thus reducing the MG content and providing the seedlings with tolerance against MG-induced glycation as well as oxidative stress. The results of the present study are corroborated by those of other researchers who reported phytoprotectant-induced increases in the Gly enzyme activity and concomitant reduction in MG content [41,42].

One of the major consequences of metal/metalloid stress is nutrient deficiency, as metal/metalloids compete with essential mineral elements for uptake into the plant [43]. Nutrient homeostasis could therefore reduce metal/metalloid accumulation, and thereby their toxicity, to enhance many physiological mechanisms under Cd stress conditions. In our study, Cd toxicity stress altered the mineral homeostasis of rice seedlings. We observed Cd toxicity-related reductions in shoot and root K, Ca, and Mg content, which is corroborated by many previous studies [44,45] that have reported a hermetic effect of Cd on mineral homeostasis in different plant species. Cd-induced oxidative stress also damages the cell membrane, which might also lower the nutrient content in roots. Vanillic acid supplementation in the presence of Cd restored the nutrient balance, in line with the results of other researchers [24]. Therefore, VA-induced nutrient availability reduced the toxic effects of Cd accumulation and restored the plant's ability to run its physiological mechanisms smoothly.

4. Materials and Methods

4.1. Plant Materials and Stress Treatments

Manually sorted surface sterilized (10 minutes with 1% sodium hypochlorite) rice (*Oryza sativa* L. cv. BRRI dhan54) seeds were soaked in deionized water (DH_2O) for 48 h and kept moist for the following24 h for sprouting. Then, the sprouted seeds were sown in plastic pots (8 cm diameter, a volume of 250 mL) hydroponically and incubated(40h). Afterward, the pots with 25 germinated seeds were facilitated growing under controlled conditions (temperature 25 ± 2 °C; relative humidity 65%–70%; light 350 μmol photon m^{-1} s^{-2}; 16/8 h day/night duration) for 12 days in a cultivation chamber. During the growing period, the seedlings were nourished with2500-fold diluted Hyponex nutrient solution (Hyponex, Japan) controlling pH 6.5–7.0. The nutrient solution was renewed after every 72 h. At 12 days, the seedlings were treated with Cd (1.0 and 2.0 mM) and VA (50 μM) for 72 h solely and in combination. Hence, the experiment consisted of six treatments fitted in a completely randomized design (CRD) with three repetitions and repeated thrice, maintaining the same conditions. Data were collected after the treatment period with standard methods described later.

4.2. Observation of Plant Growth and Biomass Accumulation

Growth and biomass accumulation under different treatments were observed by measuring the shoot and root length from bases of the seedling to the shoot and root tip, respectively, from 10randomly selected seedlings. After excision at the root and shoot junctions, the fresh weight (FW) of shoot and root were weighed and mean FW value was expressed as mg $seedling^{-1}$. Afterwards, the shoots and roots were dried separately for 48 h at 60°C and weighed again; mean dry weight (DW) value was expressed as mg $seedling^{-1}$.

We measured the relative elongation of shoot and root elongation of different treatment combinations according to the procedure stated by Song et al. [46]. Briefly, before exposure to different treatments, a set of 10 randomly selected seedlings was measured for shoot and root length, expressed as Di. After harvesting, the same seedlings were again measured for final shoot and root length, expressed as Df. The initial shoot or root and final shoot or root length of the control seedlings were measured as DCi and DCf, respectively. The following equation was employed for calculating the relative elongation of shoot or root and expressed as a percentage:

$$\text{Relative elongation} = \frac{Df - Di}{DCi - DCf} \times 100$$

4.3. Determination of Cd and Other Nutrient Contents, and Measuring BAC, BCF, and TF of Cd

After acid digestion (HNO_3:$HClO_4$ at 5:1, *v*/*v*) of dried root and shoot, an atomic absorption spectrophotometer (AAS) was employed for measuring the content of Cd, K, Ca, and Mg present in the shoot and root according to Zasoski et al. [47].

Biological accumulation coefficient indicates the ratio of Cd content of the shoots to the growing media, and BCF indicates the ratio of Cd content of the roots to the growing media, whereas TF indicates the ratio of shoot Cd and root Cd content of rice seedlings. Therefore, BAC, BCF, and TF of Cd were calculated following the equations below [48]:

$$\text{BAC} = \frac{\textit{Cd content in shoot}}{\textit{Cd content in the growing media}}$$

$$\text{BCF} = \frac{\textit{Cd content in root}}{\textit{Cd content in the growing media}}$$

$$\text{TF} = \frac{\textit{Cd content in shoot}}{\textit{Cd content in root}}$$

4.4. Determination of Stress Markers, Photosynthetic Pigment, Relative Turgidity, and Proline Content

We estimated the MDA content, following Heath and Packer [49], as the thiobarbituric acid (TBA) reactive substances. First, fresh harvested rice leaves (0.5 g) were extracted by 5% trichloroacetic acid (TCA) using a chilled mortar and pestle, and subsequent centrifugation was then conducted at 11,500× *g* for 15 min. The supernatants were mixed with TBA for reaction to obtained the MDA, which was further determined from the optical absorbance difference between 532 and 600 nm and calculated using an extinction coefficient of 155 mM^{-1} cm^{-1} expressed as nmol g^{-1} FW.

The method described by Alexieva et al. [50] with little modification was employed for the determination of H_2O_2 content. Briefly, fresh leaf samples were homogenized in 3 ml 5% TCA and centrifuged (11,500× *g* for 15 min). The supernatant was collected. An aliquot of 0.5 ml supernatant was mixed with 0.5 ml of potassium phosphate (K–P) buffer (0.5 M, pH 7.0) and 1 ml of 1M potassium iodide (KI), and the mixture was allowed for reaction to occur for 1 h in darkness. Absorbance was measured at 390 nm using standard curve and expressed as nmol g^{-1} FW.

For determining the amount of photosynthetic pigments,0.25g of fresh leaf samples were taken and emerged in 10 mL 80% ethanol and heated for 1h in a hot water bath at 60 °C. The absorbance of the colored solution was read in a spectrophotometer under 663, 645, and 470 nm. Chlorophyll *a*, *b*, (*a* + *b*), and car contents were then calculated according to the equations described by Arnon [51] and Wellburn [52], respectively.

Further, we estimated RT (%) as depicted from Barrs and Weatherly [53] using the following equation:

$$\text{RT (\%)} = \frac{\text{leaf fresh weight} - \text{Leaf dry weight}}{\text{Leaf turgid weight} - \text{leaf dry weight}} \times 100$$

Subsequently pro content was estimated according to Bates et al. [54]. Fresh leaves (100 mg) were extracted in 3% sulfosalicylic acid. To 1 mL of supernatant, 1 mL of acid ninhydrin and 1 mL of glacial acetic acid was added, followed by incubation at 100 °C for 1hour. Then, the mixture was cooled to terminate reaction and toluene was added and vortaxed to separate the toluene chromophore containing pro. Proline was assessed spectrophotometrically at 520 nm using L-proline (Wako, Japan) as standard.

Methylglyoxal estimation by *N*-acetyl-L-cysteine assay was performed following Wild et al. [55]. Fresh leaves (0.25 g) were mashed with 5% perchloric acid on an ice cold mortar and centrifuged at 11,000× *g* to remove the residue. The supernatant was mixed with charcoal for decolorization and subsequently neutralized by sodium carbonate. The neutralized supernatant was further used for *N*-acetyl-L-cysteine assay of MG estimation at a wavelength of 288 nm.

4.5. Nonenzymatic Antioxidant Assay

For the determination of the nonenzymatic antioxidants AsA and GSH, we extracted the tissue according to Kampfenkel et al. [56].The contents of total AsA, reduced AsA, total GSH, and GSSG were assayed spectrophotometrically. For the determination of AsA pool, aliquots of 200 μL of supernatants were neutralized with K–P buffer (0.5 M, pH 7.0). Further, in the aliquots of total AsA and reduced AsA, 0.1 M dithiothretitol (DTT) and DH_2O was added, respectively. Then, total and reduced AsA was determined optically at 265 nm employing a previously formed standard curve, and dehydroascorbate (DHA) was calculated by subtraction of reduced AsA from total AsA. In a similar manner GSH pool was determined after neutralizing aliquots of 200 μL supernatants with 0.5 M K–P buffer (pH 7.0) [57]. We determined GSSG after masking reduced GSH by 2-vinylpyridine. After that, total GSH and GSSG were assayed on the basis of the enzymatic recycling, where reduced GSH wasoxidized by 5,5-dithio-bis (2-nitrobenzoic acid, DTNB) in the presence of GR and reduced nicotinamide adenine dinucleotide phosphate (NADPH). The rate of absorption change was read at 412 nm optically, and previously prepared standard curves of GSH and GSSG were employed to determine the content. Finally, the reduced portion of GSH was calculated after subtracting GSSG from total GSH [58].

4.6. Protein, Antioxidant, and other Enzyme Activity Assay

Crude enzyme extract was prepared for assaying enzymatic activity where freshly harvested 0.5 g of wheat leaf samples were grinded in 1 ml ice cold extraction buffer that contained AsA (1 mM), K–P buffer (50 mM, pH 7.0), KCl (100 mM), β-mercaptoethanol (5 mM), and glycerol (10%, *w*/*v*). The homogenates were centrifuged at 11,500× *g* for 10 min, and the supernatants were collected and preserved (−60 °C). The experimental condition was maintained at 0–4°C temperature under controlled condition.

Protein quantity present in each sample was determination according to Bradford [59]. Proteins formed complexes with Coomassie Brilliant Blue dye, which was read optically at 595 nm. The amount of protein was then calculated using standard curve constructed with protein standard—bovine serum albumin (BSA).

Lipoxygenase (EC: 1.13.11.12) activity was assayed according to Doderer et al. [60] using linoleic acid as a substrate by observing the increase of absorbance at 234 nm and calculated using 25 mM^{-1} cm^{-1} as extinction coefficient.

Ascorbate peroxidase (EC: 1.11.1.11) activity was determined as stated by Nakano and Asada [61], where the assay mixture included K–P buffer (50 mM, pH 7.0), ethylenediaminetetraacetic acid (EDTA) (0.1 mM), AsA (0.5 mM), and H_2O_2 (0.1 mM). After adding H_2O_2, the reaction was started, and finally the activity of APX was computed using 2.8 mM^{-1} cm^{-1} as extinction coefficient.

Monodehydroascorbate reductase (EC: 1.6.5.4) activity was assayed following Hossain et al. [62] optically at 340 nm, using 703.4 μL of reaction mixture consistingof Tris-HCl buffer (50 mM, pH 7.5), AsA (2.5 mM), NADPH (0.2 mM), and ascorbate oxidase (AO) (0.5 unit), and was computed using 6.2 mM^{-1} cm^{-1} as extinction coefficient.

Dehydroascorbate reductase (EC: 1.8.5.1) activity wasrecorded with the method of Nakano and Asada [61], where the assay mixture contained K–P buffer (50 mM, pH 7.0), GSH (2.5 mM), EDTA (0.1 mM), and DHA (0.1 mM). Activity of DHAR was assayed from the increase in absorbance at 265 nm with the reduction of DHA for 1min. Finally, the activity of DHAR was calculated using 14 mM^{-1} cm^{-1} as extinction coefficient.

Glutathion reductase (GR; EC: 1.6.4.2) activity was measured according to Foyer and Halliwell [63] by observing the decline in absorbance at 340 nm, where reaction mixture consisted of K–P buffer (0.1 M, pH 7.0) and EDTA (1 mM). The reaction was NADPH-dependent and initiated with GSSG. Glutathione reductase activity was finally calculated using 6.2 mM^{-1} cm^{-1} as extinction coefficient.

Superoxide dismutase (SOD; EC 1.15.1.1) activity was determined on the basis of the reduction of nitro blue tetrazolium (NBT) using xanthine–xanthine oxidase system [64].

Catalase (CAT; EC: 1.11.1.6) activity was assayed following the method of Patra et al. [65] by observing the decrease in absorbance for 1 min (by conversion of H_2O_2 to water and O_2) at 240 nm, where the enzyme extract was used to initiate the reaction. The activity of enzyme was computed using 39.4 M^{-1} cm^{-1} as extinction coefficient.

Glutathione *S*-transferase (GST; EC: 2.5.1.18) activity was spectrophotometrically measured according to Booth et al. [66], where the reaction mixture contained 1.5 mM GSH, 1 mM 1-chloro-2,4-dinitrobenzene (CDNB), and enzyme. The increase of absorbance was read at 340 nm for 1min and the enzyme activity was computed using the CDNB extinction coefficient of 9.6 mM^{-1} cm^{-1}.

Glutathione peroxidase (GPX; EC: 1.11.1.9) activity was enumerated according to the procedure of Elia et al. [67], where the reaction buffer contained 100 mM K–P buffer (pH 7.0), 1 mM EDTA, 1mM sodium azide (NaN_3), 0.12 mM NADPH, 2 mM GSH, and 1unit of GR. The reaction used 0.6 mM H_2O_2 as substrate, and the activity was calculated using extinction coefficient 6.6 mM^{-1} cm^{-1}.

Glyoxalase I (Gly I; EC: 4.4.1.5) activity assay was recorded according to the method stated by Hossain et al. [68], where 700 µL of assay mixture consist of K–P buffer (100 mM, pH 7.0), magnesium sulfate (15 mM), GSH (1.7 mM), and MG (3.5 mM). After adding MG, the reaction began, and increase in absorbance was obtained at 240 nm for 1 min. The activity of Gly I was computed using 3.37 mM^{-1} cm^{-1} as extinction coefficient.

Glyoxalase II (Gly II; EC: 3.1.2.6) activity assay was performed as described by Principato et al. [69], where 500 µL of the reaction mixture contained Tris-HCl buffer (100 mM, pH 7.2), DTNB (0.2 mM), and S-D-lactoylglutathione (SLG, 1 mM). The reaction was initiated by adding SLG, and the increase in absorbance was recorded spectrophotometrically at 412 nm. Finally, the activity was computed using 13.6 mM^{-1} cm^{-1} as extinction coefficient.

4.7. Statistical Analysis

The data presented are the mean values of three replicates. The significant differences between treatments were statistically evaluated using Statistix10 software by one-way analysis of variance (ANOVA). Fisher's least significant difference (LSD) test was used for comparison of means at the $p < 0.05$ level between treatments. Results, which were significantly different at $p < 0.05$, were marked by different letters in figures. Moreover, a correlation study was performed to infer the relationship between Cd toxicity and the VA supplementation under Cd toxicity stress. Furthermore, we showed the significant difference between sole Cd and VA supplementation with Cd using asterisks.

5. Conclusions

Cadmium toxicity stress strongly reduced growth and biomass accumulation in rice seedlings. The photosynthetic pigments were damaged by Cd exposure, and the osmotic status inside the cell was altered due to the toxic effects of Cd. Cadmium exposure promoted ROS generation and impaired the antioxidant defense and glyoxalase systems in rice seedlings. Vanillic acid supplementation in the presence of Cd in the growth medium noticeably suppressed the losses in the antioxidant defense and glyoxalase system and the ROS generation and lipid peroxidation observed in cadmium-stressed plants, while improving the osmotic status and preventing the loss of photosynthetic pigments. These results indicate that VA could serve as a phytoprotectant to reduce Cd toxicity stress in rice at the early seedling stage. Further study will be conducted to elucidate the molecular mechanism of VA-induced Cd stress tolerance, as well as to evaluate the practical application of VA in greenhouse and field conditions.

Supplementary Materials: The following are available online at http://www.mdpi.com/2223-7747/9/2/188/s1, Table S1a: Correlation matrix of plant growth, osmotic status, photosynthetic pigments contents, Oxidative stress indicator, antioxidants and minerals contents, Table S1b. Correlation matrix of oxidative stress indicators, the AsA:GSH cycle and enzymatic antioxidants, Table S1c. Correlation matrix of the components of glyoxalase systems, Cd accumulation and translocation, other essential mineral components.

Author Contributions: M.H.M.B.B. conceived, designed, and performed the experiment and prepared the manuscript. K.P. and S.M.M. actively participated in conducting the experiment and edited the manuscript. J.A.M. conceived the experiment and edited the manuscript. M.H. designed the experiment, analyzed the data, and edited the manuscript. M.F. conceived, designed, and monitored the experiment. All authors read and approved the final manuscript.

Funding: This research received no external funding.

Acknowledgments: Authors M.H.M.B.B.; K.P., and S.M.M. are grateful to the Ministry of Education, Culture, Sports, Science and Technology (MEXT), Japan, for bestowing a scholarship for their Ph.D. study.

Conflicts of Interest: The authors declare that the research was conducted without any commercial or financial relationships that could be construed as a potential conflict of interest.

References

1. Di Toppi, L.S.; Gabbrielli, R. Response to cadmium in higher plants. *Environ. Exp. Bot.* **1999**, *41*, 105–130. [CrossRef]
2. Besson-Bard, A.; Gravot, A.; Richaud, P.; Auroy, P.; Duc, C.; Gaymard, F.; Taconnat, L.; Renou, J.P.; Pugin, A.; Wendehenne, D. Nitric oxide contributes to cadmium toxicity in *Arabidopsis* by promoting cadmium accumulation in roots and by up-regulating genes related to iron uptake. *Plant Physiol.* **2009**, *149*, 1302–1315. [CrossRef] [PubMed]
3. Benavides, M.P.; Gallego, S.M.; Tomaro, M.L. Cadmium toxicity in plants. *Braz. J. Plant Physiol.* **2005**, *17*, 21–34. [CrossRef]
4. Valko, M.; Jomova, K.; Rhodes, C.J.; Kuča, K.; Musílek, K. Redox-and non-redox-metal-induced formation of free radicals and their role in human disease. *Arch. Toxicol.* **2016**, *90*, 1–37. [CrossRef]
5. Keunen, E.; Remans, T.; Bohler, S.; Vangronsveld, J.; Cuypers, A. Metal-induced oxidative stress and plant mitochondria. *Int. J. Mol. Sci.* **2011**, *12*, 6894–6918. [CrossRef]
6. Sharma, P.; Jha, A.B.; Dubey, R.S.; Pessarakli, M. Reactive oxygen species, oxidative damage, and antioxidative defense mechanism in plants under stressful conditions. *J. Bot.* **2012**, *2012*, 217037. [CrossRef]
7. Gratao, P.L.; Pompeu, G.B.; Capaldi, F.R.; Vitorello, V.A.; Lea, P.J.; Azevedo, R.A. Antioxidant response of *Nicotiana tabacum* cv. Bright Yellow 2 cells to cadmium and nickel stress. *Plant Cell Tiss. Organ Cult.* **2008**, *94*, 73. [CrossRef]
8. Suhartono, E.; Triawanti, A.S.; Djati, M.S. The role of cadmium in proteins glycation by glucose: Formation of methylglyoxal and hydrogen peroxide in vitro. *J. Med. Bioeng.* **2014**, *3*, 59–62. [CrossRef]
9. Sytar, O.; Kumar, A.; Latowski, D.; Kuczynska, P.; Strzałka, K.; Prasad, M.N.V. Heavy metal-induced oxidative damage, defense reactions, and detoxification mechanisms in plants. *Acta Physiol. Plant.* **2013**, *35*, 985–999. [CrossRef]
10. Cheynier, V. Phenolic compounds: From plants to foods. *Phytochem. Rev.* **2012**, *11*, 153–177. [CrossRef]
11. Xuan, T.; Khang, D. Effects of exogenous application of protocatechuic acid and vanillic acid to chlorophylls, phenolics and antioxidant enzymes of rice (*Oryza sativa* L.) in submergence. *Molecules* **2018**, *23*, 620. [CrossRef] [PubMed]
12. Granato, D.; Shahidi, F.; Wrolstad, R.; Kilmartin, P.; Melton, L.D.; Hidalgo, F.J.; Miyashita, K.; van Camp, J.; Alasalvar, C.; Ismail, A.B.; et al. Antioxidant activity, total phenolics and flavonoids contents: Should we ban in vitro screening methods? *Food Chem.* **2018**, *264*, 471–475. [CrossRef] [PubMed]
13. Minh, L.T.; Khang, D.T.; Ha, P.T.; Tuyen, P.T.; Minh, T.N.; Quan, N.V.; Xuan, T.D. Effects of salinity stress on growth and phenolics of rice (*Oryza sativa* L.). *Int. Lett. Nat. Sci.* **2016**, *57*, 1–10. [CrossRef]
14. Quan, N.T.; Xuan, T.D. Foliar application of vanillic and p-hydroxybenzoic acids enhanced drought tolerance and formation of phytoalexin momilactones in rice. *Arch. Agron. Soil Sci.* **2018**, *64*, 1831–1846. [CrossRef]
15. Osman, H.S.; Salim, B.B. Influence of exogenous application of some phytoprotectants on growth, yield and pod quality of snap bean under NaCl salinity. *Ann. Agric. Sci.* **2016**, *61*, 1–13. [CrossRef]
16. Silva, S.; Silva, P.; Oliveira, H.; Gaivão, I.; Matos, M.; Pinto-Carnide, O.; Santos, C. Pb low doses induced genotoxicity in *Lactuca sativa* plants. *Plant Physiol.Biochem.* **2017**, *112*, 109–116. [CrossRef]
17. Novo, L.A.B.; Gonzalez, L. Germination and Early Growth of *Brassica juncea* in Copper Mine Tailings Amended with Technosol and Compost. *Sci World J.* **2014**, *2014*, 506392. [CrossRef]

18. Adrees, M.; Ali, S.; Iqbal, M.; Bharwana, S.A.; Siddiqi, Z.; Farid, M.; Ali, Q.; Saeed, R.; Rizwan, M. Mannitol alleviates chromium toxicity in wheat plants in relation to growth, yield, stimulation of anti-oxidative enzymes, oxidative stress and Cr uptake in sand and soil media. *Ecotoxicol. Environ. Saf.* **2015**, *122*, 1–8. [CrossRef]
19. Shanmugaraj, B.M.; Malla, A.; Ramalingam, S. Cadmium Stress and Toxicity in Plants: An Overview. In *Cadmium Toxicity and Tolerance in Plants: From Physiology to Remediation*; Hasanuzzaman, M., Prasad, M.N.V., Fujita, M., Eds.; Academic Press: Cambridge, MA, USA, 2019; pp. 1–17.
20. Sidhu, G.P.; Bali, A.S.; Bhardwaj, R. Role of Organic Acids in Mitigating Cadmium Toxicity in Plants. In *Cadmium Tolerance in Plants: Agronomic, Molecular, Signaling, and Omic Approaches*; Hasanuzzaman, M., Prasad, M.N.V., Nahar, K., Eds.; Academic Press: Cambridge, MA, USA, 2019; pp. 255–279.
21. Khaleda, L.; Kim, M.G.; Jeon, J.-R.; Cha, J.-Y.; Kim, W.-Y. Foliar application of humic acid or a mixture of catechol and vanillic acid enhanced growth and productivity of alfalfa. *J. Korean Soc. Grassland Forage Sci.* **2017**, *37*, 248–253. [CrossRef]
22. Zhou, X.; Wu, F. Artificially applied vanillic acid changed soil microbial communities in the rhizosphere of cucumber (*Cucumis sativus* L.). *Can. J. Soil Sci.* **2013**, *93*, 13–21. [CrossRef]
23. Zhou, X.; Wu, F. Vanillic acid changed cucumber (*Cucumis sativus* L.) seedling rhizosphere total bacterial, *Pseudomonas* and *Bacillus* spp. communities. *Sci. Rep.* **2018**, *8*, 4929. [CrossRef] [PubMed]
24. Hawrylak-Nowak, B.; Dresler, S.; Matraszek, R. Exogenous malic and acetic acids reduce cadmium phytotoxicity and enhance cadmium accumulation in roots of sunflower plants. *Plant Physiol. Biochem.* **2015**, *94*, 225–234. [CrossRef] [PubMed]
25. Rizwan, M.; Ali, S.; Akbar, M.Z.; Shakoor, M.B.; Mahmood, A.; Ishaque, W.; Hussain, A. Foliar application of aspartic acid lowers cadmium uptake and Cd-induced oxidative stress in rice under Cd stress. *Environ. Sci. Pollut. Res.* **2017**, *24*, 21938–21947. [CrossRef] [PubMed]
26. Zaheer, I.E.; Ali, S.; Rizwan, M.; Farid, M.; Shakoor, M.B.; Gill, R.A.; Najeeb, U.; Iqbal, N.; Ahmad, R. Citric acid assisted phytoremediation of copper by *Brassica napus* L. *Ecotoxicol. Environ. Saf.* **2015**, *120*, 310–317. [CrossRef]
27. Li, F.T.; Qi, J.M.; Zhang, G.Y.; Lin, L.H.; Fang, P.P.; Tao, A.F.; Xu, J.T. Effect of cadmium stress on the growth, antioxidative enzymes and lipid peroxidation in two kenaf (*Hibiscus cannabinus* L.) plant seedlings. *J. Integr Agric.* **2013**, *12*, 610–620. [CrossRef]
28. Sun, Q.; Wang, X.R.; Ding, S.M.; Yuan, X.F. Effects of exogenous organic chelators on phytochelatins production and its relationship with cadmium toxicity in wheat (*Triticum aestivum* L.) under cadmium stress. *Chemosphere* **2005**, *60*, 22–31. [CrossRef]
29. D'Alessandro, A.; Taamalli, M.; Gevi, F.; Timperio, A.M.; Zolla, L.; Ghnaya, T. Cadmium stress responses in *Brassica juncea*: Hints from proteomics and metabolomics. *J. Proteome Res.* **2013**, *12*, 4979–4997. [CrossRef]
30. Zagorchev, L.; Seal, C.E.; Kranner, I.; Odjakova, M. A central role for thiols in plant tolerance to abiotic stress. *Int. J. Mol. Sci.* **2013**, *14*, 7405–7432. [CrossRef]
31. Sun, H.W.; Ha, J.; Liang, S.X.; Kang, W.J. Protective role of selenium on garlic growth under cadmium stress. *Commun. Soil Sci. Plant Anal.* **2010**, *41*, 1195–1204. [CrossRef]
32. Wang, Z.; Zhang, Y.; Huang, Z.; Huang, L. Antioxidative response of metal-accumulator and non-accumulator plants under cadmium stress. *Plant Soil* **2008**, *310*, 137. [CrossRef]
33. Ahmad, P.; Allah, E.A.; Hashem, A.; Sarwat, M.; Gucel, S. Exogenous application of selenium mitigates cadmium toxicity in *Brassica juncea* L. (Czern & Cross) by up-regulating antioxidative system and secondary metabolites. *J. Plant Growth Regul.* **2016**, *35*, 936–950.
34. Pandey, C.; Gupta, M. Selenium and auxin mitigates arsenic stress in rice (*Oryza sativa* L.) by combining the role of stress indicators, modulators and genotoxicity assay. *J. Hazard. Mater.* **2015**, *287*, 384–391. [CrossRef] [PubMed]
35. Duman, F.; Ozturk, F.; Aydin, Z. Biological responses of duckweed (*Lemna minor* L.) exposed to the inorganic arsenic species As(III) and As(V): Effects of concentration and duration of exposure. *Ecotoxicology* **2010**, *19*, 983–993. [CrossRef] [PubMed]
36. Faraz, A.; Faizan, M.; Sami, F.; Siddiqui, H.; Hayat, S. Supplementation of salicylic acid and citric acid for alleviation of cadmium toxicity to *Brassica juncea*. *J. Plant Growth Regul.* **2019**. [CrossRef]
37. Foyer, C.F.; Noctor, G. Ascorbate and glutathione: The heart of the redox hub. *Plant Physiol.* **2011**, *155*, 2–18. [CrossRef]

38. Mhamdi, A.; Queval, G.; Chaouch, S.; Vanderauwera, S.; Breusegem, F.V.; Noctor, G. Catalase function in plants: A focus on *Arabidopsis* mutants as stress-mimic models. *J. Exp. Bot.* **2010**, *61*, 4197–4220. [CrossRef]
39. Praveen, A.; Pandey, C.; Khan, E.; Panthri, M.; Gupta, M. Silicon mediated genotoxic alterations in *Brassica juncea* under arsenic stress: Comparative study of biochemical and molecular markers. *Pedosphere* **2017**. [CrossRef]
40. Huang, Y.X.; Liao, B.H.; Wang, Z.K. Effect of cadmium stress on growth of soybeans and the contents of nutrition elements in soybean grains. *J. Saf. Environ.* **2008**, *8*, 11–15.
41. Jan, S.; Alyemeni, M.N.; Wijaya, L.; Alam, P.; Siddique, K.H.; Ahmad, P. Interactive effect of 24-epibrassinolide and silicon alleviates cadmium stress via the modulation of antioxidant defense and glyoxalase systems and macronutrient content in *Pisum sativum* L. seedlings. *BMC Plant Biol.* **2018**, *18*, 146. [CrossRef]
42. Zaid, A.; Mohammad, F.; Wani, S.H.; Siddique, K.M. Salicylic acid enhances nickel stress tolerance by up-regulating antioxidant defense and glyoxalase systems in mustard plants. *Ecotoxicol. Environ. Saf.* **2019**, *180*, 575–587. [CrossRef]
43. Yadav, V.; Arif, N.; Singh, S.; Srivastava, P.K.; Sharma, S.; Tripathi, D.K.; Dubey, N.K.; Chauhan, D.K. Exogenous mineral regulation under heavy metal stress: Advances and prospects. *Biochem. Pharmacol.* **2016**, *5*, 220. [CrossRef]
44. Gonçalves, J.F.; Antes, F.G.; Maldaner, J.; Pereira, L.B.; Tabaldi, L.A.; Rauber, R.; Rossato, L.V.; Bisognin, D.A.; Dressler, V.L.; de Moraes Flores, É.M.; et al. Cadmium and mineral nutrient accumulation in potato plantlets grown under cadmium stress in two different experimental culture conditions. *Plant Physiol. Biochem.* **2009**, *47*, 814–821. [CrossRef] [PubMed]
45. Marschner, H. Mechanisms of adaptation of plants to acid soils. *Plant Soil* **1991**, *134*, 1–20. [CrossRef]
46. Song, H.; Xu, X.; Wang, H.; Tao, Y. Protein carbonylation in barley seedling roots caused by aluminum and proton toxicity is suppressed by salicylic acid. *Russ. J. Plant Physiol.* **2011**, *58*, 653–659. [CrossRef]
47. Zasoski, R.J.; Burau, R.G. A rapid nitric-perchloric acid digestion method for multi-element tissue analysis. *Commun. Soil Sci. Plant Anal.* **1977**, *8*, 425–436. [CrossRef]
48. Malik, M.R.; Ali, S.; Fettouhi, M.; Isab, A.A.; Ahmad, S. Structural characterization of dichloridobis (N,N′-dimethylthiourea-S) cadmium (II). *J. Struct. Chem.* **2010**, *51*, 976–979. [CrossRef]
49. Heath, R.L.; Packer, L. Photoperoxidation in isolated chloroplasts: I. Kinetics and stoichiometry of fatty acid peroxidation. *Arch. Biochem. Biophys.* **1968**, *125*, 189–198. [CrossRef]
50. Alexieva, V.; Sergiev, I.; Mapelli, S.; Karanov, E. The effect of drought and ultraviolet radiation on growth and stress markers in pea and wheat. *Plant Cell Environ.* **2001**, *24*, 1337–1344. [CrossRef]
51. Arnon, D.I. Copper enzymes in isolated chloroplasts. Polyphenoloxidase in *Beta vulgaris*. *Plant Physiol.* **1949**, *24*, 1–10. [CrossRef]
52. Wellburn, A.L. The spectral determination of chlorophyllsaandb, as well as total carotenoids using various solvents with spectrophotometers of different resolutions. *J. Plant Physiol.* **1994**, *144*, 307–313. [CrossRef]
53. Barrs, H.D.; Weatherley, P.E. A re-examination of the relative turgidity technique for estimating water deficits in leaves. *Aust. J. Biol. Sci.* **1962**, *15*, 413–428. [CrossRef]
54. Bates, L.S.; Waldren, R.P.; Teare, I.D. Rapid determination of free proline for water-stress studies. *Plant Soil* **1973**, *39*, 205–207. [CrossRef]
55. Wild, R.; Ooi, L.; Srikanth, V.; Münch, G. A quick, convenient and economical method for the reliable determination of methylglyoxal in millimolar concentrations: The N-acetyl-L-cysteine assay. *Anal. Bioanal. Chem.* **2012**, *403*, 2577–2581. [CrossRef] [PubMed]
56. Kampfenkel, K.; Van Montagu, M.; Inzé, D. Extraction and determination of ascorbate and dehydroascorbate from plant-tissue. *Anal. Biochem.* **1995**, *225*, 165–167. [CrossRef] [PubMed]
57. Hewitt, E.J.; Dickes, G.J. Spectrophotometric measurements on ascorbic acid and their use for the estimation of ascorbic acid and dehydroascorbic acid in plant tissues. *Biochem. J.* **1961**, *78*, 384–391. [CrossRef]
58. Rahman, I.; Kode, A.; Biswas, S.K. Assay for quantitative determination of glutathione and glutathione disulfide levels using enzymatic recycling method. *Nat. Protoc.* **2006**, *1*, 3159–3165. [CrossRef]
59. Bradford, M.M. A rapid and sensitive method for the quantitation of microgram quantities of protein utilizingthe principle of protein-dye binding. *Anal. Biochem.* **1976**, *72*, 248–254. [CrossRef]
60. Doderer, A.; Kokkelink, I.; van der Veen, S.; Valk, B.; Schram, A.; Douma, A. Purification and characterization of two lipoxygenase isoenzymes from germinating barley. *Biochim. Biophys. Acta* **1992**, *112*, 97–104. [CrossRef]

61. Nakano, Y.; Asada, K. Hydrogen peroxide is scavenged by ascorbate-specific peroxidase in spinachchloroplasts. *Plant Cell Physiol.* **1981**, *22*, 867–880.
62. Hossain, M.A.; Nakano, Y.; Asada, K. Monodehydroascorbate reductase in spinach chloroplasts and itsparticipation in the regeneration of ascorbate for scavenging hydrogen peroxide. *Plant Cell Physiol.* **1984**, *25*, 385–395.
63. Foyer, C.H.; Halliwell, B. The presence of glutathione and glutathione reductase in chloroplasts: A proposed role in ascorbic acid metabolism. *Planta* **1976**, *133*, 21–25. [CrossRef] [PubMed]
64. El-Shabrawi, H.; Kumar, B.; Kaul, T.; Reddy, M.K.; Singla-Pareek, S.L.; Sopory, S.K. Redox homeostasis, antioxidant defense, and methylglyoxal detoxification as markers for salt tolerance in Pokkali rice. *Protoplasma* **2010**, *245*, 85–96. [CrossRef] [PubMed]
65. Patra, H.K.; Kar, M.; Mishra, D. Catalase activity in leaves and cotyledons during plant development and senescence. *Biochem. Physiol. Pflanz.* **1978**, *172*, 385–390. [CrossRef]
66. Booth, J.; Boyland, E.; Sims, A.P. An enzyme from rat liver catalysing conjugations with glutathione. *Biochem. J.* **1961**, *79*, 516–524. [CrossRef]
67. Elia, A.C.; Galarini, R.; Taticchi, M.I.; Dörr, A.J.M.; Mantilacci, L. Antioxidant responses and bioaccumulation in *Ictalurus melas* under mercury exposure. *Ecotoxicol. Environ. Safe* **2003**, *55*, 162–167. [CrossRef]
68. Hossain, M.A.; Hossain, M.Z.; Fujita, M. Stress-induced changes of methylglyoxal level and glyoxalase I activity in pumpkin seedlings and cDNA cloning of glyoxalase I gene. *Aust. J. Crop Sci.* **2009**, *3*, 53–64.
69. Principato, G.B.; Rosi, G.; Talesa, V.; Giovannini, E.; Norton, S.J. A comparative study on glyoxalase II from vertebrata. *Enzyme* **1987**, *37*, 164–168. [CrossRef]

plants

Article

Nickel Toxicity Induced Changes in Nutrient Dynamics and Antioxidant Profiling in Two Maize (*Zea mays* L.) Hybrids

Muhammad Amjad *, Hasan Raza, Behzad Murtaza, Ghulam Abbas, Muhammad Imran, Muhammad Shahid, Muhammad Asif Naeem, Ali Zakir and Muhammad Mohsin Iqbal

Department of Environmental Sciences, COMSATS University Islamabad, Vehari-Campus, Vehari 61100, Pakistan; hasanbotanist@gmail.com (H.R.); behzadmurtaza@ciitvehari.edu.pk (B.M.); Ghulamabbas@ciitvehari.edu.pk (G.A.); imranrb@ciitvehari.edu.pk (M.I.); muhammadshahid@ciitvehari.edu.pk (M.S.); asif.naeem@cuivehari.edu.pk (M.A.N.); zakirali@cuivehari.edu.pk (A.Z.); mohsiniqbal6369@gmail.com (M.M.I.)

* Correspondence: amjad_ses@yahoo.com; Tel.: +92-345-7708189

Received: 21 October 2019; Accepted: 9 December 2019; Published: 18 December 2019

Abstract: Nickel (Ni) is among the essential micronutrient heavy metals utilized by plants. However, an elevated level of Ni causes serious concerns for plants' physiology and their survival. This study evaluated the mechanisms influencing the growth, physiology, and nutrient dynamics in two commercial maize hybrids (Syngenta and Pioneer) exposed to Ni treatments in hydroponics nutrient solution (NS). Seedlings were raised in plastic trays with quartz sand, and subsequently transferred to Hoagland's NS at the two leaves stage. After three days of transplantation, Ni levels of 0, 20, and 40 mg L^{-1} were maintained in the nutrient solution. After 30 days of Ni treatments, seedlings were harvested and different growth, physiological, and nutrient concentrations were determined. The results showed that with increasing Ni concentration, the growth of maize hybrids was significantly reduced, and the maize hybrid, Pioneer, showed significantly higher growth than that of Syngenta at all levels of Ni. Higher growth in Pioneer is ascribed to elevated levels of antioxidant enzymes (SOD, CAT, GR, APX, and POX), lower damage to cellular membranes (i.e., higher MSI and lower MDA), and higher tissue nutrient concentrations (N, P, K, Ca, Mg, Fe, Mn, Zn, and Cu). Furthermore, the maize hybrids showed a difference in nutrient translocation from root to shoot which could be one of the factors responsible for differential response of these hybrids against Ni treatments.

Keywords: maize hybrids; nickel; nutrients; translocation; oxidative stress

1. Introduction

Heavy metal toxicity is among the major environmental issues reducing the yield of agricultural crops and posing serious health concerns for humans. Unlike other heavy metals, such as Cd, Pb, Hg, and Ag, Ni is an essential micronutrient that helps the urease enzyme convert urea into ammonia and carbon dioxide [1]. Ni deficiency can trigger inactivation of urease leading to urea accumulation to toxic levels which appear in the form of necrosis of leaf tips [2]. Nickel plays a very important role in plant physiology starting from germination to yield [3]. Nickel enhances yield and quality of most crops [4,5]. Some crops, such as barley (*Hordeum vulgare* L.), cannot complete their life cycle without Ni [6].

Excessive Ni, added to soil through irrigation with sewage sludge applied as a fertilizer or soil amendment, mining, and industrial processes, leads to its toxicity [7]. The worldwide average concentration of Ni in natural soils is 22 mg kg^{-1} [8,9] and has been reported up to 26.4 g kg^{-1} in soil and 0.3 mg L^{-1} in water [10,11]. Its toxicity symptoms appear between 0.19 and 0.85 mM kg^{-1} plant dry biomass [12]. Nickel toxicity inhibits the enzymatic activity, such as the Calvin cycle and chlorophyll biosynthesis which decreases photosynthetic efficiency in plants [13]. Moreover, Ni toxicity affects plants water status and the efficiency of antioxidative machinery [14,15].

Nickel toxicity shares toxic effects with other abiotic stresses such as the production of a large number of reactive oxygen species (ROS) [16]. The accumulation of reactive oxygen species in plants damages almost all the cellular components, for example, cell membrane, lipids, pigments, enzymes, chloroplasts, and nucleic acids [17,18]. The antioxidative enzymes such as superoxide dismutase (SOD), catalase (CAT), glutathione reductase (GR), and peroxidase (POX) etc., play their role in superoxide radical (O_2^-) and hydrogen peroxide (H_2O_2) and inhibit the formation of hydroxyl radicals (OH) [16,19]. Nickel, at an extremely low concentration (0.05 mM), enhances the activity of POD, SOD, and guaiacol peroxidase (GOPX) [20]. However, at elevated Ni concentrations (500 mg kg^{-1}), antioxidative enzyme activity decreases and leads to oxidative damage in plants [21].

Nickel is reported to induce the deficiency of Fe and Zn and hinders the uptake of other heavy metals such as Cd, Co, Cr, and Pb [22]. Nickel toxicity lowers the concentration of N in the roots and leaves of mungbean and chickpea [23]. Hence, a severe toxicity has been observed whenever Ni was supplemented with other heavy metals such as Cd, Cu, Pb, and Zn [24]. The protein contents and carbohydrates decreased with Ni treatment (0, 200, 400, and 800 mg kg^{-1}) in *Myplis snavelus* and sunflower [25], soyabean [26], and maize [27].

Nickel toxicity not only disrupts the uptake of important macro and micronutrients but also hinders the translocation of these nutrients from root to shoot, grain, and fruit [28]. The studies have reported that Ni toxicity tends to reduce the translocation of N from root to shoot [11,29]. Similarly, Ni interferes with translocation of micronutrients especially Fe being similar in chemical properties [30]. It has been reported that one of the causes of Fe deficiency under Ni toxicity is decrease in its translocation from root to shoot [22]. Moreover, Ni toxicity also negatively affects the assimilation of other nutrients [11].

Maize (*Zea mays* L.) is an important cereal crop which is cultivated all over the world. Maize hybrids have higher yield as compared with other varieties and genotypes. However, these hybrids need to be tested under different abiotic stresses such as salinity, drought, and metal toxicity etc. to evaluate their potential yield and survival for sustainability of the agriculture sector. Therefore, the current experiment was conducted to evaluate the physiological response, nutrient uptake, and translocation in two commercial maize hybrids (Syngenta and Pioneer) under Ni toxicity.

2. Materials and Methods

2.1. Selection of Plants and Growth Conditions

The experimental site (latitude, 30°-1.9998′ N, longitude 72°-21′ E, and altitude 184.4 m) had average a minimum and maximum temperature of 15 to 26.4 °C and a relative humidity of 57.5% to 62.7%. Two hybrids of maize, "Pioneer-32F10" and "Syngenta-8441", were selected for the experiment and seeds were purchased from the local market. These hybrids are commonly cultivated in Punjab by the farmers.

Maize seedlings were raised in a 2 inch layer of acid washed quartz sand and irrigated with distilled water. The maize seedlings were transplanted to a Styrofoam sheet floating over Hoagland's nutrient solution at the two leaves stage. The concentrations of different salts were: 5 mM KNO_3, 5 mM $Ca(NO_3)_2$, 2 mM KH_2PO_4 and 1.5 mM $MgSO_4$, 9.11 μM $MnSO_4$, 1.53 μM $ZnSO_4$, 0.235 μM $CuSO_4$, 24.05 μM H_3BO_3, 0.1 μM Na_2MoO_4, and 268.6 μM Fe-EDTA. Aeration was maintained throughout the experimental period and pH was maintained daily between 6.0 and 6.5 by 1N NaOH/HCl solution. After three days of transplantation, Ni levels of 0, 20, and 40 mg L^{-1} were maintained in the NS using $NiCl_2.6H_2O$. The nutrient solution was changed after every week throughout the experimental period.

2.2. Plant Biomass Measurements

The plants were sampled after 30 days of Ni treatments and plant growth parameters were measured. Shoot fresh weight of maize hybrids was measured with analytical balance, whereas root fresh weight was measured after paper blotting the roots to remove moisture. The shoot and root length was measured immediately, while their dry weight was measured after oven drying the samples at 65 °C for 48 h.

2.3. Determination of Antioxidant Enzymes Activity

Leaf sample (0.5 g) was homogenized with potassium phosphate buffer (pH 7.0) in a precooled mortar and pestle. The homogenate was centrifuged in a temperature-controlled centrifuge machine (4 °C) for 20 min at 10,000× *g*. The supernatant obtained was used for the determination of antioxidative enzymes.

The activity of superoxide dismutase activity (SOD) was determined by measuring its ability to inhibit the photochemical reduction of nitroblue tetrazolium (NBT). The reaction mixture was prepared along with a control (without enzyme extract) as described by Gupta et al. [31]. The reaction was started by adding riboflavin and placing the mixture under light (15W) for 15 min. The absorbance was measured at 560 nm after stopping the reaction by switching off the light. The SOD activity was calculated in mg^{-1} protein min^{-1} (1 unit = 50% reduction in absorbance).

The catalase activity (CAT) was determined spectrophotometrically by recording the decrease in absorbance at 240 nm resulting from the decomposition of H_2O_2. The reaction mixture consisted of potassium phosphate buffer, H_2O, and enzyme extract as explained by Aebi [32]. The enzyme activity was presented in mmol H_2O_2 mg^{-1} protein min^{-1}.

The activity of ascorbate peroxidase (APX) was measured by the oxidation of ascorbate and decrease in absorbance at 290 nm, according to the method described by Amako et al. [33]. The enzyme activity was expressed in mmol H_2O_2 mg^{-1} protein min^{-1}.

Glutathione reductase activity (GR) was measured by the reduction of glutathione disulfide (GSSG) to two molecules of glutathione equivalent to NADPH coenzyme expressed as µmol of NADPH mg^{-1} protein min^{-1} by recording the increase in absorbance (412 nm for 5 min) according to Sairam et al. [34].

Peroxidase (POX) activity was determined by measuring its ability to cause reduction of guaiacol (recorded at 436 nm for 90 s) in the reaction mixture prepared according to the method described by Pandey and Pathak [14]. The enzyme's activity was expressed as mmol H_2O_2 mg^{-1} protein min^{-1}.

The Bradford method was used to measure the leaf protein contents [35].

2.4. Determination of Lipid Peroxidation

Malondialdehyde contents (MDA) were measured to assess the extent of lipid peroxidation in maize leaves. Leaf sample (0.5 g) was homogenized in 5 mL of 0.1% trichloro acetic acid (TCA) and 4 mL of 0.5% thiobarbituric acid (TBA). The mixture was heated for 30 min at 95 °C and then, centrifuged at 20,000× *g* for 10 min. The supernatant was collected and the absorbance of reaction mixture was recorded at 532 and 600 nm [36]. The difference in absorbance was used to calculate the MDA content (nmol/g fresh weight).

2.5. Determination of Membrane Stability Index

The stability of leaf membrane was determined by measuring the electrical conductivities (EC) of two sets of double distilled water having 0.1 g of leaf sample at 40 °C after 30 min (EC_1) and at 100 °C after 10 min (EC_2) according to Amjad et al. [37]. The leaf membrane stability index (MSI) was calculated by the formula:

$$MSI = \left(1 - \frac{EC_1}{EC_2}\right) \times 100 \quad (1)$$

2.6. Tissues Nitrogen and Phosphorus Determination

Total nitrogen concentration in shoot and root was analyzed with Kjeldahl digestion [38], whereas tissues total phosphorus concentration was measured using the colorimetric method [39].

2.7. Atomic Absorption Spectrometric Measurements

The concentrations of Ca, Mg, K, Fe, Zn, Cu, Mn, and Ni in shoot and root were determined on an atomic absorption spectrophotometer (Thermo AA, Solaar series, Thermo Scientific) [40]. The concentrations were determined from one-gram oven-dried samples after wet digestion with a di-acid mixture of HNO_3 and $HClO_4$ (1:3).

Nickel uptake was calculated as:

$$\text{Ni uptake (mg)} = \text{tissues Ni concentration} \times \text{tissues dry mass} \quad (2)$$

While root to shoot (R-S) Ni translocation was calculated as:

$$\text{R-S translocation} = \text{concentration of nutrient in shoot/concentration of nutrient in root} \quad (3)$$

2.8. Statistical Analysis

The experiment was performed according to factorial design replicated thrice. The significance of treatment means was determined by two-way analysis of variance (ANOVA). The least significant test was applied to check significant differences (LSD) among the treatment means using the computer program Statistix 8.1.

3. Results

3.1. Growth of Maize Hybrids

The results revealed that growth parameters (shoot/root fresh and dry weight and length) decreased significantly after thirty days of exposure to 20 and 40 mg Ni L^{-1} as compared with 0 mg L^{-1} Ni treatment in both maize hybrids (Table 1). Syngenta showed significantly higher values of measured growth parameters as compared with Pioneer under 20 and 40 mg L^{-1}. However, the difference was not significant between the maize hybrids for shoot and root length at all levels of Ni and for shoot and root fresh weight at 0 mg L^{-1}.

Table 1. Effect of Ni treatments (0, 20, and 40 mg L^{-1}) on shoot fresh weight, root fresh weight, shoot dry weight, root dry weight, shoot length, and root length of two maize hybrids, Syngenta-8441 (Syngenta) and Pioneer-32F10 (Pioneer). Each value is mean of three replicates + standard error. Values in the parentheses represent the % of the values in treatment (0 mg L^{-1}) in both maize hybrids. The *p*-values show the significance for Ni levels, maize hybrids (MH), and their interaction (Ni x MH) at $p < 0.05$. Letters along the values show the results of the least significant test. Means that share the same letters show results that are not significant.

Ni Concentration	Hybrids	Shoot Fresh Weight (g) Mean ± S.E.	Root Fresh Weight (g) Mean ± S.E.	Shoot Dry Weight (g) Mean ± S.E.	Root Dry Weight (g) Mean ± S.E.	Shoot Length (cm) Mean ± S.E.	Root Length (cm) Mean ± S.E.
0 mg L^{-1}	Syngenta	3.28 ± 0.26 a	2.26 ± 0.24 a	0.316 ± 0.019 b	0.135 ± 0.018 a	38.6 ± 0.95 a	31.6 ± 0.75 a
	Pioneer	2.64 ± 0.12 b	2.33 ± 0.11 a	0.388 ± 0.002 a	0.139 ± 0.012 a	37.6 ± 1.55 a	32.5 ± 3.30 a
20 mg L^{-1}	Syngenta	2.16 ± 0.12 (66) c	2.13 ± 0.21 (94) ab	0.239 ± 0.016 (76) d	0.104 ± 0.009 (77) b	33.1 ± 1.64 (86) b	27.7 ± 1.44 (88) b
	Pioneer	2.50 ± 0.05 (95) bc	2.00 ± 0.13 (86) abc	0.351 ± 0.013 (90) c	0.131 ± 0.005 (94) ab	35.7 ± 1.61 (96) ab	28.8 ± 1.80 (89) b
40 mg L^{-1}	Syngenta	1.31 ± 0.06 (40) e	1.50 ± 0.05 (66) bc	0.190 ± 0.005 (60) f	0.094 ± 0.007 (70) c	30.4 ± 1.78 (79) c	24.2 ± 0.92 (77) c
	Pioneer	1.72 ± 0.02 (65) d	1.65 ± 0.08 (71) c	0.272 ± 0.017 (70) e	0.111 ± 0.010 (80) abc	31.1 ± 3.40 (83) bc	26.0 ± 3.07 (80) bc
p-value	Ni	0.0000	0.0018	0.0000	0.0132	0.0012	0.2104
	MH	0.0253	0.0357	0.0263	0.0345	0.0282	0.3312
	Ni x MH	0.0047	0.0772	0.0001	0.0425	0.1532	0.2461

3.2. Antioxidant Enzymes Activity

The maize hybrids showed a significant ($p < 0.05$) differential increase in antioxidant enzymes activity, i.e., SOD, CAT, GR, APX, and POX at 20 and 40 mg L^{-1} Ni as compared with the control (0 mg L^{-1} Ni) (Figure 1). At 20 mg L^{-1}, the increase in activity of SOD, CAT, GR, APX, and POX was 2.4, 1.7, 1.3, 1.6, and 2 times, respectively, as compared with the control treatment in Pioneer, whereas in Syngenta the increase was 2.0, 1.3, 1.1, 1.6, and 2.0 times, respectively. Similarly, at 40 mg L^{-1}, the increase in activity of SOD, CAT, GR, APX, and POX was 2.7, 2.4, 1.6, 3.3, and 3.6 times, respectively, as compared with the control treatment in Pioneer, whereas in Syngenta the increase was 2.5, 2.1, 1.5, 2.8, and 2.7 times, respectively.

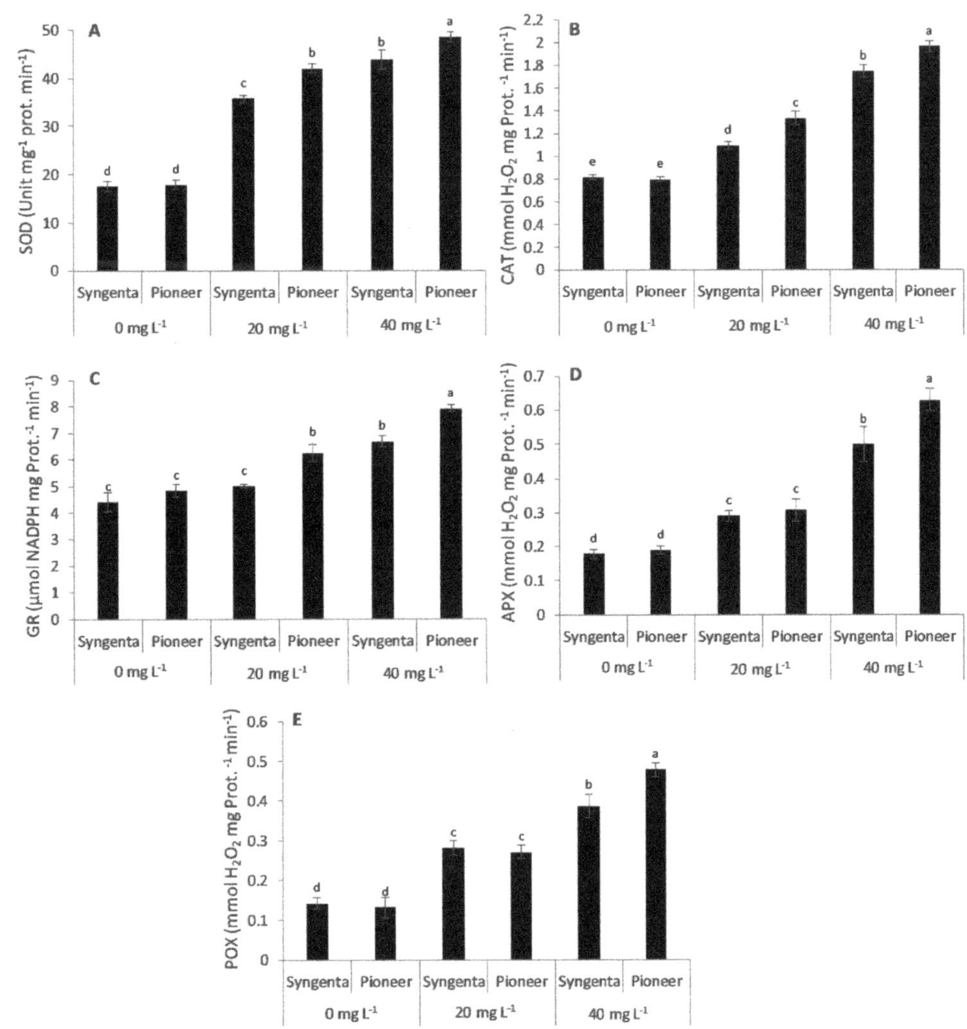

Figure 1. Effect of Ni treatments (0, 20, and 40 mg L^{-1}) on activity of antioxidant enzymes: SOD (**A**), CAT (**B**), GR (**C**), APX (**D**), and POX (**E**) in two maize hybrids, Syngenta-8441 (Syngenta) and Pioneer-32F10 (Pioneer). Different letters on bars represent the result of LSD test, bars sharing the same letter(s) show the non-significance of the means at $p < 0.05$.

3.3. Membrane Stability Index and Malondialdehyde Contents

Membrane stability in both maize hybrids against Ni treatments (20 and 40 mg L^{-1}) induced oxidative stress was measured by membrane stability index (%) and malondialdehyde concentration (Figure 2A,B, respectively). The damage caused by ROS to cellular membranes was assessed by measuring MDA contents produced as a result of lipid peroxidation. The results showed that MDA contents increased significantly with increasing Ni levels and correspondingly decreased the membrane stability index in both maize hybrids. The maize hybrids showed significant differences in terms of MDA concentration, relatively higher MDA content, and lower membrane stability index in Syngenta as compared with Pioneer.

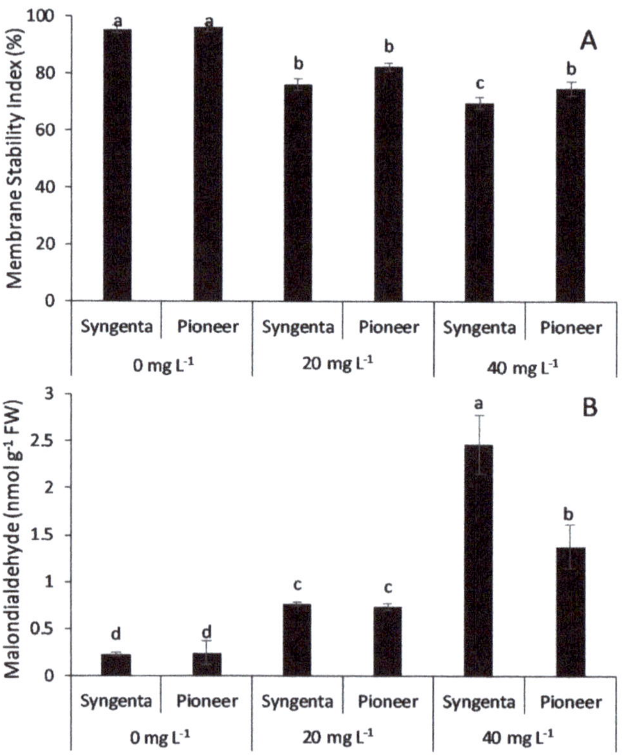

Figure 2. Membrane stability index (**A**) and lipid peroxidation (malondialdehyde) (**B**) in two maize hybrids, Syngenta (Syngenta-8441) and Pioneer (Pioneer-32F10), after 30 days of Ni treatment (0, 20, and 40 mg L^{-1}). Different letters on bars represent the result of the LSD test, bars sharing the same letter(s) show the non-significance of the means at $p < 0.05$.

3.4. Nutrient Concentration in Tissues and Root to Shoot Translocation

The results showed a significant ($p < 0.05$) decline in shoot N concentration at 20 mg L^{-1} Ni concentration as compared with 0 mg L^{-1} Ni treatment. However, there was no significant difference in shoot N concentration between the 20 and 40 mg L^{-1} Ni level (Figure 3A), whereas root N concentration increased at the 20 mg L^{-1} and decreased at the 40 mg L^{-1} Ni level. Root to shoot N translocation decreased at the 20 mg L^{-1} Ni level and remained almost constant at the 40 mg L^{-1} Ni level. Maize hybrids (Syngenta and Pioneer) showed no significant difference for both shoot and root N concentration at all levels of Ni.

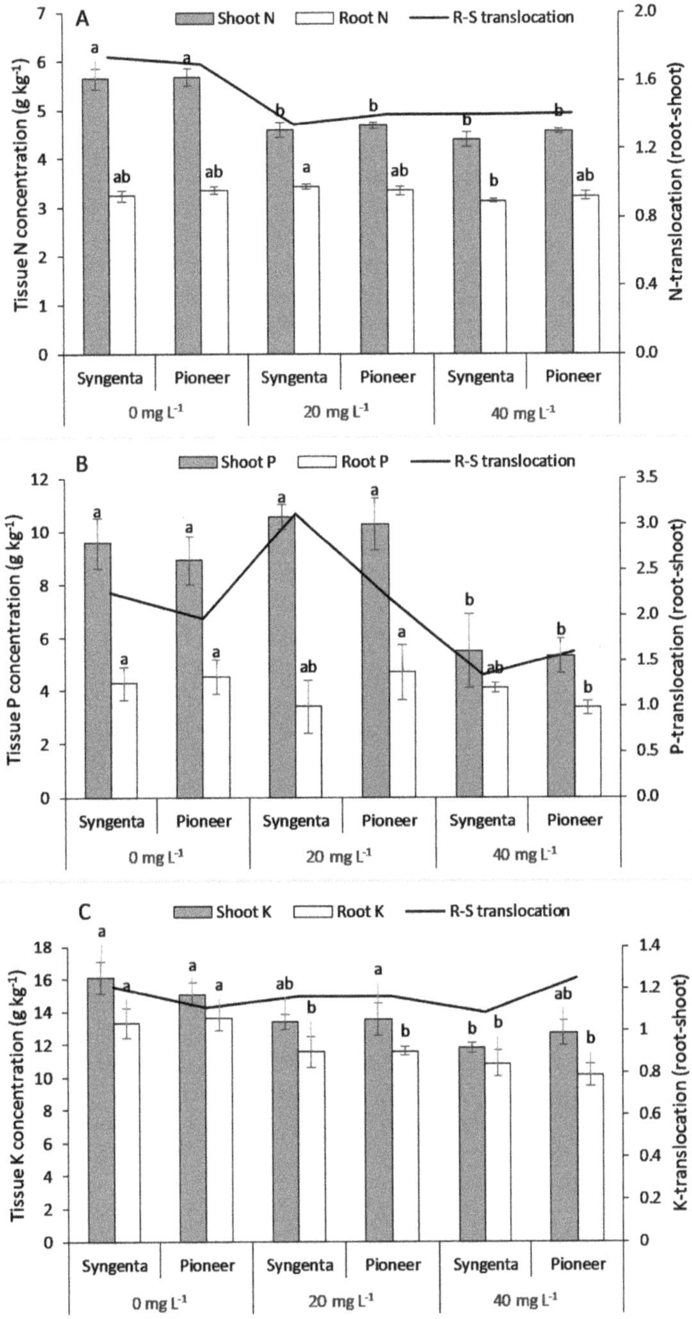

Figure 3. Concentrations of shoot and root N (**A**), P (**B**), K (**C**) and root to shoot translocation in two maize hybrids, Syngenta (Syngenta-8441) and Pioneer (Pioneer-32F10), after 30 days of Ni treatments (0, 20, and 40 mg L^{-1}). Different letters on bars represent the result of the LSD test, bars sharing the same letter(s) show the non-significance of the means at $p < 0.05$.

The results showed no significant ($p < 0.05$) increase in P concentration at 20 mg L^{-1} Ni, whereas a significant decline in P concentration was observed at 40 mg L^{-1} Ni as compared with the control (Figure 3B). Root P concentration was not significantly affected by the Ni levels. The maize hybrids showed no significant differences at all Ni levels for both shoot and root P concentration. Root to shoot translocation of P increased at the 20 mg L^{-1} Ni level and decreased at the 40 mg L^{-1} level of Ni.

Shoot and root K concentrations decreased significantly with increasing Ni toxicity (Figure 3C). The maize hybrids showed no significant differences for both shoot and root K concentrations at all levels of Ni. Root to shoot translocation of K was higher in maize hybrid Pioneer than that of Syngenta at 20 mg L^{-1} and 40 mg L^{-1} Ni.

Both root and shoot tissues showed significantly low Ca concentration at elevated levels of Ni, i.e., 20 and 40 mg L^{-1} as compared with the control for both maize hybrids (Figure 4A). Root to shoot translocation of Ca was higher in Syngenta than that of Pioneer at all the Ni levels. Both of the maize hybrids showed a significant decreasing trend regarding shoot Mg concentration with increasing Ni level, whereas there was no significant difference between them at all the levels of Ni. In addition, root Mg concentration was significantly lower at the 20 and 40 mg L^{-1} Ni levels as compared with the control with no significant differences in the maize hybrids (Figure 4B), and Mg translocation (root to shoot) was higher in Pioneer than Syngenta at the 20 mg L^{-1} Ni level and decreased at the 40 mg L^{-1} Ni level.

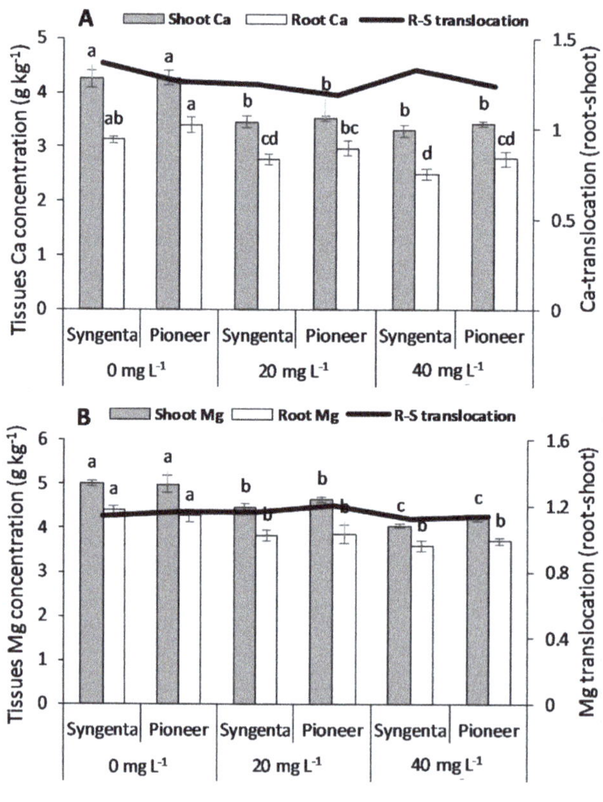

Figure 4. Concentration of shoot and root Ca (**A**), Mg (**B**), and root to shoot translocation in two maize hybrids, Syngenta (Syngenta-8441) and Pioneer (Pioneer-32F10), after 30 days of Ni treatments (0, 20, and 40 mg L^{-1}). Different letters on bars represent the result of the LSD test, bars sharing the same letter(s) show the non-significance of the means at $p < 0.05$.

Shoot Fe concentration decreased significantly, and there was no significant change in root Fe concentration in both maize hybrids with increasing Ni levels (Figure 5A). The maize hybrids showed no significant ($p < 0.05$) differences for both shoot and root Fe concentration at all the Ni levels. Root to shoot Fe translocation decreased at the 20 mg L^{-1} level of Ni as compared with the control.

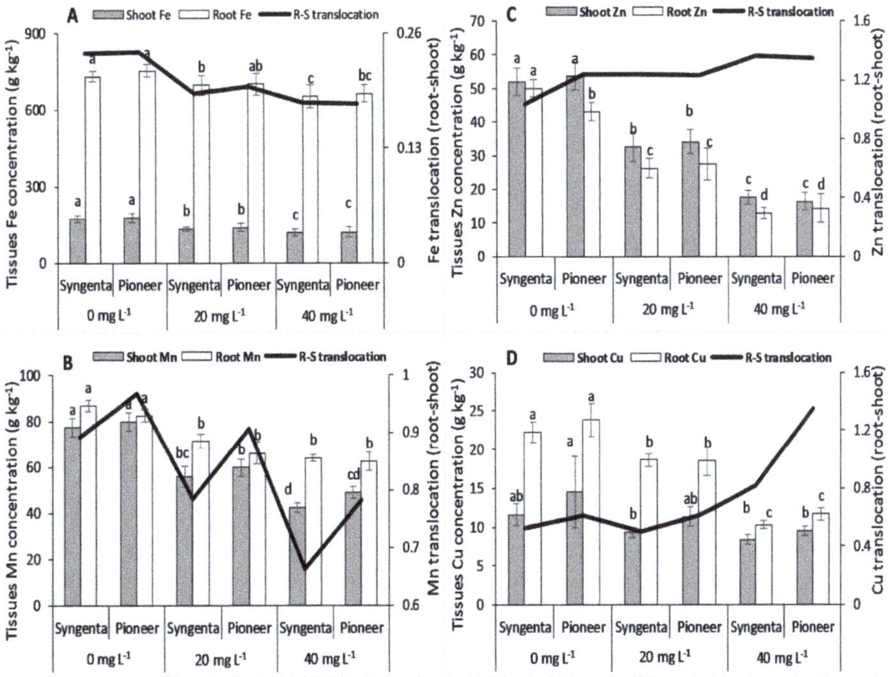

Figure 5. Concentration of shoot and root Fe (**A**), Mn (**B**), Zn (**C**), Cu (**D**), and their translocation (root to shoot) in two maize hybrids, Syngenta (Syngenta-8441) and Pioneer (Pioneer-32F10), after 30 days of Ni treatments (0, 20, and 40 mg L^{-1}). Different letters on bars represent the result of the LSD test, bars sharing the same letter(s) show the non-significance of the means at $p < 0.05$.

The results showed that Mn concentration in tissues (shoot and root) decreased significantly with increasing Ni levels in both maize hybrids (Figure 5B). Translocation (root to shoot) of Mn decreased in both maize hybrids at the 20 mg L^{-1} Ni level with higher values in Pioneer than Syngenta. There was no significant difference in Zn concentration in tissues between maize hybrids at all the Ni levels (Figure 5C). Both shoot and root Zn concentration decreased significantly at 20 and 40 mg L^{-1} as compared with 0 mg L^{-1} Ni level, whereas its root to-shoot translocation increased. Both shoot and root Cu concentration decreased significantly ($p < 0.05$) by increasing Ni (20 and 40 mg L^{-1}) levels in both maize hybrids (Figure 5D). The maize hybrids did not differ significantly in shoot and root Cu concentration at all Ni levels. Whereas, root to shoot translocation increased with increasing Ni levels.

3.5. Plant Tissues Nickel Concentration, Uptake, and Root to Shoot Translocation

The results showed that total Ni concentration in the plants increased significantly with increasing Ni levels in both maize hybrids (Figure 6A). The maize hybrids differed significantly in terms of shoot and root Ni concentration especially at elevated Ni levels, i.e., 20 and 40 mg L^{-1}. Shoot Ni concentration was significantly higher than root Ni in both maize hybrids except in Pioneer at 20 mg L^{-1}. Shoot and root Ni uptake significantly increased at increasing Ni levels with significant differences between the maize hybrids (Figure 6B). Root to shoot Ni translocation remained almost constant at 0 and 20 mg L^{-1} in both maize hybrids although higher in Syngenta than Pioneer (Figure 6C) but decreased significantly in Syngenta and increased in Pioneer at the 40 mg L^{-1} Ni level.

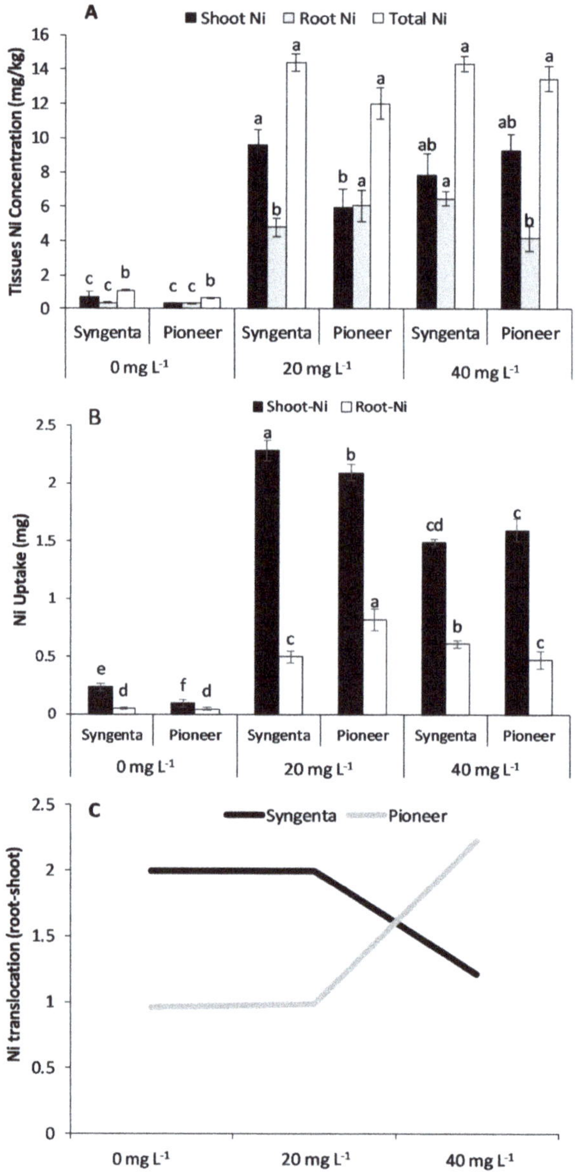

Figure 6. Concentration of shoot and root Ni concentration (**A**), uptake (**B**), and Ni root to shoot translocation (**C**) in two maize hybrids, Syngenta (Syngenta-8441) and Pioneer (Pioneer-32F10), after 30 days of Ni treatments (0, 20, and 40 mg L^{-1}). Different letters on bars represent the result of the LSD test, bars sharing the same letters show the non-significance of the means at $p < 0.05$.

4. Discussion

This study focuses on the nutrient dynamics and antioxidant defense system of two maize hybrids, i.e., Syngenta and Pioneer against Ni toxicity in nutrient solution. The decrease in growth caused by Ni toxicity could be attributed to perturbed nutrient uptake and translocation because excess Ni decreases the uptake of macro and micronutrients [15,23]. Secondarily, the decrease in growth could also be attributed to oxidative damage caused by the excess of ROS measured by lipid peroxidation (malondialdehyde) and the membrane stability index. The results showed that the maize hybrid, Syngenta, had a higher concentration of malondialdehyde and a lower membrane stability index as compared with Pioneer, and thereby showed significantly higher growth in Pioneer than Syngenta at the 20 and 40 mg L^{-1} Ni levels (Table 1). Nickel toxicity has been reported to interfere with the plant water relations, wheat growth, and plant nutrient status [41]. Different plant species and genotypes show variation in survival ability against abiotic stress induced oxidative damage, i.e., malondialdehyde concentration is responsible for their tolerance against these stresses [42]. Similarly, in this experiment, maize hybrid Pioneer showed lower malondialdehyde concentration and higher membrane stability, thus, having higher growth than Syngenta (Figure 2).

Nickel toxicity has been reported to cause damage to cellular membranes with the excessive production of ROS beyond the neutralizing capacity of the antioxidative system. Its toxicity is capable of oxidizing vital cellular membrane components including lipids, DNA, and proteins [43,44]. Nickel causes higher degrees of lipid peroxidation due to its higher mobility in plants as compared with other heavy metals' toxicity (Co, Cd, Cu, and Zn). Similarly, this study showed that Ni (20 and 40 mg L^{-1} levels) caused the lipid peroxidation (MDA concentration), thereby decreasing MSI more in Syngenta as compared with Pioneer (Figure 2A,B). Lower lipid peroxidation in Pioneer could be attributed to the higher activity of SOD, CAT, GR, APX, and POX than in Syngenta. Nickel toxicity has been reported to enhance the activity of SOD which detoxifies superoxide ions to hydrogen peroxide and molecular oxygen [45,46]. Hydrogen peroxide is subsequently detoxified by other enzymes such as CAT, APX, and POX to H_2O [47,48], whereas GR reduces oxidized glutathione (GSSG) to sustain the glutathione equilibrium in the cells [43]. The significantly higher activity of all the enzymes in the "Pioneer" maize hybrid as compared with the "Syngenta" indicated its higher Ni tolerance capability, responsible for better growth and lower lipid peroxidation (Table 1 and Figure 1).

Owing to similar characteristics with both macronutrients (Ca, Mg) and micronutrients (Fe, Cu, and Zn), Ni competes with both of them in sorption and transpiration in plants [15,49]. Because of this similarity, a high concentration of Ni causes deficiency of these nutrients in plants by hindering their sorption, uptake, and translocation [30,41]. Our results validate previous findings, that Ni toxicity significantly decreased the nutrient concentrations in both shoot and root of maize hybrids. However, the Fe concentration in tissues remained almost constant with the excessive Ni in the growth medium (Figure 5A). Previous studies have reported a synergistic effect of Ni on N and P uptake and translocation from root to shoot in different plant species within the toxic range of Ni in the growth medium [2,50,51]. Contrarily, this study showed a decrease in root to shoot translocation of N, suggesting the toxicity of the applied Ni levels. The translocation of P showed an increase at the 20 mg L^{-1} level of Ni and a decrease at the 40 mg L^{-1} level of Ni, indicating strong competition between Ni and P ions at the 40 mg L^{-1} level of Ni. A significant difference in nutrient concentration between maize hybrids suggests that genotypes and hybrids vary in their ability to counter Ni toxicity. Moreover, a higher value of Ni translocation from root to shoot in "Pioneer" was one of the factors responsible for higher growth under Ni toxicity. Higher concentration and translocation of the measured micronutrients in the maize hybrid "Pioneer" could have enhanced the activity of SOD and CAT, since these metals are present in their prosthetic groups [28,44]. Lower values of these micronutrients (Fe, Mn, Cu, and Zn) in the maize hybrid Syngenta could have reduced the biosynthesis of the measured antioxidative enzymes, and therefore caused higher oxidative damage as compared with Pioneer.

Previous reports have suggested that Ni is a highly mobile element and approximately 50% of the plant-absorbed Ni is accumulated in roots, with over 80% in the vacuoles and 20% in cortex [11,23]. However, results from this study revealed that the Ni concentration was higher in the shoot than the root tissues and the total Ni concentration (shoot and root) did not remained at significantly high Ni levels in both maize hybrids (Figure 6A–C). These results suggest that total tissues Ni concentration does not remains significant beyond a certain level owing to the limited ability of maize hybrids to accumulate and uptake Ni from the growth medium. However, increased levels of Ni toxicity cause an opposite trend of root to shoot Ni translocation as suggested by the increase in root to-shoot translocation at the 40 mg L^{-1} level of Ni in tolerant maize hybrid Pioneer and sensitive (Syngenta). The increased Ni translocation at the 40 mg L^{-1} level of Ni in Pioneer is attributed to enhanced activity of the antioxidant enzymes resulting in lower lipid peroxidation and higher growth than that in Syngenta, as shown in Figures 1 and 2.

5. Conclusions

Results from this study suggest that Ni toxicity severely affects maize plant physiology with oxidative damage and disturbs the nutrient uptake and translocation in Pioneer and Syngenta maize hybrids. There are differences in maize hybrids regarding growth, antioxidant enzymes activity, nutrient uptake, and translocation. The higher growth in the maize hybrid "Pioneer" is attributed to elevated levels of antioxidative enzymes and nutrient translocation from root to shoot which affected the normal functioning of plants.

Author Contributions: Conceptualization, M.A.; experimental work, H.R.; formal analysis, B.M. and G.A.; investigation, M.I. and M.S.; methodology, M.A.N. and A.Z.; supervision, M.A.; writing—original draft, M.A. and M.M.I.; writing—review and editing, M.I., and M.M.I. All authors have read and agreed to the published version of the manuscript.

Funding: This research received no external funding.

Acknowledgments: Authors are thankful to Ghulam Murtaza, University of Agriculture, Faisalabad for language editing and proof reading of the manuscript.

Conflicts of Interest: The authors declare that they have no conflict of interest.

References

1. Gajewska, E.; Skłodowska, M. Nickel-induced changes in nitrogen metabolism in wheat shoots. *J. Plant Physiol.* **2009**, *166*, 1034–1044. [CrossRef]
2. Fabiano, C.; Tezotto, T.; Favarin, J.L.; Polacco, J.C.; Mazzafera, P. Essentiality of nickel in plants: A role in plant stresses. *Front. Plant Sci.* **2015**, *6*, 754. [CrossRef] [PubMed]
3. Sreekanth, T.; Nagajyothi, P.; Lee, K.; Prasad, T. Occurrence, physiological responses and toxicity of nickel in plants. *Int. J. Environ. Sci. Technol.* **2013**, *10*, 1129–1140. [CrossRef]
4. Brake, S.; Jensen, R.; Mattox, J. Effects of nickel amended soils on tomato plants. *Plant Soil* **2004**, *54*, 860–869.
5. Gad, N.; El-Sherif, M.; El-Gereedly, A. Influence of nickel on some physiological aspects of tomato plants. *Aust. J. Basic Appl. Sci.* **2007**, *1*, 286–293.
6. Brown, P.H.; Welch, R.M.; Madison, J.T. Effect of nickel deficiency on soluble anion, amino acid, and nitrogen levels in barley. *Plant Soil* **1990**, *125*, 19–27. [CrossRef]
7. Ali, B.; Hayat, S.; Fariduddin, Q.; Ahmad, A. *Nickel: Essentiality, Toxicity and Tolerance in Plants*; Nickel in Relation to Plants; Narosa Publishing House: New Delhi, India, 2009; pp. 73–80.
8. McIlveen, W.D.; Negusanti, J.J. Nickel in the terrestrial environment. *Sci. Total Environ.* **1994**, *148*, 109–138. [CrossRef]
9. Kabata-Pendias, A. *Trace Elements in Soils and Plants*; CRC Press: Boca Raton, FL, USA, 2000; p. 245.
10. Zwolsman, J.J.G.; Van Bokhoven, A.J. Impact of summer droughts on water quality of the Rhine River-a preview of climate change? *Water Sci. Tech.* **2007**, *56*, 45–55. [CrossRef]
11. Chen, C.; Huang, D.; Liu, J. Functions and toxicity of nickel in plants: Recent advances and future prospects. *Clean Soil Air Water* **2009**, *37*, 304–313. [CrossRef]

12. de Queiroz Barcelos, J.P.; de Souza Osorio, C.R.W.; Leal, A.J.F.; Alves, C.Z.; Santos, E.F.; Reis, H.P.G.; dos Reis, A.R. Effects of foliar nickel (Ni) application on mineral nutrition status, urease activity and physiological quality of soybean seeds. *Aust. J. Crop Sci.* **2017**, *11*, 184. [CrossRef]
13. Sheoran, I.; Aggarwal, N.; Singh, R. Effects of cadmium and nickel on in vivo carbon dioxide exchange rate of Pigeon pea (*Cajanus cajan* L.). *Plant Soil* **1990**, *129*, 243–249. [CrossRef]
14. Pandey, N.; Pathak, G. Nickel alters antioxidative defense and water status in green gram. *Ind. J. Plant Physiol.* **2006**, *11*, 113.
15. Gajewska, E.; Bernat, P.; Długoński, J.; Skłodowska, M. Effect of nickel on membrane integrity, lipid peroxidation and fatty acid composition in wheat seedlings. *J. Agron. Crop Sci.* **2012**, *198*, 286–294. [CrossRef]
16. Gill, S.S.; Tuteja, N. Reactive oxygen species and antioxidant machinery in abiotic stress tolerance in crop plants. *Plant Physiol. Biochem.* **2010**, *48*, 909–930. [CrossRef] [PubMed]
17. Maheshwari, R.; Dubey, R. Nickel-induced oxidative stress and the role of antioxidant defence in rice seedlings. *Plant Growth Reg.* **2009**, *59*, 37–49. [CrossRef]
18. Gopal, R.; Nautiyal, N. Growth, antioxidant enzymes activities, and proline accumulation in mustard due to nickel. *Int. J. Veg. Sci.* **2012**, *18*, 223–234. [CrossRef]
19. Amjad, M.; Akhtar, J.; Haq, M.; Riaz, M.; Jacobsen, S.-E. Understanding salt tolerance mechanisms in wheat genotypes by exploring antioxidant enzymes. *Pak. J. Agric. Sci.* **2014**, *51*, 969–976.
20. Gautam, S.; Pandey, S. Growth and biochemical responses of nickel toxicity on leguminous crop (*Lens esculentum*) grown in alluvial soil. *Res. Environ. Life Sci.* **2008**, *1*, 25–28.
21. Georgiadou, E.C.; Kowalska, E.; Patla, K.; Kulbat, K.; Smolińska, B.; Leszczyńska, J.; Fotopoulos, V. Influence of heavy metals (Ni, Cu, and Zn) on nitro-oxidative stress responses, proteome regulation and allergen production in basil (*Ocimum basilicum* L.) plants. *Front Plant Sci.* **2018**, *9*, 862. [CrossRef]
22. Myśliwa-Kurdziel, B.; Prasad, M.N.V.; Strzałka, K. Photosynthesis in heavy metal stressed plants. In *Prasad MNV(ed) Heavy Metal Stress in Plants: From Biomolecules to Ecosystems*; Springer: Berlin, Germany, 2004; pp. 146–181.
23. Yusuf, M.; Fariduddin, Q.; Hayat, S.; Ahmad, A. Nickel: An overview of uptake, essentiality and toxicity in plants. *Bull. Environ. Contam. Toxicol.* **2011**, *86*, 1–17. [CrossRef]
24. Athar, R.; Ahmad, M. Heavy metal toxicity in legume-microsymbiont system. *J. Plant Nutr.* **2002**, *25*, 369–386. [CrossRef]
25. Pillay, S.V.; Rao, V.; Rao, K. Effect of nickel toxicity in *Hyptis suareeolens* (L.) Poit. and *Helianthus annuus* L. *Ind. J. Plant Physiol.* **1996**, *1*, 153–156.
26. El-Shintinawy, F.; El-Ansary, A. Differential effect of Cd^{2+} and Ni^{2+} on amino acid metabolism in soybean seedlings. *Biol. Plantar.* **2000**, *43*, 79–84. [CrossRef]
27. Baccouch, S.; Chaoui, A.; El Ferjani, E. Nickel toxicity induces oxidative damage in Zea mays roots. *J. Plant Nutr.* **2001**, *24*, 1085–1097. [CrossRef]
28. Pandey, N.; Sharma, C.P. Effect of heavy metals Co^{2+}, Ni^{2+} and Cd^{2+} on growth and metabolism of cabbage. *Plant Sci.* **2002**, *163*, 753–758. [CrossRef]
29. Ameen, N.; Amjad, M.; Murtaza, B.; Abbas, G.; Shahid, M.; Imran, M.; Asif, M.A.; Niazi, N.K. Biogeochemical behavior of nickel under different abiotic stresses: toxicity and detoxification mechanisms in plants. *Environ. Sci. Poll. Res.* **2019**, *26*, 10496–10514. [CrossRef]
30. Ahmad, P.; Jaleel, C.A.; Salem, M.A.; Nabi, G.; Sharma, S. Roles of enzymatic and nonenzymatic antioxidants in plants during abiotic stress. *Crit. Rev. Biotechnol.* **2010**, *30*, 161–175. [CrossRef]
31. Gupta, A.S.; Heinen, J.L.; Holaday, A.S.; Burke, J.J.; Allen, R.D. Increased resistance to oxidative stress in transgenic plants that overexpress chloroplastic Cu/Zn superoxide dismutase. *Proc. Natl. Acad. Sci. USA* **1993**, *90*, 1629–1633. [CrossRef]
32. Aebi, H. Catalase in vitro. In *Methods in Enzymology*; Elsevier: Amsterdam, The Netherlands, 1984; pp. 121–126.
33. Amako, K.; Chen, G.-X.; Asada, K. Separate assays specific for ascorbate peroxidase and guaiacol peroxidase and for the chloroplastic and cytosolic isozymes of ascorbate peroxidase in plants. *Plant Cell Physiol.* **1994**, *35*, 497–504.
34. Sairam, R.K.; Rao, K.V.; Srivastava, G. Differential response of wheat genotypes to long term salinity stress in relation to oxidative stress, antioxidant activity and osmolyte concentration. *Plant Sci.* **2002**, *163*, 1037–1046. [CrossRef]

35. Bradford, M. A rapid and sensitive method for the quantitation of microgram quantities of protein utilizing the principle of protein-dye binding. *Analyt. Biochem.* **1976**, *72*, 248–254. [CrossRef]
36. Smeets, K.; Cuypers, A.; Lambrechts, A.; Semane, B.; Hoet, P.; Van Laere, A.; Vangronsveld, J. Induction of oxidative stress and antioxidative mechanisms in Phaseolus vulgaris after Cd application. *Plant Physiol. Biochem.* **2005**, *43*, 437–444. [CrossRef] [PubMed]
37. Amjad, M.; Akhtar, J.; Murtaza, B.; Abbas, G.; Jawad, H. Differential accumulation of potassium results in varied salt-tolerance response in tomato (*Solanum lycopersicum* L.) cultivars. *Horti. Environ. Biotech.* **2016**, *57*, 248–258. [CrossRef]
38. Buresh, R.J.; Austin, E.R. Direct measurement of dinitrogen and nitrous oxide flux in flooded rice fields. *Soil Sci. Soci. Am. J.* **1988**, *52*, 681–688. [CrossRef]
39. Jones, D.; Smith, B.F.L.; Wilson, M.J.; Goodman, B.A. Phosphate solubilizing fungi in a Scottish upland soil. *Mycol. Res.* **1991**, *95*, 1090–1903. [CrossRef]
40. Estefan, G.; Sommer, R.; Ryan, J. *Methods of Soil, Plant, and Water Analysis: A Manual for the West, Asia and North Africa Region*; ICARDA: Beirut, Lebanon, 2013.
41. Ain, Q.; Akhtar, J.; Amjad, M.; Haq, M.; Saqib, Z. Effect of Enhanced Nickel Levels on Wheat Plant Growth and Physiology under Salt Stress. *Comm. Soil Sci. Plant Anal.* **2016**, *47*, 2538–2546. [CrossRef]
42. Rouphael, Y.; Cardarelli, M.; Rea, E.; Colla, G.J.E.; Botany, E. Grafting of cucumber as a means to minimize copper toxicity. *Environ. Exper. Bot.* **2008**, *63*, 49–58. [CrossRef]
43. Gratão, P.L.; Monteiro, C.C.; Carvalho, R.F.; Tezotto, T.; Piotto, F.A.; Peres, L.E.; Azevedo, R.A. Biochemical dissection of diageotropica and Never ripe tomato mutants to Cd-stressful conditions. *Plant Physiol. Biochem.* **2012**, *56*, 79–96. [CrossRef]
44. Feroza, K.; Rosa, M. Reactive oxygen species, abiotic stress and stress combination. *Plant J.* **2017**, *90*, 856–867.
45. Pandey, S.; Gautam, S. Effect of nickel stress on growth and physiological responses of *Trigonella foenum-graecum* L. plants grown in Gomati upland alluvial soil of Lucknow. *Ind. Botan. Soci.* **2009**, *88*, 1–3.
46. Peters, L.P.; Carvalho, G.; Martins, P.F.; Dourado, M.; Vilhena, M.B.; Pileggi, M.; Azevedo, R.A.J.P. Differential responses of the antioxidant system of ametryn and clomazone tolerant bacteria. *PLoS ONE* **2014**, *9*, e112271. [CrossRef] [PubMed]
47. Sengar, R.; Gupta, S.; Gautam, M.; Sharma, A.; Sengar, K. Occurrence, uptake, accumulation and physiological responses of nickel in plants and its effects on environment. *Res. J. Phytochem.* **2008**, *2*, 44–60. [CrossRef]
48. Sheetal, K.; Singh, S.; Anand, A.; Prasad, S. Heavy metal accumulation and effects on growth, biomass and physiological processes in mustard. *Ind. J. Plant Physiol.* **2016**, *21*, 219–223. [CrossRef]
49. Gonçalves, M.T.; Gonçalves, S.C.; Portugal, A.; Silva, S.; Sousa, J.P.; Freitas, H. Effects of nickel hyperaccumulation in Alyssum pintodasilvae on model arthropods representatives of two trophic levels. *Plant Soil* **2007**, *293*, 177–188. [CrossRef]
50. Palacios, G.; Gomez, I.; Carbonell-Barrachina, A.; Pedreño, J.N.; Mataix, J. Effect of nickel concentration on tomato plant nutrition and dry matter yield. *J. Plant Nut.* **1998**, *21*, 2179–2191. [CrossRef]
51. Kumar, P.; Rouphael, Y.; Cardarelli, M.; Colla, G. Effect of nickel and grafting combination on yield, fruit quality, antioxidative enzyme activities, lipid peroxidation, and mineral composition of tomato. *J. Plant Nutr. Soil Sci.* **2015**, *178*, 848–860. [CrossRef]

plants

MDPI

Article

Micronutrient Status and Selected Physiological Parameters of Roots in Nickel-Exposed *Sinapis alba* L. Affected by Different Sulphur Levels

Renata Matraszek-Gawron and Barbara Hawrylak-Nowak *

Department of Botany and Plant Physiology, Faculty of Environmental Biology, University of Life Sciences in Lublin, Akademicka 15, 20-950 Lublin, Poland; renata.matraszek@up.lublin.pl

* Correspondence: barbara.nowak@up.lublin.pl

Received: 3 October 2019; Accepted: 22 October 2019; Published: 23 October 2019

Abstract: An efficient method of improving the micronutrient status of Ni-treated white mustard (*Sinapis alba* L.) using intensive S-SO_4 nutrition was developed. Twelve variants of Hoagland's nutrient solution differing in the concentration of S-SO_4 (standard: 2 mM S, and elevated level: 6 or 9 mM S) and Ni (0, 0.0004, 0.04, or 0.08 mM Ni) were tested. The beneficial effect of intensive S nutrition on Ni-stressed plants was manifested by a significant rise in the content of Fe, Mn, and Zn, especially in the shoots. An increase was also found in the shoot B, Cu, and Mo content, whilst there were no changes in their root concentrations. Simultaneously, the shoot Cl concentrations dropped. The elevated level of S in the nutrient solution in general enhanced the translocation of Fe, Cu, Mo, and B in Ni-exposed plants. The beneficial effect of intensive S nutrition on the growth and micronutrient balance of Ni-exposed plants can be at least partially related to the positive changes in root surface properties, especially in cation exchange capacity (CEC). Meanwhile both reduced glutathione (GSH) and phytochelatins (PCs) probably do not significantly contribute to Ni resistance of white mustard under intensive S nutrition.

Keywords: metal toxicity; sulphur nutrition; stress mitigation; cation exchange capacity; glutathione

1. Introduction

Nickel, like other metallic micronutrients in plants, is a functional constituent of the enzyme systems and its role is primarily associated with the valence change. This element, at relatively low concentrations (0.001–0.01 mg kg^{-1} dry weight; DW) is needed for the proper N and C metabolism [1–3] as well as for producing high-vigor viable seeds and their germination [4,5]. The most common visual symptoms of Ni deficiency are growth reduction, senescence acceleration, and leaf deformation accompanied by chlorotic and necrotic lesions as a result of Fe deficiency induced in Ni-deficient plants [6,7]. However, not Ni deficiency, but its excess and strong phytotoxic effects are the serious global problem. Nickel may easily move in the environment. Of particular concern is the increasing area of Ni-contaminated agricultural soils together with rapidly rising Ni concentrations deposited in agricultural soils by airborne Ni particles. Moreover, the low soil pH as a result of reduced soil liming as well as acid rains may cause mobilization and enhance the solubility of Ni compounds [8,9]. It has been established that the Ni content in farm soils varies in a wide range from 3 to 1000 mg kg^{-1} DW. Most agricultural soils contain 25 mg kg^{-1}; however, Ni content is very often raised, up to 26,000 mg kg^{-1} or even substantially higher, due to anthropogenic activities such as mining, smelting, burning of fossil fuels (coal and oil), use of industrial and municipal wastes (sewage sludge), as well as applications of pesticides and Ni-containing fertilizers, especially phosphates [8–12]. A strongly phytotoxic effect of high Ni concentrations manifests itself as growth and development inhibition including retarded germination, yield and quality reduction, as well as disturbances in photosynthesis, respiration, water relations, and sugar transport, which cause various ultrastructural modifications. Visual symptoms induced by the Ni excess include various types of leaf chlorosis, necrosis, and wilting [6,13–15]. These chlorotic and necrotic lesions are a result of altered essential nutrient uptake and translocation. Interference with nutrient homeostasis and thus improper nutrient, especially micronutrient, status within plants is mentioned as an important mechanism of Ni phytotoxicity [16,17]. The modifications of the mineral status of Ni-stressed plants within species and even cultivars are unpredictable and contradictory. Besides, it is very difficult to study the biological role and mechanisms of Ni toxicity due to the dual character and complicated electronic chemistry of this element. Therefore, much more information concerning phytotoxicity and tolerance can be found for other widespread toxic trace metals (Cd, Pb, Cr, Cu) than for Ni [18,19]. White mustard, chosen as an object of this study, may tolerate excessive concentrations of trace metals, including Ni. This species is recognized as more sensitive to excessive amounts of Ni than to Cu or Zn [20,21].

White or yellow mustard (*Sinapis alba* L.) is believed to be native to the Mediterranean region, but nowadays is extensively cultivated throughout the world, with Canada and Nepal as the global leaders (around 52% of the world production in the 2015) as well as Ukraine, Russia, the Czech Republic, Italy, the UK, and the Netherlands as the European leaders [22]. White mustard seeds in food industry serve to produce table mustard, oil, and many kinds of spices. They are also used in the pharmaceutical and cosmetics industry. Furthermore, young fresh mustard raw leaves are used to made salad or juice [22,23]. In agriculture and horticulture, mustards are commonly used as green manure, fodder crop as well as winter or rotational cover crops in production of many species. Mustards may control weeds and a range of soil-borne pests, pathogens, and diseases. This is due to providing allelopathic compounds in the "biofumigation" process related to release of volatile toxic isothiocyanate compounds (ITCs) through the degradation of glucosinolates (GLS) [24,25].

Due to GLS synthesis in mustard, similar to the other members of the Brassicaceae family, the species is characterized by high S requirement, which are at least twice as high as that of cereal crops, especially at the flowering stage, since S is a constituent of sulfuric amino acids needed for the synthesis of seed proteins. This macronutrient is also an important constituent of lipids, polysaccharides, vitamins, and cofactors [26–33]. Besides building proteins and involvement in metabolism of secondary products, S is required for chlorophyll synthesis as well as proper cell metabolic pathways such as electron transport in Fe-S clusters, redox cycle, protein disulfide bonds. Sulphur deficiency in the environment and hence reduced yield and quality is a global problem related to progressive reduction of emissions of S compounds, common application of S-free NPK fertilizers, immobilization in soil and limited availability of S to plants, and much more intensive crop production [34,35]. An adequate S level is crucial not only for a proper plant growth and development, but also for enhanced resistance to various environmental stresses [36,37]. Plants have developed various strategies to cope with excess of trace metals. One of the strategies is induction of ligands synthesis, which are able to bind most trace metals in order to protect metal-target i.e., sensitive cellular organelles. Nickel is classified as a transition metal capable of binding to various types of naturally occurring phytocompounds. For instance, SH-containing ligands like reduced glutathione (GSH) or phytochelatins (PCs) form high-strength, durable complexes with trace metals. Nickel resistance is related mainly to GSH synthesis, since Ni is a very weak inductor of PCs synthesis. On the one hand, Ni is recognized as an important element for protecting plants against stressful conditions, among others, by participating in the regulation of the GSH pool involved in the defense against oxidative stress; on the other hand, Ni may induce oxidative stress [3,38–40].

Excessive Ni concentration disturbs the nutrient status in plants due to various unfavorable changes in their uptake and translocation [2,11–13,15], and there is evidence that S has a crucial role in enhancing tolerance to various types of stress [35]. Therefore, it may be possible to improve the nutrient status in Ni-stressed plants using S supplementation. This concerns especially species characterized by high S requirements, such as white mustard. In this study we investigated the impact of different S-SO_4 concentrations on the micronutrient status in white mustard under short-term Ni exposition. We hypothesized that intensive S nutrition may improve the growth and micronutrient status of Ni-exposed white mustard by the modulation of selected physiological parameters of roots. It is obvious that only plant species that are able to survive in an environment containing excess of trace metals at their early (juvenile) growth stages may produce resistant and healthy adult individuals. These experiments are the part of a larger project concerning the possibility of enhancing plant resistance to stress induced by trace metals with the use of intensive S nutrition.

2. Results

2.1. Micronutrient Concentrations

The concentrations of micronutrient in the roots are presented in Table 1, while micronutrients in the shoots are presented in Table 2. The root and shoot B concentrations as well as the root Zn and shoot Cu level in the intensively S-supplied (6 or 9 mM) Ni-untreated plants were higher than those at the standard S level. In turn, the changes in the shoot Zn and root Cu concentrations, similar to the changes in the content of Fe, Cl, and Mo, were generally insignificant. The Mn concentration in plants supplied with the high S increased in the roots and decreased in the shoots, and these changes were more pronounced in the roots.

Increasing Ni concentrations in the nutrient solution, irrespective of the S level, generally resulted in a significant decline in Fe and B concentrations, which was much more pronounced in the shoots than in roots, and simultaneously did not change the Mo concentrations. Moreover, the statistical analysis of the main effects showed that the root Cu and Cl concentrations of Ni-exposed white mustard remained quite stable, whilst the shoot content of both these elements was substantially reduced. An exception was the decrease in the root Cu content recorded in plants treated with 0.04 mM Ni. Simultaneously, the root and shoot Zn concentration decreased and increased, respectively, whilst the concentration of Mn increased in both roots and shoots.

Intensive S nutrition, irrespective of the Ni concentration in the nutrient solution, generally caused increases in the shoot and root Fe, B, and Zn concentrations. However, the root and shoot Fe level in plants grown at 6 mM S and the root B concentration at 9 mM S did not change significantly. At the same time, root Cu and Mo concentrations remained quite stable, whilst their shoot concentrations were markedly elevated. In turn, the Mn concentration increased in roots and Cl level decreased in shoots.

The tendencies toward changes in the micronutrient concentrations for the interactions between the S and Ni (S × Ni) were generally similar to those above described for the main effects. However, a few differences were found. For instance, noteworthy is the decrease in the shoot Mn and Mo concentrations as well as the lack of significant changes in the root B and shoot Cl and Zn concentrations in the Ni-exposed plants grown at the standard S level. In turn, the following differences between the main effects and S × Ni interactions in the Ni-exposed plants supplied with extra S, in relation to plants treated with a comparable Ni concentration grown at standard S level, were found (see Tables 1 and 2):

- no changes in the root Fe concentration under intensive S nutrition in plants under the lowest and the highest Ni exposure and the Fe increase under medium Ni concentration,
- no changes in the root B concentration under both elevated S levels,
- an increase in the shoot Mn at both the high S levels and a decrease in root Mn at the 9 mM S/0.0004 mM Ni treatment,
- an increase or no changes in the root Zn concentration at 6 and 9 mM S, respectively.

Table 1. The concentration of micronutrients in the root biomass of white mustard grown for two weeks under different sulphur and/or nickel concentrations in the nutrient solution.

Concentration in the Nutrient Solution (mM)		Concentration of the Micronutrients in the Roots (mg kg^{-1} DM)						
S	Ni	Fe	B	Cl	Cu	Mn	Mo	Zn
2		1378 ± 42.6^{a}	17.24 ± 1.58^{c}	8485 ± 53.87	5.48 ± 0.22^{ab}	47.18 ± 1.02^{e}	0.642 ± 0.093	$34.89 \pm 0.95^{d–f}$
6	0.00	1382 ± 43.0^{a}	22.13 ± 2.73^{a}	8478 ± 70.42	5.72 ± 0.35^{a}	53.37 ± 0.97^{bc}	0.585 ± 0.060	45.57 ± 1.14^{a}
9		1377 ± 52.2^{a}	20.77 ± 1.65^{b}	8498 ± 55.39	$5.13 \pm 0.56^{a–c}$	55.74 ± 0.76^{a}	0.663 ± 0.074	42.61 ± 0.99^{ab}
2		1372 ± 32.5^{a}	18.14 ± 2.51^{c}	8492 ± 97.01	$5.01 \pm 0.46^{a–d}$	$54.81 \pm 0.84^{a–c}$	0.608 ± 0.087	36.28 ± 1.22^{de}
6	0.0004	1380 ± 61.7^{a}	17.29 ± 1.67^{c}	8484 ± 41.66	5.36 ± 0.70^{ab}	$53.78 \pm 0.66^{a–c}$	0.597 ± 0.058	40.67 ± 1.48^{bc}
9		1370 ± 62.7^{a}	17.71 ± 2.49^{c}	8503 ± 84.72	$5.29 \pm 0.39^{a–c}$	51.13 ± 0.80^{d}	0.591 ± 0.051	$38.52 \pm 1.09^{b–d}$
2		1263 ± 21.9^{c}	17.05 ± 2.63^{c}	8473 ± 72.38	4.42 ± 0.31^{cd}	53.19 ± 0.75^{c}	0.617 ± 0.052	31.12 ± 1.14^{h}
6	0.04	1279 ± 42.0^{c}	18.09 ± 2.57^{c}	8488 ± 89.84	$4.73 \pm 0.66^{b–d}$	55.31 ± 0.88^{ab}	0.580 ± 0.041	35.43 ± 0.99^{de}
9		1310 ± 30.4^{b}	17.42 ± 1.66^{c}	8482 ± 75.05	4.16 ± 0.54^{d}	54.02 ± 0.76^{ac}	0.635 ± 0.049	$33.74 \pm 1.16^{e–h}$
2		1307 ± 36.1^{b}	18.59 ± 1.67^{c}	8500 ± 59.32	$4.87 \pm 0.57^{a–d}$	51.04 ± 0.64^{d}	0.659 ± 0.102	28.13 ± 0.94^{hi}
6	0.08	1299 ± 33.2^{b}	17.35 ± 1.48^{c}	8477 ± 93.41	$5.07 \pm 0.61^{a–c}$	$54.28 \pm 0.93^{a–c}$	0.668 ± 0.044	$31.37 \pm 1.21^{f–h}$
9		1313 ± 42.2^{b}	18.28 ± 2.53^{c}	8492 ± 84.85	5.34 ± 0.43^{ab}	52.95 ± 0.75^{cd}	0.584 ± 0.057	26.48 ± 0.87^{i}
Main effects								
S								
2		*1330 ± 52.4*	17.76 ± 1.48^{b}	*8488 ± 70.21*	*4.95 ± 0.37*	51.56 ± 0.43^{b}	*0.632 ± 0.027*	32.61 ± 0.73^{c}
6		*1335 ± 41.4*	18.72 ± 1.52^{a}	*8482 ± 82.08*	*5.22 ± 0.23*	54.19 ± 0.41^{a}	*0.608 ± 0.024*	38.26 ± 1.64^{a}
9		*1342 ± 59.0*	18.55 ± 1.41^{ab}	*8494 ± 70.83*	*4.98 ± 0.69*	53.46 ± 0.48^{a}	*0.618 ± 0.019*	35.34 ± 0.68^{b}
Ni								
0		1379 ± 44.8^{a}	20.05 ± 1.64^{a}	*8487 ± 91.37*	5.44 ± 0.45^{a}	52.10 ± 0.67^{c}	*0.630 ± 0.037*	41.02 ± 0.70^{a}
0.0004		1374 ± 51.3^{a}	17.71 ± 1.58^{b}	*8493 ± 72.44*	5.22 ± 0.68^{a}	53.24 ± 0.48^{ab}	*0.599 ± 0.018*	38.49 ± 0.75^{b}
0.04		1283 ± 32.4^{b}	17.52 ± 1.46^{b}	*8481 ± 63.36*	4.44 ± 0.47^{b}	54.17 ± 0.41^{a}	*0.611 ± 0.022*	33.43 ± 0.63^{c}
0.08		1306 ± 42.3^{ab}	18.07 ± 1.53^{b}	*8490 ± 70.09*	5.09 ± 0.64^{a}	52.76 ± 0.45^{bc}	*0.637 ± 0.028*	28.66 ± 0.59^{d}
Statistical significance								
S		*NS*	*	*NS*	*NS*	*	*NS*	*
Ni		*	*	*NS*	*	*	*NS*	*
S × Ni		*	*	*NS*	*	*	*NS*	*

Note: The results are presented as the mean ± SD of nine measurements (three measurements made per each of three independent repetition of the experiment over time). Means (n = 9) sharing the same letter in a column do not differ significantly according to Tukey's multiple range test at $p \leq 0.05$; significant effects for the main factors and for interaction between them are indicated with asterisks (*); NS—non-significant.

Table 2. The concentration of micronutrients in the shoot biomass of white mustard grown for two weeks under different sulphur and/or nickel concentrations in the nutrient solution.

Concentration in the Nutrient Solution (mM)		Concentration of the Micronutrients in the Shoots (mg kg^{-1} DM)						
S	Ni	Fe	B	Cl	Cu	Mn	Mo	Zn
2	0.00	114.27 ± 3.29^{ab}	42.73 ± 2.89^{cd}	12,359 ± 85.2^{a}	5.96 ± 0.19^{b}	26.72 ± 1.55^{a}	0.983 ± 0.057^{a}	42.15 ±$0.87^{c–e}$
6	0.00	111.86 ± 2.91^{ab}	49.36 ± 1.14^{b}	12,311 ± 98.5^{bc}	6.88 ± 0.15^{a}	22.45 ± 1.63^{e}	1.052 ± 0.071^{a}	40.73 ± 1.11^{e}
9	0.00	115.09 ± 3.84^{ab}	54.22 ± 2.93^{a}	12,348 ± 60.3^{ab}	7.25 ± 0.26^{a}	23.96 ± 1.72^{de}	0.928 ± $0.049^{a–d}$	41.06 ± 1.03^{e}
2	0.0004	115.24 ± 4.09^{a}	35.07 ± 1.28^{g}	12,351 ± 97.1^{ab}	4.12 ± 0.34^{d}	24.02 ± 1.53^{d}	0.770 ± $0.078^{b–d}$	43.87 ± $0.75^{b–d}$
6	0.0004	103.26 ± $3.71^{c–e}$	42.06 ± 1.05^{d}	12,301 ± 63.0^{c}	5.68 ± 0.26^{b}	25.69 ± $1.47^{a–c}$	0.981 ± 0.043^{a}	47.67 ± 1.17^{a}
9	0.0004	117.74 ± 3.35^{a}	44.85 ± 1.35^{c}	12,275 ± 60.3^{c}	5.81 ± 0.21^{b}	25.53 ± $1.61^{a–c}$	0.975 ± 0.065^{a}	45.38 ± 1.21^{ab}
2	0.04	88.22 ± 5.29^{g}	34.72 ± 1.07^{g}	12,385 ± 51.4^{a}	4.01 ± 0.28^{d}	24.72 ± 1.55^{cd}	0.759 ± 0.057^{d}	41.75 ± 1.09^{de}
6	0.04	97.29 ± 4.07^{ef}	36.94 ± 1.22^{fg}	12,149 ± 88.7^{d}	5.16 ± 0.20^{bc}	26.30 ± 1.59^{ab}	0.960 ± $0.048^{a–c}$	44.29 ± 0.86^{bc}
9	0.04	101.56 ± 5.61^{de}	38.12 ± 1.24^{f}	12,298 ± 36.0^{c}	5.92 ± 0.24^{b}	26.91 ± 1.47^{a}	0.972 ± 0.061^{a}	47.17 ± 1.14^{a}
2	0.08	93.81 ± 4.48^{fg}	35.48 ± 1.79^{fg}	12,377 ± 51.6^{a}	4.35 ± 0.21^{cd}	25.19 ± $1.62^{b–d}$	0.763 ± 0.066^{cd}	44.31 ± 0.92^{bc}
6	0.08	104.77 ± 3.02^{cd}	37.35 ± 2.14^{fg}	12,304 ± 68.3^{c}	5.37 ± 0.22^{b}	26.73 ± 1.53^{a}	0.970 ± 0.072^{ab}	46.97 ± 1.17^{a}
9	0.08	110.01 ± 5.23^{bc}	41.37 ± 1.33^{d}	12,092 ± 54.2^{e}	5.59 ± 0.31^{b}	26.70 ± 1.58^{a}	0.967 ± 0.054^{ab}	47.06 ± 1.20^{a}
Main effects								
S								
2		*102.89 ± 4.61^{b}*	*37.00 ± 1.98^{c}*	*12,368 ± 78.2^{a}*	*4.61 ± 0.14^{b}*	*25.16 ± 1.44*	*0.819 ± 0.037^{b}*	*43.02 ± 0.47^{b}*
6		*104.30 ± 3.57^{b}*	*41.43 ± 1.45^{b}*	*12,266 ± 87.1^{ab}*	*5.77 ± 0.11^{a}*	*25.29 ± 1.52*	*0.991 ± 0.029^{a}*	*44.92 ± 0.44^{a}*
9		*111.10 ± 4.79^{a}*	*44.64 ± 1.83^{a}*	*12,253 ± 56.2^{b}*	*6.14 ± 0.19^{a}*	*25.78 ± 1.47*	*0.961 ± 0.031^{a}*	*45.17 ± 0.50^{a}*
Ni								
0		*113.74 ± 2.97^{a}*	*48.77 ± 2.65^{a}*	*12,399 ± 84.3^{a}*	*6.70 ± 0.11^{a}*	*24.38 ± 1.42^{b}*	*0.988 ± 0.027*	*41.31 ± 0.78^{c}*
0.0004		*112.08 ± 3.65^{a}*	*40.66 ± 1.23^{b}*	*12,309 ± 74.4^{a}*	*5.20 ± 0.17^{b}*	*25.08 ± 1.43^{ab}*	*0.909 ± 0.022*	*45.64 ± 0.52^{b}*
0.04		*95.69 ± 4.72^{b}*	*36.59 ± 1.12^{d}*	*12,277 ± 71.1^{b}*	*5.03 ± 0.22^{b}*	*25.98 ± 1.51^{a}*	*0.897 ± 0.032*	*44.40 ± 0.47^{b}*
0.08		*102.86 ± 4.53^{a}*	*38.07 ± 1.76^{c}*	*12258 ± 59.2^{b}*	*5.10 ± 0.13^{b}*	*26.21 ± 1.48^{a}*	*0.900 ± 0.029*	*56.11 ± 0.51^{a}*
Statistical significance								
S		*	*	*	*	*NS*	*	*
Ni		*	*	*	*	*	*NS*	*
S × Ni		*	*	*	*	*	*	*

Note: For explanation see Table 1.

2.2. Micronutrient Translocation Factor (TF)

Intensive S nutrition of plants non-exposed to Ni caused an increase of TF value for Cu (Figure 1d) and its reduction for Mn and Zn (Figure 1e,g). The Mo translocation from roots to shoots increased at 6 mM S and decreased at 9 mM S (Figure 1f). Simultaneously, the B translocation decreased at 6 mM S (Figure 1b). The presence of Ni in the nutrient solution at the standard S level severely reduced the TF value of Fe, B, Cu, Mn, and Mo without notable effect on Cl translocation (Figure 1a–f). The exception was no significant changes for the TF of Fe at the 2 mM S/0.0004 mM Ni treatment. Simultaneously, the TF value of Zn increased at the highest Ni concentration used, compared to the Ni-untreated plants (Figure 1g).

Supplementation with S of the Ni-exposed plants, in relation to the standard 2 mM S level, significantly elevated the TF of B, Cu, and Mo (Figure 1b,d,f) and did not notably affect the TF value for the Cl, Mn, and Zn (Figure 1c,e,g). Only the TF values of Mn and Zn at the 9 mM S/0.0004 mM Ni and 9 mM S/0.08 mM Ni treatments, respectively, were significantly higher than TF values found for the comparable Ni concentrations in the medium under the standard S level (Figure 1e,g). The lack of significant changes in TF value of B at the 6 mM S/0.04 mM Ni treatment was an exception (Figure 1b). In turn, the TF of Fe under the lowest Ni concentration used decreased at 6 mM S and did not change at 9 mM S, but increased under intensive S nutrition at the higher Ni concentrations (0.04 and 0.08 mM; Figure 1a).

2.3. Total Surface Charge (Q_{tot}) and Cation Exchange Capacity (CEC)

The changes in Q_{tot} and CEC are presented in Figure 2a,b. The CEC values at 2 mM S were markedly lower (by 29–40%) in roots of plants treated with Ni than in those non-exposed to this metal. Under these conditions the Q_{tot} decreased significantly (by 27%) only at the highest Ni concentration. When the Ni-exposed plants were supplied with extra S at the concentration of 6 mM the CEC increased by 80–89% in comparison to plants grown at standard S level. Meanwhile, the Q_{tot} value increased significantly for plants treated with 0.04 mM Ni supplied with 6 mM S and plants treated with 0.08 mM Ni and supplied with 6 or 9 mM S. In general, the impact of the highest concentration of S on CEC and Q_{tot} was negative at the lowest and moderate Ni concentrations used and their values were significantly reduced. On the other hand, under the highest Ni concentration the intensive S nutrition at 9 mM caused an increase in both CEC and Q_{tot} values. It is worth noting that the CEC in plants exposed to 0.08 mM Ni increased almost twice under intensive S nutrition in comparison to standard S level.

2.4. GSH and PCs Accumulation in Roots

The changes in root GSH concentrations were ambiguous (Figure 3a). In general, in Ni-exposed plants an elevated GSH accumulation was found, when we compare the level of this compound in the control plants and in these grown under presence of Ni at the appropriate S levels. The intensive S nutrition of Ni-untreated plants resulted in the decrease in GSH accumulation. At 6 mM S, no significant changes in GSH content were found at any of the tested Ni concentrations, in comparison to 2 mM S. In turn, the extra S supply at the concentration of 9 mM increased the GSH concentration in the plants treated with 0.0004 or 0.08 mM Ni, but decreased in those exposed to 0.04 mM Ni.

In roots of white mustard only small amounts of PCs 2 were found. Phytochelatins were not detectable or only trace concentrations of PCs 2 appeared in the roots of plants grown at 2 mM S without Ni or at the lowest Ni concentration. The enhanced concentrations of PCs 2 in the root tissues were detected with the increasing concentration of Ni. However, the intensified S nutrition significantly enhanced PCs 2 accumulation only at the lowest and the medium Ni concentration. Under the highest Ni exposition, the PCs 2 concentration decreased in comparison to the standard S level (Figure 3b).

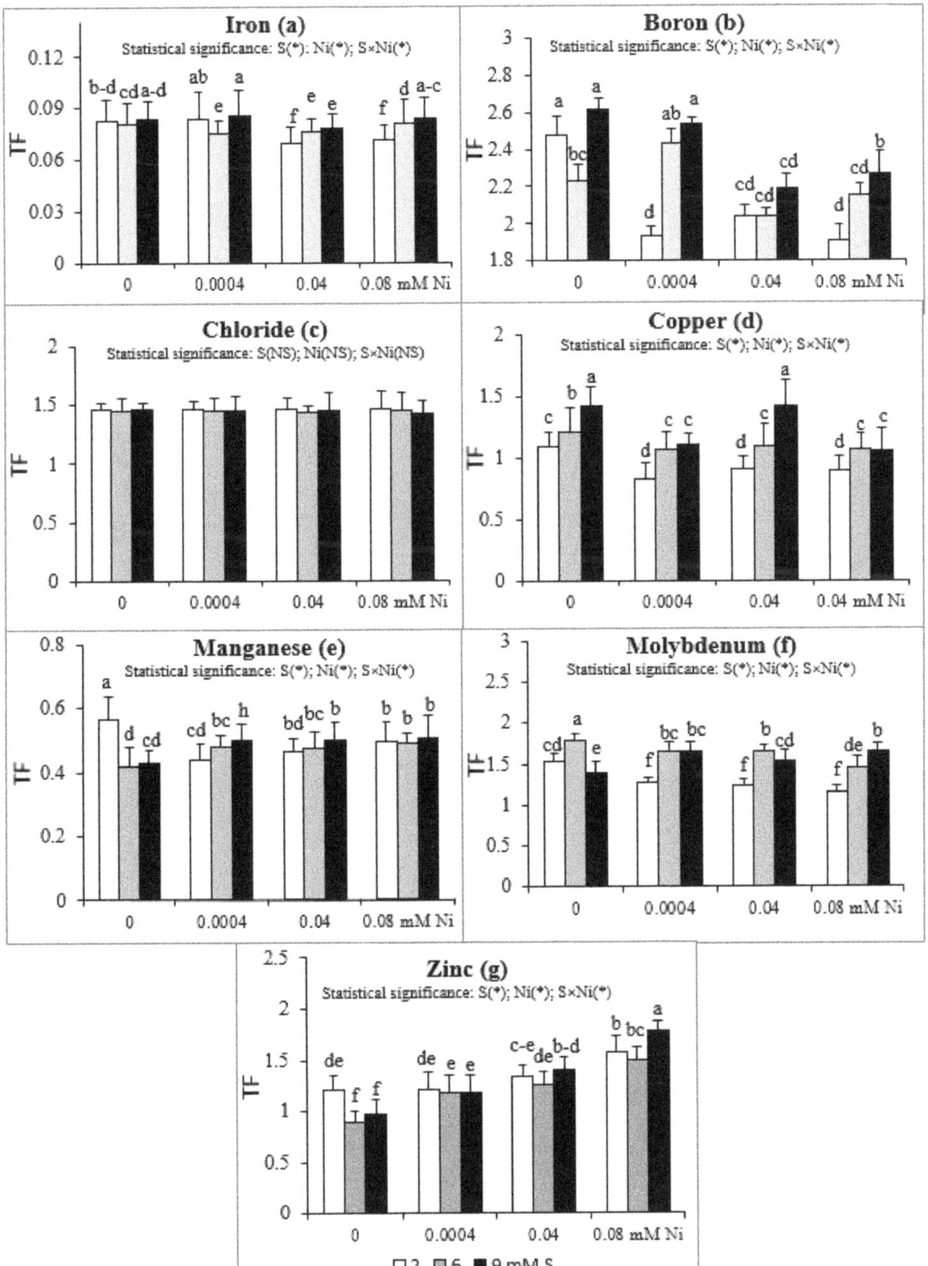

Figure 1. Translocation factor (TF) of micronutrients: (**a**) iron, (**b**) boron, (**c**) chloride, (**d**) copper, (**e**) manganese, (**f**) molybdenum, (**g**) zinc in white mustard "Rota" grown under different sulphur and/or nickel concentrations in the nutrient solution. Mean values (n = 9) followed by the same letter are not significant at 0.05 probability level based on Tukey's honestly significance tests. Asterisks indicate significant effects for main factors and interactions between them.

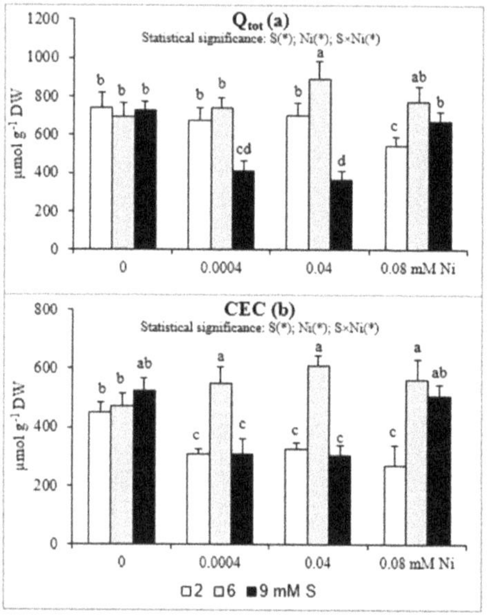

Figure 2. The (**a**) total surface charge (Q_{tot}) and (**b**) cation exchange capacity (CEC) in white mustard "Rota" grown under different sulphur and/or nickel concentrations in the nutrient solution. Mean values (n = 9) followed by the same letter are not significant at 0.05 probability level based on the Tukey's honestly significance tests. Asterisks indicate significant effects for main factors and interactions between them.

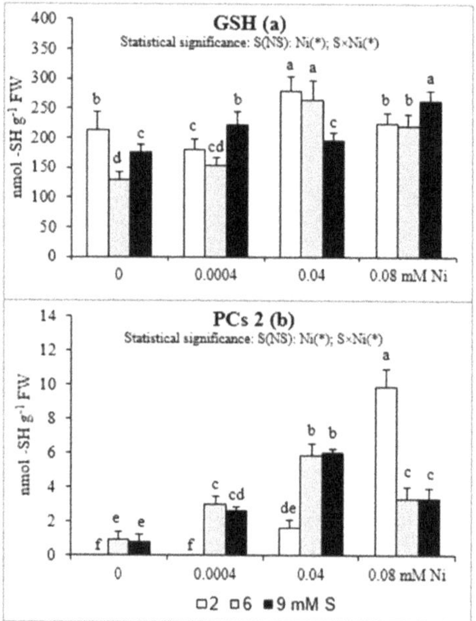

Figure 3. Concentrations of (**a**) reduced glutathione (GSH) and (**b**) phytochelatins (PCs) in white mustard "Rota" grown under different sulphur and/or nickel concentrations in the nutrient solution. Mean values (n = 9) followed by the same letter are not significant at 0.05 probability level based on Tukey's honestly significance tests. Asterisks indicate significant effects for main factors and interactions between them.

2.5. Root and Shoot Biomass

The results of the influence of differentiated Ni and S concentrations in the nutrient solution on shoot and root DW are presented in Figure 4. Both the shoot and root biomass of plants treated with 0.04 or 0.08 mM Ni was reduced. Meanwhile, the lowest concentration of this element has no effect on plant growth. In mustard exposed to 0.04 mM Ni, the intensive S nutrition at 6 mM, in comparison to the standard S level, caused an increase in the shoot DW, which was not statistically different from Ni-untreated plants. However, this phenomenon did not occur at 9 mM S and under the highest Ni concentration used (Figure 4a). In turn, in mustard not subjected to Ni, the extra S supply induced a 45–50% increase in root DW compared to the standard S level. Such a trend was also observed in the presence of Ni, but it was not statistically confirmed. The exception was a significant increase in root DW of plants treated with 0.04 mM Ni under the influence of 9 mM S (Figure 4b).

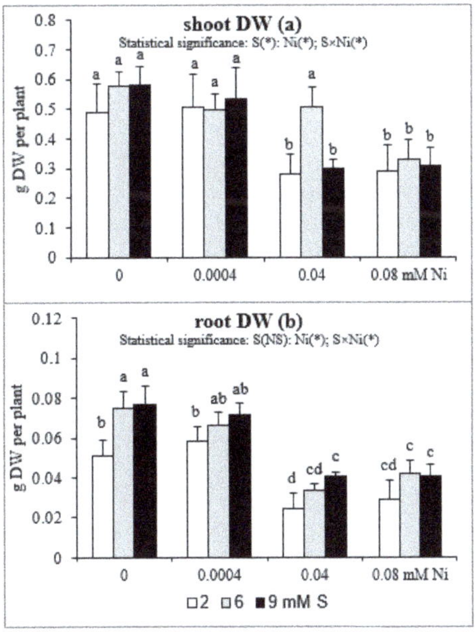

Figure 4. The dry weight (DW) of (**a**) shoots and (**b**) roots of white mustard "Rota" grown under different sulphur and/or nickel concentrations in the nutrient solution. Mean values (n = 36) followed by the same letter are not significant at 0.05 probability level based on Tukey's honestly significance tests. Asterisks indicate significant effects for main factors and interactions between them.

3. Discussion

The lowest Ni concentration used (0.0004 mM) is claimed to be the highest value acceptable for the ground water and soil solution [6,41–43]. That means that the presence of higher concentrations of this metal, including those examined in our study (0.04 and 0.08 mM), requires implementation of conservation measures. The impact of Ni on the white mustard micronutrient status under the standard S dose (2 mM) are in agreement with the widely known statement that interference with other essential metal ions are an indirect pathway of Ni phytotoxicity. It was also confirmed that Ni-induced changes in the nutrient bioconcentrations are not only species-specific, but also unpredictable and contradictory. Additionally, these changes may be different in the roots and aboveground parts of plants [13,15,44]. The phenomenon of Ni-induced alterations in essential nutrient absorption, uptake, and transport and hence the disturbance in ionic homeostasis is a consequence of competition between Ni^{2+} and other cations (Cu^{2+}, Fe^{2+}, Mn^{2+}, and Zn^{2+}) for common binding sites as a result of similar characteristics, including comparable ionic radii [44–46]. Passive diffusion and active transport are recognized as two main mechanisms of Ni ions uptake by plants. Absorption of soluble Ni compounds occurs passively via a cation transport system. Chelated Ni compounds are taken up through secondary active-transport-mediated means, i.e., permeases. In turn, endocytosis is recognized as a mechanism through which insoluble Ni compounds primarily enter plant root cells. After absorption by roots, Ni transport to the shoot occurs very easily via the xylem. The processes of Ni transport and retranslocation are strongly regulated by metal-ligand complexes (nicotianamine, histidine, and organic acids) and by some specific Ni-binding proteins [6,13,19]. Moreover, Ni uptake and translocation occur with involvement of a Zn/Fe ZRT1/IRT1–ZIP transporter and a Mn ion transporter NRAMP [19,43,44]. The antagonism between Ni^{2+} ions and Cu^{2+} and Zn^{2+} ions suggests that these three elements are absorbed at the same site on the transporter [44]. Hence, the decreased Fe, Zn, and Cu concentrations in the mustard biomass, manifested as various types of chloroses, were also recorded by other researchers in many other plant species [41,45]. The results of the present study indicate antagonism between Ni and Cl and between Ni and Mo. It should be remembered that Ni may not only compete with essential nutrients, mainly Fe, Zn, and Cu, but also inhibit their functions. Nickel may replace the essential metal of metalloproteins and bind the residues of non-metalloenzymes. The binding of Ni ions outside the catalytic site of an enzyme induces allosteric modulation and, hence, the inhibition of the enzyme. Besides the above-mentioned indirect effect of Ni on enzyme activity, i.e., competitive inhibition of nutrient absorption and transport, a direct mechanism associated with strong affinity of Ni for the functional –SH groups of proteins is also known [6,13,19]. Moreover, the Ni-induced changes in the micronutrient status recorded in the present study may be explained by the disturbances in the cell membrane permeability caused by changes in the composition of sterols and phospholipids and changes in the structure and/or activity of cell membrane enzymes, mainly the H^+-ATPase, which plays a key role in the active uptake and transport of essential nutrients [8,15,19].

The standard S concentration (2 mM) used in our experiments is recognized as a moderate level for this macronutrient. The S-SO_4^{2-} concentration in the natural environment, i.e., unpolluted with trace metals, in arid regions and in soil solutions with residues of sulfide ore mine is in the range of 0.16–7, 3–16, and 13–110 mM, respectively [46]. Our study concerning the micronutrient bioconcentration under high S-SO_4 levels (6 or 9 mM S) in mustard grown without Ni showed a synergistic effect between S and B, S and Cu, and S and Zn. Simultaneously, an antagonistic relationship between S and Cl was found. In turn, the S and Mn relationships were antagonistic at 6 mM S and synergistic at 9 mM S. It is claimed that an optimal S level increases the uptake of Mn and Zn [47]. The tendencies of changes in the root and shoot concentration of Mn, Cu, and Zn under intensive S nutrition in Ni-untreated white mustard were similar to those noted for this species in a field experiment of Jankowski et al. [48]. In their research the content of these three micronutrients in roots and shoots of Indian mustard was not affected by S fertilization, except for the Cu decrease in the shoots. However, in this study, the content of the other micronutrients in examined mustard species was not estimated. Moreover, in our previous study, an increase in the all macronutrient concentrations in roots and the S, K, and Ca level in shoots in Ni-exposed mustard supplied with extra S was found [49].

Taking into account the TF value, it may be concluded that white mustard has a strong ability to translocate B, Cl, Mo, and Zn from roots to shoots (TF > 1) and a weak ability to translocate Fe and Mn (TF < 1). This tendency was generally revealed irrespective of both the experimental factors, i.e., the S and Ni concentrations in the nutrient solution. Only the intensive S nutrition of Ni-untreated plants, compared to the standard S dose, strongly limited the translocation of Zn, reducing the Zn TF below 1. The Cu transfer within organs depended on the S and Ni concentrations in the nutrient solution. The Ni-exposed plants growing at 2 mM S showed a Cu TF value lower than 1 (about 0.9), in comparison with values exceeding 1 in plants grown without Ni. This implies that the Ni presence in the nutrient solution containing standard S levels reduced the Cu translocation. Simultaneously, the intensive S nutrition of Ni-treated mustard enhanced the ability to transfer Cu to shoots (Cu TF higher than 1). The micronutrient TF value obtained in our studies oscillated within the range of 0.070–0.086 (Fe), 1.91–2.611 (B), 1.42–1.46 (Cl), 0.82–1.41 (Cu), 0.42–0.57 (Mn), 1.16–1.80 (Mo), and 0.89–1.58 (Zn).

The electric charge is most frequently studied root surface property to describe the root CEC, its changes with soil pH, balance of plant cations of different valence and toxicity of trace metals. Electric charge of the root compartments is dominated by negatively charged groups, thus positively charged cations, including essential nutrients, accumulate near the roots surface [50]. In our study we have noticed a positive effect of intensive S nutrition of Ni-exposed plants on the studied properties of roots (Q_{tot} and CEC), especially at 6 mM S. The beneficial effect of extra S supply of Ni-treated mustard on the changes in the studied root properties may contribute to a better uptake of micronutrients and thus positively affect the mineral status of plants, which consequently stimulates their growth. On the other hand, in plants grown at 9 mM S, the values of Q_{tot} and CEC decreased at 0.004 and 0.04 mM Ni, but increased at 0.08 mM Ni in comparison to those non-treated with Ni. We suppose that under high Ni bioconcentration, the higher S levels can be required to abolish toxic effect of Ni to the studied root parameters.

The antioxidant defense is believed to play a key role in the Ni tolerance, as the oxidative stress induction and the disturbances in the nutrient status are the major mechanism of the phytotoxicity of this element [13]. Also, there is no doubt that Ni-tolerance is linked to S metabolism, primarily with high levels of O-acetyl-L-serine (OAS), Cys, and GSH associated with the high activity of Ser acetyl transferase (SAT). Nickel is recognized as an element with strong ability to bind to various types of chelating agents, especially S-donor ligands rich in highly reactive S functional groups. However, Ni is recognized as a weak PCs inductor [4,6,51–53], which has been also confirmed in our study. Although the amount of PCs in plants exposed to Ni was not high, but in the presence of this metal the level PCs 2 significantly increased and the intensive S nutrition enhanced PCs 2 accumulation under the lowest and moderate Ni exposure. On the other hand, there are strong evidence that Ni may play a role in plant response to stressful conditions by decreasing the level of methylglyoxal (MG; a toxic, mutagenic alpha-ketoaldehyde) as well as participating in the regulation of the GSH homeostasis. In the degradation pathway of MG are involved glyoxalases I and II (GLY-I and II) and recently it was found that Ni ions may activate GLY-I in plants. The GLY-I dependence on Ni may play an additional role in the regeneration of GSH [3,54]. In our study the concentrations of GSH at the standard S level increased under moderate Ni exposure but not under both the lowest and the highest concentrations of this metal. We suppose that the concentration of 0.0004 mM Ni could be too low to effectively induce GSH accumulation but the concentration 0.08 mM Ni could have an adverse effect on GLY-I activity and therefore no increase in GSH concentration was found. The intensive S nutrition at 9 mM S caused enhanced accumulation of GSH only under the lowest and the highest Ni exposure. This effect was not observed at 6 mM S. Therefore, the beneficial effect of intensive S fertilization of Ni-stressed white mustard, which manifested itself as a rise in the Fe, Mn, and Zn bioconcentration, especially in the shoot biomass, is probably related to positive changes in root surface property as CEC, but not with an increase in GSH or PC synthesis in root tissues.

4. Materials and Methods

4.1. Plant Material and Growth Conditions

The biological object of the study was white mustard (*Sinapis alba* L.) "Rota" (Brassicaceae). Seven-day-old homogenous, healthy seedlings obtained from seeds germinated on quartz sand moistened with distilled water were transferred to 1 L glass jars (two plants per each jar) filled with full-strength Hoagland's solution No. 2 with different levels of S and Ni. A combination of three S levels (standard: 2 mM S, and intensive: 6 or 9 mM S; sulfate VI (S-SO_4)) and four Ni concentrations (0, 0.0004, 0.04, and 0.08 mM Ni; $NiCl_2$) was used to arrange the experimental treatments. In all experimental treatments, the standard S dose (2 mM) was supplied as $MgSO_4$ and supplemented with corresponding amounts of Na_2SO_4. The dose of S applied as Na_2SO_4 for the level of 2, 6, and 9 mM were 0, 4, and 7 mM, respectively. In each treatment, the levels of Na and Cl were equalized and the pH of the nutrient solution was set at 5.8–6.0 by adding appropriate amounts of diluted solutions of NaCl or HCl. The plants were cultured in a controlled-climate vegetation room at 14-h day length, PPF of 250–270 μmol × m^{-2} × s^{-1} at the level of the tops of the plants, temperature 25/20 °C (day/night), and relative air humidity of 50–60%. The nutrient solution was aerated for 15 min. every three days and replenished with a fresh nutrient solution when the medium level was depleted to ca. 70% of the initial level. After 14 days of vegetation under the differentiated S and Ni concentrations, the plants were harvested and subjected to the analysis.

4.2. Determination of Biomass and Micronutrient Concentrations

The roots and shoots of twelve randomly selected plants were dried in a forced air oven at 105 °C for 48 h, their dry weight (DW) were determined, and the samples were subjected to the analysis of the micronutrient concentration. The dry plant samples were ground to form a powder using a laboratory grinding mill. The total content of Fe, Mn, Zn, Mo, and Cu in roots and shoots were analyzed by atomic-absorption spectrophotometry (AAS) after wet mineralization with sulfuric acid (VI) and perhydrol [55,56]. To measure the B concentration, the Azomethine-H method was employed and the absorbance was read by spectrophotometry at 410 nm [57]. The Cl concentration was determined by the nephelometric method using nitric acid and silver nitrate [58]. The data obtained were used to calculate the value of the translocation factor (TF) of micronutrients (defined as a quotient of concentration of a given element in shoots and its concentration in the roots).

4.3. Determination of CEC and Q_{tot} by Potentiometric Titration

The values of CEC under differentiated experimental conditions were determined using potentiometric titration described in detail by Szatanik-Kloc et al. [50]. In brief, the fresh roots were placed in a ventilated room at 30 °C for 48 h. Then, a suspension of plant roots equilibrated overnight with 1 M L^{-1} NaCl was adjusted to pH = 3.0 until the pH was stable over the next 5 min and titrated automatically (Titrino 702 MS, Metrohm AG, Switzerland) by 60 s increments of 1 μL 0.100 M L^{-1} sodium hydroxide solution to pH = 10. The surface charge at pH = 7 was taken as the root CEC and the charge at pH = 10 was considered as the total surface charge (Q_{tot}).

4.4. Determination of γ-Glu-Cys Peptides by HPLC Method

The determine the GSH and PCs concentrations the root samples were weighted and ground in an ice-cooled mortar with a double volume of 0.1 M HCl. The crude assay solution was obtained by homogenate centrifugation at 14 000 rpm at 4 °C (3 times by 5 min). Beckman chromatograph (model 126/166) with Supelco precolumn (4.6 × 10 mm) and column (4.6 × 250 mm) (both filled with Ultrasphere C18) were used. The peptide solution (100 μL) was separated in a linear gradient (0–20%) of acetonitrile (ACN) in 0.05% trifluoroacetic acid (TFA) and was subjected to a post-column reaction with 200 μM 5,5′-dithiobis-2-nitrobenzoic acid (DTNB) in the 0.05 M potassium-phosphate buffer (pH = 7.6). The absorbance of the resulted reaction products was measured at 405 nm using a Beckman detector (model 166). The chromatograms were analysed using Karat 7.0 software (Beckman).

4.5. Statistical Analysis

The results were processed statistically using analysis of variance (ANOVA) for a two-factor experiment (3 levels of S and 4 concentrations of Ni in the nutrient solution) established in a completely randomized design, using Statistica 9.0 software. Each of the twelve experimental treatments included twenty replications (20 jars with 2 plants in each) and the whole experiment was repeated independently three times under the same conditions. This means that each of the twelve experimental treatments included in total 60 jars and 120 plants. The main effects of the S level and Ni concentration were compared using Tukey's multiple comparison test at the significance level $p \leq 0.05$. The comparison of the values within the same treatment as well as the mean values in each of the twelve treatments collected from each of the three independent replicates of the experiment over the time did not show statistically proven differences.

5. Conclusions

The results from the present study show that it is possible to prevent to some extent unfavorable changes in the micronutrient status of the white mustard "Rota" exposed to Ni (0.0004–0.08 mM) with the use of intensive S-SO_4 nutrition (6 or 9 mM S). Generally, an increase in the Fe, Mn, and Zn bioconcentrations, especially in the shoots, were found in Ni-treated plants supplied with extra S. The elevated concentrations of shoot B, Cu, and Mn were also revealed, without changes in their root concentrations. Simultaneously, the shoot Cl concentrations decreased. Furthermore, intensive S nutrition of Ni-exposed mustard, in relation to the standard 2 mM S dose, in general, enhanced the translocation of Fe, Cu, Mo, and B from roots to shoots. The improved micronutrient status of Ni-treated mustard supplied with extra S can be related to positive changes in total surface charge and cation exchange capacity of roots. However, an increase in the shoot biomass was noted only at 6 mM S in plants exposed to 0.04 mM Ni. These results contribute to the knowledge concerning mechanisms employed by plants intensively supplied with S to cope with Ni stress. The present studies offer an opportunity to increase the resistance of white mustard to Ni excess using an intensified S nutrition. These methods are quite promising and effective, easy to apply, as well as sustainable and safe for the environment. It may find practical application, which is especially important for farmers and horticulturalists, but needs further confirmation under field conditions.

Author Contributions: Conceptualization, R.M.-G., B.H.-N.; methodology, R.M.-G.; software, R.M.-G.; validation, R.M.-G., B.H.-N.; investigation, R.M.-G.; writing—original draft preparation, R.M.-G.; writing—review and editing, B.H.-N.; visualization, B.H.-N.; funding acquisition, R.M.-G., B.H.-N.

Funding: This research was financially supported by the statutory funds (OKA/DS/3, Department of Plant Physiology, University of Life Sciences in Lublin) from the Polish Ministry of Science and Higher Education.

Acknowledgments: We thank the employees of Department of Plant Physiology of Maria Curie Sklodowska University (Prof. Małgorzata Wójcik, Prof. Sławomir Dresler) and Institute of Agrophysics PAS (Prof. Alicja Szatanik-Kloc) for their technical support.

Conflicts of Interest: The authors declare no conflict of interest.

References

1. Ragsdale, S.W. Nickel-based enzyme systems. *J. Biol. Chem.* **2009**, *284*, 18571–18575. [CrossRef] [PubMed]
2. Poonkothai, M.; Shyamala Vijayavathi, B. Nickel as an essential element and a toxicant. *Int. J. Environ. Sci.* **2012**, *1*, 285–288.
3. Fabiano, C.C.; Tezotto, T.; Favarin, J.L.; Polacco, J.C.; Mazzafera, P. Essentiality of nickel in plants: A role in plant stresses. *Front. Plant Sci.* **2015**, *6*, 754. [CrossRef] [PubMed]
4. Bhatia, N.P.; Nkang, A.E.; Walsh, K.B.; Baker, A.J.; Ashwath, N.; Midmore, D.J. Successful seed germination of the nickel hyperaccumulator *Stackhousia tryonii*. *Ann. Bot.* **2005**, *96*, 159–163. [CrossRef]
5. Polacco, J.C.; Mazzafera, P.; Tezotto, T. Opinion: Nickel and urease in plants: Still many knowledge gaps. *Plant Sci.* **2013**, *199*, 79–90. [CrossRef]
6. Ahmad, M.S.A.; Ashraf, M. Essential roles and hazardous effects of nickel in plants. *Rev. Environ. Contam. Toxicol.* **2011**, *214*, 125–167. [CrossRef]
7. López, M.Á.; Magnitski, S. Nickel: The last of the essential micronutrients. *Agron. Colomb.* **2011**, *29*, 49–56.
8. Bhalerao, S.A.; Sharma, A.S.; Poojari, A.C. Toxicity of nickel in plants. *Int. J. Pure Appl. Biosci.* **2015**, *3*, 345–355.
9. Pavlovkin, J.; Fiala, R.; Čiamporová, M.; Martinka, M.; Repka, V. Impact of nickel on grapevine (*Vitis vinifera* L.) root plasma membrane, ROS generation, and cell viability. *Acta Bot. Croat.* **2016**, *75*, 25–30. [CrossRef]
10. Nieminen, T.M.; Ukonmaanaho, L.; Rausch, N.; Shotyk, W. Biogeochemistry of nickel and its release into the environment. In *Nickel and Its Surprising Impact in Nature*, 1st ed.; Sigel, A., Sigel, H., Sigel, R.K.O., Eds.; John Wiley & Sons: Chichester, UK, 2007; pp. 1–30. [CrossRef]

11. Sreekanth, T.V.M.; Nagajyothi, P.C.; Lee, K.D.; Prasad, T.N.V.K.V. Occurrence, physiological responses and toxicity of nickel in plants. *Int. J. Environ. Sci. Technol.* **2013**, *10*, 1129–1140. [CrossRef]
12. Harasim, P.; Filipek, T. Nickel in the environment. *J. Elem.* **2015**, *20*, 525–534. [CrossRef]
13. Chen, C.; Huang, D.; Liu, J. Functions and toxicity of nickel in plants: Recent advances and future prospects. *Clean Soil Air Water* **2009**, *37*, 304–313. [CrossRef]
14. Shafeeq, A.; Butt, Z.A.; Muhammad, S. Response of nickel pollution on physiological and biochemical attributes of wheat (*Triticum aestivum* L.) var. Bhakar-02. *Pak. J. Bot.* **2012**, *44*, 111–116.
15. Hussain, M.B.; Ali, S.; Azam, A.; Hina, S.; Farooq, M.A.; Ali, B.; Bharwana, S.A.; Gill, M.B. Morphological, physiological and biochemical responses of plants to nickel stress: A review. *Afr. J. Agric. Res.* **2013**, *8*, 1596–1602. [CrossRef]
16. Ghasemi, R.; Ghaderian, S.M.; Krämer, U. Interference of nickel with copper and iron homeostasis contributes to metal toxicity symptoms in the nickel hyperaccumulator plant *Alyssum inflatum*. *New Phytol.* **2009**, *184*, 566–580. [CrossRef] [PubMed]
17. Yadav, S.S.; Shukla, R.; Sharma, Y.K. Nickel toxicity on seed germination and growth in radish (*Raphanus sativus*) and its recovery using copper and boron. *J. Environ. Biol.* **2009**, *30*, 461–466.
18. Sengar, R.S.; Gupta, S.; Gautam, M.; Sharma, A.; Sengar, K. Occurrence, uptake, accumulation and physiological responses of nickel in plants and its effects on environment. *Res. J. Phytochem.* **2008**, *2*, 44–60. [CrossRef]
19. Yusuf, M.; Fariduddin, Q.; Hayat, S.; Ahmad, A. Nickel: An overview of uptake, essentiality and toxicity in plants. *Bull. Environ. Contam. Toxicol.* **2011**, *86*, 1–17. [CrossRef]
20. Stanisławska-Głubiak, E.; Korzeniowska, J. Tolerance of white mustard (*Sinapis alba* L.) to soil pollution with several heavy metals. *Ecol. Chem. Eng. Ser. A* **2011**, *18*, 445–450.
21. Putnik-Delić, M.I.; Maksimović, I.V.; Gani-Novaković, I.; Zeremski, T.; Marjanović-Jaromela, A. The effect of Ni concentration of the most abundant essential cations in several *Brassica* species. *J. Nat. Sci.* **2014**, *126*, 15–23. [CrossRef]
22. Manohar, P.; Ram; Pushpan, R.; Rohini, S. Mustard and its uses in *Ayurveda*. *Indian J. Tradit. Knowl.* **2009**, *8*, 400–404.
23. Damian, C. Physical properties of mustard seeds (*Sinapis alba* L.). *Lucr. Ştiinţifice Ser. Zooteh.* **2014**, *61*, 39–44.
24. Paunović, D.; Šolević Knudsen, T.; Krivokapić, M.; Zlatković, B.; Antić, M. Sinalbin degradation products in mild yellow mustard paste. *Hem. Ind.* **2012**, *66*, 29–32. [CrossRef]
25. Young-Mathews, A. Plant guide for field mustard (*Brassica rapa* ssp. *rapa*). USDA-Natural Resources Conservation Service, Plant Materials Center, Corvallis. Available online: https://plants.usda.gov/plantguide/pdf/pg_brrar.pdf (accessed on 20 September 2019).
26. Ryant, P. Effect of sulphur fertilisation on yield and quality of white mustard seeds. *Acta Univ. Agric. Silvic. Mendel. Brun.* **2009**, *57*, 95–104. [CrossRef]
27. Bloem, E.; Haneklaus, S.; Daniels, R.; Schnug, E. Influence of sulfur fertilization on floral scent patterns of crops in full bloom. *Landbauforsch Volk. Appl. Agric. For. Res.* **2010**, *60*, 45–50.
28. Filipek-Mazur, B.; Gondek, K. Yielding and sulphur content in white mustard as the effect of application of multi-component fertilizers containing sulphur. *Acta Agrophys.* **2005**, *6*, 343–351.
29. Barczak, B.; Kozera, W.; Knapowski, T.; Ralcewicz, M. Selected field components in white mustard (*Sinapis alba*) versus sulfur fertilization. *J. Cent. Eur. Agric.* **2011**, *12*, 380–389. [CrossRef]
30. Shekhawat, K.; Kandpal, B.K.; Chand, G.; Singh, D. Sulphur management for increased productivity of Indian mustard: A review. *Ann. Plant. Soil Res.* **2015**, *17*, 1–12.
31. Ray, K.; Sengupta, K.; Pal, A.K.; Banerjee, H. Effects of sulphur fertilization on yield, S uptake and quality of Indian mustard under varied irrigation regimes. *Plant Soil Environ.* **2015**, *61*, 6–10. [CrossRef]
32. Hawkesford, M.J. Plant responses to sulphur deficiency and the genetic manipulation of sulphate transporters to improve S-utilization efficiency. *J. Exp. Bot.* **2000**, *51*, 131–138. [CrossRef]
33. Mašauskiene, A.; Mašauskas, V. Soil sulphur problems and management. In *Sustainable Agriculture. Series Ecosystem Health and Sustainable Agriculture*; Jakobsson, C., Ed.; Baltic University Press: Uppsala, Sweden, 2012; pp. 113–116.
34. Eriksen, J. Sulphur Cycling in Agrosystems. Ph.D. Thesis, Aarhus University, Tjele, Denmark, 2010.
35. Procházková, D.; Pavlíková, D.; Pavlík, M. Sulphur: Role in alleviation of environmental stress in crop plants. In *Plant-Environment Interaction: Responses and Approaches to Mitigate Stress*; Azooz, M.M., Ahmad, P., Eds.; John Wiley & Sons: Hoboken, NJ, USA, 2016; pp. 84–96. [CrossRef]

36. Prasad, M.N.V. *Heavy Metal. Stress in Plants: From Biomolecules to Ecosystems*, 2nd ed.; Springer: Berlin/Heidelberg, Germany, 2004; p. 462. [CrossRef]
37. Maheshwari, R.; Dubey, R.S. Nickel-induced oxidative stress and the role of antioxidant defense in rice seedlings. *Plant Growth Regul.* **2009**, *59*, 37–49. [CrossRef]
38. Yadav, N.; Sharma, S. An account of nickel requirement, toxicity and oxidative stress in plants. *Biol. Forum Int. J.* **2016**, *8*, 414–419.
39. Barker, A.V.; Pilbeam, D.J. *Handbook of Plant Nutrition*, 2nd ed.; CRC Press: Boca Raton, FL, USA, 2015; p. 773.
40. Rahman, H.; Sabreen, S.; Alam, S.; Kawai, S. Effects of nickel on growth and composition of metal micronutrients in barley plants grown in nutrient solution. *J. Plant Nutr.* **2005**, *28*, 393–404. [CrossRef]
41. Yadav, S. Toxicity of Nickel in Plants and Its Interaction with Certain Micronutrient. Ph.D. Thesis, University of Lucknow, Lucknow, India, 2013.
42. Fageira, N.K. *Mineral. Nutrition of Rice*, 1st ed.; CRC Press: Boca Raton, FL, USA, 2013; p. 586. [CrossRef]
43. Aibara, I.; Miwa, K. Strategies for optimization of mineral nutrient transport in plants: Multilevel regulation of nutrient-dependent dynamics of root architecture and transporter activity. *Plant Cell Physiol.* **2014**, *55*, 2027–2036. [CrossRef]
44. Mitra, G.N. *Regulation of Nutrient Uptake by Plants*, 1st ed.; Springer: New Delhi, India, 2015; p. 195. [CrossRef]
45. Moosavi, A.A.; Mansouri, S.; Zahedifar, M. Effect of soil water stress and nickel application on micronutrient status of canola grown on two calcareous soils. *Plant. Prod. Sci.* **2015**, *18*, 377–387. [CrossRef]
46. Ernst, W.H.O.; Krauss, G.J.; Verkleij, J.A.C.; Wesenberg, D. Interaction of heavy metals with the sulphur metabolism in angiosperms from an ecological point of view. *Plant Cell Environ.* **2008**, *31*, 123–143. [CrossRef]
47. Ranade-Malvi, U. Interaction of micronutrients with major nutrients with special reference to potassium. *Karnataka J. Agric. Sci.* **2011**, *24*, 106–109.
48. Jankowski, K.J.; Budzyński, W.S.; Kijewski, Ł.; Klasa, A. Concentrations of cooper, zinc and manganese in the roots, straw and oil cake of white mustard (*Sinapis alba* L.) and Indian mustard (*Brassica juncea* (L.) Czern. et Coss) depending on sulphur fertilization. *Plant Soil Environ.* **2014**, *60*, 364–371. [CrossRef]
49. Matraszek, R.; Hawrylak-Nowak, B.; Chwil, M.; Chwil, S.; Rudaś, M. Effect of the interaction of nickel stress and sulphur supplementation on the content and accumulation of macronutrients in white mustard (*Sinapis alba* L.). *Electron. J. Pol. Agric. Univ.* **2017**, *20*, 1. [CrossRef]
50. Szatanik-Kloc, A.; Szerement, J.; Józefaciuk, G. The role of cell walls and pectins in cation exchange and surface area of plant roots. *J. Plant Physiol.* **2017**, *215*, 85–90. [CrossRef]
51. Freeman, J.L.; Persans, M.W.; Nieman, K.; Albrecht, C.; Peer, W.; Pickering, I.J.; Salt, D.E. Increased glutathione biosynthesis plays a role in nickel tolerance in *Thlaspi* nickel hyperaccumulators. *Plant Cell* **2004**, *16*, 2176–2191. [CrossRef] [PubMed]
52. Hossain, M.A.; Piyatida, P.; Texteira da Silva, J.A.; Fujita, M. Molecular mechanism of heavy metal toxicity and tolerance in plants: Central role of glutathione in detoxification of Reactive Oxygen Species and methylglyoxal and in heavy metal chelation. *J. Bot.* **2012**, *2012*, 872875. [CrossRef]
53. Viehweger, K. How plants cope with heavy metals? *Bot. Stud.* **2014**, *55*, 35. [CrossRef] [PubMed]
54. Hasanuzzaman, M.; Alam, M.M.; Nahar, K.; Mohsin, S.M.; Bhuyan, M.H.M.B.; Parvin, K.; Hawrylak-Nowak, B.; Fujita, M. Silicon-induced antioxidant defense and methylglyoxal detoxification works coordinately in alleviating nickel toxicity in *Oryza sativa* L. *Ecotoxicology* **2019**, *28*, 261. [CrossRef] [PubMed]
55. Jones, J.B., Jr. *Laboratory Guide for Conducting Soil Tests and Plant Analysis*, 1st ed.; CRC Press: Boca Raton, FL, USA, 2001; p. 363.
56. Aref, F. Concentration and uptake of zinc and boron in corn leaf as affected by zinc sulfate and boric acid fertilizers in a deficient soil. *Life Sci. J.* **2011**, *8*, 26–31.
57. Wolf, B. Improvements in the azomethine-H method for determination of boron. *Commun. Soil Sci. Plant Anal.* **1974**, *5*, 39–44. [CrossRef]
58. Nowosielski, O. *Methods for the Determination of Fertilisation Requirements*, 1st ed.; PWRiL: Warszawa, Poland, 1974. (In Polish)

plants

MDPI

Article

Modulation of the Functional Components of Growth, Photosynthesis, and Anti-Oxidant Stress Markers in Cadmium Exposed *Brassica juncea* L.

Dhriti Kapoor [1,2,†], Mahendra P. Singh [2,†], Satwinderjeet Kaur [1], Renu Bhardwaj [1], Bingsong Zheng [3,*] and Anket Sharma [1,3,*]

1 Department of Botanical & Environmental Sciences, Guru Nanak Dev University, Amritsar 143005, Punjab, India

2 School of Bioengineering and Biosciences, Lovely Professional University, Delhi-Jalandhar Highway, Phagwara 144411, Punjab, India

3 State Key Laboratory of Subtropical Silviculture, Zhejiang A&F University, Hangzhou 311300, China

* Correspondence: bszheng@zafu.edu.cn (B.Z.); anketsharma@gmail.com or anketsharma@zafu.edu.cn (A.S.); Tel.: +86-(0)571-63730936 (B.Z.)

† Contributed equally.

Received: 16 July 2019; Accepted: 29 July 2019; Published: 31 July 2019

Abstract: Heavy metals (including Cadmium) are being entered into the environment through various sources and cause toxicity to plants. Response of *Brassica juncea* L. var. RLC-1 was evaluated after exposing them to different concentration of cadmium (Cd) for seven days. Seeds of *B. juncea* were treated with different concentrations of Cd like 0.2–0.6 mM for 7 days, allowing them to grow in Petri-dishes, and seedlings were examined for different physiological responses. Following exposure to Cd, in the seedlings of *B. juncea*, growth parameters (root and shoot length), stress markers (lipid peroxidation and H_2O_2 content), secondary metabolites, photosynthetic pigments, and ion analysis, were estimated along with enzymatic and non-enzymatic antioxidants. We observed a significant reduction in root and shoot length after Cd treatment as compared to control seedlings. Malondialdehyde and H_2O_2 contents were increased accompanied by enhanced Cd uptake. Activities of antioxidative enzymes were also significantly altered following Cd exposure to the seedlings of *B. juncea*. Conclusively, we suggest that Cd exposure to the seedlings triggered an induction of several defense responses in *B. juncea* including major metabolites.

Keywords: cadmium toxicity; oxidative stress; antioxidative defense system; photosynthetic pigments

1. Introduction

Plants require optimum level of light, temperature, nutrients, and moisture for their growth and development. Any major deviation from these required optimum conditions affects the plants adversely and may lead to death of the plants. The damage to plants results in retardation of growth, fresh weight, and food production [1–10]. Heavy metals are one of the major environmental toxicants. Enhancing toxicity of these metals is dangerous for nutritional, biological, and environmental aspects. Chemical nature of metal, its doses, composition, pH of soil, and types of plants are certain factors on which metal toxicity depends [11]. Heavy metal pollution occurs through both natural and anthropogenic sources. Reports suggested that major areas of world have been contaminated by heavy metals through the activities like mining, smelting operations, and agriculture [12,13]. The natural sources like volcanic eruptions, rock outcropping or geologic parent material also contribute their release in the environment. The toxicity of these metals to plants depends upon the oxidation states [14]. The metal toxicity at cellular and molecular level changes various vital activities of plants, which includes deactivation and degradation of enzymes, proteins, displacement/substitution of important metal ions, blocking of functional groups of metabolically important molecules, structural changes, and membrane destabilization [5,14–18]. As a consequence, change in plant metabolism, decrease in respiration, photosynthesis, and change in enzyme activities is observed. In addition, they also stimulate the generation of free oxidants such as hydroxyl radicals (OH), singlet oxygen ($^1O_2$), hydrogen peroxide (H_2O_2), and superoxide radicals (O_2^-), and thus disturb the redox homeostasis [19,20].

Cadmium (Cd) is present in the soil as free hydrated ions in the form of soil solution or can also be composited with organic or inorganic ligands [21]. Visible effects of high Cd doses in the plants showed reduced growth, leaf chlorosis, closure of stomata, disturbance in water balance, and photosynthetic apparatus of plants [22,23]. According to Al-Yemens [24], reductions in the weights of roots and shoots of *Vigna ambacensis* occur under low Cd concentration. In *Brassica juncea*, application of Cd caused reduction in root and shoot biomass, total chlorophyll content, and total protein content [25]. Increasing concentrations of Cd in soil inhibited germination and seedling ratio of maize seeds [26]. Plants under cadmium stress have decreased net photosynthesis due to inhibition of Rubisco to fix the carbon dioxide [27]. Cd also changes the activities of different metabolic enzymes [28–30]. Furthermore, it also binds with sulphydryl groups (SH) of proteins and causes displacement of essential elements, deficiency effects, disruption of structure, and inhibition of enzyme activities [31]. Cd stress stimulates lipid peroxidation, stimulates the synthesis of oxidants like hydrogen peroxide, superoxide, and singlet oxygen and hydrogen radicals, and also alters the activities of antioxidative enzymes [32]. *Brassica juncea*, an important oilseed crop, when cultivated in heavy metals polluted sites, is affected severely towards growth, development, and also the yield. *B. juncea* is mainly grown as a food crop, also used for its medicinal purposes. Mustard green possesses antiseptic, aperitif, diuretic, emetic, and rubefacient properties. It is rich source of antioxidants like flavonoids, indoles, carotenes, lutein, and zea-xanthin [33]. Cadmium toxicity occurs at different concentration and keeping this in mind, we designed an experiment to with different Cd concentrations (0.2, 0.4, and 0.6 mM) to investigate whether it has toxic effect on *B. juncea* plants or it may tolerate lower doses. To assess this, we studied physiological and biochemical parameters using *B. juncea* as plant model system.

2. Material and Methods

2.1. Plant Material

Certified seeds of *Brassica juncea* L. cv. 'RLC-1' used for the current investigation were purchased from Punjab Agriculture University (PAU), Ludhiana, Punjab, India. Seeds of *B. juncea* were surface sterilized, followed by treatment with 0, 0.2, 0.4, and 0.6 mM Cd in form of $CdCl_2$ solution, and grown on Whatman No. 1 filter paper placed in the autoclaved Petri plates. For each treatment, 3 Petri plates were used with 20 seeds each. Controlled conditions like 25 ± 2 °C, 16 h photoperiod and 175 μmol m^{-2} s^{-1} light intensity were provided in a seed germinator. Three mL of $CdCl_2$ solution was given after every 3 days. Full seedlings were collected on the 7th day for the measurements of morphological parameters and stored at −20 °C for further analysis. Sample collection and further analysis were done using three biological replicates.

2.2. Morphological Parameters

Seven-day-old Cd treated or control (untreated) seedlings were harvested and their root length and shoot length were measured.

2.3. Oxidative Stress Markers

2.3.1. Malondialdehyde Content

Membrane damage due to exposure to Cd was judged in terms of total malondialdehyde (MDA) content by following the method given by Heath and Packer [34]. Extraction of 1 g seedlings was performed in 0.1% (*w*/*v*) trichloroacetic acid (TCA) and then centrifuged at 5000 rpm. To the supernatant, 20% (*w*/*v*) TCA containing 0.5% (*w*/*v*) TBA was added. Optical density of the supernatant was measured at 532 nm. Extinction coefficient of 155 mM cm^{-1} was taken to measure MDA content.

2.3.2. H_2O_2 Content

Hydrogen peroxide (H_2O_2) content was estimated by method mentioned by Velikova et al. [35]. Briefly, 500 mg of plant material was taken in which 2 mL of TCA was added and then centrifuged at 12,000 rpm for 15 min. 0.5 mL of 10 mM PBS (phosphate buffered saline) and 1 mL of 1 M potassium iodide was added to 0.5 mL of supernatant. Optical density was taken at 390 nm.

2.4. *Pigment Analysis*

2.4.1. Chlorophyll Content

Level of chlorophyll was measured by using method given by Arnon [36] method. One gram plant tissue was extracted in 80% acetone. The material was then centrifuged at 4 °C for the duration of 20 min at 13,000 rpm. The supernatant was taken to measure chlorophyll content. The optical density was measured at 645 and 663 nm.

2.4.2. Anthocyanin Content

Anthocyanin content was estimated by method described by Mancinelli [37]. Plant tissue (one gram) was crushed in 3 mL of acidified methanol and centrifuged at 4 °C for the time period of 20 min at 13,000 rpm. Then the optical density of supernatant was measured at 530 and 657 nm.

2.5. *Determination of Sodium and Potassium Ions*

Flame emission photometer (Systronics 128) was used to estimate sodium and potassium ion content as mentioned in Chapman [38]. Plant material (0.5 g) was digested in nitric acid (HNO_3) and perchloric acid ($HClO_4$) and total volume was made up to 50 mL by diluting with distilled water. Then the extract was filtered and ion contents were estimated.

2.6. *Determination of Cd Content and Bioconcentration Factor (BCF)*

Atomic Absorption Spectrophotometer (AAS) technique was used to estimate Cd content. Half a gram (0.5 g) dried plant samples were digested in nitric acid and perchloric acid as mentioned in Chapman [38]. Digested samples were filtered and diluted up to 50 mL with distilled water. $CdCl_2$ were used as standard solution for the standardization in AAS. Calculation of BCF was done as mentioned by Retamal-Salgado [39].

2.7. *Estimation of Antioxidant Enzyme Activities*

Activities of antioxidative enzymes were determined by the standard methods of Aebi [40] for catalase (CAT) (EC 1.11.1.6), Kono [41] for superoxide dismutase (SOD) (EC 1.15.1.1), Carlberg and Mannervik [42] for glutathione reductase (GR) (EC 1.6.4.2), Dalton et al. [43] for dehydroascorbate reductase (DHAR) (EC 1.8.5.1), and Hossain et al. [44] for monodehydroascorbate reductase (MDHAR) (EC 1.15.1.1).

2.7.1. CAT

The decomposition rate of hydrogen peroxide was followed by reduction in optical density at 240 nm in reaction mixture including 1.5 mL phosphate buffer, 0.7 mL of hydrogen peroxide, and 300 µL of enzyme extract.

2.7.2. SOD

Reaction mixture including 1.73 mL sodium carbonate buffer, 500 μL NBT, and 100 μL Triton X-100 was collected in the test cuvettes. Adding of 100 μL hydroxylamine hydrochloride started the reaction. A total of 70 μL of the plant extract was added after 2 min. Percentage reduction with the rate of NBT decrease was observed as rise in absorbance at 540 nm.

2.7.3. GR

The reaction mixture comprised 50 mM K-phosphate buffer (pH 7.6) (1.45 mL), 3 mM EDTA (0.3 mL), 0.1 mM NADPH (0.3 mL), 1 mM oxidized glutathione (0.3 mL), and 150 μL enzyme extract. The absorbance was read at 340 nm.

2.7.4. DHAR

For execution of this assay, reaction mixture was prepared containing 50 mM phosphate buffer (1.2 mL), 0.1 mM EDTA (0.3 mL), 1.5 mM glutathione (0.3 mL), 0.2 mM dehydroascorbate (0.3 mL), and 400 μL plant extract. A rise in the optical density was estimated at 265 nm.

2.7.5. MDHAR

The assay mixture was consisted of 50 mM of phosphate buffer (1.4 mL), 0.1 mM EDTA (0.3 mL), 0.3 mM NADH (0.3 mL), 0.25 units of ascorbate oxidase, and plant extract (0.5 mL). Decrease in absorbance was measured at 340 nm.

2.8. Non-Enzymatic Antioxidants

2.8.1. Ascorbic Acid Content

Level of ascorbic acid was measured by Roe and Kuether [45]. The assay mixture consisted of 100 mg charcoal, 4 mL of double distilled water, 0.5 mL extract of plant, and 0.5 mL of 50% TCA. Then, 0.4 mL of DNPH was taken and the mixture was incubated at 37 °C for 3 h by chilling on ice bath. Sulphuric acid (65% purity, 1.6 mL) was taken and added into it and placed at room temperature for 30 min. Absorbance was taken at 520 nm and 1 mg 100 mL^{-1} ascorbic acid was taken as standard.

2.8.2. Total Phenolic Content

Phenolic content was estimated by Singleton and Rossi [46]. In 0.4 g of dry plant material, 40 mL of 60% ethanol was mixed. Then it was shaken in water at 60 °C for 10 min. Filtered extract was taken and diluted to 100 mL with 60% ethanol. From diluted plant sample, 2.5 mL was taken and re-diluted with 25 mL of distilled water. A two milliliter sample was added with 10 mL of FC reagent and after 5 min, 7.5% of sodium carbonate was added into the reaction mixture. To the mixture, 2 h incubation was given. The optical density was measured at 765 nm. As a reference standard, Gallic acid was taken.

2.9. Statistical Analysis

All the experiments were carried out in triplicate and performed three times independently. Data is expressed in mean ± SE. To test the statistical significant difference between the treatments, one way-analysis of variance (ANOVA) was carried out followed by post hoc Tukey's test using SPSS software. For all the analyses, a p value less than 0.05 were considered statistically significant.

3. Results

3.1. Morphological Parameters

A steep decline in root length was observed with increasing cadmium concentration as compared to control. Approximately three-fold decrease in root length was evident in the seedlings exposed to the highest concentration of cadmium. Slight inhibition in shoot length was found in Cd stressed seedlings. Maximum fall in shoot length was noticed in 0.6 mM Cd stressed seedlings (Table 1).

Table 1. Effect of Cadmium (Cd) metal on growth parameters of 7 days old seedlings of *B. juncea*.

Cd Conc. (mM)	Root Length (cm)	Shoot Length (cm)
0.0	11.7 ± 1.57 a	4.37 ± 0.22 a
0.2	7.63 ± 0.41 b	4.17 ± 0.2 ab
0.4	5.97 ± 0.12 b	3.43 ± 0.2 ab
0.6	3.93 ± 0.5 b	3.33 ± 0.26 b

Data represents mean ± SE, and the experiment was repeated three times independently. Means with same letters are not significantly different from each other at $p < 0.05$.

3.2. MDA and H_2O_2 Content

Malondiaeldehyde (MDA) and H_2O_2 contents were found lowest in control seedlings. However, with increasing doses of Cd, levels of MDA and H_2O_2 were found to increase. Maximum increase in these oxidative stress markers was observed in the seedlings exposed to 0.6 mM Cd (Figure 1).

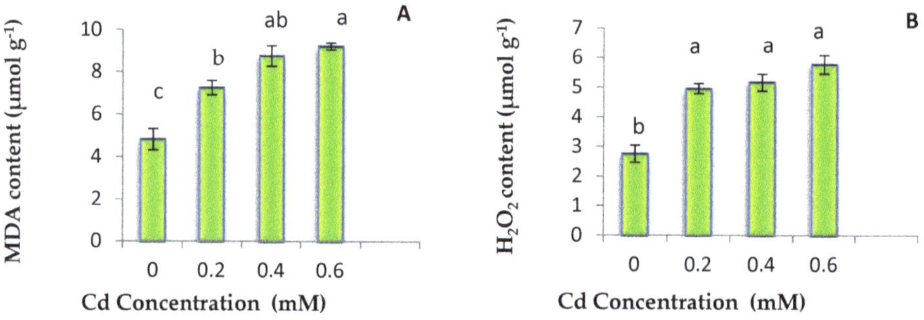

Figure 1. Effects of different Cd treatments on Malondiaeldehyde (MDA) content (**A**) and H_2O_2 content (**B**) in 7 days old seedlings of *B. juncea*. Data presented as mean ± SE and means with same letters are not significantly different from each other at $p < 0.05$.

3.3. Pigments

3.3.1. Chlorophyll Content

Cadmium toxicity decreased the total chlorophyll, chl a, and chl b content as compared to control plants. Highest Cd toxicity decreased the level of total chlorophyll 1.99 folds in seven-days old seedlings of *B. juncea* as compared to control. Moreover, a small but significant variation in 0.2 mM and 0.4 mM Cd stressed seedlings was observed for chl a and total chlorophyll. A continuous decline was noticed with increased Cd concentration, where control seedlings possessed highest levels of chl a and chl b (Table 2).

Table 2. Effect of Cd metal on pigment contents of 7 days old *B. juncea* seedling.

Cd Conc. (mM)	Total Chl (mg g^{-1} FW)	Chl a (mg g^{-1} FW)	Chl b (mg g^{-1} FW)	Anthocyanin (mg g^{-1} FW)
0.0	26.75 ± 0.77 a	16.78 ± 0.56 a	7.54 ± 0.47 a	2.41 ± 0.28 b
0.2	21.69 ± 0.9 b	15.04 ± 0.93 a	5.33 ± 1.29 ab	3.6 ± 0.32 b
0.4	18.56 ± 0.39 c	11.59 ± 0.44 b	6.12 ± 0.41 a	5.09 ± 0.41 a
0.6	13.38 ± 0.39 d	7.1 ± 0.23 c	2.24 ± 0.19 b	6.02 ± 0.11 a

Data represents mean ± SE, $n = 3$ and experiment was repeated three times independently. Means with same letters are not significantly different from each other at $p < 0.05$.

3.3.2. Anthocyanin Content

Application of Cd metal raised the anthocyanin level 2.5 folds from control to 0.6 mM Cd. Moreover, a continuous increase in anthocyanin content was noticed with Cd treatment from 0.2 to 0.6 mM (Table 2).

3.4. Cd Uptake and Bioconcentration Factor (BCF) Studies

Results revealed significant accumulation of Cd in *B. juncea* plants (Table 3). It was noticed that Cd accumulation in seedlings was increased with the increasing concentration of Cd. Maximum uptake of Cd was noticed in 0.6 mM treated plants. Control plants were also examined for Cd uptake, but Cd was not detected. Moreover, BCF analysis revealed that Cd uptake efficiency was reduced with increasing Cd treatments (Table 3).

Table 3. Data showing Cd accumulation and bioconcentration factor in 7 days old seedlings of *B. juncea* grown under different Cd treatments.

Cd Conc. (mM)	Cd Accumulation in Seedlings (μg g^{-1} DW)	BCF
0.2	69.13 ± 5.14 ^{b}	1.85
0.4	83.39 ± 2.35 ^{ab}	1.13
0.6	89.68 ± 2.87 ^{a}	0.82

Data represents mean ± SE, $n = 3$, and the experiment was repeated three times independently. Means with same letters are not significantly different from each other at $p < 0.05$.

3.5. Sodium and Potassium Ion Analysis

Sodium and potassium ions were found to decrease with increasing Cd concentrations (Table 4). Concentration of sodium ion was recorded maximum in control seedlings. Maximum reduction in sodium ions was noticed in the seedlings exposed to highest Cd concentration. Similarly, highest concentration of potassium ions was noticed in the control seedlings. However, significant reduction in potassium content was only noticed in those seedlings which were treated with highest Cd concentration.

Table 4. Effect of Cd on contents of sodium and potassium ions in 7 days old seedlings of *B. juncea*.

Cd Conc. (mM)	Sodium Ion (ppm)	Potassium Ion (ppm)
0.0	6.38 ± 0.57 ^{a}	5.82 ± 0.59 ^{a}
0.2	5.89 ± 0.59 ^{a}	5.25 ± 0.29 ^{ab}
0.4	4.99 ± 0.49 ^{b}	5.5 ± 0.62 ^{ab}
0.6	4.42 ± 0.36 ^{b}	4.24 ± 0.79 ^{b}

Data represents mean ± SE, $n = 3$, and the experiment was repeated three times independently. Means with same letters are not significantly different from each other at $p < 0.05$.

3.6. Antioxidant Enzyme Activities

Antioxidative defense system got activated after Cd stress and activities of enzymatic antioxidants were noticed to increase. Catalase activity was found minimum in control seedlings, whereas maximum under highest Cd treatment. However, change in CAT activity was not statistically significant among different Cd concentrations. A continuous increase in activity of SOD was noticed due to Cd treatment as compared to control seedlings and 1.54-folds increased SOD activity was observed in 0.6 mM Cd stressed seedlings in comparison to control. Activity of GR enzyme was recorded maximum in 0.4 mM Cd treated seedlings. However, among different Cd treatments, GR activity was not significantly different. Maximum DHAR and MDHAR activities were observed in seedlings grown under maximum Cd concentration. For MDHAR, almost 1.63 folds rise in enzyme activity was noted under highest Cd treatment as compared to control (Table 5).

Table 5. Effect of Cd metal on activities of enzymatic antioxidants in 7 days old seedlings of *B. juncea*.

Cd Conc. (mM)	CAT (UA mg^{-1} Protein)	SOD (UA mg^{-1} Protein)	GR (UA mg^{-1} Protein)	DHAR (UA mg^{-1} Protein)	MDHAR (UA mg^{-1} Protein)
0.0	6.43 ± 0.19 ^{b}	5.43 ± 0.35 ^{b}	7.63 ± 0.27 ^{b}	8.33 ± 0.49 ^{c}	11.29 ± 0.33 ^{c}
0.2	8.27 ± 0.35 ^{a}	6.38 ± 0.21 ^{b}	9.66 ± 0.40 ^{a}	10.25 ± 0.33 ^{b}	14.65 ± 0.44 ^{b}
0.4	8.51 ± 0.53 ^{a}	7.75 ± 0.25 ^{a}	10.87 ± 0.54 ^{a}	11.25 ± 0.26 ^{ab}	16.94 ± 0.62 ^{a}
0.6	8.94 ± 0.32 ^{a}	8.39 ± 0.24 ^{a}	9.93 ± 0.47 ^{a}	12.21 ± 0.29 ^{a}	18.46 ± 0.55 ^{a}

Data represents Mean ± SE, $n = 3$, and the experiment was repeated three times independently. Means with same letters are not significantly different from each other at $p < 0.05$.

3.7. Level of Non-Enzymatic Antioxidants

A maximum ascorbic acid content was estimated by 1.76 folds in 0.6 mM Cd treated seedlings as compared to control. Moreover, ascorbic acid content was found to be significantly different among all Cd treatments. Similarly, total phenolic content was also increased with Cd treatment and maximum content was found in seedlings grown under highest Cd treatment. However, no significant difference in the phenolic contents of control and 0.2 mM treated seedlings was observed (Table 6).

Table 6. Effect of Cd metal on non-enzymatic antioxidants in 7 days old seedlings of *B. juncea*.

Cd Conc. (mM)	Ascorbic Acid (mg g^{-1} FW)	Total Phenolic Content (mg g^{-1} FW)
0.0	8.63 ± 0.25 ^{d}	6.86 ± 0.20 ^{c}
0.2	10.34 ± 0.49 ^{c}	7.89 ± 0.49b ^{c}
0.4	12.97 ± 0.31 ^{b}	8.89 ± 0.35 ^{ab}
0.6	15.23 ± 0.41 ^{a}	9.96 ± 0.15 ^{a}

Data represents mean ± SE, $n = 3$, and the experiment was repeated three times independently. Means with same letters are not significantly different from each other at $p < 0.05$.

3.8. Correlation Analysis

Correlation analysis between accumulated Cd and other physiological parameters revealed that deposition of Cd in seedlings had a negative impact on growth and stimulated the antioxidative defense system to counterattack oxidative damage and ultimately helps in enhancing the tolerance against Cd (Table 7).

Table 7. Data showing correlation between Cd accumulation and different parameters.

Parameter	Correlation Coefficient	p Value
Root length	−0.8738	0.0002
Shoot length	−0.6936	0.0123
MDA content	0.9189	2.39×10^{-5}
H_2O_2 content	0.9347	8.38×10^{-6}
Total chlorophyll content	−0.8791	0.0001
Anthocyanin content	0.8566	0.0003
Sodium ion content	−0.6271	0.0291
Potassium ion content	−0.3413	0.2775
CAT activity	0.8703	0.0002
SOD activity	0.8365	0.0006
GR activity	0.8729	0.0002
DHAR activity	0.8856	0.0001
MDHAR activity	0.9227	1.89×10^{-5}
Ascorbic acid content	0.8186	0.0011
Total phenolic content	0.8241	0.0009

4. Discussion

Cadmium limits the crop productivity worldwide as this metal deposits within plant organs and adversely affects the essential physiological processes. Retardation in growth, accumulation of lipid peroxides, alteration in the level of pigments, antioxidants and activities of antioxidative enzymes and generation of ROS were observed due to Cd toxicity in various plant species [47–52]. The current study revealed a statistically significant decrease in the root length and shoot length under Cd stress. Due to adverse effects on the roots, nutrition and water supply is also negatively altered, which further affects the growth and physiology of aerial parts of plants. These observations are in agreement with the results of Pandey and Tripathi [53], where Cd treatment to *Albizia procera* seedlings showed retardation of growth in morphological parameters. Cadmium stress adversely affected root length, shoot length, leaf area, and biomass of *A. procera*. Similarly, in another study supplement of 10 ppm dose of Cd^{2+} and Cr^{6+} to *Medicago sativa* plants drastically inhibited growth as compared to the control plants. Further, when their doses were raised to 20 ppm, growth was diminished by 62.0% and 65.0%, respectively [54]. Inhibitory effect of Cd was more evident in the growth of roots as compared to others as roots are the first position of exposure to the ions of Cd through apoplastic transport that leads to maximum accumulation [55,56] The accumulation of Cd suppresses elongation of root and also responsible for deformation of root due to direct inhibition of root and shoot metabolism [57]. Correlation analysis of accumulated Cd in seedlings with other physiological parameters also confirmed the negative impact of Cd upon those parameters (Table 7).

Chlorophyll rapidly declined with increasing the doses of Cd in the present study. Reduction of Fe due to Cd metal causes chlorosis in leaves and also negatively affects chlorophyll metabolism. Heavy metals cause degradation of the micronutrients required for the plants and thus lead to inhibition of the level of pigments like chlorophyll [58]; this has been found as one of major reasons for the impairment of photosynthesis and inhibition of growth in plants under the effect by this metal. In the current study, higher Cd concentration showed negative impact on the sodium and potassium contents. Photosynthesis is the basic phenomena of the plant which regulates the development and growth of plants [59], and previous reports have documented that Cd can significantly degrade the chlorophylls and inhibit the photosynthesis [60]. Similarly, our results also showed statistically significant down-regulated chlorophyll levels thereby depicting degradation, or inhibition of their biosynthesis following Cd exposure even at the lowest the concentration used (0.2 mM).

Other important components that are present in the plants are anthocyanins, these are bioactive secondary metabolites of the plants wherein biological functions act as powerful antioxidant machinery and can also lead to the chelation of metal ions [61]. Stimuli due to unsuitable environment like exposure to Cd are well documented to stimulate anthocyanins accumulation in the plants [62]. Interestingly, in this study, Cd exposure had statistically significant higher anthocyanins in *B. Juncea* even at 0.2 mM Cd treated groups.

The treatment of Cd metal to *B. juncea* plants in the present study revealed that metal stress enhanced the level of antioxidants and activities of antioxidative enzymes. There was a significant increase in the ascorbic acid, total phenolic content, and activities of CAT, SOD, GR, DHAR, and MDHAR in Cd stressed *B. juncea* plants. SOD has a significant contribution in the protection against reactive oxygen species, produced as by-products of biological oxidations. These observations have also been documented by Zhang et al. [63] and Nehnevajova et al. [64]. A previous report by Najeeb et al. [65] evidenced a Cd modulated antioxidant defense system of the seedlings of *B. juncea*. The exposure of Cd inhibited the activities of antioxidant markers like SOD, POD, GST, and PPO and also significantly stimulated the activities of CAT, APOX, GPOX, DHAR, and GR. The stress caused by Cd inhibits enzymatic markers may be due to impairment of enzyme synthesis or due to inhibition of growth along with accumulation of H_2O_2 [65]. In this context, superoxide dismutase (Cu-Zn SOD) acts as a tier of cellular defense machinery that converts superoxide anion to H_2O_2. Then, these H_2O_2 is converted to water molecule majorly through antioxidant enzymes like catalases and peroxidases. Interestingly, ascorbate peroxidase reduces hydrogen peroxide through ascorbic acid-glutathione pathway that uses antioxidant enzymatic markers [66]. The glutathione peroxidases and peroxiredoxins also eliminate H_2O_2 and other organic hydroperoxides using non-enzymatic antioxidant markers like gluathione, thioredoxin, and glutaredoxin as nucleophiles [67]. The synthesis of phenolic compounds acts efficiently on metal decontamination as antioxidants [68]. The exposure to Cd reduces intracellular GSH contents and Cd exposure has also been reported to reduce GSH content in a number of plant species [69,70]. Nazar and his colleagues [71] explained that the reduction of GSH levels may be due to stress imposed by Cd, as it is known to prompt the formation of phytochelatins (oligomers of reduced glutathione) through increasing phytochelatin synthase activity, and these phytochelatins can conjugate with free ions of Cd and get sequestered into the vacuoles.

The present investigation revealed the increased level of MDA, H_2O_2 and also enhanced production of free radicals and oxidative damage due to membrane destabilization in plants being exposed to metal. MDA is a cytotoxic by product of lipid peroxidation and also acts as a marker of generation of free radical [63]. High doses of heavy metals lead to the production of ROS like H_2O_2, and it is the primary response of plants under stress. ROS generation occurs in the presence of oxidative stress or directly through Haber–Weiss reactions [72] and this supports the present study as we have found similar results in this context.

5. Conclusions

From the present study it is concluded that Cd toxicity adversely affects the seedlings even at low concentration as reported earlier (0.2 mM) in terms of increased production of ROS, higher lipid peroxidation, and by negatively affecting the photosynthetic pigment system. While in response to the detrimental effects of metal, defense responses of *B. juncea* seedlings got activated by the action of metabolites like phenols and compatible solutes. Besides, antioxidative defense system of seedlings helped in scavenging of free radicals generated during metal stress. It was also noticed that the ROS content was initially increased with Cd treatment, but no significant change was noticed between middle and highest Cd treatment. This might be due to the enhanced ROS scavenging as a result of activated antioxidative system. Furthermore, growth of seedlings was reduced with Cd treatment but there was also no statistical difference in growth parameters among different Cd treatments. These facts revealed that activated antioxidant system resulted in providing better Cd tolerance to seedlings.

Author Contributions: Conceptualization, R.B., B.Z. and A.S.; methodology, investigation, formal analysis, D.K. and M.P.S.; resources, B.Z. and A.S.; writing original draft, D.K. and M.P.S.; editing and revisions D.K., R.B., S.K. and A.S.

Funding: We acknowledge the different funding agencies: This study was supported by National Key Research and Development Program of China (2018YFD1000600); Independent research topics of the State Key Laboratory of Subtropical Silviculture (ZY20180208, ZY20180308); Key Research and Development Program of Zhejiang Province (2018C02004); Key Project of Zhejiang Provincial Natural Science Foundation (LZ18C160001); Fruit Innovation Team Project of Zhejiang Province (2016C02052-12); National Undergraduate Innovation and Entrepreneurship Training Project (201610341010); Undergraduate Science and Technology Innovation Plan of Zhejiang Province (2017R412006); Undergraduate Research Training Program in Zhejiang A & F University (102-2013200005, 102-2013200041, 102-2013200042,KX20180047, KX20180043, KX20180065).

Acknowledgments: Authors are thankful to the Department of Botanical and Environmental Sciences, Guru Nanak Dev University, Amritsar, India, for providing laboratory facilities. Dhriti Kapoor is grateful for UGC, New Delhi, for providing financial assistance.

Conflicts of Interest: The authors declare no conflict of interest.

Abbreviations

H_2O_2	hydrogen peroxide
CAT	catalase
SOD	superoxide dismutase
GR	glutathione reductase
DHAR	dehydroascorbate reductase
MDHAR	monodehydroascorbate reductase

References

1. Di Salvatore, M.; Carafa, A.M.; Carratu, G. Assessment of heavy metals phytotoxicity using seed germination and root elongation tests: A comparison of two growth substrates. *Chemosphere* **2008**, *73*, 1461–1464. [CrossRef] [PubMed]
2. Shahzad, B.; Tanveer, M.; Rehman, A.; Cheema, S.A.; Fahad, S.; Rehman, S.; Sharma, A. Nickel; whether toxic or essential for plants and environment-A review. *Plant Physiol. Biochem.* **2018**, *132*, 641–651. [CrossRef] [PubMed]
3. Kohli, S.K.; Handa, N.; Sharma, A.; Gautam, V.; Arora, S.; Bhardwaj, R.; Wijaya, L.; Alyemeni, M.N.; Ahmad, P. Interaction of 24-epibrassinolide and salicylic acid regulates pigment contents, antioxidative defense responses, and gene expression in *Brassica juncea* L. seedlings under Pb stress. *Environ. Sci. Pollut. Res. Int.* **2018**, *25*, 15159–15173. [CrossRef] [PubMed]
4. Shahzad, B.; Tanveer, M.; Che, Z.; Rehman, A.; Cheema, S.A.; Sharma, A.; Song, H.; ur Rehman, S.; Zhaorong, D. Role of 24-epibrassinolide (EBL) in mediating heavy metal and pesticide induced oxidative stress in plants: A review. *Ecotoxicol. Environ. Saf.* **2018**, *147*, 935–944. [CrossRef] [PubMed]
5. Sharma, A.; Soares, C.; Sousa, B.; Martins, M.; Kumar, V.; Shahzad, B.; Sidhu, G.P.S.; Bali, A.S.; Asgher, M.; Bhardwaj, R.; et al. Nitric oxide-mediated regulation of oxidative stress in plants under metal stress: A review on molecular and biochemical aspects. *Physiol. Plant.* **2019**. [CrossRef] [PubMed]
6. Tomescu, D.; Şumălan, R.; Copolovici, L.; Copolovici, D. The influence of soil salinity on volatile organic compounds emission and photosynthetic parameters of *Solanum lycopersicum* L. varieties. *Open Life Sci.* **2017**, *12*, 135–142. [CrossRef]
7. Tang, W.; Luo, C. Overexpression of Zinc Finger Transcription Factor ZAT6 Enhances Salt Tolerance. *Open Life Sci.* **2018**, *13*, 431–445. [CrossRef]
8. Guo, H.; Feng, X.; Hong, C.; Chen, H.; Zeng, F.; Zheng, B.; Jiang, D. Malate secretion from the root system is an important reason for higher resistance of *Miscanthus sacchariflorus* to cadmium. *Physiol. Planta* **2017**, *159*, 340–353. [CrossRef]
9. Guo, H.; Chen, H.; Hong, C.; Jiang, D.; Zheng, B. Exogenous malic acid alleviates cadmium toxicity in *Miscanthus sacchariflorus* through enhancing photosynthetic capacity and restraining ROS accumulation. *Ecotoxicol. Environ. Saf.* **2017**, *141*, 119–128. [CrossRef]
10. Chen, X.; Qiu, L.; Guo, H.; Wang, Y.; Yuan, H.; Yan, D.; Zheng, B. Spermidine induces physiological and biochemical changes in southern highbush blueberry under drought stress. *Braz. J. Bot.* **2017**, *40*, 841–851. [CrossRef]
11. Nagajyoti, P.C.; Lee, K.D.; Sreekanth, T. Heavy metals, occurrence and toxicity for plants: A review. *Environ. Chem. Lett.* **2010**, *8*, 199–216. [CrossRef]
12. Herawati, N.; Suzuki, S.; Hayashi, K.; Rivai, I.; Koyama, H. Cadmium, copper, and zinc levels in rice and soil of Japan, Indonesia, and China by soil type. *Bull. Environ. Contam. Toxicol.* **2000**, *64*, 33–39. [CrossRef] [PubMed]
13. Kumar, V.; Sharma, A.; Kaur, P.; Sidhu, G.P.S.; Bali, A.S.; Bhardwaj, R.; Thukral, A.K.; Cerda, A. Pollution assessment of heavy metals in soils of India and ecological risk assessment: A state-of-the-art. *Chemosphere* **2018**, *216*, 449–462. [CrossRef] [PubMed]
14. Villiers, F.; Ducruix, C.; Hugouvieux, V.; Jarno, N.; Ezan, E.; Garin, J.; Junot, C.; Bourguignon, J. Investigating the plant response to cadmium exposure by proteomic and metabolomic approaches. *Proteomics* **2011**, *11*, 1650–1663. [CrossRef] [PubMed]
15. Kohli, S.K.; Handa, N.; Sharma, A.; Gautam, V.; Arora, S.; Bhardwaj, R.; Alyemeni, M.N.; Wijaya, L.; Ahmad, P. Combined effect of 24-epibrassinolide and salicylic acid mitigates lead (Pb) toxicity by modulating various metabolites in *Brassica juncea* L. seedlings. *Protoplasma* **2018**, *255*, 11–24. [CrossRef] [PubMed]
16. Handa, N.; Kohli, S.K.; Sharma, A.; Thukral, A.K.; Bhardwaj, R.; Abd_Allah, E.F.; Alqarawi, A.A.; Ahmad, P. Selenium modulates dynamics of antioxidative defence expression, photosynthetic attributes and secondary metabolites to mitigate chromium toxicity in *Brassica juncea* L. plants. *Environ. Exp. Bot.* **2019**, *161*, 180–192. [CrossRef]
17. Kaur, R.; Yadav, P.; Sharma, A.; Kumar Thukral, A.; Kumar, V.; Kaur Kohli, S.; Bhardwaj, R. Castasterone and citric acid treatment restores photosynthetic attributes in *Brassica juncea* L. under Cd(II) toxicity. *Ecotoxicol. Environ. Saf.* **2017**, *145*, 466–475. [CrossRef] [PubMed]

18. Kaur, P.; Bali, S.; Sharma, A.; Vig, A.P.; Bhardwaj, R. Effect of earthworms on growth, photosynthetic efficiency and metal uptake in *Brassica juncea* L. plants grown in cadmium-polluted soils. *Environ. Sci. Pollut. Res. Int.* **2017**, *24*, 13452–13465. [CrossRef]
19. Anjum, N.A.; Ahmad, I.; Mohmood, I.; Pacheco, M.; Duarte, A.C.; Pereira, E.; Umar, S.; Ahmad, A.; Khan, N.A.; Iqbal, M. Modulation of glutathione and its related enzymes in plants' responses to toxic metals and metalloids—A review. *Environ. Exp. Bot.* **2012**, *75*, 307–324. [CrossRef]
20. Kaur, P.; Bali, S.; Sharma, A.; Kohli, S.K.; Vig, A.P.; Bhardwaj, R.; Thukral, A.K.; Abd-Allah, E.F.; Wijaya, L.; Alyemeni, M.N. Cd induced generation of free radical species in *Brassica juncea* is regulated by supplementation of earthworms in the drilosphere. *Sci. Total Environ.* **2019**, *655*, 663–675. [CrossRef]
21. Singh, R.; Gautam, N.; Mishra, A.; Gupta, R. Heavy metals and living systems: An overview. *Indian J. Pharmacol.* **2011**, *43*, 246–253. [CrossRef] [PubMed]
22. Hasan, S.A.; Ali, B.; Hayat, S.; Ahmad, A. Cadmium-induced changes in the growth and carbonic anhydrase activity of chickpea. *Turk. J. Biol.* **2007**, *31*, 137–140.
23. Mishra, S.; Srivastava, S.; Tripathi, R.D.; Govindarajan, R.; Kuriakose, S.V.; Prasad, M.N. Phytochelatin synthesis and response of antioxidants during cadmium stress in *Bacopa monnieri* L. *Plant Physiol. Biochem.* **2006**, *44*, 25–37. [CrossRef] [PubMed]
24. Al-Yemeni, M.N. Effect of cadmium, mercury and lead on seed germination and early seedling growth of *Vigna ambacensis* L. *Indian J. Plant Physiol.* **2001**, *6*, 147–151.
25. Seth, C.S.; Chaturvedi, P.K.; Misra, V. The role of phytochelatins and antioxidants in tolerance to Cd accumulation in *Brassica juncea* L. *Ecotoxicol. Environ. Saf.* **2008**, *71*, 76–85. [CrossRef] [PubMed]
26. Gozubenli, H. Seed vigor of maize grown on the contaminated soils by cadmium. *Asian J. Plant Sci.* **2010**, *9*, 168. [CrossRef]
27. Wahid, A.; Ghani, A.; Ali, I.; Ashraf, M. Effects of cadmium on carbon and nitrogen assimilation in shoots of mungbean [*Vigna radiata* (L.) Wilczek] seedlings. *J. Agron. Crop Sci.* **2007**, *193*, 357–365. [CrossRef]
28. Kovalchuk, O.; Titov, V.; Hohn, B.; Kovalchuk, I. A sensitive transgenic plant system to detect toxic inorganic compounds in the environment. *Nat. Biotechnol.* **2001**, *19*, 568–572. [CrossRef]
29. Anjum, S.A.; Tanveer, M.; Hussain, S.; Bao, M.; Wang, L.; Khan, I.; Ullah, E.; Tung, S.A.; Samad, R.A.; Shahzad, B. Cadmium toxicity in Maize (*Zea mays* L.): Consequences on antioxidative systems, reactive oxygen species and cadmium accumulation. *Environ. Sci. Pollut. Res.* **2015**, *22*, 17022–17030. [CrossRef]
30. Anjum, S.A.; Tanveer, M.; Hussain, S.; Shahzad, B.; Ashraf, U.; Fahad, S.; Hassan, W.; Jan, S.; Khan, I.; Saleem, M.F. Osmoregulation and antioxidant production in maize under combined cadmium and arsenic stress. *Environ. Sci. Pollut. Res.* **2016**, *23*, 11864–11875. [CrossRef]
31. Amari, T.; Ghnaya, T.; Abdelly, C. Nickel, cadmium and lead phytotoxicity and potential of halophytic plants in heavy metal extraction. *S. Afr. J. Bot.* **2017**, *111*, 99–110. [CrossRef]
32. Anuradha, S.; Rao, S.S.R. Effect of 24-epibrassinolide on the photosynthetic activity of radish plants under cadmium stress. *Photosynthetica* **2009**, *47*, 317–320. [CrossRef]
33. Shekhawat, K.; Rathore, S.; Premi, O.; Kandpal, B.; Chauhan, J. Advances in agronomic management of Indian mustard (*Brassica juncea* (L.) Czernj. Cosson): An overview. *Int. J. Agron.* **2012**, *2012*. [CrossRef]
34. Heath, R.L.; Packer, L. Photoperoxidation in isolated chloroplasts. I. Kinetics and stoichiometry of fatty acid peroxidation. *Arch. Biochem. Biophys.* **1968**, *125*, 189–198. [CrossRef]
35. Velikova, V.; Yordanov, I.; Edreva, A. Oxidative stress and some antioxidant systems in acid rain-treated bean plants: Protective role of exogenous polyamines. *Plant Sci.* **2000**, *151*, 59–66. [CrossRef]
36. Arnon, D.I. Copper enzymes in isolated chloroplasts. Polyphenoloxidase in beta vulgaris. *Plant Physiol.* **1949**, *24*, 1–15. [CrossRef]
37. Mancinelli, A.L. Photoregulation of anthocyanin synthesis: VIII. Effect of light pretreatments. *Plant Physiol.* **1984**, *75*, 447–453. [CrossRef] [PubMed]
38. Chapman, S.B. *Methods in Plant Ecology*; Blackwell Scientific Publications: London, UK, 1976.
39. Retamal-Salgado, J.; Hirzel, J.; Walter, I.; Matus, I. Bioabsorption and bioaccumulation of cadmium in the straw and grain of maize (*Zea mays* L.) in growing soils contaminated with cadmium in different environment. *Int. J. Environ. Res. Public Health* **2017**, *14*, 1399. [CrossRef]
40. Aebi, H. Catalase. In *Methods of Enzymatic Analysis*; Elsevier: Amsterdam, The Netherlands, 1974; pp. 673–684.
41. Kono, Y. Generation of superoxide radical during autoxidation of hydroxylamine and an assay for superoxide dismutase. *Arch. Biochem. Biophys.* **1978**, *186*, 189–195. [CrossRef]

42. Carlberg, I.; Mannervik, B. Purification and characterization of the flavoenzyme glutathione reductase from rat liver. *J. Biol. Chem.* **1975**, *250*, 5475–5480.
43. Dalton, D.A.; Russell, S.A.; Hanus, F.; Pascoe, G.A.; Evans, H.J. Enzymatic reactions of ascorbate and glutathione that prevent peroxide damage in soybean root nodules. *Proc. Natl. Acad. Sci. USA* **1986**, *83*, 3811–3815. [CrossRef] [PubMed]
44. Hossain, M.A.; Nakano, Y.; Asada, K. Monodehydroascorbate reductase in spinach chloroplasts and its participation in regeneration of ascorbate for scavenging hydrogen peroxide. *Plant Cell Physiol.* **1984**, *25*, 385–395.
45. Roe, J.H.; Kuether, C.A. The determination of ascorbic acid in whole blood and urine through the 2, 4-dinitrophenylhydrazine derivavative of dehydroascorbic acid. *J. Biol. Chem.* **1943**, *147*, 399–407.
46. Singleton, V.L.; Rossi, J.A. Colorimetry of total phenolics with phosphomolybdic-phosphotungstic acid reagents. *Am. J. Enol. Vitic.* **1965**, *16*, 144–158.
47. Hashem, H. Cadmium toxicity induces lipid peroxidation and alters cytokinin content and antioxidant enzyme activities in soybean. *Botany* **2013**, *92*, 1–7. [CrossRef]
48. Kaur, P.; Bali, S.; Sharma, A.; Vig, A.P.; Bhardwaj, R. Role of earthworms in phytoremediation of cadmium (Cd) by modulating the antioxidative potential of *Brassica juncea* L. *Appl. Soil Ecol.* **2018**, *124*, 306–316. [CrossRef]
49. Khanna, K.; Jamwal, V.L.; Sharma, A.; Gandhi, S.G.; Ohri, P.; Bhardwaj, R.; Al-Huqail, A.A.; Siddiqui, M.H.; Ali, H.M.; Ahmad, P. Supplementation with plant growth promoting rhizobacteria (PGPR) alleviates cadmium toxicity in *Solanum lycopersicum* by modulating the expression of secondary metabolites. *Chemosphere* **2019**, *230*, 628–639. [CrossRef]
50. Zhang, K.; Wang, G.; Bao, M.; Wang, L.; Xie, X. Exogenous application of ascorbic acid mitigates cadmium toxicity and uptake in Maize (*Zea mays* L.). *Environ. Sci. Pollut. Res.* **2019**, *26*, 19261–19271. [CrossRef]
51. Kaya, C.; Okant, M.; Ugurlar, F.; Alyemeni, M.N.; Ashraf, M.; Ahmad, P. Melatonin-mediated nitric oxide improves tolerance to cadmium toxicity by reducing oxidative stress in wheat plants. *Chemosphere* **2019**, *225*, 627–638. [CrossRef]
52. Wang, H.; Zhao, S.; Liu, R.; Zhou, W.; Jin, J. Changes of photosynthetic activities of maize (*Zea mays* L.) seedlings in response to cadmium stress. *Photosynthetica* **2009**, *47*, 277–283. [CrossRef]
53. Pandey, P.; Tripathi, A. Effect of heavy metals on morphological and biochemical characteristics of *Albizia procera* (Roxb.) Benth. seedlings. *Int. J. Environ. Sci.* **2011**, *1*, 1009.
54. Aydinalp, C.; Marinova, S. The effects of heavy metals on seed germination and plant growth on alfalfa plant (*Medicago sativa*). *Bulg. J. Agric. Sci.* **2009**, *15*, 347–350.
55. Verma, K.; Shekhawat, G.S.; Sharma, A.; Mehta, S.K.; Sharma, V. Cadmium induced oxidative stress and changes in soluble and ionically bound cell wall peroxidase activities in roots of seedling and 3–4 leaf stage plants of *Brassica juncea* (L.) czern. *Plant Cell Rep.* **2008**, *27*, 1261–1269. [CrossRef]
56. Chen, X.; Wang, J.; Shi, Y.; Zhao, M.; Chi, G. Effects of cadmium on growth and photosynthetic activities in pakchoi and mustard. *Bot. Stud.* **2011**, *52*, 41–46.
57. Andresen, E.; Küpper, H. Cadmium toxicity in plants. In *Cadmium: From Toxicity to Essentiality*; Springer: New York, NY, USA, 2013; pp. 395–413.
58. Sandalio, L.M.; Dalurzo, H.C.; Gomez, M.; Romero-Puertas, M.C.; del Rio, L.A. Cadmium-induced changes in the growth and oxidative metabolism of pea plants. *J. Exp. Bot.* **2001**, *52*, 2115–2126. [CrossRef]
59. Ashraf, M.; Harris, P.J. Photosynthesis under stressful environments: An overview. *Photosynthetica* **2013**, *51*, 163–190. [CrossRef]
60. Ying, R.-R.; Qiu, R.-L.; Tang, Y.-T.; Hu, P.-J.; Qiu, H.; Chen, H.-R.; Shi, T.-H.; Morel, J.-L. Cadmium tolerance of carbon assimilation enzymes and chloroplast in Zn/Cd hyperaccumulator Picris divaricata. *J. Plant Physiol.* **2010**, *167*, 81–87. [CrossRef]
61. Keilig, K.; Ludwig-Mueller, J. Effect of flavonoids on heavy metal tolerance in *Arabidopsis thaliana* seedlings. *Bot. Stud.* **2009**, *50*, 311–318.
62. Dai, L.-P.; Dong, X.-J.; Ma, H.-H. Antioxidative and chelating properties of anthocyanins in *Azolla imbricata* induced by cadmium. *Pol. J. Environ. Stud.* **2012**, *21*, 837–844.
63. Zhang, H.; Xia, Y.; Wang, G.; Shen, Z. Excess copper induces accumulation of hydrogen peroxide and increases lipid peroxidation and total activity of copper–zinc superoxide dismutase in roots of *Elsholtzia haichowensis*. *Planta* **2008**, *227*, 465–475. [CrossRef]

64. Nehnevajova, E.; Lyubenova, L.; Herzig, R.; Schröder, P.; Schwitzguébel, J.-P.; Schmülling, T. Metal accumulation and response of antioxidant enzymes in seedlings and adult sunflower mutants with improved metal removal traits on a metal-contaminated soil. *Environ. Exp. Bot.* **2012**, *76*, 39–48. [CrossRef]
65. Najeeb, U.; Jilani, G.; Ali, S.; Sarwar, M.; Xu, L.; Zhou, W. Insights into cadmium induced physiological and ultra-structural disorders in *Juncus effusus* L. and its remediation through exogenous citric acid. *J. Hazard. Mater.* **2011**, *186*, 565–574. [CrossRef]
66. Mittler, R. Oxidative stress, antioxidants and stress tolerance. *Trends Plant Sci.* **2002**, *7*, 405–410. [CrossRef]
67. Noctor, G.; Mhamdi, A.; Foyer, C.H. The roles of reactive oxygen metabolism in drought: Not so cut and dried. *Plant Physiol.* **2014**, *164*, 1636–1648. [CrossRef]
68. Zheng, G.; Lv, H.; Gao, S.; Wang, S. Effects of cadmium on growth and antioxidant responses in *Glycyrrhiza uralensis* seedlings. *Plantsoil Environ.* **2010**, *56*, 508–515. [CrossRef]
69. Foyer, C.H. Regulation of glutathione synthesis and its role in abiotic and biotic stress defence. In *Sulfur Nutrition and Sulfur Assimilation in Higher Plants*; SPB Academic Publishing: The Hague, The Netherlands, 2000; pp. 127–153.
70. Romero-Puertas, M.C.; Corpas, F.J.; Rodríguez-Serrano, M.; Gómez, M.; Luis, A.; Sandalio, L.M. Differential expression and regulation of antioxidative enzymes by cadmium in pea plants. *J. Plant Physiol.* **2007**, *164*, 1346–1357. [CrossRef]
71. Nazar, R.; Iqbal, N.; Masood, A.; Khan, M.I.R.; Syeed, S.; Khan, N.A. Cadmium toxicity in plants and role of mineral nutrients in its alleviation. *Am. J. Plant Sci.* **2012**, *3*, 1476. [CrossRef]
72. Mithofer, A.; Schulze, B.; Boland, W. Biotic and heavy metal stress response in plants: Evidence for common signals. *Febs Lett.* **2004**, *566*, 1–5. [CrossRef]

plants

MDPI

Article

Bentonite and Biochar Mitigate Pb Toxicity in *Pisum sativum* by Reducing Plant Oxidative Stress and Pb Translocation

Muhammad Zulqurnain Haider [1], Sabir Hussain [2,†], Pia Muhammad Adnan Ramzani [3], Mutahar Iqbal [2,†], Muhammad Iqbal [2], Tanvir Shahzad [2], Maryam Fatima [2], Shahbaz Ali Khan [2], Imran Khan [4], Muhammad Shahid [5], Muhammad Ibrahim [2], Hafiz Syed Tanzeem Ull Haq [2] and Faisal Mahmood [2,*]

1 Department of Botany, Government College University, Faisalabad 38000, Pakistan; drmzhaider@gcuf.edu.pk

2 Department of Environmental Sciences and Engineering, Government College University, Faisalabad 38000, Pakistan; Sabir.hussain@gcuf.edu.pk (S.H.); mutahariqbal70@gmail.com (M.I.); iqbal.farhad@gmx.at (M.I.); tanvirshahzad@gcuf.edu.pk (T.S.); mfvirgo000@gmail.com (M.F.); shahbaz_2010@live.com (S.A.K.); ebrahem.m@gmail.com (M.I.); tanzeem_syed@yahoo.com (H.S.T.U.H.)

3 Cholistan Institute of Desert Studies, The Islamia University of Bahawalpur, Bahawalpur 63100, Pakistan; dr.piamuhammad@iub.edu.pk

4 Department of Agronomy, University of Agriculture, Faisalabad 38040, Pakistan; agronomist786@hotmail.com

5 Department of Bioinformatics and Biotechnology, Government College University, Faisalabad 38000, Pakistan; mshahid@gcuf.edu.pk

* Correspondence: faisalmahmood@gcuf.edu.pk; Tel.: +0092-(0)3330613047

† Co-first Authors.

Received: 30 October 2019; Accepted: 3 December 2019; Published: 5 December 2019

Abstract: Lead (Pb)-polluted soils pose a serious threat to human health, particularly by transmitting this heavy metal to the food chain via the crops grown on them. The application of novel amendments in Pb-polluted soils can significantly reduce this problem. In this research, we report the effects of various organic and inorganic amendments i.e., bentonite (BN), biochar (BR), lignin (LN), magnesium potassium phosphate cement (CM) and iron hydroxyl phosphate (FeHP), on the Pb bioavailability in Pb-polluted soil, upon Pb distribution in shoots, roots, grain, the translocation factor (TF) and the bioconcentration factor (BCF) of Pb in pea (*Pisum sativum* L.) grain. Furthermore, effects of the said amendments on the plant parameters, as well as grain biochemistry and nutritional quality, were also assessed. Lead pollution significantly elevated Pb concentrations in roots, shoots and grain, as well as the grain TF and BCF of Pb, while reducing the nutritional quality and biochemistry of grain, plant height, relative water content (RWC), chlorophyll contents (chl a and chl b) and the dry weight (DW) of shoot, root and grain. The lowest Pb distribution in shoots, roots and grain were found with BN, FeHP and CM, compared to our control. Likewise, the BN, FeHP and CM significantly lowered the TF and BCF values of Pb in the order FeHP > CM > BN. Similarly, the highest increase in plant height, shoot, root and grain DW, RWC, chl a and chl b contents, grain biochemistry and the micronutrient concentrations, were recorded with BR amendment. Biochar also reduced grain polyphenols as well as plant oxidative stress. Given that the BR and BN amendments gave the best results, we propose to explore their potential synergistic effect to reduce Pb toxicity by using them together in future research.

Keywords: lead pollution; antioxidants; bentonite; translocation factor; grain biochemistry; biochar; oxidative stress

1. Introduction

Urbanization and industrialization has led to the contamination of world soils with heavy metals [1]. Lead (Pb) is an important pollutant in soil among other heavy metals. It can enter into soils through the disposal of effluents from industries like batteries and paints, mining and smelting, and the burning of fossil fuels, i.e., coal and leaded gasoline [2,3]. Similarly, several natural sources causing soil pollution with Pb are the weathering of rocks, volcanic eruption, forest fires and sea sprays [4].

Lead pollution has an adverse effect on humans, especially the health of teenagers. Consumption of food grown within Pb-contaminated soil may cause neurological effects and cognitive disorders [5,6]. Soil pollution with Pb is not only threatening human health, it is also damaging the environment by polluting the ground and surface water resources [6].

Given the disadvantages of traditional mitigation practices deployed for heavy metals-polluted soils [7,8], gentle remediation practices like phytoremediation and in-situ immobilization are being promoted. Phytoremediation is a process in which plants are used in combination with suitable agronomic practices to remove heavy metals from the environment, or at least mitigate their toxicity [9]. It involves the use of many approaches individually or in combination. For instance, phytoextraction involves the uptake of toxic metals by plants in their shoots, whereas in phytostabilization, the mobility of pollutants in soils is restricted by using suitable plant cover on contaminated sites [10]. The phytoextraction of heavy metals can be stimulated by using different organic chelants like ethylenediaminetetraacetic acid (EDTA), ethylenediamine-N,N′-disuccinic acid (EDDS) etc., which solubilize heavy metals in the rhizosphere, thereby enhancing their uptake by the plants. However, these chelating agents are slowly biodegradable in nature, and cause ground water contamination [11]. Similar to phytoextraction, phytostabilization can also be enhanced by amending soils with suitable materials that can immobilize metals in the presence of selected plant species. Organic amendments can immobilize certain heavy metals through enhanced organo-metal complexes, chemisorption, ion exchange, complexation and adsorption. Moreover, increase in pH as the result of amendments may also immobilize heavy metals [9].

The recent trend to use magnesium potassium phosphate cement (CM) for Pb immobilization in Pb-polluted soil is more advantageous than ordinary Portland cement (OPC) [6]. The CM converts Pb into highly insoluble pyromorphite ($Pb_5(PO_4)_3X$, X = Cl^-, OH^-, F^-) and Pb-phosphate ($Pb_3(PO_4)_2$) while OPC makes $Pb(OH)_2$ in a Pb-rich environment. Compared to the $Pb(OH)_2$ that has a comparatively high solubility [solubility product constant (Ksp) ≅ 10^{-4}], the pyromorphite and Pb-phosphate have much lower solubility (Ksp ≅ 10^{-60}–10^{-85} and 10^{-6}, respectively). Furthermore, CM has vital properties like fast-setting, high early strength and resistance to soil alkaline or acidic conditions, which all make it an excellent Pb immobilizing agent compared to OPC in Pb-polluted soils [6,12]. Biochar (BR) is prepared by the pyrolyzing organic waste under oxygen-limited conditions [13]. It, when derived from alkaline feedstocks, raises the soil pH similar to liming materials. Moreover, it has a large surface area and high sorption capacity that enable it to effectively lessen the Pb bioavailability in this Pb-polluted soil [3,13,14]. Lignin (LN), a waste product of the paper industry, has abundant oxygen-containing groups, carboxyl, lactonic and phenolic hydroxyl groups [15,16]. Due to these functional groups, LN has been widely used as an effective amendment for the immobilization of Pb in Pb-polluted water and soil [15,16]. Bentonite (BN), an expandable clay mainly comprised of montmorillonite, has high durable negative charges and large definite surface area [7]. Numerous studies have reported that amending metal-polluted soils with BN has significantly reduced the bioavailability of a variety of metals, especially Pb [1,5,7]. Likewise, iron hydroxyl phosphate (FeHP) forms stable metal-phosphate precipitates e.g., $Pb_5(PO_4)_3$ (OH, F, Cl) in Pb-polluted soils [17].

A lot of research in the past few decades has been carried out on the assessment of various organic and inorganic amendments to reduce the Pb bioavailability in Pb-polluted soils and Pb uptake by different crops. However, there exists no research on the efficacy of different cost-effective and innovative amendments on reducing Pb distribution in pea grain, plant oxidative stress and any improvement in the grain biochemistry and nutritional value and plant agronomic and biophysical traits grown on Pb-polluted soil. Therefore, the key objectives of this research were to (i) evaluate the efficacy of BN, BR, LN, CM and FeHP to reduce Pb bioavailability in Pb-polluted soil, (ii) evaluating the positive effects of Pb immobilization on pea grain quality, productivity and Pb translocation in grain and (iii) to determine the level of Pb uptake and changes in plant oxidative stress and antioxidant defense machinery in response to the said amendments.

2. Results

2.1. Pb Allocation in Plant Parts and Soil

The data regarding the concentrations of Pb in grain, shoots and roots were in the range of 56.0 to 129.8, 220.9 to 365.1 and 658.6 to 843.8 mg kg^{-1} DW, respectively. The Pb concentration was 3.31 to 6.09 mg kg^{-1} soil in the DTPA extract (Figure 1). Amending the Pb-polluted soil with BN 5%, BR 2%, LN 2%, CM 0.5% and FeHP 2% significantly reduced the Pb concentrations in grain, shoots, roots and DTPA extract, compared to the control. In the BN 5%, CM 0.5% and FeHP 2% treatments, the Pb concentrations in grain were reduced by 57%, 53% and 50%; in roots by 22%, 19% and 18% and in DTPA extract by 46%, 42% and 40%, respectively. Likewise, the BN 5% and CM 0.5% treatments showed the highest reduction in the concentrations of Pb in shoots by 39% and 33%, respectively (Figure 1).

The TF values for Pb were in the range 0.34–0.43 in all treatments (Table 1). All treatments significantly reduced the TF values for Pb when compared to control. The BN 5%, CM 0.5%, and FeHP 2% showed the highest significant decrease in TF values for Pb, compared to control. Overall, the TF values for Pb among various treatments was in the following order: Control > LN 2% > BR 2% > FeHP 2% > CM 0.5% > BN 5%. Similarly, the BCF values for Pb for all treatments were in the range 0.22–0.37 (Table 1). Results revealed that all treatments were able to significantly reduce the BCF values for Pb, compared to control.

In this context, the BN 5%, FeHP 2% and CM 0.5% treatments exhibited the highest significant reduction in BCF values for Pb, compared to control. The order of reduction in the BCF values for Pb was Control > LN 2% > BR 2% > FeHP 2% > CM 0.5% > BN 5%.

Soil pH after plant harvest ranged from 8.01 to 8.92, with the highest pH values found in BR 2%, CM 0.5% and BN 5%, compared to control. Likewise, the lowest value of pH was found in the LN 2% treatment, compared to control (Figure 1).

Table 1. Influences of bentonite (BN), biochar (BR), lignin (LN), cement (CM) and iron hydroxyl phosphate (FeHP) on growth and biomass, chlorophyll contents and relative water content (RWC) in leaves of pea, translocation factor (TF) and the bioconcentration factor (BCF) values for Pb in plant and Pb contents in shoot, root and grain grown on a Pb-contaminated soil are illustrated. Numerical values represent means (from three replicates i.e., $n = 3$) along bars sharing identical alphabetic letters which are statistically (at $p < 0.05$) non-significant to each other. These values are mean of three replicates ± SE.

Treatments	Growth Parameters					Chlorophyll Contents		Pb Translocation		Pb Contents		
	Shoot DW (g pot^{-1})	Root DW (g pot^{-1})	Grain DW (g pot^{-1})	Plant Height (cm plant^{-1})	RWC (%)	Chl a (mg g^{-1} FW)	Chl b (mg g^{-1} FW)	TF	BCF	Shoot (mg kg^{-1})	Root (mg kg^{-1})	Grain (mg kg^{-1})
Control	3.3 ± 0.1 e	1.3 ± 0.0 e	1.3 ± 0.0 d	47.5 ± 1.7 d	60.7 ± 2.2 d	32.5 ± 1.2 e	26.9 ± 1.0 d	0.432 ± 0.02 a	0.37 ± 0.01 a	1.21 ± 0.02 b	1.08 ± 0.02 cd	0.16 ± 0.001 a
BN 5%	4.7 ± 0.2 b	1.7 ± 0.1 b	1.8 ± 0.1 ab	64.3 ± 2.3 ab	73.4 ± 2.7 ab	47.5 ± 1.7 b	40.9 ± 1.5 ab	0.339 ± 0.01 d	0.22 ± 0.01 d	1.04 ± 0.02 c	1.13 ± 0.02 c	0.10 ± 0.001 d
BR 2%	5.4 ± 0.2 a	2.0 ± 0.1 a	1.7 ± 0.1 a	70.9 ± 2.6 a	78.8 ± 2.9 a	56.0 ± 2.0 a	48.5 ± 1.8 a	0.361 ± 0.01 bc	0.27 ± 0.01 bc	1.45 ± 0.03 a	1.49 ± 0.03 a	0.11 ± 0.001 c
LN 2%	4.4 ± 0.2 bc	1.6 ± 0.1 be	1.6 ± 0.1 bc	59.8 ± 2.2 bc	69.6 ± 2.5 bc	44.0 ± 1.6 bc	38.6 ± 1.4 bc	0.379 ± 0.01 b	0.28 ± 0.01 b	1.24 ± 0.02 b	1.21 ± 0.02 b	0.12 ± 0.001 b
CM 0.5%	3.8 ± 0.1 d	1.4 ± 0.1 de	1.5 ± 0.1 c	53.4 ± 1.9 cd	64.0 ± 2.3 cd	37.6 ± 1.4 d	30.4 ± 1.1 d	0.360 ± 0.01 cd	0.24 ± 0.01 cd	0.94 ± 0.02 d	0.95 ± 0.02 e	0.09 ± 0.001 e
FeHp 2%	4.2 ± 0.2 cd	1.5 ± 0.1 cd	1.6 ± 0.1 bc	57.4 ± 2.1 c	67.1 ± 2.4 bcd	41.8 ± 1.5 cd	35.1 ± 1.3 c	0.371 ± 0.01 bcd	0.26 ± 0.01 bcd	1.08 ± 0.02 c	1.04 ± 0.02 d	0.10 ± 0.001 d

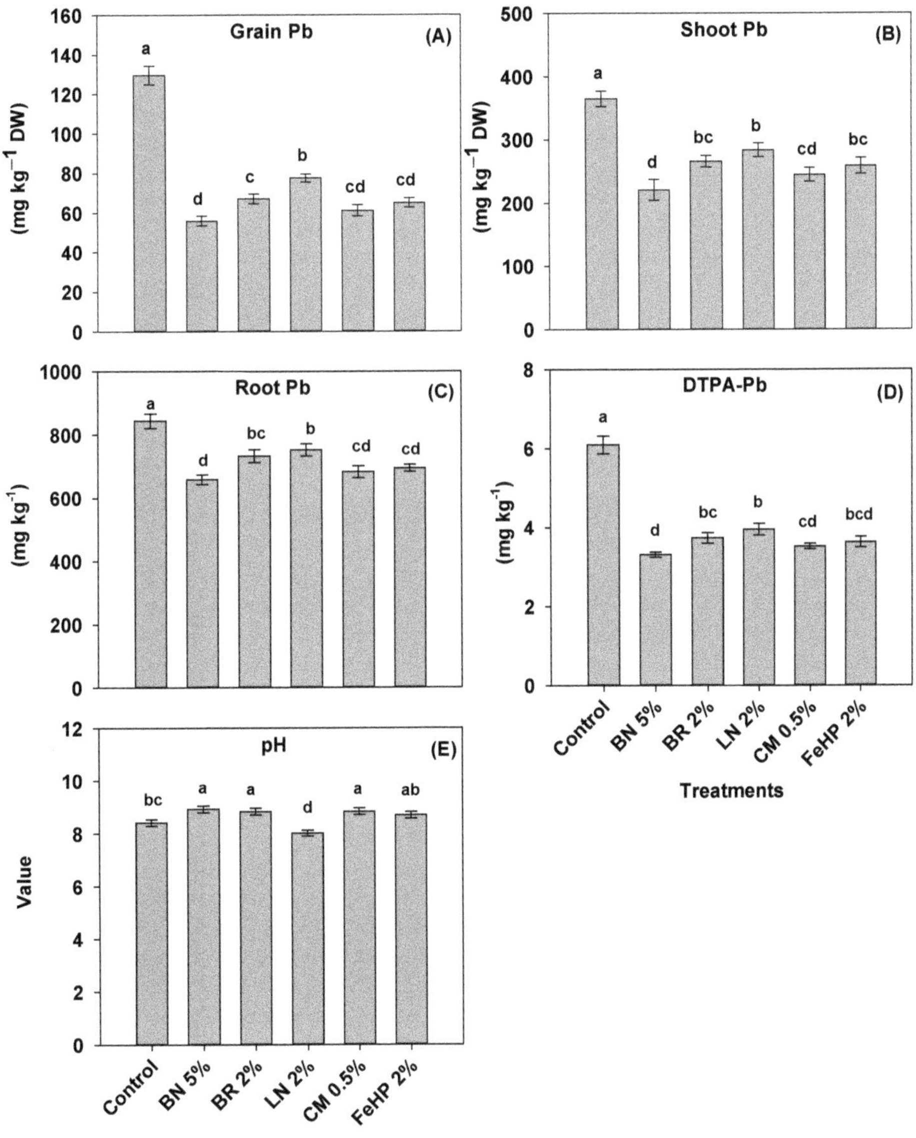

Figure 1. Effects of amendments in Pb-polluted soil on the concentrations of Pb in pea grain (**A**), in the shoot (**B**), the root (**C**) and diethylenetriaminepentaacetic acid (DTPA)-extractable fraction (**D**), as well as soil pH after plant harvest (**E**). Values are the means of three replicates, the error bars represent the standard error of means and the lower case alphabets indicate significant differences ($p \leq 0.05$) among treatments based on one way Analysis of Variance (ANOVA) (using LSD test, $p = 0.05$ at $df = 5$ and $n = 3$). DW: dry weight; BN: bentonite, BR: biochar; LN: lignin; CM: magnesium potassium phosphate cement; FeHP: iron hydroxyl phosphate.

2.2. *Agronomic, Photosynthetic and Biophysical Parameters of Pea Plant as Influenced by Amendments*

The obtained data for plant height, shoot, root and grain DW were in the range from 47.5 to 70.9 cm, 3.31 to 5.45 g pot^{-1}, 1.28 to 2.03 g pot^{-1} and 1.26 to 1.83 g pot^{-1}, respectively (Table 1). Relative to the control, all treatments significantly improved the shoot and grain DW. However, except for CM 0.5%, the rest of the treatments significantly improved plant height and root DW, in comparison to the control treatment. The highest improvement in root and shoot DW by 58% and 64%, respectively, was observed in the BR 2% treatment. The BR 2% and BN 5% treatments showed the highest improvement in grain DW by 36% and 45% and plant height by 49% and 35%, respectively, relative to the control.

Likewise, the values of Chl a, Chl b, and RWC were in the range from 32.5 to 56 mg g^{-1} fresh weight (FW) and 26.9 to 48.5 mg g^{-1} FW and 60.7% to 78.8%, respectively (Table 1). All treatments significantly improved the Chl a and Chl b contents, with the exception of CM 0.5% only in case of Chl b, compared to the control. Likewise, with the exception of CM 0.5% and FeHP 2%, the rest of the treatments significantly improved the values of RWC compared to the control (Table 1). The BR 2% and BN 5% treatments showed the highest improvement in RWC values by 30% and 21%, and Chl b contents by 80% and 52%, respectively, relative to the control. Likewise, the highest improvement in Chl a contents was observed in BR 2%, and that was 72%, in comparison to the control.

2.3. *Status of Micronutrients, Antinutrient and Biochemical Compounds in Pea Grain as Influenced by Amendments*

The data concerning protein, fat, fiber and carbohydrate contents in pea grain across treatments were in the ranges from 15.9% to 19.7%, 1.78% to 2.19%, 6.94% to 9.93% and 48% to 64.5%, respectively (Figure 2). With different exceptions, amending Pb-polluted soil with the selected amendments significantly improved the contents of grain biochemical compounds, compared to control. The highest improvement in protein contents by 16%, 24% and 19%, and fat contents by 13%, 23% and 17%, respectively, were observed in the BN 5%, BR 2% and LN 2% treatments. Similarly, BR 2% and LN 2% exhibited the highest significant improvement in grain carbohydrate contents by 34% and 24%, respectively, whereas fiber contents by 43% in BR 2%, relative to the control (Figure 2).

The Fe, Zn and Mn concentrations in grain were in the ranges from 18.4 to 27.6 mg kg^{-1} DW, 14.7 to 18.0 mg kg^{-1} DW and 11.3 to 15.1 mg kg^{-1} DW respectively. The polyphenols were in the range 6.88 to 10.5 mg g^{-1} DW across all treatments (Figure 2). All amendments significantly decreased the grain polyphenol contents, while improvement in the grain Fe and Zn concentrations were found significant only in the BR 2%, LN 2% and FeHP 2% treatments. Likewise, only these BN 5%, BR 2% and LN 2% treatments significantly improved grain Mn concentrations (Figure 2). In this context, the BR 2%, LN 2% and FeHP 2% treatments reduced polyphenol contents up to 34%, 31%, and 29%, respectively, while the highest improvement in the grain Zn concentrations by 22% and 16% and Mn by 26% and 33% were observed in the BR 2% and LN 2% treatments, respectively. Likewise, the highest improvement in the grain Fe concentration by 50% was found in the FeHP 2% treatment.

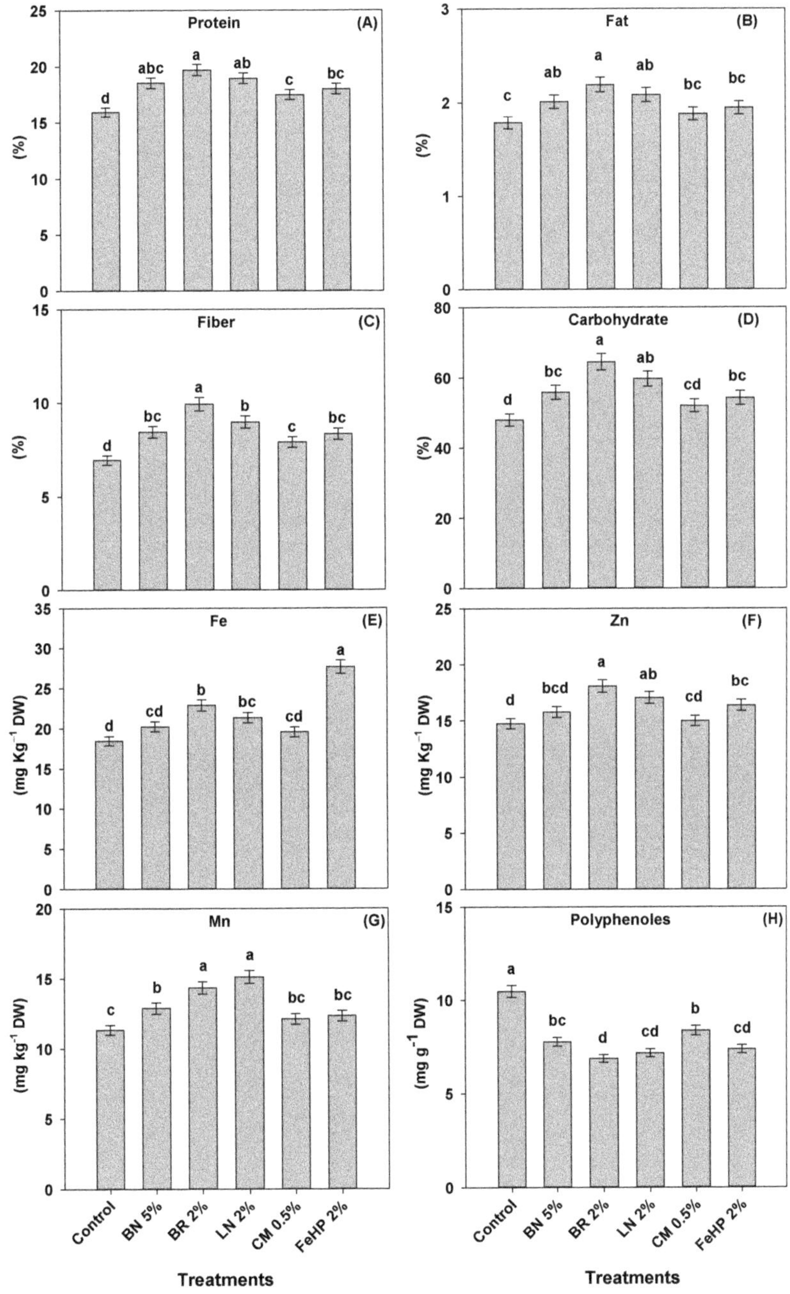

Figure 2. Effects of amendments in Pb-polluted soil on the contents of protein (**A**), fat (**B**), fiber (**C**) carbohydrate (**D**), and the concentrations of Fe (**E**), Zn (**F**), Mn (**G**) and polyphenols (**H**) in pea grain. Values are means of three replicates, error bars represent the standard error of the means, and the lower case alphabets indicate significant differences ($p \leq 0.05$) among treatments based on one-way ANOVA (using LSD test, $p = 0.05$ at $df = 5$ and $n = 3$).

2.4. Effect of Amendments on Antioxidant Defense Machinery and Oxidative Stress in Pea Plants

In all treatments, the activities of ascorbate peroxidase (APX), superoxide dismutase (SOD), catalase (CAT) and dehydroascorbate reductase (DHAR) were in the range from 0.38 to 0.80, 51.5 to 108.5, 38.7 to 53.0 and 31.8 to 72.9 μmol min^{-1} mg^{-1} protein, respectively (Figure 3). All treatments significantly improved the APX, SOD, CAT and DHAR activities, compared to the control. The topmost improvement in the APX, SOD, CAT and DHAR activities was found in BR 2% by 111%, 111%, 106% and 129%, respectively when compared to control.

The contents of malondialdehyde (MDA) and hydrogen peroxide (H_2O_2) were in the ranges from 35.1 to 58.1 and 29.0 to 74.4 nmol g^{-1} FW, while O_2^- generation rate from 15.2 to 37.8 nmol min^{-1} g^{-1} FW, respectively (Figure 3). Interestingly, all treatments were capable of significantly reducing the contents of MDA and H_2O_2 as well as the O_2^- generation rate, compared to the control. The obtained data for the BR 2% and LN 2% treatments showed the highest reduction in MDA contents up to 39% and 36%; H_2O_2 contents up to 61% and 56%, and O_2^- generation rates up to 60% and 55%, respectively, compared to the control.

2.5. Principal Component Analysis and Pearson Coefficient Correlation (r^2) Among Studied Attributes

The values of Pearson correlation with their significance at probability levels (p) < 0.05, 0.01, 0.001 are presented in Table 2. It clearly shows that the plant biomass has positive significant correlation ($p < 0.001$) with studied physiological attributes like Chl a and Chl b, as well as antioxidant activities such as SOD, CAT, APX and DHAR. In addition, shoot dry weight is also positively significantly correlated ($p < 0.001$) with carbohydrates contents, total soluble proteins, Zn, fiber and fat contents of pea grain, as well as grain Mn ($p < 0.01$). However, shoot biomass was negatively correlated ($p < 0.001$) with the MDA and H_2O_2 contents, O_2 generation and grain polyphenols. Moreover, the shoot dry weight has also negative significant correlation with BCF shoot, TF shoot, DTPA Pb, grain Pb and shoot Pb. Correlation studies presented in Figure 4 generated through PCA show that the studied attributes are categorized in two major groups. The first component of the PCA explained 76.6%, while the second component explained 13.93%, of the variance. The details of component loadings and communalities for each parameter is described in Tables S1 and S2, while the scree plot of PCA is represented in Figure S1

Table 2. Pearson coefficient correlation (r^2) values of studied attributes of pea showing significance differences grown under Pb stress with various soil amendments.

	SDW	GDW	Chl-a	Chl-b	RWC	DTPA-Pb	S-Pb	G-Pb	TF-S	BCF-S	Prot	Carb	Fat	Fiber	Mn	Zn	PPs	APX	CAT	SOD	DHAR	MDA	H_2O_2	O_2-ge
SDW	1.000											5												
GDW	0.852 ***	1.000																						
Chl-a	0.999 ***	0.845 ***	1.000																					
Chl-b	0.994 ***	0.841 ***	0.996 ***	1.000																				
RWC	0.994 ***	0.860 ***	0.996 ***	0.995 ***	1.000																			
DTPA-Pb	−0.634 **	−0.837 ***	−0.613 **	−0.581 **	−0.593 **	1.000																		
S-Pb	−0.573 *	−0.839 ***	−0.552 *	−0.513 *	−0.546 *	0.973 ***	1.000																	
G-Pb	−0.631 **	−0.832 ***	−0.611 **	−0.573 *	−0.590 *	0.998 ***	0.981 ***	1.000																
TF-S	−0.684 **	−0.894 ***	−0.665 **	−0.630 **	−0.663 **	0.965 ***	0.986 ***	0.972 ***	1.000															
BCF-S	−0.590 **	−0.840 ***	−0.568 *	−0.533 *	−0.561 *	0.982 ***	0.995 ***	0.985 ***	0.988 ***	1.000														
Prot	0.940 ***	0.852 ***	0.934 ***	0.939 ***	0.920 ***	−0.743 ***	−0.636 **	−0.721 ***	−0.723 ***	−0.674 **	1.000													
Carb	0.939 ***	0.720 ***	0.939 ***	0.949 ***	0.922 ***	−0.553 *	−0.426 ns	−0.531 *	−0.539 *	−0.470 *	0.964 ***	1.000												
Fat	0.949 ***	0.766 ***	0.949 ***	0.963 ***	0.939 ***	−0.576 **	−0.460 *	−0.554 *	−0.572 *	−0.502 *	0.974 ***	0.996 ***	1.000											
Fiber	0.949 ***	0.751 ***	0.947 ***	0.948 ***	0.923 ***	−0.629 **	−0.504 *	−0.612 **	−0.605 **	−0.542 *	0.976 ***	0.992 ***	0.987 ***	1.000										
Mn	0.723 ***	0.569 *	0.722 ***	0.762 ***	0.717 ***	−0.434 ns	−0.280 ns	−0.386 ns	−0.379 ns	−0.353 ns	0.864 ***	0.896 ***	0.899 ***	0.856 ***	1.000									
Zn	0.867 ***	0.603 **	0.874 ***	0.891 ***	0.845 ***	−0.422 ns	−0.266 ns	−0.397 ns	−0.367 ns	−0.302 ns	0.894 ***	0.960 ***	0.948 ***	0.952 ***	0.853 ***	1.000								
PPs	−0.826 ***	−0.827 ***	−0.818 ***	−0.815 ***	−0.786 ***	0.846 ***	0.724 ***	0.823 ***	0.764 ***	0.754 ***	−0.947 ***	−0.860 ***	−0.868 ***	−0.901 ***	−0.771 ***	−0.823 ***	1.000							
APX	0.891 ***	0.691 **	0.891 ***	0.904 ***	0.863 ***	−0.576 **	−0.423 ns	−0.548 *	−0.517 *	−0.468 *	0.960 ***	0.983 ***	0.976 ***	0.984 ***	0.904 ***	0.976 ***	−0.906 ***	1.000						
CAT	0.850 ***	0.684 **	0.849 ***	0.857 ***	0.811 ***	−0.633 **	−0.471 *	−0.606 **	−0.542 *	−0.510 *	0.944 ***	0.944 ***	0.935 ***	0.961 ***	0.856 ***	0.958 ***	−0.943 ***	0.986 ***	1.000					
SOD	0.867 ***	0.685 **	0.868 ***	0.878 ***	0.833 ***	−0.594 **	−0.435 ns	−0.567 *	−0.514 *	−0.472 *	0.942 ***	0.955 ***	0.947 ***	0.966 ***	0.860 ***	0.975 ***	−0.926 ***	0.991 ***	0.997 ***	1.000				
DHAR	0.864 ***	0.642 **	0.867 ***	0.882 ***	0.835 ***	−0.518 *	−0.356 ns	−0.491 *	−0.446 ns	−0.396 ns	0.927 ***	0.965 ***	0.954 ***	0.966 ***	0.877 ***	0.991 ***	−0.885 ***	0.993 ***	0.987 ***	0.995 ***	1.000			
MDA	−0.833 ***	−0.741***	−0.831 ***	−0.839 ***	−0.796 ***	0.706 ***	0.554 *	0.677 **	0.610 **	0.590 **	−0.948 ***	−0.913 ***	−0.912 ***	−0.937 ***	−0.839 ***	−0.919 ***	0.975 ***	−0.964 ***	−0.990 ***	−0.983 ***	−0.961 ***	1.000		
H_2O_2	−0.787 ***	−0.648 **	−0.787 ***	−0.801 ***	−0.747 ***	0.612 **	0.440 ns	0.580 **	0.497 *	0.479 *	−0.909 ***	−0.902 ***	−0.894 ***	−0.920 ***	−0.847 ***	−0.939 ***	0.938 ***	−0.965 ***	−0.992 ***	-0.987 ***	-0.973 ***	0.990 ***	1.000	
O_2-ge	−0.804 ***	−0.722 ***	−0.802 ***	−0.811 ***	−0.765 ***	0.700 ***	0.544 *	0.671 **	0.593 **	0.578 **	−0.928 ***	−0.889 ***	−0.888 ***	−0.916 ***	−0.821 ***	−0.908 ***	0.972 ***	−0.951 ***	−0.985 ***	−0.976 ***	−0.952 ***	0.998 ***	0.992 ***	1.000

*, ** and *** = significant at 0.05, 0.01, 0.001 levels respectively; ns = non-significant. SDW: Shoot dry weight; GDW; Grain dry weight; GDW: Chl-a: Chlorophyll a contents; Chl-b: Chlorophyll b contents; RWC: Relative water contents; R-Pb: Root Pb contents; S-Pb: Shoot Pb contents; G-Pb: Grain Pb contents; TF-S: TF shoot; BCF-S: BCF-Shoot; Prot: Protein; Carb: Carbohydrates; PP: Polyphenols; O2-ge: O2 generation.

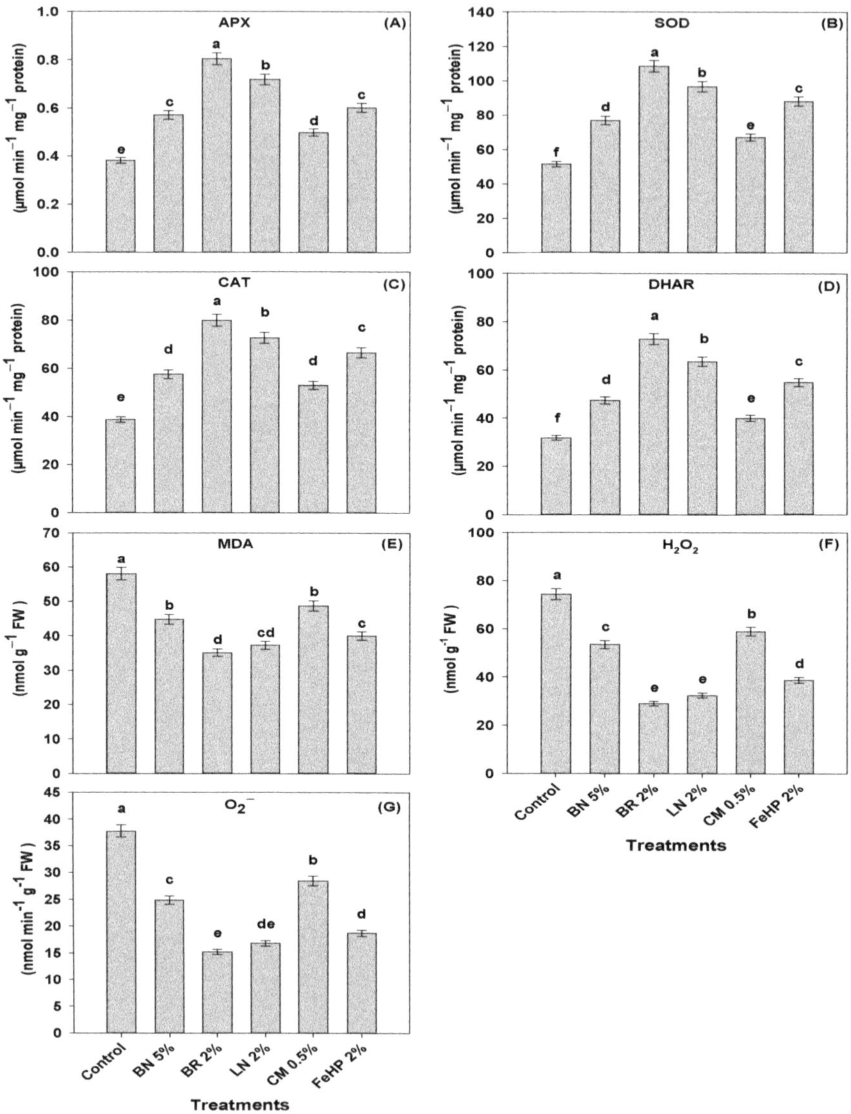

Figure 3. Effects of amendments inclusion in Pb-polluted soil on the activities of ascorbate peroxidase (APX) (**A**), superoxide dismutase (SOD) (**B**), catalase (CAT) (**C**) and dehydroascorbate reductase (DHAR) (**D**), as well as the contents of the malondialdehyde (MDA) (**E**), hydrogen peroxide (H_2O_2) (**F**) and O_2^- generation rate (**G**) in pea leaves. Values are means of three replicates, error bars represent standard error of means, and the lower case alphabets indicate significant differences ($p \leq 0.05$) among treatments based on one-way ANOVA (using LSD test, $p = 0.05$ at $df = 5$ and $n = 3$).

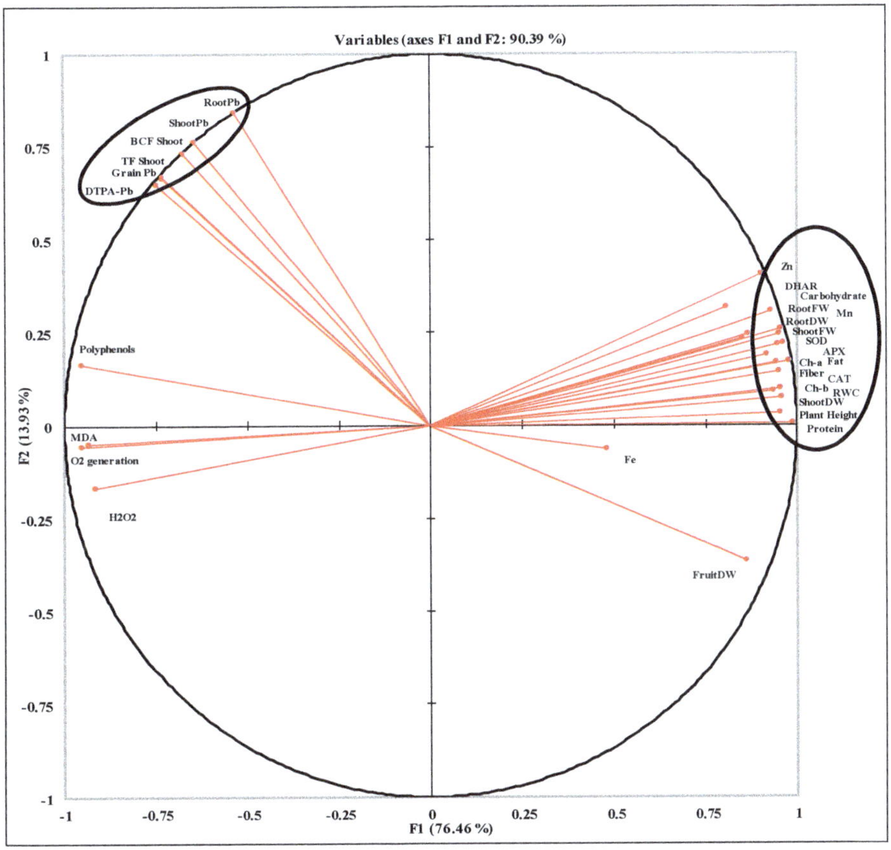

Figure 4. Principal component analysis (PCA) of studied attributes of pea grown under Pb stress with various soil amendments.

3. Discussion

3.1. Speciation of Pb in Pea Shoots, Roots, and Grain, Related BCF and TF Values and Pb Bioavailability in Post-Harvest Soil as Influenced by Amendments

All amendments noticeably reduced Pb concentration in pea roots, shoots, grain and bioavailable Pb (DTPA-extractable fraction), compared to control. The lowest Pb concentrations in roots, shoots, grain and DTPA-extract were found in BN 5% and CM, compared to control (Figure 1). Our results are in agreement with the results of previous studies, where the application of BN in Pb-polluted soil significantly reduced the concentrations of Pb in the shoots of pak choi and maize, as well as DTPA-extractable fraction [7]. The application of BN in Pb-polluted soil has also been reported to significantly reduce the Pb concentrations in rapeseed [5], rice grain and straw [1]. It has been reported that Pb concentrations were significantly reduced in the leachates after amending a Pb-polluted soil with CM [6]. The reduction in Pb concentrations in roots, shoots, grain and DTPA-extractable Pb fraction can be attributed to versatile characteristics of BN and CM. Application of CM in Pb-polluted soil reduces the bioavailability of Pb via three ways i.e., (1) sequestration of Pb by physical encapsulation by hydration products (struvite-K), (2) conversion of Pb into highly insoluble pyromorphite ($Pb_5(PO_4)_3X$, X = Cl^-, OH^-, F^-) and Pb-phosphate ($Pb_3(PO_4)_2$) compounds and (3) raising soil pH after their application in the soil. It has been reported that pyromorphite and Pb-phosphate have extremely low solubility (Ksp ≅ 10^{-60}–10^{-85} and 10^{-6}, respectively) and thrive in extreme soil alkaline and acidic conditions [6,9].

Moreover, BN has vital characteristics that can alter the physicochemical characteristics of the soil. Application of BN increases soil pH which enhances sorption and precipitation of Pb in the soil. Furthermore, the larger surface of BN increases the CEC of the soil, which in turn reduces the Pb bioavailability to the plants [1,5,7].

The TF and BCF values for Pb were significantly reduced with all amendments, compared to the control. However, the least values of TF and BCF were observed after amending Pb-polluted soil with BN 5%, FeHP 2% and CM 0.5% (Table 1). The lower TF values of Pb are attributed to its accumulation in the cell membrane and vacuoles of the root, which is associated with the stable attribute of Pb in the soil-plant system [18]. The results of a previous study have explained that Pb is early recognized as a toxic compound by the roots of plants and is vacuolated either in the cell wall or vacuole, which in turn leads to its low translocation to the aerial parts [19]. The BCF stands for the transport potential of Pb from the soil to the plant body, and is dependent on the characteristics of soils and Pb speciation [20]. The BN 5%, FeHP 2% and CM 0.5% treatments showed the least BCF values of Pb, compared to the control. These lower BCF values of Pb indicate a very low transfer of Pb in pea showing its least bioavailability in the BN 5%, FeHP 2% and CM 0.5% amended soil, compared to control treatment [21].

3.2. Agronomic, Photosynthetic and Biophysical Parameters of Pea Plant as Influenced by Amendments

With few exceptions, all treatments significantly improved shoot, root and grain DW, as well as plant height, RWC and chlorophyll contents in pea, compared to control. However, BR treatment showed the highest shoot and root DW, and Chl-a contents, while BN and BR treatments showed highest grain DW, plant height, Chl-b and RWC contents, compared to control (Table 1). An improvement in shoot, root and grain DW, as well as plant height of brinjal [13], the contents of Chl-a and Chl-b in menthol [3] and RWC in chicory were observed after BR application in Pb-polluted soil [22]. Likewise, the application of BN in Pb-polluted soil significantly improved shoot DW and Chl contents in Chinese cabbage [7]. Improvement in the plant DW, RWC and Chl contents could be attributed to the various characteristics of BR like provision of essential nutrients, improvement in WHC and alteration in redox condition of the soil [23–25]. Likewise, improvement in the agronomic and biophysical parameters of plants is also due to the alleviation of Pb toxicity to them after the Pb has been immobilized onto the BR surface [3,12,13]. Moreover, improvement in soil CEC after the application of BN is responsible for the reduced phytoavailability of Pb. It has been reported that improvement in plant biomass and associated parameters are mainly due to the alleviation of Pb toxicity to the plants [7,26].

3.3. Status of Micronutrients, Antinutrient and Biochemical Compounds in Pea Grain as Influenced by Amendments

With few exceptions, the application of all amendments significantly improved the contents of protein, fat, fiber and carbohydrates, while it decreased the contents of polyphenols, compared to control. However, the best results regarding these parameters were observed in BR and LN treatments (Figure 2). It has been reported that Pb toxicity reduced the contents of biochemical compounds while there was an increase in the polyphenol contents in brinjal [13], sesame [27] and *Conocarpus erectus* [10]. The results of our study are in line with the findings of previous investigations where improvement in the contents of biochemical compounds and reduction in polyphenol contents of different plants grown in BR amended metal-polluted soils was reported [13,27]. Likewise, BR application in Pb-polluted soil improved the biochemical compounds of spinach [14]. Improvement in the contents of biochemical compounds in pea grain is due to the vital characteristics of BR and LN. Biochar and LN have very high WHC, and are rich in organic matter. Application of BR and LN in soil improves the activities of rhizosphere microbes due to the decomposition of these organic materials and the enhanced release of dissolved organic carbon, which in turn are responsible for enhancing the plant protein contents [15,27]. Amending soil with BR and LN increases the WHC of soil and improves the availability of water to the plants through xylem, and augments the metabolic activity in the plant. Improvement in the plant metabolic activity enhances the plant biochemical compounds and reduces polyphenol contents [15,16].

Enhancement in the contents of biochemical compounds could also be associated with the alleviation of Pb toxicity to the plants [10,14]. Application of LN in soil improves the soil organic matter content. The hydroxyl and carboxyl groups present on organic matter react with OH^- in soil and make them electronegative. This special feature of LN increases the variable negative charges in the soil which promote the adsorption of Pb ions on the soil colloids, thereby reducing Pb toxicity to the plants [15,16]. Likewise, BR also alleviates Pb toxicity to the plants via adsorbing Pb ions onto the larger surface area [13,14].

In our experiment, the highest grain Mn and Zn concentrations were found with BR 2% and LN 2% treatments, compared to control. Likewise, the highest significant Fe concentration was observed with FeHP treatment, compared to control (Figure 2). It has been reported that amending metal-polluted soil with BR significantly improved the concentrations of Mn and Zn in brinjal [13] and sunflower [28]. Likewise, an improvement in the Fe concentrations of rice grain was reported after the application of iron compounds [29]. The higher concentrations of Fe in pea grain could be attributed to the presence of Fe in the FeHP amendment. Furthermore, BR application in the soil also provides essential nutrients to the plants and increases the concentrations of Zn and Mn [16]. Similarly, LN contains several functional groups that help to strongly adsorb Pb ions which reduce the Pb bioavailability in the soil [15,16]. Therefore, the reduced bioavailability of Pb in the soil improved the phyto-availability of micronutrients, especially Mn and Zn and their uptake by the plant roots due to their antagonistic behavior with Pb ions [30]. It has been reported that the sum of released cations like K, Na, Ca and Mg from amendments is almost equal to the quantity of adsorbed metal ions, explaining the primary function of cation exchange in Pb sorption by BH [31].

3.4. Status of Antioxidant Defense Machinery and Oxidative Stress in Pea Plant as Influenced by Amendments

In our experiment, all of the amendments noticeably reduced the MDA, H_2O_2 contents and the O_2^- generation rate, as well as improved the activities of APX, SOD, CAT and DHAR in pea plant relative to the control treatment. However, the most pronounced results regarding these parameters were found in BR treatment followed by LN treatment (Figure 3). Our results are in line with the findings of previous studies where the activities of SOD, CAT, peroxidase (PER), ascorbic acid (AsA) and APX in brinjal [15], SOD, CAT and PER in *Mentha arvensis* and CAT, PER and polyphenol oxidase (PPO) in chicory [22], were enhanced with the incorporation of BR in Pb-polluted soil. Whereas, the contents of the H_2O_2 and O_2^- generation rate were significantly reduced in chicory with BR application in a Pb-polluted soil [22]. Plants produce antioxidant enzymes like APX, SOD, CAT and DHAR upon the exposure to heavy metals stress which acts as a defense system against oxidative stress; the latter is the higher production of ROS as a result of oxidizing chain reactions [16]. The mechanism responsible for this improvement in antioxidant activities and the reduction of ROS contents in the plants is due to the fundamental characteristics of BR; i.e., promotion of plant health, provision of essential nutrients, resistance against metal stress after they get immobilized on BR, having the larger surface area and high cation exchange capacity [16]. Similarly, LN adsorbs metal ions [15] and decreases the metal stress to the plant, which in turn increases the mobility of essential nutrients to the plant. Higher mobility of essential nutrients to plants is known to improve their vigor and thereby reduce the contents of reactive oxygen species in them [16].

4. Materials and Methods

4.1. Collection of Experiment Soil and its Characterization

Experimental soil was purchased from a plant shop named "Evergreen nursery, Faisalabad, Pakistan". The soil was air-dried and sieved through a 2 mm sieve to remove stones and debris. The physicochemical properties of the soil were determined by employing standard methods. Soil pH was determined by the method of McLean, [32] by preparing a suspension (soil:deionized water, 1:1) followed by shaking for approximately 1 h and measurement on a calibrated pH meter (model WTW7110, Weilheim, Germany). Similarly, soil texture was determined using the hydrometer method, soil organic matter using the Walkley-Black method and Cation exchange capacity (CEC) was determined by the approaches of Gee and Bauder, [33], Jackson [34], and Rhoades [35], respectively. Determination of total phosphorus, exchangeable potassium and calcium carbonate, were carried out according to the methods of Watanabe and Olsen [36], Richards [37] and Allison and Moodie [38], respectively. Likewise, the bioavailable fraction of Pb in the Pb-polluted soil was measured on the atomic absorption spectrophotometer (AAS, PerkinElmer AAnalyst™ 800, Shelton, CT, USA) after extracting the soil with 5 mM diethylenetriaminepentaacetic acid (DTPA extractant (soil to DTPA extractant, 1:2) [39]. The properties of experimental soil are presented in Table 3.

Table 3. Physiochemical characteristics of experimental soil.

Characteristics	Units	Amount
Clay	%	29.7 ± 1.07
Silt	%	27 ± 0.97
Sand	%	40.3 ± 0.80
Organic matter content (OMC)	%	0.84 ± 0.03
Bicarbonate (HCO3)	%	0.17 ± 0.01
pH	-	8.4 ± 0.30
Cation exchange capacity (CEC)	cmolc kg^{-1}	29.2 ± 1.06
Electrical conductivity (EC)	DSm^{-1}	3.8 ± 0.14
Content of calcium carbonate (CaCO3)	%	2.9 ± 0.11
Phosphorus (P)	mg kg^{-1}	8.3 ± 0.30
Potassium (K)	mg kg^{-1}	81 ± 2.94
Nitrogen (N)	mg kg^{-1}	174 ± 6.31
Total Pb	mg kg^{-1}	1000 ± 36.2
diethylenetriaminepentaacetic acid (DTPA)-extractable Pb	mg kg^{-1}	6.1 ± 0.22

4.2. Soil Spiking with Pb

Soil was amended with $Pb(NO_3)_2$ to obtain a 1000 mg kg^{-1} Pb concentration in the soil. For this purpose, a known amount of $Pb(NO_3)_2$ was dissolved in distilled water. Later, this $Pb(NO_3)_2$ solution was poured in the soil and properly mixed. The soil was homogenized and packed in plastic sacks that were later kept at 25 °C for 60 days in a dark room. The soil in plastic sacks was manually mixed with a spatula twice a week for a homogeneous distribution of moisture. During this incubation period, distilled water was used to maintain the moisture at 65% water holding capacity (WHC) of the soil. The soil was air-dried after the end of the incubation.

4.3. Addition of Immobilizing Agents

Five reducing agents, including BN, BR, LN, CM 0.5% and FeHP were used for the immobilization of Pb in Pb-spiked soil. The BN was acquired from Jinan Yuansheng Chemical Technology Co., Ltd., Licheng District, Jinan, Shandong, China. The preparation and characterization of BR used in this experiment have been described in Shahbaz et al. [23]. Lignin was procured from the Jinan Yuansheng Chemical Technology Co., Ltd., Licheng District, Jinan, Shandong, China and CM from a local store. Likewise, a mixture of [Sodium phosphate (1000 mL, 0.5 mol L^{-1}) after a reaction with ferric chloride (1000 mL, 0.3 mol L^{-1})] was prepared and placed in a hermetic container at 35 °C for 24 h. The pH of the aforementioned reaction mixture was fixed at 4, using 1 M NaOH solution. After the completion of the reaction, the pH of the reaction mixture rose to 4.3. Ultimately, the resultant product i.e., (FeHP), was wiped out with distilled water and desiccated at 65 °C for 15 h.

4.4. Pot Experiment

Overall, the pot experiment consisted of six treatments i.e., BN, BR, LN, CM and FeHP and control Pb-spiked soil without any amendment (Table 4). The percentage of each immobilizing amendment used in this study was selected after carefully reviewing the previous studies i.e., BR 2% [40], BN 5% [1], LN 2% [15], CM 0.5% [6] and FeHP 2% [17], respectively. The resulting uniform mixture was incorporated with leftover soil via mechanical shaker, while maintaining the moisture at 65% WHC followed by incubation at 25 °C for six weeks in darkness. After the incubation, the treated soil was transferred to the plastic pots (height 33 cm, diameter 25.4 cm). The relocation of the experiment was accomplished in the field area of Government College University Faisalabad, Faisalabad, Pakistan with intense care, considering the appropriate environmental conditions, for instance, suitable temperature 25 °C, proper light 8–10 h and moisture 50%.

Table 4. Overview of treatment plan considered in this pot experiment using BN = Bentonite, BR = Biochar, LN = Lignin CM = Cement and FeHP = Iron hydroxyl phosphate.

Treatments	Abbreviations	Input Amounts of Both Amendments (g pot^{-1})
Control	Control	-
Bentonite (5%)	BN	150
Biochar (2%)	BR	60
Lignin (2%)	LN	60
Cement (0.5%)	CM	15
Iron Hydroxyl phosphate (2%)	FeHP	60

Pots were moistened to achieve suitable sowing conditions. Pea seeds were procured from Ayub Agriculture Research Institute, Faisalabad, Pakistan and soaked in water for 8 h preceding sowing in the pots. Eight seeds were sown per pot, which germinated in about 10 days within sowing. The pea plants were fertilized using balanced plant fertilizer [Grow Fertilizer (18-18-18), White Flower Farm, Litchfield, CT, USA] at the age of three weeks. Some parameters like plant height, length of plant and biomass (root and shoot), were assessed by using a portable measuring stick after 60 days of pea plant growth. After 60 d of growth, plants were harvested at the base and divided into root and shoot parts.

4.5. Plant and Soil Analysis

4.5.1. Estimating the Pb Concentrations in Plant Parts and DTPA Extract

The harvested soil was extracted from the pot with intense care, air-dried and sieved through a 2 mm sieve. An aliquot of harvested soil was used to measure its pH, as described earlier.

DTPA-extractable Pb was extracted by using the standard methodology of Lindsay and Norvell, [39], and the extracts were analyzed on AAS. Subsequently, plants (root and shoot) were rinsed to remove the adhered dust and dirt. The pea plant biomass was desiccated in an oven (Memmert, Beschickung-loading, model 100–800, Schwabach, Germany) at 70 °C for 24 h to get constant dry weight. The dried plant material was pulverized in a grinder (IKAWerke, MF 10 Basic, Staufen, Germany) followed by di-acid digestion (HNO_3:$HClO_4$ = 2:1) after sieving at 0.5mm as devised by Jones and Case [41]. Eventually, the Pb concentration in plant digest was analyzed on AAS.

4.5.2. Chlorophyll Contents, Antioxidant Enzymes Activities and Reactive Oxygen Species (ROS) Contents in Barley Leaf

The chlorophyll a (Chl-a) and chlorophyll b (Chl-b) contents in the leaves were assessed following the methods of Hiscox and Israelstam [42]. To this end, 1 g of fresh leaf sample was homogenized in 20 mL of methanol, chloroform and water (12:5:3 ratio). The contents of Chl-a and Chl-b were assessed by measuring the absorbance on a spectrophotometer at 664.5 and 647.4 nm, respectively.

The antioxidant enzymes [SOD, CAT, APX and DHAR] in pea plant were determined by the methodology of El-Shabrawi et al. [43], Aebi [44], Nakano and Asada [45], respectively. The reaction mixture for SOD was prepared by mixing supernatant (1 mL) with potassium phosphate (K-P) buffer (50 mM), CAT (0.1 u), nitro blue tetrazolium (NBT, 2.24 mM), xanthine (2.36 mM, pH 7) and xanthine oxidase (0.1 u), as it follows the xanthine–xanthine oxidase system. The mixture for APX was prepared by mixing the supernatant (0.5 mL) with [AsA (0.5 mM), H_2O_2 (0.1 mM), sodium phosphate buffer (pH 7) and ethylenediaminetetraacetic acid (EDTA) (0.25 mL)]. The [dehydroascorbate (DHA, 0.1 mM), glutathione (GSH, 2.5 mM) and K-P buffer (50 mM, pH 7)] were mixed with supernatant [0.5 mL] for the mixture of DHAR. Similarly, the CAT reaction was commenced with H_2O_2 and its degradation was analyzed. Finally, the variations in absorbance were recorded via spectrophotometer at 560 nm, 240 nm, 290 nm and 265 nm for SOD, CAT, APX and DHAR, respectively. The specific activity was observed for the extraction coefficient at 40 mM cm to obtain CAT and extinction coefficient at 2.8 mM cm and 14 mM cm to obtain APX and DHAR.

The contents of MDA, H_2O_2 and O_2^- generation rate were measured by the methods of Jambunathan, [46], Velikova et al. [47] and Yang et al. [48], respectively. The reaction mixtures were prepared with fresh leaf tissue (500 mg) and assimilated with trichloroacetic acid (TCA, 5 mL, 0.1%) for MDA, K-P buffer [(5 mL) for H_2O_2 and (12 mL, 65 mM, pH 7.8)] for O_2. Afterward, the mixtures were centrifuged at 10,000× *g* for 15 min and supernatants (2.5 mL, 0.5 mL & 5 mL) for MDA, H_2O_2 and O_2, respectively, were used. Subsequently, supernatants were assorted with thiobarbituric acid (TBA, 1 mL, 0.5% *w/v*) in TCA (20%), heated at 95 °C for 30 min and later chilled in an ice bath for MDA. Similarly, the mixture (TCA, 5 mL, 0.1% *w/v*) was prepared in K-P buffer (10 mM, pH 7) using potassium iodide (1 M, 1 mL) for H_2O_2 determination. For the estimation of O_2^- generation, K-P buffer (0.9 mL, 65 mM, pH 7.8) was mixed with hydroxylamine hydrochloride (1 mL, 10 mM), sulfanilamide (1 mL, 17 mM) and naphthylamine (1 mL, 7 mM) before incubation at 25 °C for 20 min. The variation in absorbance was deliberated via Beer and Lambert's equation to calculate MDA at 532 nm–600 nm. While the absorbance for the H_2O_2 and O_2^- generation rate were scrutinized via spectrophotometer at 390 nm and 530 nm, respectively [49].

4.5.3. Estimation of grain Biochemical Compounds and Micronutrients

Various standard protocols were used to analyze the protein, fat, fiber, carbohydrate and polyphenols in the ground pea grain samples. A standard approach was developed for the estimation of plant protein by using the protein dye-binding method involving an equal volume of sample buffer (impeded by Bradford assay) added into the protein reagent to compensate for the interference. Bovine serum albumin was used to accomplish the reaction [50]. The association of official analytical chemists (AOAC) [51] methods were used for the estimation of fat, fiber and carbohydrates in pea grain.

Polyphenols in pea grain were determined by using the Folin-Ciocalteu method proposed by Singleton et al. [52]. Following this method, the calibration curve of the standard for gallic acid was plotted for the estimation of phenolic compounds in pea grain, and represented as mg g^{-1} equivalent to gallic acid (GAE) after recording the absorbance at 760 nm on a spectrophotometer.

Some of the pea grain were crushed in a grinder (IKAWerke, MF 10 Basic, Staufen, Germany) followed by sieving at 0.5 mm. The grounded grains were subjected to di-acid (HNO_3:$HClO_4$, 2:1) digestion. Afterwards, further characterizations of pea grain micronutrients (Ca, Zn, Fe, Mg, Mn, Ni) were performed spectrophotometrically (PerkinElmer, AAnalyst 100, Waltham, MA, USA). The aforementioned process was developed by Jones and Case [41].

4.5.4. Computation of Translocation Factor (TF) and Bioconcentration Factor (BCF) of Pb

The values of BCF and TF of Pb were calculated by Equations (1) and (2) respectively as recommended by Salazar and Pignata [53].

$$BCF = Cshoot/Csoil \tag{1}$$

$$TF = Cshoot/Croot \quad (2)$$

where, the Cshoot, Croot and Csoil are the concentrations of Pb in the shoots (mg kg^{-1} DW), roots (mg kg^{-1} DW) and soil (mg kg^{-1} DW soil), respectively.

4.6. Statistical Analysis

A completely randomized design was used to execute this pot experiment, and the results were interpreted by using one-way Analysis of Variance (ANOVA) with the help of the Statistix 8.1 software package (Copyright 2005, Analytical software, Tallahassee, FL, USA). The described means are the average of three replicates, and are stated with their standard error (SE). A least significant difference (LSD) test was carried out to detect the significant difference ($p < 0.05$) between treatment means. The principle component analysis (PCA) and Pearson coefficient correlation (r^2) among studied attributes was computed by using the xlstat software version 4.15 (Addinsoft, Paris, France). The table of communalities for each parameter was generated with the IBM SPSS Statistics software windows version 25 (IBM Corp, Armonk, NY, USA).

5. Conclusions

In this pot experiment, Pb-polluted soil was amended with different amendments i.e., BC 2%, BN 5%, LN 2%, CM 0.5% and FeHP 2%, and their effects on different parameters of pea were observed. Results showed that BN 5%, FeHP 2% and CM 0.5% significantly reduced the concentrations of Pb in shoots, roots and grain, as well as the TF and BCF values of Pb in the order FeHP 2% > CM 0.5% > BN 5%. The agronomic (plant height, shoot, root and grain DW), biochemical (chl a and chl b contents and grain biochemistry), biophysical (RWC) parameters and grain micronutrient concentrations were improved with BR 2%. However, the 2% BR rate that we used in our pot experiment may not be economically viable for field scale application.

Supplementary Materials: The following are available online at http://www.mdpi.com/2223-7747/8/12/571/s1, Table S1: Factor loadings for PCA (Principle component analysis) of studied attributes of pea grown under Pb stress with various soil amendments. Table S2: Communalities of studied attributes of pea grown under Pb stress with various soil amendments. Figure S1: Scree plot representing the Eigen values and cumulative variability (%) in relation to factors used in PCA

Author Contributions: Conceptualization, M.I. (Mutahar Iqbal) and F.M.; Data curation, M.I. (Mutahar Iqbal), S.H. and S.A.K.; Formal analysis, S.H., P.M.A.R., M.I. (Muhammad Iqbal) and M.F.; Methodology, P.M.A.R., M.Z.H., M.F., S.A.K. and F.M.; Resources, M.Z.H. and T.S.; Writing—original draft, M.I. (Mutahar Iqbal) and M.I. (Muhammad Iqbal); Writing—review & editing, T.S., F.M., I.K., K.S., M.I. (Muhammad Ibrahim) and H.S.T.U.H.

Acknowledgments: The authors of this manuscript are thankful to Veysel Turan, Bingöl University, Turkey for carrying out innumerable analyses involved in this study.

Conflicts of Interest: The authors declare no conflict of interest.

References

1. Sun, Y.; Li, Y.; Xu, Y.; Liang, X.; Wang, L. In situ stabilization remediation of cadmium (Cd) and lead (Pb) co-contaminated paddy soil using bentonite. *Appl. Clay Sci.* **2015**, *105*, 200–206. [CrossRef]
2. López-Orenes, A.; Dias, M.C.; Ferrer, M.Á.; Calderón, A.; Moutinho-Pereira, J.; Correia, C.; Santos, C. Different mechanisms of the metalliferous Zygophyllum fabago shoots and roots to cope with Pb toxicity. *Environ. Sci. Pollut. Res.* **2018**, *25*, 1319–1330. [CrossRef] [PubMed]
3. Nigam, N.; Khare, P.; Yadav, V.; Mishra, D.; Jain, S.; Karak, T.; Panja, S.; Tandon, S. Biochar-mediated sequestration of Pb and Cd leads to enhanced productivity in Mentha arvensis. *Ecotoxicol. Environ. Saf.* **2019**, *172*, 411–422. [CrossRef] [PubMed]
4. Tchounwou, P.B.; Yedjou, C.G.; Patlolla, A.K.; Sutton, D.J. Heavy Metal Toxicity and the Environent. In *Molecular, Clinical and Environmental Toxicology*; Springer: Berlin, Germany, 2012; pp. 133–164.
5. Shaheen, S.M.; Rinklebe, J. Impact of emerging and low cost alternative amendments on the (im) mobilization and phytoavailability of Cd and Pb in a contaminated floodplain soil. *Ecol. Eng.* **2015**, *74*, 319–326. [CrossRef]

6. Wang, Y.-S.; Dai, J.-G.; Wang, L.; Tsang, D.C.W.; Poon, C.S. Influence of lead on stabilization/solidification by ordinary Portland cement and magnesium phosphate cement. *Chemosphere* **2018**, *190*, 90–96. [CrossRef]
7. Lahori, A.H.; Zhang, Z.; Shaheen, S.M.; Rinklebe, J.; Guo, Z.; Li, R.; Mahar, A.; Wang, Z.; Ren, C.; Mi, S. Mono-and co-applications of Ca-bentonite with zeolite, Ca-hydroxide, and tobacco biochar affect phytoavailability and uptake of copper and lead in a gold mine-polluted soil. *J. Hazard. Mater.* **2019**, *374*, 401–411. [CrossRef]
8. Wang, L.; Yu, K.; Li, J.-S.; Tsang, D.C.W.; Poon, C.S.; Yoo, J.-C.; Baek, K.; Ding, S.; Hou, D.; Dai, J.-G. Low-carbon and low-alkalinity stabilization/solidification of high-Pb contaminated soil. *Chem. Eng. J.* **2018**, *351*, 418–427. [CrossRef]
9. Wiszniewska, A.; Hanus-Fajerska, E.; MUSZYŃSKA, E.; Ciarkowska, K. Natural organic amendments for improved phytoremediation of polluted soils: a review of recent progress. *Pedosphere* **2016**, *26*, 1–12. [CrossRef]
10. Tauqeer, H.M.; Hussain, S.; Abbas, F.; Iqbal, M. The potential of an energy crop "Conocarpus erectus" for lead phytoextraction and phytostabilization of chromium, nickel, and cadmium: An excellent option for the management of multi-metal contaminated soils. *Ecotoxicol. Environ. Saf.* **2019**, *173*, 273–284. [CrossRef]
11. Fernández, D.A.; Roldán, A.; Azcón, R.; Caravaca, F.; Bååth, E. Effects of water stress, organic amendment and mycorrhizal inoculation on soil microbial community structure and activity during the establishment of two heavy metal-tolerant native plant species. *Microb. Ecol.* **2012**, *63*, 794–803. [CrossRef]
12. Wang, L.; Iris, K.M.; Tsang, D.C.W.; Li, S.; Li, J.; Poon, C.S.; Wang, Y.-S.; Dai, J.-G. Transforming wood waste into water-resistant magnesia-phosphate cement particleboard modified by alumina and red mud. *J. Clean. Prod.* **2017**, *168*, 452–462. [CrossRef]
13. Turan, V.; Khan, S.A.; Iqbal, M.; Ramzani, P.M.A.; Fatima, M. Promoting the productivity and quality of brinjal aligned with heavy metals immobilization in a wastewater irrigated heavy metal polluted soil with biochar and chitosan. *Ecotoxicol. Environ. Saf.* **2018**, *161*, 409–419. [CrossRef] [PubMed]
14. Břendová, K.; Zemanová, V.; Pavlíková, D.; Tlustoš, P. Utilization of biochar and activated carbon to reduce Cd, Pb and Zn phytoavailability and phytotoxicity for plants. *J. Environ. Manage.* **2016**, *181*, 637–645. [CrossRef] [PubMed]
15. Li, T.; Lü, S.; Zhang, S.; Gao, C.; Liu, M. Lignin-based multifunctional fertilizer for immobilization of Pb (II) in contaminated soil. *J. Taiwan Inst. Chem. Eng.* **2018**, *91*, 643–652. [CrossRef]
16. Yan, M.; Li, Z. Microwave-assisted functionalized lignin with dithiocarbamate for enhancing adsorption of Pb (II). *Mater. Lett.* **2016**, *170*, 135–138. [CrossRef]
17. Yuan, Y.; Chai, L.; Yang, Z.; Yang, W. Simultaneous immobilization of lead, cadmium, and arsenic in combined contaminated soil with iron hydroxyl phosphate. *J. soils sediments* **2017**, *17*, 432–439. [CrossRef]
18. Castaldi, P.; Melis, P.; Silvetti, M.; Deiana, P.; Garau, G. Influence of pea and wheat growth on Pb, Cd, and Zn mobility and soil biological status in a polluted amended soil. *Geoderma* **2009**, *151*, 241–248. [CrossRef]
19. Di Toppi, L.S.; Gabbrielli, R. Response to cadmium in higher plants. *Environ. Exp. Bot.* **1999**, *41*, 105–130. [CrossRef]
20. Singh, A.; Agrawal, M.; Marshall, F.M. The role of organic vs. inorganic fertilizers in reducing phytoavailability of heavy metals in a wastewater-irrigated area. *Ecol. Eng.* **2010**, *36*, 1733–1740. [CrossRef]
21. Guo, G.; Lei, M.; Chen, T.; Yang, J. Evaluation of different amendments and foliar fertilizer for immobilization of heavy metals in contaminated soils. *J. soils sediments* **2018**, *18*, 239–247. [CrossRef]
22. El-Banna, M.F.; Mosa, A.; Gao, B.; Yin, X.; Ahmad, Z.; Wang, H. Sorption of lead ions onto oxidized bagasse-biochar mitigates Pb-induced oxidative stress on hydroponically grown chicory: Experimental observations and mechanisms. *Chemosphere* **2018**, *208*, 887–898. [CrossRef] [PubMed]
23. Shahbaz, A.K.; Lewińska, K.; Iqbal, J.; Ali, Q.; Iqbal, M.; Abbas, F.; Tauqeer, H.M.; Ramzani, P.M.A. Improvement in productivity, nutritional quality, and antioxidative defense mechanisms of sunflower (*Helianthus annuus* L.) and maize (*Zea mays* L.) in nickel contaminated soil amended with different biochar and zeolite ratios. *J. Environ. Manage.* **2018**, *218*, 256–270. [CrossRef] [PubMed]
24. Joseph, S.; Husson, O.; Graber, E.R.; Van Zwieten, L.; Taherymoosavi, S.; Thomas, T.; Nielsen, S.; Ye, J.; Pan, G.; Chia, C. The electrochemical properties of biochars and how they affect soil redox properties and processes. *Agronomy* **2015**, *5*, 322–340. [CrossRef]
25. Mukherjee, A.; Lal, R. Biochar impacts on soil physical properties and greenhouse gas emissions. *Agronomy* **2013**, *3*, 313–339. [CrossRef]

26. Shaheen, S.M.; Rinklebe, J.; Selim, M.H. Impact of various amendments on immobilization and phytoavailability of nickel and zinc in a contaminated floodplain soil. *Int. J. Environ. Sci. Technol.* **2015**, *12*, 2765–2776. [CrossRef]
27. Mehmood, S.; Saeed, D.A.; Rizwan, M.; Khan, M.N.; Aziz, O.; Bashir, S.; Ibrahim, M.; Ditta, A.; Akmal, M.; Mumtaz, M.A. Impact of different amendments on biochemical responses of sesame (*Sesamum indicum* L.) plants grown in lead-cadmium contaminated soil. *Plant Physiol. Biochem.* **2018**, *132*, 345–355. [CrossRef]
28. Turan, V.; Ramzani, P.M.A.; Ali, Q.; Abbas, F.; Iqbal, M.; Irum, A.; Khan, W.-D. Alleviation of nickel toxicity and an improvement in zinc bioavailability in sunflower seed with chitosan and biochar application in pH adjusted nickel contaminated soil. *Arch. Agron. Soil Sci.* **2018**, *64*, 1053–1067. [CrossRef]
29. Zhou, Q.; Lin, L.; Qiu, W.; Song, Z.; Liao, B. Supplementation with ferromanganese oxide–impregnated biochar composite reduces cadmium uptake by indica rice (*Oryza sativa* L.). *J. Clean. Prod.* **2018**, *184*, 1052–1059. [CrossRef]
30. Kabata-Pendias, A. *Trace Elements in Soils and Plants*; CRC Press, Taylor and Francis Group: Boca Raton, FL, USA, 2011.
31. Li, H.; Dong, X.; da Silva, E.B.; de Oliveira, L.M.; Chen, Y.; Ma, L.Q. Mechanisms of metal sorption by biochars: biochar characteristics and modifications. *Chemosphere* **2017**, *178*, 466–478. [CrossRef]
32. McLean, E.O. Soil pH and lime requirement. *Methods soil Anal. Part 2. Chem. Microbiol. Prop.* **1982**, 199–224.
33. Gee, G.W.; Bauder, J.W. Particle-size analysis 1. *Meth. soil Anal. Part 1—Phys. Mineral. Methods* **1986**, 383–411.
34. Jackson, M.L. Soil chemical analysis, constable and Co. *Ltd. London* **1962**, *497*.
35. Rhoades, J.D. Cation Exchange Capacity 1. *Methods soil Anal. Part 2. Chem. Microbiol. Prop.* **1982**, 149–157.
36. Watanabe, F.S.; Olsen, S.R. Test of an ascorbic acid method for determining phosphorus in water and $NaHCO_3$ extracts from soil 1. *Soil Sci. Soc. Am. J.* **1965**, *29*, 677–678. [CrossRef]
37. Richards, L.A. *Diagnosis and Improvement of Saline and Alkali Soils*; LWW: Philadelphia, PA, USA, 1954; Volume 78, ISBN 0038-075X.
38. Allison, L.E.; Moodie, C.D. Carbonate. *Methods Soil Anal. Part 2. Chem. Microbiol. Prop.* **1965**, 1379–1396.
39. Lindsay, W.L.; Norvell, W.A. Development of a DTPA soil test for zinc, iron, manganese, and copper 1. *Soil Sci. Soc. Am. J.* **1978**, *42*, 421–428. [CrossRef]
40. Nejad, Z.D.; Jung, M.C. The effects of biochar and inorganic amendments on soil remediation in the presence of hyperaccumulator plant. *Int. J. Energy Environ. Eng.* **2017**, *8*, 317–329. [CrossRef]
41. Jones Jr, J.B.; Case, V.W. Sampling, handling and analyzing plant tissue samples. *Sampling, Handl. Anal. Plant Tissue Samples.* **1990**, 389–427.
42. Hiscox, J.D.; Israelstam, G.F. A method for the extraction of chlorophyll from leaf tissue without maceration. *Can. J. Bot.* **1979**, *57*, 1332–1334. [CrossRef]
43. El-Shabrawi, H.; Kumar, B.; Kaul, T.; Reddy, M.K.; Singla-Pareek, S.L.; Sopory, S.K. Redox homeostasis, antioxidant defense, and methylglyoxal detoxification as markers for salt tolerance in Pokkali rice. *Protoplasma* **2010**, *245*, 85–96. [CrossRef]
44. Aebi, H. [13] Catalase in vitro. In *Methods in Enzymology*; Elsevier: Amsterdam, The Netherlands, 1984; Volume 105, pp. 121–126. ISBN 0076-6879.
45. Nakano, Y.; Asada, K. Hydrogen peroxide is scavenged by ascorbate-specific peroxidase in spinach chloroplasts. *Plant cell Physiol.* **1981**, *22*, 867–880.
46. Jambunathan, N. Determination and detection of reactive oxygen species (ROS), lipid peroxidation, and electrolyte leakage in plants. In *Plant Stress Tolerance*; Springer: Berlin, Germany, 2010; pp. 291–297.
47. Velikova, V.; Yordanov, I.; Edreva, A. Oxidative stress and some antioxidant systems in acid rain-treated bean plants: protective role of exogenous polyamines. *Plant Sci.* **2000**, *151*, 59–66. [CrossRef]
48. Yang, H.; Wu, F.; Cheng, J. Reduced chilling injury in cucumber by nitric oxide and the antioxidant response. *Food Chem.* **2011**, *127*, 1237–1242. [CrossRef] [PubMed]
49. Nahar, K.; Rahman, M.; Hasanuzzaman, M.; Alam, M.M.; Rahman, A.; Suzuki, T.; Fujita, M. Physiological and biochemical mechanisms of spermine-induced cadmium stress tolerance in mung bean (*Vigna radiata* L.) seedlings. *Environ. Sci. Pollut. Res.* **2016**, *23*, 21206–21218. [CrossRef]
50. Bradford, M.M. A rapid and sensitive method for the quantitation of microgram quantities of protein utilizing the principle of protein-dye binding. *Anal. Biochem.* **1976**, *72*, 248–254. [CrossRef]
51. Horwitz, W.; Chichilo, P.; Reynolds, H. Official methods of analysis of the Association of Official Analytical Chemists. *Off. methods Anal. Assoc. Off. Anal. Chem.* **1970**.

52. Singleton, V.L.; Orthofer, R.; Lamuela-Raventós, R.M. [14] Analysis of total phenols and other oxidation substrates and antioxidants by means of folin-ciocalteu reagent. In *Methods in Enzymology*; Elsevier: Amsterdam, The Netherlands, 1999; Volume 299, pp. 152–178, ISBN 0076-6879.
53. Salazar, M.J.; Pignata, M.L. Lead accumulation in plants grown in polluted soils. Screening of native species for phytoremediation. *J. Geochemical Explor.* **2014**, *137*, 29–36. [CrossRef]

plants MDPI

Article

Assessing the Correlations between Different Traits in Copper-Sensitive and Copper-Resistant Varieties of Jute (*Corchorus capsularis* L.)

Muhammad Hamzah Saleem [1], Shafaqat Ali [2,*], Mahmoud F. Seleiman [3], Muhammad Rizwan [2], Muzammal Rehman [5], Nudrat Aisha Akram [6], Lijun Liu [1,*], Majed Alotaibi [3], Ibrahim Al-Ashkar [3,7] and Muhammad Mubushar [3]

1 MOA Key Laboratory of Crop Ecophysiology and Farming System in the Middle Reaches of the Yangtze River, College of Plant Science and Technology, Huazhong Agricultural University, Wuhan 430070, China; saleemhamza312@webmail.hzau.edu.cn
2 Department of Environmental Sciences and Engineering, Government College University, Allama Iqbal Road, Faisalabad 38000, Pakistan; mrazi1532@yahoo.com
3 Plant Production Department, College of Food and Agriculture Sciences, King Saud University, P.O. Box 2460, Riyadh 11451, Saudi Arabia; mseleiman@ksu.edu.sa (M.F.S.); malotaibia@ksu.edu.sa (M.A.); ialashkar@ksu.edu.sa (I.A.-A.); 43910662@ksu.edu.sa (M.M.)
4 Department of Crop Sciences, Faculty of Agriculture, Menoufia University, Shibin El-kom 32514, Egypt
5 School of Agriculture, Yunnan University, Kunming 650504, China; muzammal@ynu.edu.cn
6 Department of Botany, Government College University Allama Iqbal Road, Faisalabad 38000, Pakistan; nudrataauaf@yahoo.com
7 Agronomy Department, Faculty of Agriculture, Al-Azhar University, Cairo 11651, Egypt
* Correspondence: shafaqataligill@yahoo.com (S.A.); liulijun@mail.hzau.edu.cn (L.L.)

Received: 4 November 2019; Accepted: 20 November 2019; Published: 26 November 2019

Abstract: The current study was conducted to explore the potential for phytoremediation in different varieties of jute grown under toxic concentrations of copper (Cu). For this purpose, a Petri dish experiment was conducted under controlled conditions using four varieties of jute, i.e., HongTieGuXuan, C-3, GuBaChangaJia, and ShangHuoMa, grown in double filter paper under 50 μmol L^{-1} of artificially spiked copper (Cu) using $CuSO_4.H_2O$. The results of the present study revealed that jute varieties C-3 and HongTieGuXuan were able to survive under high concentrations of Cu without a significant decrease in plant height, plant fresh and dry weights, total chlorophyll content, or seed germination, while varieties GuBaChangaJia and ShangHuoMa exhibited a significant reduction in their growth and biomass. Furthermore, high concentrations of Cu in the medium resulted in lipid peroxidation. This could be due to the oxidative damage induced in the roots and leaves of the jute varieties, which might be a result of by hydrogen peroxide (H_2O_2) and electrolyte leakage. Reactive oxygen species (ROS) generated due to Cu toxicity can be overcome by the increasing activity of antioxidants, and it was also noted that GuBaChangaJia and ShangHuoMa exhibited high Cu stress, while C-3 and HongTieGuXuan showed some resistance to Cu toxicity. Contrastingly, Cu accumulation and uptake was higher in C-3 and HongTieGuXuan, while a little Cu was accumulated in the roots and leaves of GuBaChangaJia and ShangHuoMa. On the basis of these findings, it can be suggested that C-3 and HongTieGuXuan have the potential to cope with Cu stress and can be considered Cu-resistant varieties, while GuBaChangaJia and ShangHuoMa are considered Cu-sensitive varieties. Moreover, C-3 and HongTieGuXuan have the potential to revoke large amounts of Cu, and can be cultivated as phytoremediation tools in Cu-contaminated soil.

Keywords: jute varieties; copper stress; phytoremediation; bioaccumulation factor; translocation factor; oxidative stress; growth

1. Introduction

"Heavy metals" is a general term which applies to metals and metalloids with atomic densities greater than 4000 kg m^{-3}. Almost all the heavy metals are toxic to human beings even at low metal ion concentrations [1–3]. Heavy metals such as copper (Cu), mercury (Hg), lead (Pb), cadmium (Cd), and zinc (Zn) are harmful to the living organisms, including plants, when their concentrations increase beyond the permissible limits. Copper (Cu) is an important micronutrient present in many minerals and rocks, and also required for many metabolic processes in all living organisms, i.e., prokaryotes and eukaryotes [3–6]. Furthermore, Cu functions in many different enzymes including hemocynins, oxidases, and reductases, and also takes part in many biological and physiological processes in plants [5,7–10]. Contrastingly, excess Cu in the soil causes adverse effects on plants such as alteration in DNA structure, affected photosynthesis and respiration, stunted growth, chlorosis, and poor root development in higher plants [11–16]. Furthermore, phytotoxicity of Cu causes injury at the cellular level to many cellular organelles, which causes the generation of reactive oxygen species (ROS) [4,6]. Moreover, high concentrations of Cu in the soil enhance the production of ROS by producing superoxide radicals (O^{-1}) and singlet oxygen ($^1O_2$) in the Mehler reaction [5,8], which leads to lipid peroxidation and the production of high contents of malondialdehyde (MDA) in plant tissues/cells, which induces oxidative damage in plants due to the phytotoxicity of Cu [7,13,17,18]. ROS production in plants is removed by a variety of antioxidant enzymes such as SOD, POD, CAT, and APX. Plant responses to oxidative stress also depend upon plant species and cultivars [7,9,19,20]. Previously, increasing antioxidant activity has been observed in *Raphanus sativus* [20], *Vitis vinifera* [18], and *Brassica napus* [14] under toxic concentrations of Cu in the soil.

Phytoremediation is the use of green plants) to remove, contain, and render harmful or toxic pollutants from the soil or medium. Furthermore, it is cost-effective, eco-friendly or with few environmental side effects, and is a publicly accepted technique which can also apply on specific sites or over large areas [8,13,21,22]. At present, 25 different plant species have been used as Cu accumulators (able to accumulate large amounts of metal in their above-ground parts rather than in below-ground parts) and Cu excluders (able to accumulate large amounts of metal in their below-ground parts rather than in above-gound parts) have been identified [7,8]. Previously, many different fibrous species, e.g., *Boehmeria nivea*, *Linum usitatissimum*, and *Hibiscus cannabinus*, been used for the phytoextraction of different heavy metals (Cu, Cd, and Pb) [11,23,24]. Among different fibrous species, jute (*Corchorus capsularis* L.) has been found to be relatively tolerant to heavy metal stress, possibly due to its huge biomass and specific physiological and biochemical activities [25–27]. Jute is a C-3 plant belonging to the family Malvaceae, and is cultivated as a natural fiber and also known as the golden fiber throughout the world. It is the cheapest and strongest fiber among all the natural fibrous crops, and main exporters of its fibers are India, Bangladesh, China, Thailand, and many other countries of southwest Asia [28]. Raw jute has various uses in making twine, matting, bales of cotton, sacks, coarse clothes, ropes, chair coverings, carpets, area rugs, curtains, and backing for linoleum [25,28].

Keeping in view the importance of jute, the present experiment was planned to explore the effect of Cu toxicity on the germination, seedling growth, biomass, antioxidant capacity, and Cu uptake of different varieties of jute. Sufficient literature is available on the phytotoxicity of Cu in *Raphanus sativus*, *Brassica juncea*, and *Phyllostachys pubescens* [8,15,20], but very few studies have been conducted on jute to study its different morphological and physiological traits. Findings from the present study will add to our understanding as to which varieties of jute can tolerate the toxic effects of Cu when grown in Cu-contaminated soil. The results from this study will add to our knowledge about (i) phytoremediation potential and (ii) the morpho-physiological traits and biochemical response of different varieties of jute grown in Cu-contaminated soil.

2. Materials and Methods

2.1. Growth Conditions and Experimental Treatment

The seeds of different jute varieties were collected from Bast and Fiber Research Center of Huazhong Agricultural University, Hubei Province, China. The seeds of C-3, HongTieGuXuan, GuBaChangaJia, and ShangHuoMa (which are types of white jute) were used for the Petri dish experiment, and 40 seeds were planted in each Petri dish. Two filter papers (90 mm in diameter) were used in each Petri dish and 5 mL of 50 µmol L^{-1} Cu solution was added. Cu solutions were prepared with pure distilled water using copper sulfate ($CuSO_4.5H_2O$) (99% purity). After quantification of Cu, as percent availability in $CuSO_4.5H_2O$, 50 µmol L^{-1} doses of this compound were taken. Cu solution was applied every alternate day for the prevention of fungal infection and other contamination [29]. After washing carefully with distilled water, seeds were tested on filter paper (What-man No. 1). Seeds were surface sterilized with 0.1% $HgCl_2$ for the prevention of surface fungal/bacterial contamination [30]. The experiment was conducted in March 2018 at Huazhong Agricultural University, Wuhan China, in a growth chamber under white lights (100 W, Guangdong PHILIPS Co., Guangdong, China) with a day/night temperature of 25 ± 2 °C and day/night humidity of 80%. The nutrient solution was provided once in a week, replacing the Cu treatment for 24 h. The experiment was conducted in a complete randomized design with nine replications for each treatment. The seed germination was recorded at 4 days after sowing (DAS). The seeds were considered to have germinated when the shoots were more than 2 mm [10], and plant height, plant fresh and dry weight, chlorophyll content, and different antioxidants were analyzed at 14 DAS. Furthermore, the Cu concentration of roots and shoots were also measured in this study. All chemicals used were of analytical grade, procured from Sinopharm Chemical Reagent Co., Ltd (Shanghai, China).

2.2. Growth and Morphological Traits

Different morphological attributes were noted after the harvest of all plants. Plants in each treatment were harvested and separated into roots and shoots for growth and morphology traits. Total length was defined as the length of the plant from the surface growth medium line of the Petri dish to the tip of the uppermost shoot. Total fresh weight was measured by measuring the weight of roots and shoots with the help of a digital weighing balance. After that, plant samples were oven dried for 1 h at 105 °C, then 65 °C for 72 h until the weight become uniform and dry biomass was recorded. Roots were washed with distilled water and dipped in 20 mM Na_2EDTA for 15–20 min, washed thrice with distilled water and finally with de-ionized water, and then oven dried for further analysis. Chlorophyll contents were determined following the method of Arnon [31] and are expressed as mg g^{-1} FW.

2.3. Determination of Oxidative Stress

The method used to describe the concentration of lipid peroxidation was presented by Heath and Packer [32] and expressed as µmoles g^{-1} FW.

For the estimation of H_2O_2 content, an H_2O_2 Assay Kit (Suzhou Comin Biotechnology Co., Ltd.) was used.

The electrolyte leakage (EL) was measured according to the standard procedure of Dionisio-Sese and Tobita [33].

2.4. Analysis of Antioxidant Enzyme Activities

The activity of SOD was measured by the method of Chen and Pan [34] and expressed in Ug^{-1} FW.

The activity of POD was measured by the method of Sakharov and Aridilla [35] and expressed in Ug^{-1} FW.

The enzymatic activity of CAT was measured by the method of Aebi [36] and expressed in Ug^{-1} FW.

Ascorbate peroxidase activity was measured according to Nakano and Asada [37] and expressed in Ug^{-1} FW.

2.5. Determination of Cu Concentration

The dried samples were ground into powdered form using stainless steel and 0.1 g of root and shoot samples was taken for digestion in $HNO_3/HClO_4$ (4:1) solution. Final readings were taken from an atomic absorption spectrophotometer (AAS) model Agilent 240 FS-AA [7].

Bioaccumulation factor (BAF) was measured as the proportion of Cu concentration in plant tissues and Cu concentration in medium using the following formula:

$$\text{BAF} = \frac{\text{Cu concentration in plant tissues}}{\text{Cu concentration in the medium}} \quad (1)$$

Translocation factor (TF) was evaluated as the proportion of Cu concentration in shoots with respect to the roots:

$$\text{TF} = \frac{\text{Cu concentration in shoots}}{\text{Cu concentration in the roots}} \quad (2)$$

2.6. Statistical Analysis

The data recorded were statistically analyzed using Statistix 8.1 (Analytical Software, Tallahassee, FL, USA). Testing showed that all the data were approximately normally distributed. Thus, the differences between treatments were determined using analysis of variance, and the least significant difference test ($p \leq 0.05$) was used for multiple comparisons between treatment means. Graphical representation was carried out using SigmaPlot 12 and R studio.

3. Results

3.1. Effects of Cu Toxicity on Seed Germination, Growth, and Chlorophyll Content

Jute varieties showed various germination behaviors in response to Cu exposure (Table 1). Germination rate ranging from 75–100% in different varieties of jute under Cu toxicity. Based on the observations of seed germination, HongTieGuXuan and C-3 showed a better germination rate than GuBaChangaJia and ShangHuoMa. Maximum germination percentage was observed in HongTieGuXuan (100%) and C-3 (100%), followed by GuBaChangaJia (77.5%) and ShangHuoMa (75%) under toxic concentrations of Cu in the medium.

Table 1. Effect of Cu stress on plant height (cm), plant fresh weight (g), plant dry weight (g), total chlorophyll content (mg g^{-1} FW), and seed germination (%) in different varieties of jute.

Varieties	Plant Height	Plant Fresh Weight	Plant Dry Weight	Total Chlorophyll	Seed Germination
HongTieGuXuan	3.72 ± 0.05 a	0.255 ± 0.05 b	0.103 ± 0.005 a	2.64 ± 0.09 a	100
C-3	3.73 ± 0.85 a	0.3 ± 0.01 a	0.100 ± 0.01 a	2.82 ± 0.03 a	100
GuBaChangaJia	0.87 ± 0.81 b	0.103 ± 0.003 c	0.045 ± 0.015 b	1.51 ± 0.03 b	77.5
ShangHuoMa	0.61 ± 0.13 b	0.105 ± 0.003 c	0.040 ± 0.02 b	1.47 ± 0.08 b	75

Values in the table are from one harvest ± SD (n = 5). Different letters within a column indicate significant difference between the treatments ($p < 0.05$).

In the present study, growth in terms of plant height, plant fresh weight, and plant dry weight was also measured in different varieties of jute when cultivated under high concentrations of Cu in the medium. Results regarding growth and biomass of different varieties of jute are presented in Table 1. Maximum plant height and fresh and dry biomass were observed in HongTieGuXuan and C-3 under toxic levels of Cu in the medium. Moreover, the minimum plant height (0.61 ± 0.13), plant fresh weight (0.101 ± 0.003), and plant dry weight (0.040 ± 0.02) was observed in ShangHuoMa, while maximum plant height (3.73 ± 0.85), plant fresh weight (0.30 ± 0.01), and plant dry weight (0.100 ± 0.01) were observed in C-3 under toxic concentrations of Cu in the medium.

In the present study, total chlorophyll contents of the leaves of different varieties of jute are presented in Table 1. Different chlorophyll contents were observed in different varieties to jute after exposure of high concentrations of Cu in the medium. Maximum contents of chlorophyll were observed in HongTieGuXuan and C-3. The maximum contents of chlorophyll were observed in C-3 (2.82 ± 0.03 mg g^{-1} FW) followed by HongTieGuXuan (2.64 ± 0.09 mg g^{-1} FW), while minimum contents of chlorophyll were observed in ShangHuoMa (1.47 ± 0.08 mg g^{-1} FW) followed by GuBaChangaJia (1.51 ± 0.03 mg g^{-1} FW).

3.2. *Effect of Cu Toxicity on Oxidative Stress*

In this study, the effects of Cu toxicity on different varieties of jute on malondialdehyde (MDA), hydrogen peroxide (H_2O_2), and electrolyte leakage (EL) from the roots and shoots were also investigated (Figure 1). Exposure of Cu concentration to different varieties of jute significantly increased the MDA, H_2O_2, and EL in the roots and leaves of different varieties of jute (Figure 1). The contents of MDA, H_2O_2, and EL were higher in the roots when compared to the above-ground parts of the plant. According to the results, maximum contents of MDA (17.7 ± 0.8 μmoles g^{-1}), H_2O_2 (539 ± 6 μmoles g^{-1}), and EL (70 ± 2%) were observed in the roots of ShangHuoMa. Similarly, in the shoots, the maximum contents of MDA (11.9 ± 0.2 μmoles g^{-1}), H_2O_2 (445 ± 5 μmoles g^{-1}), and EL (56 ± 2%) were also observed in of ShangHuoMa compared to other varieties of jute.

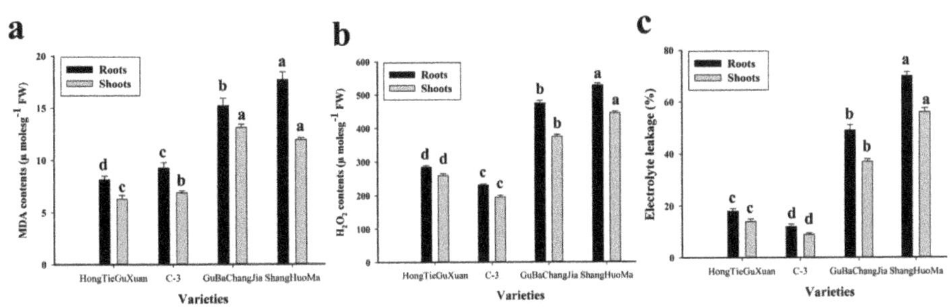

Figure 1. Effect of Cu stress on malondialdehyde (MDA) (**a**), H_2O_2 (**b**) and electrolyte leakage (EL) (**c**) in different varieties of jute. Values in the figures represent just one harvest. Different letters within a column indicate significant differences between the treatments ($p < 0.05$). Bars indicate the mean ± SD ($n = 3$).

3.3. *Effect of Cu Toxicity on Antioxidant Activities*

In the present study, the antioxidants superoxidase dismutase (SOD), peroxidase (POD), catalase (CAT), and ascorbate peroxidase (APX) were also investigated in different varieties of jute under toxic concentrations of Cu in the medium (Figure 2). Activities of antioxidant were higher in GuBaChangaJia and ShangHuoMa than in C-3 and HongTieGuXuan. The maximum activity of SOD (90 ± 1.4 U g^{-1} FW), POD (3400 ± 138 U g^{-1} FW), CAT (222 ± 5 U g^{-1} FW), and APX (441 ± 4 U g^{-1} FW) was observed in the roots of ShangHuoMa. Similarly, in the shoots, the maximum activity of SOD (86 ± 1.4 U g^{-1} FW), POD (2920 ± 23 U g^{-1} FW), CAT (191 ± 3 U g^{-1} FW), and APX (375 ± 4 U g^{-1} FW) were also observed in ShangHuoMa.

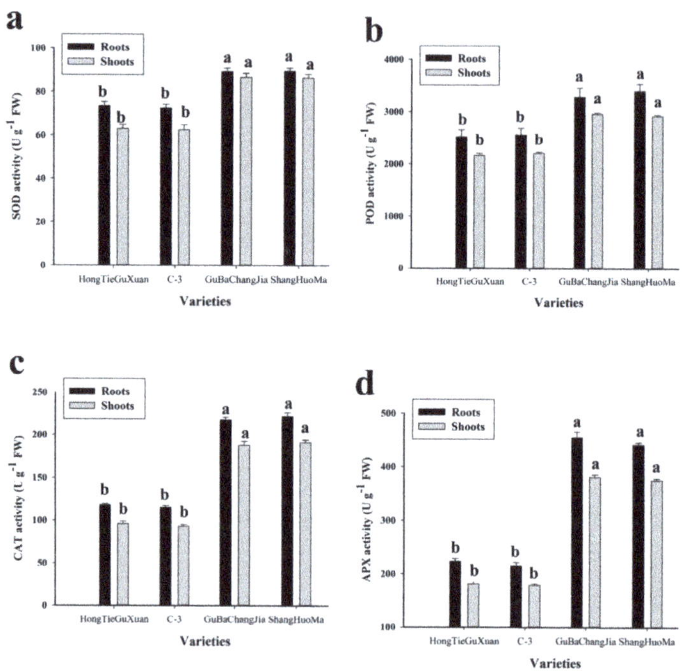

Figure 2. Effect of Cu stress on superoxidase dismutase (SOD) (**a**), peroxidase (POD) (**b**), catalase (CAT) (**c**), and ascorbate peroxidase (APX) (**d**) in different varieties of jute. Values in the figures represent just one harvest. Different letters within a column indicate significant differences between the treatments ($p < 0.05$). Bars indicate the mean ± SD ($n = 3$).

3.4. Cu Uptake and Bioaccumulation in the Roots and Shoots

In this study, the concentrations of Cu from different parts (roots and shoots) of jute varieties were also determined under toxic concentrations of Cu in the medium. The concentrations of Cu in the roots and shoots of different varieties of jute are presented in Table 2. These results suggested that Cu concentration was higher in the varieties that exhibited significantly better growth than the varieties that were more affected by Cu stress. The concentration of Cu in the roots ranged from 37 to 60 mg kg^{-1}, while in the shoots, Cu concentration ranged from 38 to 61 mg kg^{-1} (Table 2). The maximum Cu was accumulated in the roots of C-3 (60 ± 0.8 mg kg^{-1}), followed by HongTieGuXuan (56 ± 1.4 mg kg^{-1}) and GuBaChangaJia (40 ± 0.8 mg kg^{-1}). Similarly, in the shoots, maximum Cu concentration was accumulated in C-3 (61 ± 0.9 mg kg^{-1}), followed by HongTieGuXuan (57 ± 1.1 mg kg^{-1}) and GuBaChangaJia (41 ± 0.6 mg kg^{-1}).

Table 2. Accumulation of Cu in roots (mg kg^{-1}) and shoots (mg kg^{-1}) of different varieties of jute.

Varieties	Cu Concentration in Roots	Cu Concentration in Shoots
HongTieGuXuan	56 ± 1.4 b	57 ± 1.1 b
C-3	60 ± 0.8 a	61 ± 0.9 a
GuBaChangaJia	40 ± 0.8 c	41 ± 0.6 c
ShangHuoMa	37 ± 1 d	38 ± 0.9 d

Values in the table represent just one harvest (mean ± SD) (n =3). Different letters within a column indicate significant differences between the treatments ($p < 0.05$). Relative radiance of plastic filter used: HongTieGuXuan, C-3, GuBaChangaJia, and ShangHuoMa.

In this study, bioaccumulation factor (BAF) and translocation factor (TF) were also calculated (Figure 3). It was noticed that the values of BAF and TF of HongTieGuXuan and C-3 were greater than 1 while the values of BAF and TF of GuBaChangaJia and HongTieGuXuan were less than 1 (Figure 3). The maximum BAF value was shown by C-3 in the roots (1.19) and shoots (1.20), compared to other varieties of jute. Contrastingly, the minimum BAF value was shown by HongTieGuXuan in the roots (0.74) and shoots (0.74), compared to other varieties of jute. Similarly, the maximum value of TF was also shown by C-3 (1.01), followed by HongTieGuXuan (1.01) and GuBaChangaJia (0.99).

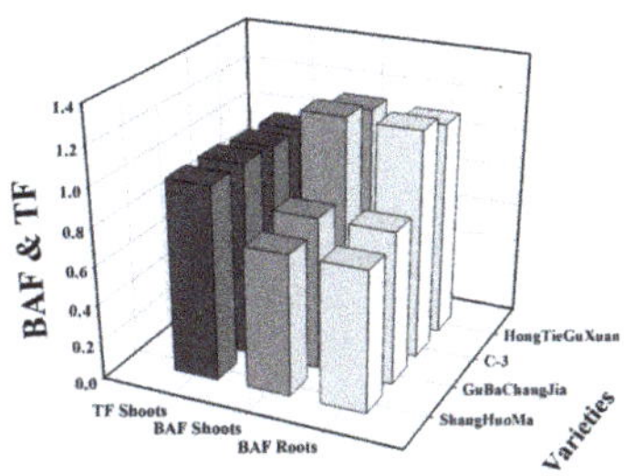

Figure 3. Effect of Cu stress on bioaccumulation factor (BAF) and translocation factor (TF) in different varieties of jute.

3.5. Correlation between Growth, Biomass, Total Chlorophyll Content, and Cu Uptake

A Pearson's correlation analysis was carried out to quantify the relationship between growth, biomass, total chlorophyll content, and Cu uptake in the roots and shoots of different varieties of jute (Figure 4). Cu concentration in the roots was positively correlated with Cu concentration in the shoots, and also positively correlated with growth parameters and chlorophyll content. In the same way, Cu concentration in the shoots was also strongly positively correlated with plant height, fresh and dry biomass, and chlorophyll content. This correlation reflected the close connection between Cu uptake and growth in different varieties of jute.

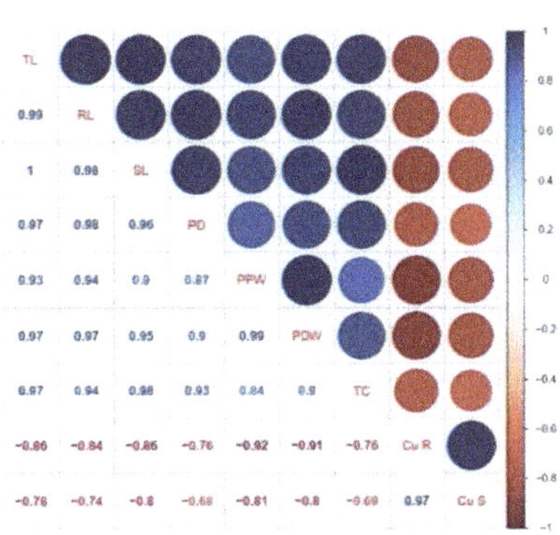

Figure 4. Correlation of Cu uptake, with growth parameters and total chlorophyll content. Cu R: Cu concentration in the roots; Cu S: Cu concentration in the shoots; SL: shoot length; RL: root length; TL: total length; PD: plant diameter; PFW: plant fresh weight; PDW: plant dry weight; and TC: total chlorophyll content.

4. Discussions

In the last few decades, soil concentrations of Cu have surpassed toxic levels (>30 mg kg^{-1}) due to overpopulation and industrialization [5,6,15,38]. It has been observed that a reduction in plant growth and biomass in a common response of plants to Cu stress [4,8,39]. For the successful production of jute when cultivated in the Cu-contaminated soil of China, development and selection of tolerant varieties through screening is crucial. For better growth and development of jute for fiber extraction and natural resources, it is necessary to cultivate the most suitable variety of jute [25,26]. However, the resistance or tolerance mechanism of a plant depend upon its specific physiological and biochemical activities [14,40,41]. Therefore, a preliminary experiment was conducted to expose Cu-sensitive varieties and Cu-resistant varieties to toxic concentrations of Cu in the medium.

The germination assay is a basic method to observe the effects of heavy metal stress on plant seedlings. Moreover, seed germination is one of the most important biological parameters in the life cycle of a plant [10,16,42,43]. Seed germination of different varieties of jute exhibited different responses to exposure to Cu stress via the medium (Table 1). It was also noticed that HongTieGuXuan and C-3 showed the maximum germination percentage, while GuBaChangaJia and ShangHuoMa showed the minimum germination percentage, which might be due to high toxicity caused by the Cu in the medium. Previously, it has also been suggested that germination percentage at seedling stage can be affected by the phytotoxicity of Cu, which might be due to the accumulation of carbon partitioning in the tissues of the plant [10,16]. Furthermore, the minimum germination rate in GuBaChangaJia and ShangHuoMa might be due to water deficiency due to excess Cu in the medium inhibiting cell expansion and reducing carbon assimilation [10,44]. These results coincided with Nizam et al. [42], who found that high concentrations of As reduced the germination percentage of some varieties, while others showed a better germination rate compared to the control.

High concentrations of Cu are extremely toxic for growth and biomass of plants. It was noticed that dominant plants collected from Cu mining sites showed more resistance than normal plants [8,39]. In the present study, HongTieGuXuan and C-3 showed better growth in terms of plant height and plant fresh and dry biomass compared to GuBaChangaJia and ShangHuoMa (Table 1). The poor growth and composition in GuBaChangaJia and ShangHuoMa under high concentrations of Cu in the medium can be attributed to insufficient uptake of nutrients, low availability of water, perturbed root architecture, and poor stomata regulation of plant metabolic processes [45]. Furthermore, nutrient acquisition, stimulation of the defense system (antioxidants), structural integrity of metabolites, and considerable water use efficiency are positively associated with heavy metal stress tolerance in different plant species [10,46,47]. Similar findings were shown by Uddin et al. [48] when they studied different varieties of jute and noticed that different varieties exhibited different responses to exposure to high concentrations of Pb in their soil.

Copper is a micronutrient that provides support to the shaping and function of chloroplasts in plant cells [4], but excess Cu affects chloroplast structure and ultimately reduces the photosynthetic pigment of the plants [49,50]. In the present study, excess Cu in the medium caused a drastic reduction in chlorophyll content in Cu-sensitive varieties (Table 1). The reduction of chlorophyll content in GuBaChangaJia and ShangHuoMa under toxic concentration of Cu in the medium might be due to the inhibited activities of various enzymes associated with chlorophyll biosynthesis [11,50]. Furthermore, the accumulation of Cu concentrations in the tissues of GuBaChangaJia and ShangHuoMa also caused low chlorophyll content in the leaves of these jute varieties.

Heavy metal stress can alter the equilibrium of reactive oxygen species (ROS) production, which promotes membrane lipid peroxidation and ROS accumulation, and disturbs the function and structure of the cell membrane [20,23,39]. ROS have also been shown to play a role in ABA mediated root growth in Arabidopsis [50,51]. Moreover, the generation of ROS under high concentrations of Cu in the soil is enhanced by cuprous and cupric Cu ions, which induce oxidative damage in plant cells/tissues [9,18,52]. MDA is an oxidized product of membrane lipids and is supported by leakage of plasma membrane and cell turgor loss [39,53]. The production of ROS in plant cells/tissues is very dangerous, and plants have evolved special defense systems such as SOD, POD, CAT, and APX to scavenge ROS [7,8,14]. Plant responses to oxidative stress depend on the plant species and variety [47,54]. It was noticed that Cu-sensitive (GuBaChangaJia and ShangHuoMa) varieties underwent high oxidative stress, as shown by high contents of MDA, H_2O_2, and EL (Figure 1). Similar results were suggested by Khan et al. [47]:

that tolerant varieties undergo less oxidative stress when compared to sensitive varieties. Furthermore, it was also observed that to overcome oxidative stress, plants have a special defense system to scavenge ROS generation (Figure 2). GuBaChangaJia and ShangHuoMa showed more antioxidant activity than HongTieGuXuan and C-3, which might be due to more oxidative stress in these species. The difference in antioxidant activities might be due to species-specific biochemical responses, as shown by Akram et al. [45].

Metal accumulation in different parts of a plant depends upon plant species, growth stage, fertilizer application, and growth conditions [4,7]. Based on tolerance mechanisms, plant species can been divided into two types: (1) Metal excluders accumulate heavy metals from the substrate in their roots, but restrict their transport and entry into their aerial parts; (2). Hyperaccumulators are able to accumulate large amounts of metals in their above-ground parts rather than in belowground parts [55]. Furthermore, bioaccumulation factor (BAF) and translocation factor (TF) are important in screening hyperaccumulators for phytoremediation of heavy metals. Screening of hyperaccumulators depends upon BAF and TF (both should be greater than 1) for evaluation and selection of plants for phytoremediation [8,25,56–58]. In many previous studies, jute has been considered as a hyperaccumulator species for different heavy metals; for example, Ahmed and Salima [25] studied jute under different heavy metal exposures and noticed that is a hyperaccumulator for different heavy metals including Cd, Cu, Cr, and Pb. Our results suggested that HongTieGuXuan and C-3 are able to accumulate large amounts of Cu in their above-ground parts and can be considered hyperaccumulator varieties, while GuBaChangaJia and ShangHuoMa are able to accumulate low Cu concentrations in their above-ground parts and can be considered Cu-excluder species. Furthermore, the values of BAF and TF of HongTieGuXuan and C-3 were greater than 1, while the values of BAF and TF for GuBaChangaJia and ShangHuoMa were less than 1. Phytoremediation potential of different varieties of jute has already been studied by Uddin et al. [48] under lead-contaminated soil, and it was found the similar results that Pb-tolerant varieties were good hyperaccumulator species (showed values of BAF and TF greater than 1) and were able to accumulate large amount of Pb in their above-ground parts rather than in the below-ground parts of the plant.

5. Conclusions

Overall, Cu stress significantly decreased growth and development (seed germination, plant height, fresh and dry biomass, chlorophyll content) in all varieties of jute, but Cu tolerance index was higher in HongTieGuXuan and C-3 than in GuBaChangaJia and ShangHuoMa. However, jute has an active antioxidative defense system to scavenge the ROS produced due to high concentrations of Cu in the medium. The results suggested that HongTieGuXuan and C-3 are Cu-tolerant varieties, while GuBaChangaJia and ShangHuoMa are Cu-sensitive varieties. Furthermore, the accumulation of Cu in different parts (roots and shoots) of the plants also indicated that HongTieGuXuan and C-3 can be considered Cu hyperaccumulator species, while GuBaChangaJia and ShangHuoMa can be considered Cu excluder species. However, future research is needed to study the effects of Cu stress on different varieties of jute for fiber extract in field areas.

Author Contributions: Conceptualization, M.H.S., S.A., M.F.S. and L.L.; Data curation, M.H.S., M.R. (Muhammad Rizwan) and N.A.A.; Formal analysis, M.H.S. and M.A.; Funding acquisition, M.F.S. and M.A.; Investigation, M.R. (Muzammal Rehman) and M.M.; Methodology, M.H.S., S.A., M.F.S. and I.A.-A.; Project administration, M.F.S. and M.R. (Muzammal Rehman); Resources, M.F.S. and L.L.; Software, M.R. (Muhammad Rizwan), M.R. (Muzammal Rehman) and M.A.; Supervision, S.A. and L.L.; Validation, M.R. (Muhammad Rizwan), N.A.A., I.A.-A. and M.M.; Visualization, M.F.S., M.R. (Muhammad Rizwan), M.R. (Muzammal Rehman) and M.M.; Writing—original draft, M.H.S. and N.A.A.; Writing—review and editing, S.A., M.F.S., M.A., L.L. and I.A.-A.

Funding: This research was supported by China Agriculture Research System project (CARS-16-E10) and the National Natural Science Foundation of China (31571717). The authors extend their appreciation to the Deanship of Scientific Research at King Saud University for funding this work through Research Group No. (RG-1441-301). The authors thank the Researchers Support & Services Unit at King Saud University for their technical support.

Conflicts of Interest: The authors declare that there is no conflict of interest.

References

1. Vardhan, K.H.; Kumar, P.S.; Panda, R.C. A review on heavy metal pollution, toxicity and remedial measures: Current trends and future perspectives. *J. Mol. Liq.* **2019**, *290*, 111197. [CrossRef]
2. Daud, M.K.; Ali, S.; Abbas, Z.; Zaheer, I.E.; Riaz, M.A.; Malik, A.; Hussain, A.; Rizwan, M.; Zia-ur-Rehman, M.; Zhu, S.J. Potential of duckweed (*Lemna minor*) for the Phytoremediation of Landfill Leachate. *J. Chem.* **2018**, *2018*. [CrossRef]
3. Nagajyoti, P.C.; Lee, K.D.; Sreekanth, T. Heavy metals, occurrence and toxicity for plants: A review. *Environ. Chem. Let.* **2010**, *8*, 199–216. [CrossRef]
4. Rehman, M.; Liu, L.; Wang, Q.; Saleem, M.H.; Bashir, S.; Ullah, S.; Peng, D. Copper environmental toxicology, recent advances, and future outlook: A review. *Environ. Sci. Pol. Res.* **2019**, *26*, 18003–18016. [CrossRef] [PubMed]
5. Li, L.; Zhang, K.; Gill, R.A.; Islam, F.; Farooq, M.A.; Wang, J.; Zhou, W. Ecotoxicological and Interactive Effects of Copper and Chromium on Physiochemical, Ultrastructural, and Molecular Profiling in *Brassica napus* L. *BioMed Res. Inter.* **2018**, *2018*. [CrossRef]
6. Adrees, M.; Ali, S.; Rizwan, M.; Ibrahim, M.; Abbas, F.; Farid, M.; Zia-ur-Rehman, M.; Irshad, M.K.; Bharwana, S.A. The effect of excess copper on growth and physiology of important food crops: A review. *Environ. Sci. Pol. Res.* **2015**, *22*, 8148–8162. [CrossRef]
7. Rehman, M.; Maqbool, Z.; Peng, D.; Liu, L. Morpho-physiological traits, antioxidant capacity and phytoextraction of copper by ramie (*Boehmeria nivea* L.) grown as fodder in copper-contaminated soil. *Environ. Sci. Pol. Res.* **2019**, *26*, 5851–5861. [CrossRef]
8. Chen, J.; Shafi, M.; Li, S.; Wang, Y.; Wu, J.; Ye, Z.; Peng, D.; Yan, W.; Liu, D. Copper induced oxidative stresses, antioxidant responses and phytoremediation potential of Moso bamboo (*Phyllostachys pubescens*). *Sci. Rep.* **2015**, *5*, 13554. [CrossRef]
9. Liu, J.; Wang, J.; Lee, S.; Wen, R. Copper-caused oxidative stress triggers the activation of antioxidant enzymes via *ZmMPK3* in maize leaves. *PLoS ONE* **2018**, *13*, e0203612. [CrossRef]
10. Ahsan, N.; Lee, D.-G.; Lee, S.-H.; Kang, K.Y.; Lee, J.J.; Kim, P.J.; Yoon, H.-S.; Kim, J.-S.; Lee, B.-H. Excess copper induced physiological and proteomic changes in germinating rice seeds. *Chemosphere* **2007**, *67*, 1182–1193. [CrossRef] [PubMed]
11. Rehman, M.; Liu, L.; Bashir, S.; Saleem, M.H.; Chen, C.; Peng, D.; Siddique, K.H. Influence of rice straw biochar on growth, antioxidant capacity and copper uptake in ramie (*Boehmeria nivea* L.) grown as forage in aged copper-contaminated soil. *Plant Phy. Biochem.* **2019**, *138*, 121–129. [CrossRef] [PubMed]
12. Gao, G.; Liu, Y.; Li, X.; Feng, Z.; Xu, Z.; Wu, H.; Xu, J. Expected CO_2-induced ocean acidification modulates copper toxicity in the green tide alga *Ulva prolifera*. *Environ. Exp. Bot.* **2017**, *135*, 63–72. [CrossRef]
13. Habiba, U.; Ali, S.; Farid, M.; Shakoor, M.B.; Rizwan, M.; Ibrahim, M.; Abbasi, G.H.; Hayat, T.; Ali, B. EDTA enhanced plant growth, antioxidant defense system, and phytoextraction of copper by *Brassica napus* L. *Environ. Sci. Pol. Res.* **2015**, *22*, 1534–1544. [CrossRef] [PubMed]
14. Zaheer, I.E.; Ali, S.; Rizwan, M.; Farid, M.; Shakoor, M.B.; Gill, R.A.; Najeeb, U.; Iqbal, N.; Ahmad, R. Citric acid assisted phytoremediation of copper by *Brassica napus* L. *Ecotox. Environ. Safe.* **2015**, *120*, 310–317. [CrossRef]
15. Nair, P.M.G.; Chung, I.M. Study on the correlation between copper oxide nanoparticles induced growth suppression and enhanced lignification in Indian mustard (*Brassica juncea* L.). *Ecotox. Environ. Safe.* **2015**, *113*, 302–313. [CrossRef]
16. Ashagre, H.; Shelema, M.; Kedir, R.; Ebsa, S. Seed germination and seedling growth of haricot bean (*Phaseolus vulgaris* L.) cultivars as influenced by copper sulphate. *World J. Agric. Sci.* **2013**, *1*, 312–317.
17. Farooq, M.A.; Li, L.; Ali, B.; Gill, R.A.; Wang, J.; Ali, S.; Gill, M.B.; Zhou, W. Oxidative injury and antioxidant enzymes regulation in arsenic-exposed seedlings of four *Brassica napus* L. cultivars. *Environ. Sci. Pol. Res.* **2015**, *22*, 10699–10712. [CrossRef]
18. Quartacci, M.; Ranieri, A.; Sgherri, C. Antioxidative defence mechanisms in two grapevine (*Vitis vinifera* L.) cultivars grown under boron excess in the irrigation water. *VITIS-J. Grape. Res.* **2015**, *54*, 51–58.
19. Das, K.; Roychoudhury, A. Reactive oxygen species (ROS) and response of antioxidants as ROS-scavengers during environmental stress in plants. *Front. Environ. Sci.* **2014**, *2*, 53. [CrossRef]

20. Sun, B.-Y.; Kan, S.-H.; Zhang, Y.-Z.; Deng, S.-H.; Wu, J.; Yuan, H.; Qi, H.; Yang, G.; Li, L.; Zhang, X.-H. Certain antioxidant enzymes and lipid peroxidation of radish (*Raphanus sativus* L.) as early warning biomarkers of soil copper exposure. *J. Hazard. Mater.* **2010**, *183*, 833–838. [CrossRef]
21. Ashraf, S.; Ali, Q.; Zahir, Z.A.; Ashraf, S.; Asghar, H.N. Phytoremediation: Environmentally sustainable way for reclamation of heavy metal polluted soils. *Ecotox. Environ. Saf.* **2019**, *174*, 714–727. [CrossRef] [PubMed]
22. Lajayer, B.A.; Moghadam, N.K.; Maghsoodi, M.R.; Ghorbanpour, M.; Kariman, K. Phytoextraction of heavy metals from contaminated soil, water and atmosphere using ornamental plants: Mechanisms and efficiency improvement strategies. *Environ. Sci. Pollut. Res.* **2019**, *26*, 8468–8484. [CrossRef] [PubMed]
23. Li, F.; Qi, J.; Zhang, G.; Lin, L.; Fang, P.; Tao, A.; Xu, J. Effect of cadmium stress on the growth, antioxidative enzymes and lipid peroxidation in two kenaf (*Hibiscus cannabinus* L.) plant seedlings. *J. Integr. Agric.* **2013**, *12*, 610–620. [CrossRef]
24. Hosman, M.E.; El-Feky, S.S.; Mohamed, M.; Shaker, E.M. Mechanism of phytoremediation potential of flax (*Linum usitatissimum* L.) to Pb, Cd and Zn. *Asian J. Plant Sci. Res.* **2017**, *7*, 30–40.
25. Ahmed, D.A.; Slima, D.F. Heavy metal accumulation by *Corchorus olitorius* L. irrigated with wastewater. *Environ. Sci. Pollut. Res.* **2018**, *25*, 14996–15005. [CrossRef]
26. Ogoko, E. Accumulation of heavy metal in soil and their transfer to leafy vegetables with phytoremediation potential. *Am. J. Chem.* **2015**, *5*, 125–131.
27. Abubakari, M.; Moomin, A.; Nyarko, G.; Dawuda, M. Heavy metals concentrations and risk assessment of roselle and jute mallow cultivated with three compost types. *Ann. Agric. Sci.* **2017**, *62*, 145–150. [CrossRef]
28. Faruk, O.; Bledzki, A.K.; Fink, H.-P.; Sain, M. Biocomposites reinforced with natural fibers: 2000–2010. *Prog. Polym. Sci.* **2012**, *37*, 1552–1596. [CrossRef]
29. Nath, K.; Saini, S.; Sharma, Y.K. Chromium in tannery industry effluent and its effect on plant metabolism and growth. *J. Environ. Biol.* **2005**, *26*, 197–204.
30. Young, H.H.; Johnson, F.P. *Practice of Urology: Based on a Study of 12,500 Cases*; WB Saunders: Philadelphia, PA, USA, 1926.
31. Arnon, D.I. Copper enzymes in isolated chloroplasts. Polyphenoloxidase in *Beta vulgaris*. *Plant Physiol.* **1949**, *24*, 1. [CrossRef]
32. Heath, R.L.; Packer, L. Photoperoxidation in isolated chloroplasts: I. Kinetics and stoichiometry of fatty acid peroxidation. *Arch. Biochem. Biophys.* **1968**, *125*, 189–198. [CrossRef]
33. Dionisio-Sese, M.L.; Tobita, S. Antioxidant responses of rice seedlings to salinity stress. *Plant Sci.* **1998**, *135*, 1–9. [CrossRef]
34. Chen, C.-N.; Pan, S.-M. Assay of superoxide dismutase activity by combining electrophoresis and densitometry. *Bot. Bull. Acad. Sin.* **1996**, *37*, 107–111.
35. Sakharov, I.Y.; Ardila, G.B. Variations of peroxidase activity in cocoa (*Theobroma cacao* L.) beans during their ripening, fermentation and drying. *Food Chem.* **1999**, *65*, 51–54. [CrossRef]
36. Aebi, H. Catalase in vitro. In *Methods in Enzymology*; Elsevier: Amsterdam, The Netherlands, 1984; pp. 121–126.
37. Nakano, Y.; Asada, K. Hydrogen peroxide is scavenged by ascorbate-specific peroxidase in spinach chloroplasts. *Plant Cell Physiol.* **1981**, *22*, 867–880.
38. Celis-Plá, P.S.; Brown, M.T.; Santillán-Sarmiento, A.; Korbee, N.; Sáez, C.A.; Figueroa, F.L. Ecophysiological and metabolic responses to interactive exposure to nutrients and copper excess in the brown macroalga *Cystoseira tamariscifolia*. *Mar. Pollut. Bull.* **2018**, *128*, 214–222. [CrossRef]
39. Thounaojam, T.C.; Panda, P.; Mazumdar, P.; Kumar, D.; Sharma, G.; Sahoo, L.; Sanjib, P. Excess copper induced oxidative stress and response of antioxidants in rice. *Plant Physiol. Biochem.* **2012**, *53*, 33–39. [CrossRef]
40. Kamran, M.; Malik, Z.; Parveen, A.; Zong, Y.; Abbasi, G.H.; Rafiq, M.T.; Shaaban, M.; Mustafa, A.; Bashir, S.; Rafay, M. Biochar alleviates Cd phytotoxicity by minimizing bioavailability and oxidative stress in pak choi (*Brassica chinensis* L.) cultivated in Cd-polluted soil. *J. Environ. Manag.* **2019**, *250*, 109500. [CrossRef]
41. Kamran, M.; Malik, Z.; Parveen, A.; Huang, L.; Riaz, M.; Bashir, S.; Mustafa, A.; Abbasi, G.H.; Xue, B.; Ali, U. Ameliorative Effects of Biochar on Rapeseed (*Brassica napus* L.) Growth and Heavy Metal Immobilization in Soil Irrigated with Untreated Wastewater. *J. Plant Growth Regul.* **2019**, 1–16. [CrossRef]

42. Uddin Nizam, M.; Mokhlesur Rahman, M.; Kim, J.-E. Phytoremediation Potential of Kenaf (*Hibiscus cannabinus* L.), Mesta (*Hibiscus sabdariffa* L.), and Jute (*Corchorus capsularis* L.) in Arsenic-contaminated Soil. *Korean J. Environ. Agric.* **2016**, *35*, 111–120. [CrossRef]
43. Gang, A.; Vyas, A.; Vyas, H. Toxic effect of heavy metals on germination and seedling growth of wheat. *J. Environ. Res. Dev.* **2013**, *8*, 206.
44. Singh, K.P.; Mohan, D.; Sinha, S.; Dalwani, R. Impact assessment of treated/untreated wastewater toxicants discharged by sewage treatment plants on health, agricultural, and environmental quality in the wastewater disposal area. *Chemosphere* **2004**, *55*, 227–255. [CrossRef] [PubMed]
45. Akram, N.A.; Iqbal, M.; Muhammad, A.; Ashraf, M.; Al-Qurainy, F.; Shafiq, S. Aminolevulinic acid and nitric oxide regulate oxidative defense and secondary metabolisms in canola (*Brassica napus* L.) under drought stress. *Protoplasma* **2018**, *255*, 163–174. [CrossRef] [PubMed]
46. Schmidthoffer, I.; Szilák, L.; Molnár, P.; Csontos, P.; Skribanek, A. Drought tolerance of European barley (*Hordeum vulgare* L.) varieties. *Agriculture (Pol'nohospodárstvo)* **2018**, *64*, 137–142. [CrossRef]
47. Khan, M.N.; Zhang, J.; Luo, T.; Liu, J.; Ni, F.; Rizwan, M.; Fahad, S.; Hu, L. Morpho-physiological and biochemical responses of tolerant and sensitive rapeseed cultivars to drought stress during early seedling growth stage. *Acta Physiol. Plant.* **2019**, *41*, 25. [CrossRef]
48. Uddin, M.N.; Wahid-Uz-Zaman, M.; Rahman, M.M.; Islam, M.S.; Islam, M.S. Phytoremediation Potentiality of Lead from Contaminated Soils by Fibrous Crop Varieties. *Am. J. Appl. Sci. Res.* **2016**, *2*, 22. [CrossRef]
49. Wodala, B.; Eitel, G.; Gyula, T.; Ördög, A.; Horváth, F. Monitoring moderate Cu and Cd toxicity by chlorophyll fluorescence and P 700 absorbance in pea leaves. *Photosynthetica* **2012**, *50*, 380–386. [CrossRef]
50. Ma, X.; Zhang, X.; Yang, L.; Tang, M.; Wang, K.; Wang, L.; Bai, L.; Song, C. Hydrogen peroxide plays an important role in PERK4-mediated abscisic acid-regulated root growth in Arabidopsis. *Funct. Plant Biol.* **2018**, *46*, 165–174. [CrossRef]
51. Liu, H.; Guo, S.; Lu, M.; Zhang, Y.; Li, J.; Wang, W.; Wang, P.; Zhang, J.; Hu, Z.; Li, L.; et al. Biosynthesis of DHGA12 and its roles in Arabidopsis seedling establishment. *Nat. Commun.* **2019**, *10*, 1768. [CrossRef]
52. Brahim, L.; Mohamed, M. Effects of copper stress on antioxidative enzymes, chlorophyll and protein content in *Atriplex halimus*. *Afr. J. Biotech.* **2011**, *10*, 10143–10148.
53. Farooq, M.A.; Ali, S.; Hameed, A.; Ishaque, W.; Mahmood, K.; Iqbal, Z. Alleviation of cadmium toxicity by silicon is related to elevated photosynthesis, antioxidant enzymes; suppressed cadmium uptake and oxidative stress in cotton. *Ecotoxicol. Environ. Saf.* **2013**, *96*, 242–249. [CrossRef] [PubMed]
54. Jiang, J.; Qin, C.; Shu, X.; Chen, R.; Song, H.; Li, Q.; Xu, H. Effects of copper on induction of thiol-compounds and antioxidant enzymes by the fruiting body of *Oudemansiella radicata*. *Ecotoxicol. Environ. Saf.* **2015**, *111*, 60–65. [CrossRef] [PubMed]
55. Muszynska, E.; Hanus-Fajerska, E. Why are heavy metal hyperaccumulating plants so amazing? *BioTechnol. J. Biotech. Comput. Biol. Bionanotech.* **2015**, *96*, 265–271. [CrossRef]
56. Saleem, M.H.; Fahad, S.; Khan, S.U.; Ahmar, S.; Khan, M.H.U.; Rehman, M.; Maqbool, Z.; Liu, L. Morpho-physiological traits, gaseous exchange attributes, and phytoremediation potential of jute (*Corchorus capsularis* L.) grown in different concentrations of copper-contaminated soil. *Ecotoxicol. Environ. Safe.* **2019**. [CrossRef] [PubMed]
57. Jan, R.; Khan, M.A.; Asaf, S.; Lee, I.-J.; Kim, K.M. Metal resistant endophytic bacteria reduces cadmium, nickel toxicity, and enhances expression of metal stress related genes with improved growth of *Oryza Sativa*, via regulating its antioxidant machinery and endogenous hormones. *Plants* **2019**, *8*, 363. [CrossRef]
58. Li, L.; Hou, M.; Cao, L.; Xia, Y.; Shen, Z.; Hu, Z.J.E. Glutathione S-transferases modulate Cu tolerance in *Oryza sativa*. *Environ. Exp. Bot.* **2018**, *155*, 313–320. [CrossRef]

plants

MDPI

Article

Protein Carbonylation As a Biomarker of Heavy Metal, Cd and Pb, Damage in *Paspalum fasciculatum* Willd. ex Flüggé

Manuel Salas-Moreno [1,2], Neyder Contreras-Puentes [2], Erika Rodríguez-Cavallo [2], Jesús Jorrín-Novo [3], José Marrugo-Negrete [4,*] and Darío Méndez-Cuadro [2,*]

1 Biosistematic Research Group, Biology Department, Faculty of Naturals Sciences, Technological University of Chocó, Quibdó, Colombia; hasamo49@hotmail.com

2 Analytical Chemistry and Biomedicine Group, Exacts and Natural Sciences Faculty, University of Cartagena, 130015 Cartagena, Colombia; neidercontreras11@gmail.com (N.C.-P.); erodriguezc1@unicartagena.edu.co (E.R.-C.)

3 Department of Biochemistry and Molecular Biology, University of Cordoba, 14014 Cordoba, Spain; bf1jonoj@uco.es

4 Universidad de Córdoba, Carrera 6 No. 77-305, Montería, Córdoba, Colombia

* Correspondence: jmarrugo@correo.unicordoba.edu.co (J.M.-N.); dmendezc@unicartagena.edu.co (D.M.-C.); Tel.: +57-3233016454 (J.M.-N.); +57-3015584887 (D.M.-C.)

Received: 22 October 2019; Accepted: 14 November 2019; Published: 16 November 2019

Abstract: Heavy metal tolerant plants have phytoremediation potential for the recovery of contaminated soils, and the characterization of their metabolic adaptation processes is an important starting point to elucidate their tolerance mechanisms at molecular, biochemical and physiological levels. In this research, the effects of Cd and Pb on growth and protein carbonylation in tissues of *Paspalum fasciculatum* exposed to 30 and 50 mg·Kg^{-1} Cd and Pb respectively were determined. *P. fasciculatum* seedlings exposed to metals grew more than controls until 60 days of cultivation and limited their oxidative effects to a reduced protein group. Carbonyl indexes in leaf and root proteins reached a significant increase concerning their controls in plants exposed 30 days to Cd and 60 days to Pb. From the combined approach of Western Blot with Sodium Dodecyl Sulfate-Polyacrylamide Gel Electrophoresis (SDS-PAGE) and protein analysis by Matrix Asisted Laser Desorption/Ionisation-Time Of Flight (MALDI-TOF/TOF) mass spectrometry, chloroplastic proteins were identified into the main oxidative stress-inducible proteins to Cd and Pb, such as subunits α, γ of ATP synthetase, Chlorophyll CP26 binding protein, fructose-bisphosphate aldolase and long-chain ribulose bisphosphate carboxylase (RuBisCO LSU). Cd generated damage in the photosynthetic machinery of the leaves of *P. fasciculatum* into the first 30 days of treatment; five of the oxidized proteins are involved in photosynthesis processes. Moreover, there was a proteolytic fragmentation of the RuBisCO LSU. Results showed that intrinsic tolerance of *P. fasciculatum* to these metals reached 60 days in our conditions, along with the bioaccumulating appreciable quantities of metals in their roots.

Keywords: *P. fasciculatum*; heavy metals; tolerant plant; protein carbonylation; photosynthesis proteins; mining soils

1. Introduction

Cadmium (Cd) and lead (Pb) are toxic elements, which at high concentrations have various effects on the biochemistry, morphology and physiology of the plants. Lead toxicity in plants includes a wide range of metabolic alterations, damage to biomolecules, changes in membrane permeability, inhibition in activities of many enzymes, reduction in photosynthesis and transpiration and enhanced generation of reactive oxygen species (ROS) [1]. While oxidative stress induced by Cd can hinder physiological processes such as photosynthesis and respiration by inhibition of defense antioxidants systems [2,3], chlorophyll content reduction and decreases the activity of enzymes involved in CO_2 fixation [4]. In leaves, its toxic effects include chlorosis, photosynthesis inhibition, structural and functional damages of photosystem II, disturb of mineral nutrition, intracellular redox equilibrium and water balance [2,5].

Oxidative stress promoted by Cd and Pb induces structural and functional alterations on target proteins [6]; thus, for example, induction of modification in amino-acid side chains and the main residues such as histidine, arginine and lysine, which are converted to aldehyde or keto groups [7]. The increase in the amount of these carbonyl groups per protein molecule is called protein carbonylation; this damage is irreversible because the cells are incapable to repair changes in these proteins through enzymatic reactions [7]. Carbonylation is frequently used as an oxidative stress biomarker, which is a non-enzymatic post-translational modification. Heavy metals are indirect causes of oxidative stress in plants, which is a direct cause of increased carbonylation of proteins; this has been observed in leaves exposed at Cd of species like *Arabidopsis thaliana* [8].

Plants have a series of strategies for heavy metal detoxification, thus reducing the adverse effects of exposure and their accumulation; this implies a complex response at molecular, biochemical, physiological and cellular levels [9]. Recent advances in the comprehension of tolerance mechanisms have been reached with proteomics, metabolomics and transcriptomics studies [10].

In particular, some species of genus *Paspalum* have been assayed by its potential in metals remediation such as Cd and Pb, where they have shown moderate uptake capabilities [11]. Paspalum species are distributed in tropical and subtropical regions of America, with few species in Africa and Asia. The genus presents a greater species diversity in South American countries such as Brazil, Paraguay, Uruguay, Colombia, and Argentina. In Colombia, 91 species of the genus Paspalum have been recorded [12], of which more than 20 of these species are disseminated in the Colombian Caribbean region [13], including those as *Paspalum fasciculatum* that grows in mining soils or flooded soils with a high metal load.

P. fasciculatum Willd. ex Flüggé is a plant with phytoremediation potential, Cd-phytostabilizer and Pb-phytoextractor that can bioaccumulate large amounts of Cd and Pb in the roots [14]; however, toxic effects of these metals on its proteome are unknown. In order to gain insights into the effects and strategies of tolerance in this plant, we used proteomics redox approaches to measure the oxidative effects of Cd and Pb exposure during the growth of *P. fasciculatum* under controlled greenhouse conditions. Thus, the carbonyl index was used as a biomarker to quantify the global oxidative damage suffered by leaves and roots proteomes of *P. fasciculatum* grown in mining soils doped with Cd and Pb along with its effects on biomass and accumulation patterns. Moreover, the carbonylated proteins in the leaves were identified as oxidative stress targets by the action of Cd and Pb in mining soils, using Western Blot analysis and tandem mass spectrometry.

2. Results

2.1. Plant Growth

P. fasciculatum seedlings were cultured in mining soils doped with Cd and Pb metals and growth was monitored for 90 days and expressed in terms of biomass production (Figure 1), as was previously reported [14]. It was observed that the plants in the first 60 days of growth, showed a more significant growth than the plants in control soil ($p \leq 0.05$). By contrast, at 90 days a significant reduction in biomass in all organs (leaves, stem and roots) was recorded for plants exposed at TP50, and leaves of TC30, compared with control plants ($p \leq 0.05$). It was found that the growth behavior of the plants examined in this study differs from that of the plants used in most studies of this type [15].

Plants of exposed groups suffered adverse effects on their biomass with evident symptoms of necrosis, chlorosis in leaves and a decrease in thickness in the stem between 60 to 90 days of growth;

indicating that these plants have a good tolerance to exposure to Cd and Pb, at least into the first 60 days of exposure. Toxic effects observed in leaves were higher in plants exposed to Pb than in those exposed to Cd, which may be related to the high accumulation of Cd in the roots and the greatest translocation of Pb (Table S1).

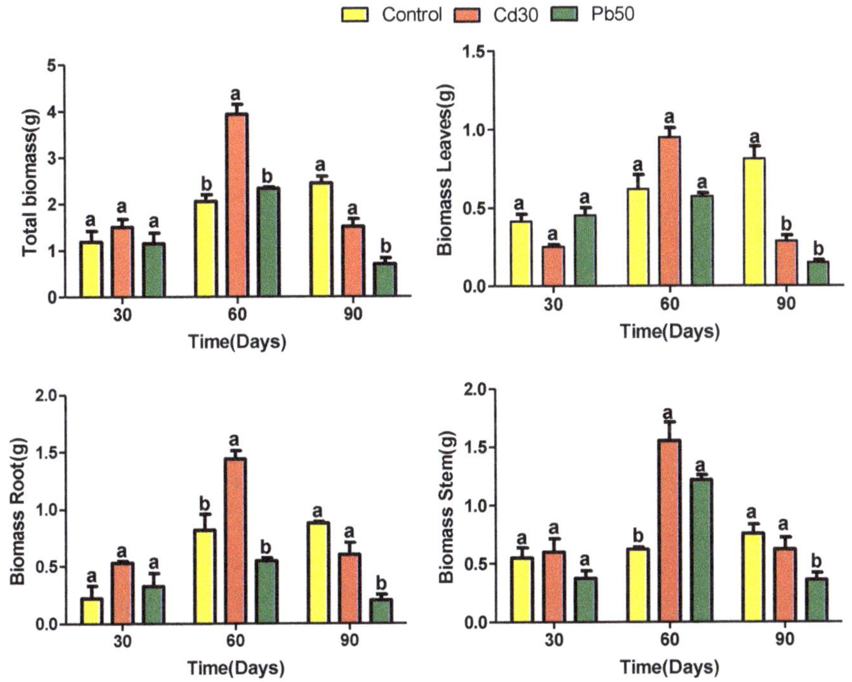

Figure 1. Amount of dry biomass in different tissues of *Paspalum fasciculatum* exposed to Cd and Pb in mining soils. Treatments Pb50 = 50 mg kg^{-1} Pb, Cd30 = 30 mg kg^{-1} Cd, ($n = 3$). The letters are the statistical significance ($p < 0.05$) between the control and treatments-Cd30 and Pb50.

2.2. Cd and Pb Concentrations in Plants

As was expected, bioaccumulation of Cd and Pb in the organs of *P. fasciculatum* increased with the concentrations of these metals in the soil, following the order stems < leaves < roots ($p \leq 0.05$) (Table 1). During the first 30 days of both exposed groups, plant organs accumulated more metal than during the subsequent 60 days of phytoremediation, with the exception of the roots in Pb50, which accumulated more Pb at 90 days. This could be interpreted as a dilution of the metals in the plant, due to significant growth concerning the control observed in some organs at 60 days.

In terms of uptake, *P. fasciculatum* showed a larger capacity to bioaccumulate Cd than Pb; its capacity to accumulate Cd in the roots is remarkable. However, this uptake is not continuous during the 90 days of exposure (except in roots of Pb50), which indicates that this species can be used to phytoremediation during short periods of time (no more than 2 or 3 months)

Table 1. The concentration of Cd and Pb (mg kg^{-1}) in the organs of *P. fasciculatum* grown in doped mining soils.

Treatments	Plant Tissue	Cd30 (mg kg^{-1})			Pb50 (mg kg^{-1})		
		30 Days	60 Days	90 Days	30 Days	60 Days	90 Days
	Roots	190.5 ± 8 ^{a}	107.1 ± 22.7 ^{a}	130.8 ± 22.7 ^{a}	36.7 ± 6.9 ^{a}	20.8 ± 2.2 ^{a}	45.7 ± 1.9 ^{a}
	Stems	23.2 ± 3.8 ^{a}	12.7 ± 3.67 ^{a}	7.6 ± 0.7 ^{a}	5.4 ± 0.6 ^{a}	1.1 ± 0.02 ^{a}	3.5 ± 0.1 ^{a}
	Leaves	27.6 ± 5.6 ^{a}	16.4 ± 7.7 ^{a}	16.1 ± 0.74 ^{a}	4.8 ± 2.6 ^{a}	1.3 ± 0.71 ^{a}	2.9 ± 0.3 ^{a}
Control	Roots	2.7 ± 0.5 ^{b}	3.0 ± 0.4 ^{b}	2.1 ± 0.1 ^{b}	0.6 ± 0.4 ^{b}	1.2 ± 0.05 ^{b}	0.9 ± 0.7 ^{b}
	Stems	0.6 ± 0.5 ^{b}	0.5 ± 0.02 ^{b}	0.9 ± 0.2 ^{b}	ND ^{b}	ND ^{a}	ND ^{b}
	Leaves	1.4 ± 1.1 ^{b}	1.2 ± 0.2 ^{b}	1.3 ± 0.2 ^{b}	ND ^{b}	ND ^{a}	ND ^{b}

The letters are the statistical significance ($p < 0.05$) between the control and treatments-Cd30 and Pb50 in three periods of time. ND = not detectable, ($n = 3$).

2.3. Leaves and Roots Proteins Extract Obtained

Trichlorocaetic acid(TCA)-acetone precipitation protocol [16] was used to obtain proteins from the leaves and roots of P. fasciculatum. Based on previous studies in P. fasciculatum, samples from the Cd30 and Pb50 treatments with 30, 60 and 90 days of exposure, were chosen because of the significant differences observed in biomass yielding in regard to corresponding controls. Protein extracts were used in each treatment to determine the carbonyl indexes and protein carbonylation patterns. Yielding extracts of leaves and root proteins were in the range of 0.3 to 1.0 and 0.05 to 0.6 μg/μL in TC30, respectively. For Pb50 were 0.3–1.1 μg/μL, while in controls values were between 0.60 to 1.55 and 0.08 to 0.8 μg/μL. Yielding is condensed in the supplementary material (Table S2).

2.4. Oxidative Damage Induced by Cd and Pb in the Protein of *P. fasciculatum*

The determination of carbonyl indexes was carried out by building calibration curves with a standard protein of BSA and its linearity was evaluated. Figure 2 shows the dot blot membrane obtained.

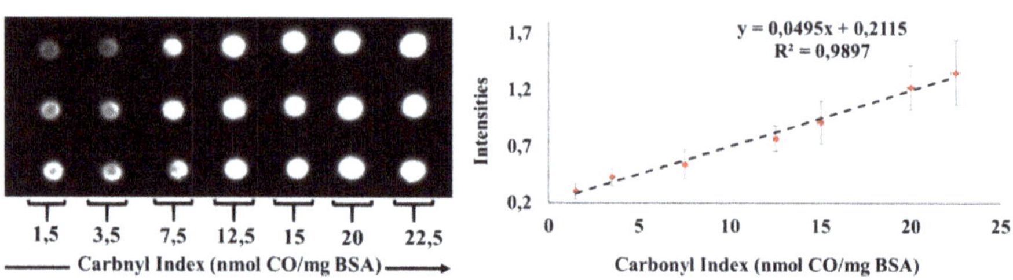

Figure 2. Curve of BSA for determination of carbonyl indexes employed Dot Blot assay. 200 ng of derivatized Bovine Serum Albumin (BSA) marked with 2,4-Dinitrophenylhydrazine (DNPH) were spotted by triplicate using Polyvinylidene fluoride (PVDF) membranes. Two curves were analyzed on two different days. The curves show linearity in the interval of the carbonyl index between 1.5 and 22.5 nmol of carbonyl/mg protein ($R^2 > 0.99$).

Carbonyl indexes were determined for proteins of roots and leaves of *P. fasciculatum* using the dot blots shown in Figure S1 and its values are listed in Table 2. In general, it was observed an increase of protein carbonylation on roots and leaves proteomes in all samples analyzed from plants exposed to Cd and Pb (Figure 3). In particular, roots and leaves of the Cd30 treatment showed the greatest differences in oxidative damage at proteins, due to the carbonyl indexes values that increased 4.7 and 9.9 folds compared to the control, respectively. These results are congruent with the highest values of Cd absorption reached during growing experiments. For Pb50 treatment, oxidative damage in leaves was larger as the exposure time elapsed, which was coincident with the progressive biomass reduction observed. The relative increase in the carbonyl indexes for organs at the different treatments are listed in Table S3.

Table 2. Protein carbonyl indexes of *Paspalum fasciculatum* Willd (Ex Flugue) (Poaceae) organs.

Days	Carbonyl Index of Protein Exposed to Cd30				Carbonyl Index of Protein Exposed to Pb50	
	Roots		Leaves		Leaves	
	Control	Cd30	Control	Cd30	Control	Pb50
30	0.6 ± 0.2	2.9 ± 0.3	0.3 ± 0.2	3.2 ± 0.6	ND	1.4 ± 0.5
60	0.7 ± 0.5	2.08 ± 0.1	1.2 ± 0.4	2.8 ± 0.6	0.5 ± 0.01	1.5 ± 0.8
90	1.6 ± 0.9	3.1 ± 0.8	ND	0.5 ± 0.04	1.4 ± 0.8	2.9 ± 0.5

C.I. calculated are expressed as means ± SD (n = 3). Treatments Pb50 = 50 mg kg^{-1} Pb, Cd30 = 30 mg kg^{-1} Cd; ND: non-determinate. Cd: cadmium. Pb: lead.

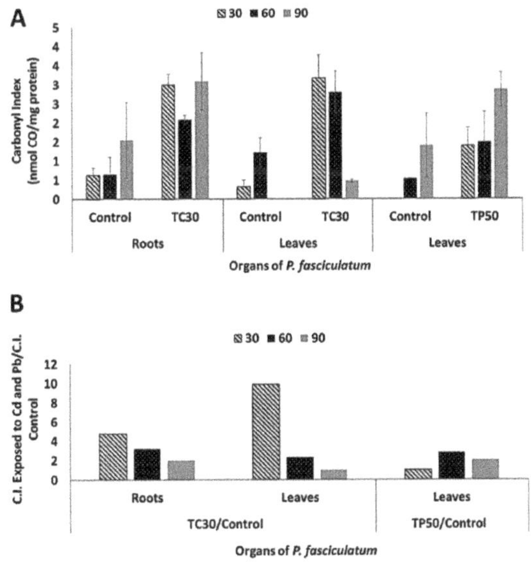

Figure 3. Carbonyl indexes in roots and leaves proteins of *P. fasciculatum* exposed to Cd and Pb at dissimilar days. (**a**) Evidence variation of C.I [means ± Standard Desviation (SD)] for each control, roots and leaves in exposition assays to 30 mg kg^{-1} Cd and leaves exposed to 50 mg kg^{-1} Pb. (**b**) Relation increase of C.I values for samples exposed to Cd and Pb/Control.

2.5. *Carbonylation Patterns of Roots and Leaves Proteins of P. fasciculatum Exposed to Cd and Pb*

The qualitative pattern of carbonylated proteins was obtained for roots and leaves of *P. fasciculatum* exposed to metals, in all conditions assayed. In fact, protein carbonylated bands were not detected by the anti-DNPH antibody for the controls in experimental conditions. By contrast, carbonylated bands were detected as an effect of its exposition to Cd and Pb, which were notorious in roots and leaves exposed to Cd and leaves in contact with Pb (Figure 4), confirming the obtained by Dot blot assay.

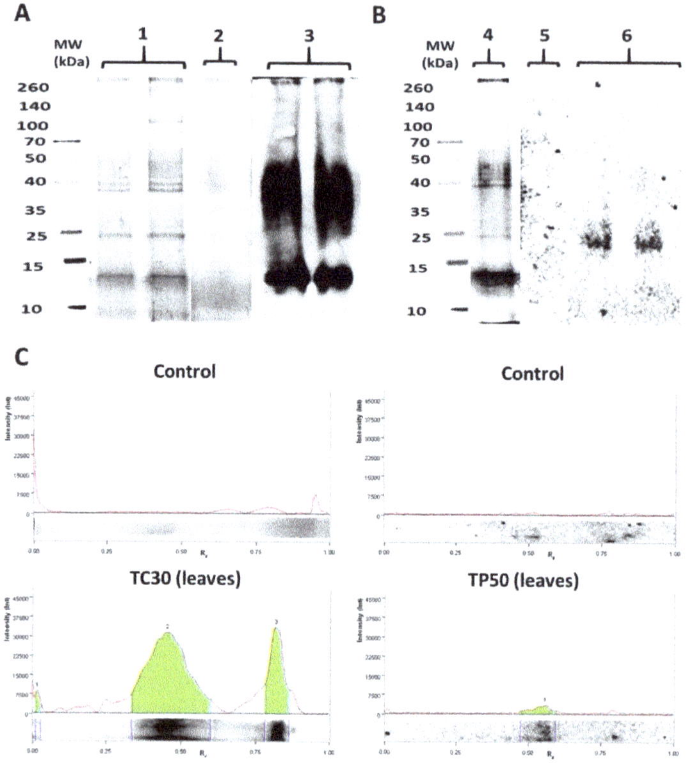

Figure 4. Carbonylation patterns in samples of *P. fasciculatum* exposed to Cd and Pb. (**a**) Protein profiles obtained for roots of *P. fasciculatum* exposed to 30 mg kg^{-1} Cd for 30 days: electrophoregram of Blotting of Coomassie Blue (CBB) stained proteins on SDS-PAGE (1), oxyblots for 30 days' control (2) and exposed samples (3). (**b**) Protein profiles obtained for leaves P. fasciculatum exposed to 50 mg kg^{-1} Pb during 60 days: electrophoregram of CBB stained proteins on SDS-PAGE (4), Oxyblots for 60 days' control (5) and exposed samples (6). (**c**) Densitograms with the intensity of the chemiluminescent signal for each carbonylated protein band obtained in oxyblots.

2.6. Identification of Proteins in Carbonylated Bands

To identify potentially carbonylated proteins as a consequence of Cd and Pb exposition, profiles of oxidized proteins obtained by Western blot were matched with SDS-PAGE electrophoregrams. For this, leaves protein extracts of Cd30 and Pb50 treatments exposed for 30 days and 60 days were chosen, respectively, because they showed the largest carbonyl indexes (Figure 4).

Qualitative analysis of the oxyblots obtained exhibited intense and wide protein carbonylated bands in the proteome of *P. fasciculatum* exposed to Cd and Pb. These results are consistent with the observations made in the dot blot assay. Preparative duplicates of SDS-PAGE were matched against oxyblots to identify carbonylated protein bands (Figure 5). Thus, seven protein bands were removed, which were digested with trypsin and subsequently analyzed with a MALDI-TOF-TOF spectrometer.

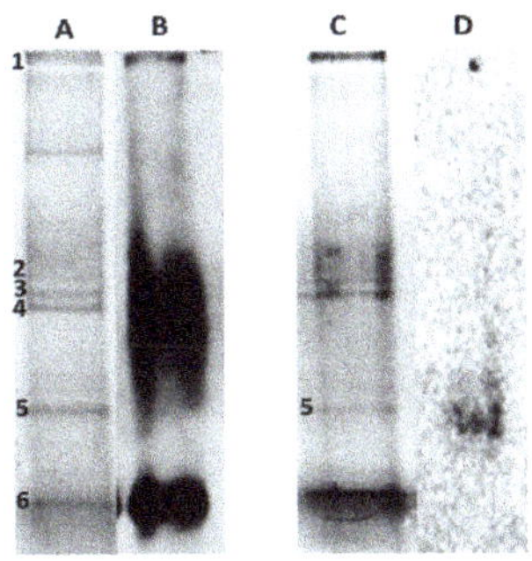

Figure 5. Carbonylated proteins profiles in leaves of *P. fasciculatum*. (**A**) and (**C**). Protein leaves of *P. fasciculatum* exposed Cd and Pb, electrophoresed at gel polyacrylamide staining with Coomassie brilliant blue. (**B**) and (**D**). Oxyblots of protein roots and leaves of *P. fasciculatum* exposed Cd and Pb.

Band 5, sited around 25 KDa, was common for Cd30 and Pb50, while bands 1 to 4 and 6 were only observed in TC30. A total of five proteins were identified and classified according to their function, thus: 2 ATP synthase subunit, Chlorophyll a-b binding protein, Fructose-bisphosphate aldolase, the Ribulose bisphosphate carboxylase large chain (RuBisCO LSU) and four fragments or subunits of this same protein were identified. Proteins identified are listed in Table 3 along with their accession number, mass (kDa), isoelectric point, score, submitted name, encoded on and biological process (annotated in Uniprot-Swissprot database). The proteins identified as targets of oxidative damage due to exposure to Cd and Pb are found in proteins mostly accumulated in this species due to the stress by these metals; it has been observed that these proteins are also major targets of adverse effects according to other similar research [3,17–19].

Furthermore, Fructose-bisphosphate aldolase and ATP synthase subunit gamma were identified in carbonylated protein bands by Cd30 and Pb50 (bands 5). Fructose-bisphosphate aldolase allosteric enzyme which plays a key role in glycolysis and gluconeogenesis. While ATP synthase subunit gamma has an essential role both in regulating the activity of ATPase and in the currents of protons across of the CF_0 complex in the chloroplast thylakoid membrane. In the case protein bands that exhibited oxidative damage, only in treatment-Cd30 were all subunits of RuBisCO identified, which catalyzes the carboxylation of D-ribulose 1, 5-bisphosphate. This step is important for CO_2 fixation and oxidative fragmentation of the pentose substrate in the photorespiration process. ATP synthase subunit alpha and beta, produces ATP from ADP in the presence of a proton gradient across the membrane, the alpha chain is a regulatory subunit and the beta subunits are hosted primarily in the catalytic sites. Chlorophyll a-b binding protein 1, 2, 8 and CP26, light-harvesting complex (LHC), playan important role in the reception of light, apprehension and delivery of excitation energy to photosystems, with which it has a direct relationship.

Table 3. Carbonylated proteins identified from *P. fasciculatum* Willd (Ex Flugue) leaves exposed at Cd (30 days) and Pb (60 days).

Band	Submitted Name		Score	Accession	Biological Process	Encoded on	Condition of Exposure
1	Ribulose bisphosphate carboxylase large chain [*Hordeum vulgare*]	52.0	104.0	RBL_HORVU	Catalyzes: CO_2 fixation, oxidative fragmentation of the pentose substrate in the photorespiration process.	Plastic, Chloroplast	Cd30
2	ATP synthase subunit alpha [*Oryza nivara*]	55.6	157.0	ATPA_ORYNI	Translocase, ATP synthesis, Hydrogen ion transport, Ion transport	Plastic, chloroplastic	Cd30
	Ribulose bisphosphate carboxylase large chain [*Cuscuta sandwichiana*]	53.4	99.4	RBL_CUSSA	Catalyzes: carbon dioxide fixation, oxidative fragmentation of the pentose substrate in the photorespiration process	Plastic, chloroplastic	Cd30
3	Ribulose bisphosphate carboxylase large chain [*Avena sativa*]	52.9	150.0	RBL_AVESA	Catalyzes: carbon dioxide fixation, oxidative fragmentation of the pentose substrate in the photorespiration process	Plastic, chloroplastic	Cd30
4	Ribulose bisphosphate carboxylase large chain [*Avena sativa*]	52.9	150.0	RBL_AVESA	Catalyzes: carbon dioxide fixation, oxidative fragmentation of the pentose substrate in the photorespiration process	Plastic, chloroplastic	Cd30
5	Fructose-bisphosphate aldolase [*Oryza sativa* subsp. Japonica]	42.0	145.0	ALFP_ORYSJ	Allosteric enzyme, kinase, transferase, photosynthesis, Glycolysis; Plays a key role in glycolysis and gluconeogenesis	Cytoplasm	Cd30 and Pb50
	ATP synthase subunit gamma [*Zea mays*]	39.8	67.5	ATPG_MAIZE	ATP synthesis, Hydrogen ion transport, Ion transport, Transport, proton-transporting ATP synthase activity, rotational mechanism;	Chloroplast; chloroplast thylakoid membrane, Peripheral membrane protein	Cd30 and Pb50
	Chlorophyll a-b binding protein CP26 [*Arabidopsis thaliana*]	30.1	79.1	CB5_ARATH	light-harvesting in photosystem I, The light-harvesting complex (LHC) functions as a light receptor	Chloroplast, chloroplast thylakoid membrane	Cd30

3. Discussion

Paspalum genus count with evidence of tolerance to heavy metals for some species such as *P. dischum* and *P. vaginatum*. For the first one, an exclusion strategy that restricts the translocation of the metals to the aerial parts has been proposed, favoring their accumulation in roots, but decreasing the availability in the soil [20]. This phytostabilizing mechanism allows it to grow well in tailings sites contaminated with Cu, Pb and Zn; hence, it has been proposed as a promising species for the repopulation of wastelands contaminated with these metals [21]. For the second one, Cd-tolerance genes have been described which could be related to the regulation of pathways involved in the synthesis of phytochelatins, heat shock factor transcription (HSFA4), protection against stress, cytochrome P450 complex (CYP450) and sugar metabolism [22].

For *P. fasciculatum*, we previously measured its capabilities to change the bioavailability of Cd and Pb in doped mining soils with these metals, its growth and development in these soils contaminated and the metals bioaccumulation in their tissues [14]. As a result, *P. fascicl ulatum* is a heavy metal tolerant plant that grows in mining, flooded and dry soils, which can bioaccumulate significant amounts of Cd and Pb and increases the content of organic matter and the pH in the rhizosphere. Therefore, it can improve the quality of the soil, aiding the revegetation and colonization of other plants in such a way that it can have a significant ecological effect on the soil [14].

To deepen the physiological and biochemical aspects of the metal tolerance for this species;, in this study, we evaluated the effects of exposure to high concentrations of Cd and Pb on the development and oxidative damage on the proteome of different organs of *P. fasciculatum*. Aspects such as the growth behavior, absorption of metals in tissues and the values for carbonyl index ratios by exposure to Cd and Pb in *P. fasciculatum* (Table 2) showed differential characteristics into the time period of tolerance of the plant (30 and 60 days). The results obtained showed that the plant limited the oxidative effects to a small group of major proteins, which could be associated with the translocations of metals to the shoots where Cd and Pb induced the carbonylation of important proteins of photosynthesis, energy production, glycolysis and gluconeogenesis (Table 3).

This implies the presence of adverse effects in the energy processes, the balance of sugars, a decrease of chlorophyll and photosynthesis. Nonetheless, plants grew even more than controls up to 60 days, due maybe to the intrinsic tolerance of this plant. These plants demonstrated a marked tendency to store larger amounts of heavy metals in their roots, especially Cd. Pb presented a little more translocation to shoots, thus avoiding more severe damage to photosynthetic organs. Therefore, the mechanism of tolerance of *P. fasciculatum* would imply a decrease in the concentration of Cd and Pb in the cytoplasm.

In similar stress conditions, the synthesis of glutathione and/or phytochelatins [22–24], followed by compartmentalization and sequestration of metals in vacuoles, have been described along with the activation of the protective antioxidant systems [25,26].

In leaves of *P. fasciculatum*, proteins of photosynthesis were identified as the main target of the oxidative damage induced by Cd and Pb. This has been observed by other authors, who employed a diversity of abiotic stressors, including heavy metals [17,27]. Thereby, oxidative damage induced that heavy metals could force the cell to strengthen its tolerance mechanisms at the cost of growth [17], as observed in our experiment in the last 90 days. Once the tolerance period is exceeded, eventually the cells could slip into early irreversible senescence, as it was observed in the heavy metal-treated *Medicago sativa* [23], *Brassica rapa* [28] and *Nicotiana tabacum* [29]. It has been described that heavy metal effects can be reflected in the carboxylation phase of photosynthesis, mainly on enzymes of CO_2 fixation, as RuBisCO [30]. In fact, ribulose bisphosphate of *P. fasciculatum* was identified in different protein carbonylated bands, showing similar behavior to RuBisCO of rice leaves exposed to metal stress, which increased the number of proteolytic fragments in protein electrophoresis as a consequence of the oxidative damage induced by the metals [31]. Furthermore, proteomics studies have observed that the Cd drastically reduced the amounts of RuBisCO LSU in SDS-PAGE by displacing the magnesium cofactor from its structure [19,31]. Mg^{+2} ion is the most important co-factor in carboxylation reactions catalyzed by photosynthetic enzymes as ribulose 1,5 bisphosphate carboxylase and phosphoenolpyruvate carboxylase and its removal inhibit their activity [3].

Likewise, in this study, other proteins affected by oxidative stress were chlorophyll (Chl) a/b binding proteins, which showed oxidative damage with Cd30 at 30 days of exposure. The light-harvesting chlorophyll a/b-binding (LHCB) proteins, are proteins with an important structural and functional role in the process of photosynthesis, it is part of the light-harvesting complex of photosystem II (PSII), which is always attached with chlorophyll A and B, and xanthophyll, thus forming the antenna complex [32]. These proteins are an important structural piece of the major light-harvesting complex, it is the membrane protein with greater distribution by a large number of plants on the planet [33]. Experimental studies show that Pb-stress is a potential inhibitor in the activity of PS II and I; being the sensitivity to Pb-stress greater in PSII than that of PSI [27]. In addition, this produces suppression of Chl biosynthesis, causing a decrease in pigment content and adverse effects on the fluorescence of Chl [27].

As a consequence, Cd showed greater alterations on light-harvesting chlorophyll a/b protein complex II, which causes adverse effects in the process of photosynthesis by decreasing PSII activity. This in turn causes an inhibition of quantum yield and electron transport, thus, negatively affecting the photo-assimilatory pathways, such as CO_2 fixation, for alterations in enzymes such as RuBisCO [34]. Some studies conducted in spinach have shown differential accumulation of light-harvesting complex I and II (LHCI and LHCII) proteins in processes of adaptation to stress-by heavy metals [35]. Therefore, Cd and Pb are indirect causes of ROS, which have the capacity to produce structural and organizational damage to the lipids of the thylakoid membrane, creating an alteration in the organization of the structure of granum stacks and chloroplast ultrastructure [34,36]. Some studies indicate that PSI photoinhibition is produced by both oxygen radical (O^{2-}) and singlet oxygen ($1O^{*2}$) generated inside the thylakoid membranes; Cd and Pb causes decrease in a size and number of grana stacks, which increases the generation of ROS within the cells, directly influencing the alteration of the activity of antioxidant enzymes, and therefore altering multiproteic complexes such as PSI, PSII and ATPase in the thylakoid membrane [18,37]. Again, these toxic effects are resisted by the *P. fasciculatum* during the tolerance period.

On the other hand, it has been demonstrated that the CF_1 CF_0-ATP synthase (F_1F_0) is an important enzyme in the synthesis of ATP, starting from ADP; this protein has a crucial role in the production of energy in the cell. Structurally, it is composed of an integral membrane protein (CF_0) and another component exposed to the stroma (CF_1); the latter is made up of several subunits, $\alpha3\beta3\gamma1\delta1\varepsilon1$ [38,39]. In this research, several subunits of the ATP synthase showed oxidative damage, in Cd30 α and γ. In Pb50 only the γ subunit showed oxidative damage, which shows an effect of Cd and Pb in the alteration of energy production processes at the level of the chloroplast. Studies have shown that some ROS alter the structural components of ATP synthase, the peroxide hydrogen (H_2O_2) produces adverse effects on α and β-subunits; $1O^{*2}$ in the γ-subunit, also decreases critically the hydrolysis of ATP and the flow of protons [40,41].

Obtaining energy is an important physiological factor in all forms of life; therefore, glycolysis and gluconeogenesis are important biochemical processes in the development of organisms, including plants. Oxidative stress generated for both Cd30 and Pb50, oxidative damage in fructose-1,6-bisphosphate aldolase, which has a crucial role in energy metabolism, because it catalyzes rupture of β-fructose-1,6-phosphate to glyceraldehyde-3-phosphate and dihydroxyacetone in glycolysis, the process is reversed in gluconeogenesis [40]. There are reports of the adverse effects of Cd on fructose-1,6-bisphosphate aldolase enzyme activity [42]. This adverse effect of heavy metals on the important proteins of photosynthesis and energy metabolism has been observed as intolerant, hyperaccumulating and phytoremediating in plants [25,29].

Finally, our results show that *P. fasciculatum* has an intrinsic tolerance to the presence of Cd and Pb. In this, the intense oxidation induced by metals is limited to a group of major proteins in the first 30 days of exposure. This event perhaps activates cellular responses of the tolerance mechanism that attenuates oxidative damage and allows the plant to grow up to 60 days under the conditions of adverse cultures used.

Whereas the carbonylated proteins identified by oxidative damage in this study were those with the highest accumulation, as observed by gel electrophoresis; its carbonylation could be part of the cellular responses involved in the mechanisms of tolerance to stress by Cd and Pb in the leaves of *P. fasciculatum*.

4. Materials and Methods

4.1. Sampling and Preparation of Soils and Growing Conditions of the Plants

The soil samples (50 Kg) were taken 30 cm from the surface at the tailings of the gold mine "El Alacrán", located in the northwest of Colombia, between coordinates 7°44′29.01″ North and 75°44′10.8″ West, in the municipality of Puerto Libertador (Córdoba). Soils were stored in labeled polyethylene bags, transported to the laboratory, dried at 40 °C and sieved. Cd and Pb content was determined by atomic absorption spectroscopy (see below) and listed in Table 4, three samples were taken for each treatment. Next, three experimental groups were established. Two exposed groups, Cd30 and Pb50, corresponding to plants grown in mine soil samples doped with enough $CdCl_2$ and $PbSO_4$ solutions in order to prepare individual samples with 30 and 50 mg of Cd and Pb per kg of soil, respectively. Whereas a third group corresponding to plants in mine soils without doping was used as control. Cd and Pb concentrations were chosen considering the designation of unpolluted Colombian soil, which is 0.012 and 0.008 mg kg^{-1} of Cd and Pb, respectively [43]. Worldwide, these concentrations are 27 mg kg^{-1} Cd and 0.41 mg kg^{-1} Pb [44], and in countries such as the USA and Sweden, the permitted limit of heavy metal concentration in soil is 15 and 40 mg kg^{-1} Pb; 1.4 and 0.4 mg kg^{-1} Cd, respectively [45].

Table 4. Physical and chemical characteristics of the soil from "El Alacrán" gold mine.

Soil Properties					
Properties of Bioavailability		**Texture**		**Metals**	
pH	3.67 ± 0.03	Sand (%)	27.5 ± 0.03	Cd (mg kg^{-1})	7.27 ± 0.1
OM (%)	1.54 ± 0.09	Clay (%)	4.4 ± 0.09	Pb (mg kg^{-1})	2.72 ± 0.4
CEC	13.1 ± 0.01	Silt (%)	68.1 ± 0.06		

[1] OM (%) = Organic matter percentage; CEC = cation exchange capacity.

P. fasciculatum plants from soils free of Cd and Pb were cut, its taxonomic determination was carried out in the herbarium of the University of Cordoba (code HUC-8132). For the processing in the laboratory, establishment, growth conditions and monitoring of plants in doped soils and control in greenhouse conditions, we rely on what is supported by Salas-Moreno and Marrugo-Negrete [14]. To determine the amount of biomass generated (g), metal concentrations in the soil and plant organs (leaf, stem and root) were measured at three growth periods (30, 60 and 90 days).

4.2. Analysis of Plants and Soil Samples

Soil and plant samples were taken once a month to measure the concentrations of Cd and Pb. Organs of the plants were subjected to washing with sterile water, then they were stored in paper bags in an oven for several days at 40 °C. After that, the produced biomass was determined by weighing the exposed and control samples using an OHAUS Adventurer digital balance. For the evaluation of the symptoms of toxicity in plants such as necrosis and chlorosis, we rely on periodic observations every day, we also calculate the length and thickness of the stem every week, as well as the number of fallen leaves or with symptoms of toxicity.

For measurements of Cd and Pb concentrations in plant tissues (three samples were taken per organ of each biological replica, in each repetition), 0.5 g of plant material (dry weight) was taken, then digested with a mixture of 5 mL/2 mL of HNO_3/H_2O_2 in a microwave [46]. In the same way, 0.5 g of soil (dry weight) was subjected to microwave digestion using 10 mL of 65% HNO_3 solution according to EPA method 3051A [47]. In this process, a Milestone ETHOS TOUCH 127697 series microwave oven was carried out. The total measurements of Cd and Pb were determined by graphical oven atomic absorption spectroscopy (GF-AAS) using a Thermo Scientific iCE 3000 series analyzer. For the analytical control, certified materials were used for plants and soils (lichen: IAEA-336, 0.117 mg Cd kg^{-1} and 4.9 mg Pb kg^{-1} and soil/sediment: CRM008-050, 0.82 mg Cd kg^{-1} and 95.3 mg Pb kg^{-1}).

The pipette method was used to establish the soil texture, and the texture triangle was used to establish the soil type. Cation exchange capacity (CEC) was determined by establishing the total number of removable cations (Ca^{2+} Mg^{2+} K^+ Na^+ Fe^{3+} Al^{3+} Mn^{4+}), which were extracted with 1.0 ammonium acetate. Organic matter content (% OM) was established as the resulting sample quantity (measured from the weight of a sample), after a calcination loss of 2.0 g of soil at 450 °C for 4 h [48], while pH was determined with a WTW 330i pH meter (Table 4). All measurements were performed on three soil samples per treatment group.

4.3. Protein Extraction of Roots and Leaves from *P. fasciculatum*

Protein extraction from roots and leaves samples were obtained for control and exposed plants Cd30 and Pb50 at 30, 60 and 90 days of growth (Figure 6), in total 27 samples were analyzed by each repetition, three samples for exposure time in each organ. Organ samples were rinsed with distilled water, dried with blotting paper and cut into small pieces using a clean scissor. Samples were weighed and ground to a fine powder in liquid nitrogen using pestle and mortar [16]. 500 mg of tissue powder was macerated at 4 °C for 1 h with 2 mL of solution 10% *w/v* TCA in acetone supplemented with 0.07% of 2-mercaptoethanol/protease inhibitor cocktail mini-complete (Roche, BSL, Switzerland). Pellets were recovered by centrifuging (9000× *g* for 10 min at 4 °C) and washed twice with one ml of ice-cold (−20 °C) acetone containing 0.07%, 2-mercaptoethanol/protease inhibitor cocktail mini-complete. Pellets were dried at 40 °C for 10 min. Finally, proteins were extracted in 200 μL buffer lysis (5M urea/2M thiourea/1% triton-x100/50 mM Tris-HCl, pH 8.8 and 1% DTT/protease inhibitor cocktail mini-complete), vortexing for five minutes and incubated during 4 °C for 1 h. Supernatants were recovered by centrifuging at 10,000× *g* for 10 min at 4 °C. To quantify the extracted proteins, the Quick Start Bradford assay microplate (Bio-Rad, Hercules, CA, USA) was used, by generating a standard curve from BSA [49].

Figure 6. Seedlings of *P. fasciculatum* in a greenhouse under to Cd and Pb-stress for 90 days. Treatments: Cd30 = 30 mg kg^{-1}; Pb50 = 50 mg kg^{-1}.

4.4. Measurement of Carbonyl Index

Irreversible oxidative damage on roots and leaves proteins of *P. fasciculatum* was quantified by Dot blot immunoassays proposed by Wehr and Levine [50], 27 samples were analyzed by repetition. To determine this, a calibration curve was built using BSA with a variety of carbonyl index values following strictly the procedure of Contreras-Puentes [51]. We prepared a solution of 2 mg/mL BSA, which was subjected to oxidation with 10 mM $FeSO_4$ for 2h at 37 °C. A part of the initial BSA solution was stored to establish the basal carbonyl index value. Subsequently, the basal and oxidized solutions were derivatized with 10 mM DNPH in 0.5 M phosphoric acid, and then they were incubated for 10 min. Then, these solutions were made alkaline with 6M NaOH, incubated for 10 min and read at 450 nm [52]. The molar absorption coefficient of 22308 M-1 cm^{-1} of DNPH was determined to calculate the nmol of carbonyl/mg of protein. Stoichiometric mixing of BSA basal with BSA oxidized was performed to obtain several points on the calibration curve. Protein freshly solutions (1 mg/mL) of *P. fasciculatum* roots and leaves from treatments and control samples, were derivatized with the DNPH probe (10 mM DNPH and 2 M HCl) [53]. Derivatized BSA and samples were spotted by triplicate on PVDF membrane (Immun-Blot® PVDF Membrane for Protein Blotting. Cat. #162-0177) and the membrane was immersed in a solution of Phosphate Buffered Saline (PBS) (5% skimmed milk powder dissolved in phosphate buffered saline) for 2 h at room temperature. Afterwards, they were incubated with rabbit polyclonal anti-DNPH antibody 1:10.000 (Sigma, St Louis, USA) in PBS–Tween-20 0.05% and milk 5%, for 2 h at room temperature with constant mixing movement. Then, the membranes were subjected to washing and incubation with peroxidase linked anti-rabbit IgG antibody at 1:10,000 (Amersham Bio sciences; 1h at room temperature). Chemiluminescence signals were developed using the Western maxTM rabbit IgG detection kit from Amresco and captured in a ChemiDoc System (Bio-Rad, Hercules, CA, USA). The intensity of each analyzed point was calculated through optical densitometry with Image Lab software (Bio-Rad, Hercules, CA, USA). To preserve similar measurements in the quantitative analysis, an area of 11.0 mm^2 was determined for each protein point and the intensity measurement was organized in a matrix table in Microsoft Excel version 2013.

4.5. Identification of Proteins in Carbonylated Bands

Protein identification in carbonylated bands was only performed on leaves of exposed samples (Cd30 and Pb50) and control at 30 and 60-days of growth (three samples of protein bands for each treatment) because of the greatest differences in carbonyls indexes values between groups were found at these times. The use of proteomics techniques such as SDS-PAGE, Western blot and Tandem Mass Spectrometry were necessary for this investigation. In this way, DNPH-derivatized leaves proteins were electrophoresed by duplicate in 10% acrylamide/bisacrylamide gels. One of the gels was transferred to a PVDF membrane semi-dry for 30 min with a Trans-Blot turbo instrument (Bio-Rad, Hercules, CA, USA). Once the membrane was transferred, it was blocked and subjected to an incubation process with polyclonal anti-DNPH antibody, and immediately revealed by chemiluminescence, as described above [51,54]. On the other hand, the second gel was subjected to fixation in methanol 50%: 2% phosphoric acid solution and visualization of the separated proteins, which was performed by staining with Coomassie Blue Brilliant G-250. Finally, using a ChemiDoc system, immunoblot images and stained gel were visualized; in this way, the oxidized protein bands were compared and selected.

Protein bands that were chosen for the analysis were in-gel reduced, alkylated and digested with trypsin, as described in [55]. In summary, the stains were washed twice with water, reduced for 15 min with 100% acetonitrile (ACN) and dried in a Savant Speed Vac for 30 min. Afterwards, the samples were reduced with 10 mM DTT in 25 mM NH_4 HCO_3 (ammonium bicarbonate) for 30 min at 56 °C and then alkylated with 55 mM iodoacetamide in 25 mM NH_4 HCO_3 during 15 min in the dark. Samples were subjected to digestion with 12.5 ng/μL trypsin (sequencing grade, Roche Molecular Biochemicals) in 25 mM NH_4HCO_3 (pH 8.5) overnight to 37 °C. Finished digestion, 1 μL was spotted onto a MALDI target plate and allowed to air dry at room temperature. Then, 0.4 μL of a 3 mg/mL of the α-cyano-4-hydroxy-cinnamic acid matrix (Sigma, St Louis, USA) in 50% ACN was supplemented to the dried peptide digest spots and allowed again to air-dry at room temperature.

For MALDI-TOF/TOF, samples were analysed using an Analyzer MALDI-TOF/TOF/MS (4800 Plus Proteomics) (Applied Biosystems, MDS Sciex, Toronto, Canada), The spectrometer was operated in positive mode, the voltage acceleration used in this spectrometry was of 20 kV. All mass spectra were internally calibrated using peptides from the autodigestion of trypsin (m/z = 842.509 and 2211.104) and the peptides observed with a signal-to-noise greater than 10 were collated and represented as a list of monoisotopic molecular weights.

Through MS/MS sequencing analysis, the proteins identified by peptide mass fingerprints were processed. Taking into account the MS expectros, the most abundant precursors were chosen to undergo MS/MS analysis with collision-induced dissociation (CID) on (atmospheric gas was used) operated in 1 kV positive ion reflector mode and precursor mass windows ± 4 Da. The analysis and processing of the spectra were carried out based on the optimization of the plate model and default calibration.

The protein identification process was carried out with the help of databases, a search of each spectrum was performed in MASCOT engine v. 2.6. of the NCBI nr-database (Viridiplantae) (date 23_2018), carried out by Global Protein Server software v.3.6 from ABSciex. The parameters taken into account for the identification of proteins were the following: the modifications fixe, carbamidomethyl cysteine; the modifications variable, oxidized methionine.; peptide mass tolerance: 50 ppm for PMF, or 80–100 ppm for MS/MS or combined searches; 1 missed trypsin cleavage site; peptide charge state: +1; and MS/MS fragments tolerance: 0.3 Da. During the identification process, all proteins presented probability scores greater than the score set by MASCOT as significant (p-value < 0.05).

4.6. Statistical Analysis

The statistical analysis applied was completely random factorial, the results are presented in the experimental unit as the mean ± the standard deviation of the triplicate determinations of accumulated biomass, metals concentrations and Dot Blot assays. For data on heavy metals concentrations in the plant, organs were assessing normality using the Shapiro-Wilk test and homogeneity of variance using the Bartlett test. Later, the data were subjected to the ANOVA test, and the means comparisons were made when it was necessary using the Bonferroni tests. Statistical software GraphPad PRISM version 6.0c was used for all analyzes. A level of significance of 0.05 was selected. The significant differences between experimental groups were performed by analysis of variance in GraphPad PRISM 6.01, or between two groups by Tukey Test. Dot blot assay was triplicated.

5. Conclusions

P. fasciculatum demonstrated a tolerance capacity to stress by Cd and Pb during the process of phytoremediation in doped mining soils; this plant developed a defense mechanism reducing the concentration of metals in the cytoplasm and photosynthetic tissues and bioaccumulating almost exclusively on their roots. Good tolerance was limited in time at 60 days. The oxidative stress induced by Cd and Pb caused an increase in the carbonylation of proteins in leaves and roots of *P. fasciculatum* in treatments TC30 and TP50, at 30 and 60 days, respectively. Due to the translocation a shoots, Cd caused oxidative damage in proteins of leaves such as RuBisCO LSU, alpha and gamma units of ATP synthetase and chlorophyll binding protein CP26, exclusively affecting the photosynthetic machinery of the leaves in *P. fasciculatum* in 30 days of exposure; in addition, a proteolytic fragmentation in the subunits of the RuBisCO LSU associated with oxidative stress.

Supplementary Materials: The following are available online at http://www.mdpi.com/2223-7747/8/11/513/s1, Figure S1: Oxidative dot-blots of roots and leaves proteins from *Paspalum fasciculatum* Willd (ex Flüggé) (Poaceae), Table S1: Bioaccumulation (BAF) and Translocation (TF) Factor in *P. fasciculatum*, Table S2: Yielding of protein from *Paspalum fasciculatum* leaves and root exposed at Cd and Pb in mining soil, Table S3: Relation values C.I of exposition to *Paspalum fasciculatum* by Cd and Pb with control.

Author Contributions: M.S.-M.: Performed the experiments, and data/evidence collection. Wrote the initial draft. N.C.-P.: Performed the experiments of immune assays for protein carbonylation and wrote these results in the initial draft. E.R.-C.: Developed and designed the methodology for Dot blot Western blot assays and performed verification, whether as a part of the activity or separate, of the overall replication/reproducibility of results/experiments and other research outputs. J.J.-N.: Provided reagents, materials, instrumentation and computing resources for mass spectrometry analysis and did the critical review, commentary and revision for the pre-publication stage. J.M.-N.: Developed and designed the methodology for plant growth in greenhouse conditions and formulated the evolution of overarching research goals and aims. D.M.-C.: Oversight and leadership responsibility for the research activity planning and execution, including mentorship external to the core team.

Funding: Our research was carried out with the support of the Colombian Institute for the Development of Science and Technology 'Francisco José de Caldas' (COLCIENCIAS), the Technological University of Chocó 'DLC', and the University of Córdoba by the financing of the codified project 1112-489-25604 contract 472. DMC and ERC were funded by COLCIENCIAS and the University of Cartagena, Grant 1107-711-50102, and project Acta 134-2017.

Acknowledgments: Authors are grateful to the members of the Research Group "Biochemistry, Proteomics and Biology of Vegetal and Agricultural Systems" of the University of Córdoba (Spain), particularly to María Ángeles Castillejo Sánchez.

Conflicts of Interest: The authors declare that there is no conflict of interest.

References

1. Verma, S.; Dubey, R.S. Lead toxicity induces lipid peroxidation and alters the activities of antioxidant enzymes in growing rice plants. *Plant Sci.* **2003**, *164*, 645–655. [CrossRef]
2. Belkadhi, A.; De Haro, A.; Soengas, P.; Obregon, S.; Cartea, M.E.; Djebali, W.; Chaïbi, W. Salicylic acid improves root antioxidant defense system and total antioxidant capacities of flax subjected to cadmium. *OMICS J. Integr. Biol.* **2013**, *17*, 398–406. [CrossRef] [PubMed]
3. Hossain, M.A.; Piyatida, P.; da Silva, J.A.T.; Fujita, M. Molecular mechanism of heavy metal toxicity and tolerance in plants: Central role of glutathione in detoxification of reactive oxygen species and methylglyoxal and in heavy metal chelation. *J. Bot.* **2012**, *2012*, 1–37. [CrossRef]
4. Romero-Puertas, M.C.; Palma, J.M.; Gómez, M.; Del Río, L.A.; Sandalio, L.M. Cadmium causes the oxidative modification of proteins in pea plants. *Plant Cell Environ.* **2002**, *25*, 677–686. [CrossRef]
5. Zhang, F.; Wan, X.; Zhong, Y. Nitrogen as an important detoxification factor to cadmium stress in poplar plants. *J. Plant Interact.* **2014**, *9*, 249–258. [CrossRef]
6. Bagheri, R.; Bashir, H.; Ahmad, J.; Baig, A.; Qureshi, M.I. Effects of Cadmium Stress on Plants. In *Environmental Sustainability: Concepts, Principles, Evidences and Innovations*; Kishangarh: New Delhi, India, 2014; pp. 271–277.
7. Rogowska-Wrzesinska, A.; Wojdyla, K.; Nedić, O.; Baron, C.P.; Griffiths, H.R. Analysis of protein carbonylation—Pitfalls and promise in commonly used methods. *Free Radic. Res.* **2014**, *48*, 1145–1162. [CrossRef]
8. Polge, C.; Jaquinod, M.; Holzer, F.; Bourguignon, J.; Walling, L.; Brouquisse, R. Evidence for the existence in *Arabidopsis thaliana* of the proteasome proteolytic pathway: Activation in response to cadmium. *J. Biol. Chem.* **2009**, *284*, 35412–35424. [CrossRef]
9. Yadav, S.K. Heavy metals toxicity in plants: An overview on the role of glutathione and phytochelatins in heavy metal stress tolerance of plants. *South Afr. J. Bot.* **2010**, *76*, 167–179. [CrossRef]
10. Singh, S.; Parihar, P.; Singh, R.; Singh, V.P.; Prasad, S.M. Heavy metal tolerance in plants: Role of transcriptomics, proteomics, metabolomics, and ionomics. *Front. Plant Sci.* **2016**, *6*, 1143. [CrossRef]
11. Zhang, X.; Xia, H.; Li, Z.; Zhuang, P.; Gao, B. Potential of four forage grasses in remediation of Cd and Zn contaminated soils. *Bioresour. Technol.* **2010**, *101*, 2063–2066. [CrossRef]
12. Giraldo-Cañas, D. CatáLogo de la familia poaceae en colombia. *Darwiniana* **2011**, *49*, 139–247.
13. Giraldo-Cañas, D.A. *Las Gramíneas en Colombia: Riqueza, Distribución, Endemismo, Invasión, Migración, Usos y Taxonomías Populares*; Instituto de Ciencias Naturales, Universidad Nacional de Colombia: Bogotá DC, Colombia, 2013; p. 382.
14. Salas-Moreno, M.; Marrugo-Negrete, J. Phytoremediation potential of Cd and Pb-contaminated soils by *Paspalum fasciculatum* Willd. ex Flugge. *Int. J. Phytoremediation* **2019**, 1–11. [CrossRef] [PubMed]
15. Zhan, J.; Li, T.; Zhang, X.; Yu, H.; Zhao, L. Rhizosphere characteristics of phytostabilizer *Athyrium wardii* (Hook.) involved in Cd and Pb accumulation. *Ecotoxicol. Environ. Saf.* **2018**, *148*, 892–900. [CrossRef]

16. Deng, G.; Liu, L.; Wang, H.; Lao, C.; Wang, B.; Zhu, C.; Peng, D. Establishment and optimization of two-dimensional electrophoresis (2-DE) technology for proteomic analysis of ramie. *Int. J. Agric. Biol.* **2013**, *15*, 570–574.
17. Bashir, N.Y.; Lockwood, P.; Chasteen, A.L.; Nadolny, D.; Noyes, I. The ironic impact of activists: Negative stereotypes reduce social change influence. *Eur. J. Soc. Psychol.* **2013**, *43*, 614–626. [CrossRef]
18. Muradoglu, F.; Gundogdu, M.; Ercisli, S.; Encu, T.; Balta, F.; Jaafar, H.Z.; Zia-Ul-Haq, M. Cadmium toxicity affects chlorophyll a and b content, antioxidant enzyme activities and mineral nutrient accumulation in strawberry. *Biol. Res.* **2015**, *48*, 3–9. [CrossRef]
19. Polatajko, A.; Feldmann, I.; Hayen, H.; Jakubowski, N. Combined application of a laser ablation-ICP-MS assay for screening and ESI-FTICR-MS for identification of a Cd-binding protein in *Spinacia oleracea* L. after exposure to Cd. *Metallomics* **2011**, *3*, 1001–1008. [CrossRef]
20. Bhattacharya, T.; Chakraborty, S.; Banerjee, D.K. Heavy metal uptake and its effect on macronutrients, chlorophyll, protein, and peroxidase activity of *Paspalum distichum* grown on sludge-dosed soils. Heavy metal uptake and its effect on P. distichum. *Environ. Monit. Assess.* **2010**, *169*, 15–26. [CrossRef]
21. Shu, W.S.; Ye, Z.H.; Lan, C.Y.; Zhang, Z.Q.; Wong, M.H. Lead, zinc and copper accumulation and tolerance in populations of *Paspalum distichum* and *Cynodon dactylon*. *Environ. Pollut.* **2002**, *120*, 445–453. [CrossRef]
22. Chen, Y.; Chen, C.; Tan, Z.; Liu, J.; Zhuang, L.; Yang, Z.; Huang, B. Functional identification and characterization of genes cloned from halophyte seashore paspalum conferring salinity and cadmium tolerance. *Front. Plant Sci.* **2016**, *7*, 102. [CrossRef]
23. Gutsch, A.; Keunen, E.; Guerriero, G.; Renaut, J.; Cuypers, A.; Hausman, J.F.; Sergeant, K. Long-term cadmium exposure influences the abundance of proteins that impact the cell wall structure in *Medicago sativa* stems. *Plant Boil. (Stuttg)* **2018**, *20*, 1023–1035. [CrossRef] [PubMed]
24. Pochodylo, A.L.A.L.; Aristilde, L. Molecular dynamics of stability and structures in phytochelatin complexes with Zn, Cu, Fe, Mg, and Ca: Implications for metal detoxification. *Environ. Chem. Lett.* **2017**, *15*, 495–500. [CrossRef]
25. Li, X.; Zhou, Y.; Yang, Y.; Yang, S.; Sun, X.; Yang, Y. Physiological and proteomics analyses reveal the mechanism of *Eichhornia crassipes* tolerance to high-concentration cadmium stress compared with *Pistia stratiotes*. *PLoS ONE* **2015**, *10*, e0124304. [CrossRef] [PubMed]
26. Lin, Y.F.; Aarts, M.G. The molecular mechanism of zinc and cadmium stress response in plants. *Cell. Mol. Life Sci.* **2012**, *69*, 3187–3206. [CrossRef]
27. Prasad, A.; Kumar, A.; Suzuki, M.; Kikuchi, H.; Sugai, T.; Kobayashi, M.; Pospíšil, P.; Tada, M.; Kasai, S. Detection of hydrogen peroxide in Photosystem II (PSII) using catalytic amperometric biosensor. *Front. Plant Sci.* **2015**, *6*, 1–10. [CrossRef]
28. Khan, K.Y. Effect of humic acid amendment on cadmium bioavailability and accumulation by pak choi (*Brassica rapa* ssp. *chinensis* L.) to alleviate dietary toxicity risk. *Arch. Agron. Soil Sci.* **2017**, *63*, 1431–1442. [CrossRef]
29. Stefanic, P.P.; Cvjetko, P.; Biba, R.; Domijan, A.M.; Letofsky-Papst, I.; Tkalec, M.; Šikić, S.; Cindrić, M.; Balen, B. Physiological, ultrastructural and proteomic responses of tobacco seedlings exposed to silver nanoparticles and silver nitrate. *Chemosphere* **2018**, *209*, 640–653. [CrossRef]
30. Takahashi, S.; Murata, N. How do environmental stresses accelerate photoinhibition? *Trends Plant Sci.* **2008**, *13*, 178–182. [CrossRef]
31. Hajduch, M.; Rakwal, R.; Agrawal, G.K.; Yonekura, M.; Pretova, A. High-resolution two-dimensional electrophoresis separation of proteins from metal-stressed rice (*Oryza sativa* L.) leaves: Drastic reductions/fragmentation of ribulose-1,5-bisphosphate carboxylase/oxygenase and induction of stress-related proteins. *Electrophoresis* **2001**, *22*, 2824–2831. [CrossRef]
32. Wientjes, E.; Philippi, J.; Borst, J.W.; van Amerongen, H. Imaging the Photosystem I/Photosystem II chlorophyll ratio inside the leaf. *Biochim. Biophys. Acta Bioenerg.* **2017**, *1858*, 259–265. [CrossRef]
33. Gururani, M.A.; Venkatesh, J.; Tran, L.S.P. Regulation of photosynthesis during abiotic stress-induced photoinhibition. *Mol. Plant* **2015**, *8*, 1304–1320. [CrossRef] [PubMed]
34. Bashir, H.; Qureshi, M.I.; Ibrahim, M.M.; Iqbal, M. Chloroplast and photosystems: Impact of cadmium and iron deficiency. *Photosynthetica* **2015**, *53*, 321–335. [CrossRef]
35. Perreault, F.; Dionne, J.; Didur, O.; Juneau, P.; Popovic, R. Effect of cadmium on photosystem II activity in *Chlamydomonas reinhardtii*: Alteration of O-J-I-P fluorescence transients indicating the change of apparent activation energies within photosystem II. *Photosynth. Res.* **2011**, *107*, 151–157. [CrossRef] [PubMed]

36. Pribil, M.; Labs, M.; Leister, D. Structure and dynamics of thylakoids in land plants. *J. Exp. Bot.* **2014**, *65*, 1955–1972. [CrossRef] [PubMed]
37. Sigfridsson, K.G.V.; Bernát, G.; Mamedov, F.; Styring, S. Molecular interference of Cd^{2+} with Photosystem II. *Biochim. Biophys. Acta Bioenerg.* **2004**, *1659*, 19–31. [CrossRef] [PubMed]
38. Seelert, H.; Dencher, N.A. ATP synthase superassemblies in animals and plants: Two or more are better. *Biochim. Biophys. Acta Bioenerg.* **2011**, *1807*, 1185–1197. [CrossRef] [PubMed]
39. Yoshida, K.; Matsuoka, Y.; Hara, S.; Konno, H.; Hisabori, T. Distinct redox behaviors of chloroplast thiol enzymes and their relationships with photosynthetic electron transport in arabidopsis thaliana. *Plant Cell Physiol.* **2014**, *55*, 1415–1425. [CrossRef]
40. Buchert, F.; Forreiter, C. Singlet oxygen inhibits ATPase and proton translocation activity of the thylakoid ATP synthase CF1CFo. *FEBS Lett.* **2010**, *584*, 147–152. [CrossRef]
41. Mahler, H.; Wuennenberg, P.; Linder, M.; Przybyla, D.; Zoerb, C.; Landgraf, F.; Forreiter, C. Singlet oxygen affects the activity of the thylakoid ATP synthase and has a strong impact on its γ subunit. *Planta* **2007**, *225*, 1073–1083. [CrossRef]
42. Shah, K.; Dubey, R.S. A18 kDa cadmium inducible protein Complex: Its isolation and characterisation from rice (*Oryza sativa* L.) seedlings. *J. Plant Physiol.* **1998**, *152*, 448–454. [CrossRef]
43. Marrugo-Negrete, J.; Pinedo-Hernández, J.; Díez, S. Assessment of heavy metal pollution, spatial distribution and origin in agricultural soils along the Sinú River Basin, Colombia. *Environ. Res.* **2017**, *154*, 380–388. [CrossRef] [PubMed]
44. Kabata-pendias, A.; Pendias, H. *Trace Elements in Soils and Plants*, 3rd ed.; CRC Press: Boca Raton, FL, USA; London, UK, 2001; p. 403.
45. ECDGE. *Heavy Metals and Organic Compounds from Wastes Used as Organic Fertilisers*; ECDGE: Perchtoldsdorf, Austria, 2004.
46. Jedrzejczak, R. Determination of total mercury in foods of plant origin in Poland by cold vapour atomic absorption spectrometry. *Food Addit. Contam.* **2002**, *19*, 996–1002. [CrossRef] [PubMed]
47. United States Environmental Protection Agency. *Method 3051A (SW-846): Microwave Assisted Acid Digestion of Sediments, Sludges, and Oils*; United States Environmental Protection Agency: Washington, DC, USA, 2007.
48. Coquery, M.; Welbourn, P.M. The relationship between metal concentration and organic matter in sediments and metal concentration in the aquatic macrophyte Eriocaulon septangulare. *Water Res.* **1995**, *29*, 2094–2102. [CrossRef]
49. Bradford, M.M. A rapid and sensitive method for the quantitation of microgram quantities of protein utilizing the principle of protein-dye binding. *Anal. Biochem.* **1976**, *72*, 248–254. [CrossRef]
50. Wehr, N.B.; Levine, R.L. Quantitation of protein carbonylation by dot blot. *Anal. Biochem.* **2012**, *423*, 241–245. [CrossRef] [PubMed]
51. Contreras-Puentes, N.; Rodriguez-Cavallo, E.; Mendez-Cuadro, D. Membrane protein carbonylation of *Plasmodium falciparum* infected erythrocytes under conditions of sickle cell trait and G6PD deficiency. *Mol. Biochem. Parasitol.* **2019**, *227*, 5–14. [CrossRef] [PubMed]
52. Mesquita, C.S.; Oliveira, R.; Bento, F.; Geraldo, D.; Rodrigues, J.V.; Marcos, J.C. Simplified 2,4-dinitrophenylhydrazine spectrophotometric assay for quantification of carbonyls in oxidized proteins. *Anal. Biochem.* **2014**, *458*, 69–71. [CrossRef]
53. Levine, R.L.; Garland, D.; Oliver, C.N.; Amici, A.; Climent, I.; Lenz, A.G.; Ahn, B.W.; Shaltiel, S.; Stadtman, E.R. Determination of carbonyl content in oxidatively modified proteins. *Methods Enzymol.* **1990**, *186*, 464–478.
54. Mendez, D.; Linares, M.; Diez, A.; Puyet, A.; Bautista, J.M. Stress response and cytoskeletal proteins involved in erythrocyte membrane remodeling upon *Plasmodium falciparum* invasion are differentially carbonylated in G6PD A-deficiency. *Free Radic. Biol. Med.* **2011**, *50*, 1305–1313. [CrossRef]
55. Sechi, S.; Chait, B.T. Modification of cysteine residues by alkylation. A tool in peptide mapping and protein identification. *Anal. Chem.* **1998**, *70*, 5150–5158. [CrossRef]

plants

MDPI

Article

Silicon and Iron Differently Alleviate Copper Toxicity in Cucumber Leaves

Dragana Bosnić [1], Predrag Bosnić [2], Dragana Nikolić [1], Miroslav Nikolić [2] and Jelena Samardžić [2,*]

[1] Laboratory for Plant Molecular Biology, Institute of Molecular Genetics and Genetic Engineering, University of Belgrade, Vojvode Stepe 444a, 11000 Belgrade, Serbia; dragana.bosnic@imgge.bg.ac.rs (D.B.); dragana.nikolic.imgge@rcub.bg.ac.rs (D.N.)

[2] Department of Plant Nutrition, Institute for Multidisciplinary Research, University of Belgrade, Kneza Višeslava 1, 11000 Belgrade, Serbia; predrag.bosnic@imsi.bg.ac.rs (P.B.); mnikolic@imsi.bg.ac.rs (M.N.)

* Correspondence: jelena.samardzic@imgge.bg.ac.rs; Tel.: +381-11-3976-414

Received: 1 November 2019; Accepted: 20 November 2019; Published: 28 November 2019

check for updates

Abstract: Copper (Cu) toxicity in plants may lead to iron (Fe), zinc (Zn) and manganese (Mn) deficiencies. Here, we investigated the effect of Si and Fe supply on the concentrations of micronutrients and metal-chelating amino acids nicotianamine (NA) and histidine (His) in leaves of cucumber plants exposed to Cu in excess. Cucumber (*Cucumis sativus* L.) was treated with 10 μM Cu, and additional 100 μM Fe or/and 1.5 mM Si for five days. High Cu and decreased Zn, Fe and Mn concentrations were found in Cu treatment. Additional Fe supply had a more pronounced effect in decreasing Cu accumulation and improving the molar ratio between micronutrients as compared to the Si supply. However, the simultaneous supply of Fe and Si was the most effective treatment in alleviation of Cu-induced deficiency of Fe, Zn and Mn. Additional Fe supply increased the His but not NA concentration, while Si supply significantly increased both NA and His whereby the NA:Cu and His:Cu molar ratios exceeded the control values indicating that Si recruits Cu-chelation to achieve Cu tolerance. In conclusion, Si-mediated alleviation of Cu toxicity was directed toward Cu tolerance while Fe-alleviative effect was due to a dramatic decrease in Cu accumulation.

Keywords: copper toxicity; micronutrient deficiency; silicon; iron; nicotianamine; histidine; Cu-chelation

1. Introduction

Micronutrients such as copper (Cu), iron (Fe), zinc (Zn) and manganese (Mn), are essential to all plants, since they participate in numerous metabolic processes. The interaction between the elements becomes apparent when the excessive level of one of the micronutrients in the growth medium is present so that it may interfere with the uptake, transport and accumulation of the other nutrients, consequently disturbing the overall plant nutrient balance [1]. Copper present at high concentration has phytotoxic effects such as inhibition of plant growth and development, and generation of oxidative stress [2]. Plants exposed to overload of Cu are highly prone to the induced deficiency of other essential ions thus suffering a disturbance of essential metabolic processes [1]. Being a transition metal, Cu in excess competes for specific binding sites of other metal ions and consequently displace them from their binding sites [2]. Most of the plant metal transporters have limited ion specificity, therefore there are no efficient mechanisms for plants to differentially regulate the uptake of specific metal ion and avoid competition [3]. Divalent cation uptake transporters such as: Iron-regulated transporter 1 (IRT1), Zrt/IRT-like protein (ZIP) transporter family, natural resistance-associated macrophage proteins (NRAMPs) are able to unselectively mediate the transport of a broad range of metals (Fe, Zn, Cu, Mn and Cd) depending on their availability [3–5]. The copper and Fe competition is of particular concern since they are redox active counterparts and their metabolic crossroads are in tight relationships [6]. It has been proposed that high Cu induces alterations in Fe nutrition by decreasing the root Fe availability and Fe acquisition processes along with inhibition of Fe-deficiency responses [7–10]. Excess Cu competes with Fe for the reduction sites, i.e., Fe (III) reductase (FRO) activity, which precedes the uptake of both metals [11]. Accordingly, inhibition of root FRO activity with increased Cu supply has been found in cucumber [7,12]. However, it has also been noticed that additional Fe supply can be used for mitigation of the deleterious effects caused by high Cu and for correcting Fe deficiency induced by excess of Cu [13–15].

Silicon (Si) is not essential for most of the plant species; however, its presence in plants may influence plant response to both mineral toxicity and deficiency stress [16,17]. For example, addition of Si to Cu-stressed cucumber plants enhances accumulation of Cu-binding molecules (proteins and organic acids) that could buffer excess Cu, thus diminishing the level of oxidative stress [18]. However, there are only a few studies addressing nutrient imbalances in plant leaves under Cu toxicity and Si nutrition. Lower leaf Cu and increased calcium (Ca) concentration was found to be one of the main beneficial effects of Si nutrition for both *Zinnia elegans* and *Erica andevalensis* exposed to toxic Cu [19,20]. This was attributed to the well-known antagonistic effect of Cu and Ca, which was alleviated by Si supplementation. In addition, Si may alleviate the excess of Cu by increasing the uptake of other essential metals such as Mn and Zn [21].

Apart from the metal uptake, the tissue metal partition and distribution may also be affected by metal imbalances. Nicotianamine (NA) is a non-proteinogenic amino acid and a major metal chelator involved in the homeostasis of metals such as Cu, Fe, Zn and Mn. NA is essential for metal distribution in leaves via phloem and seed loading of Cu, Fe, Zn and Mn [22–24]. However, NA has an exclusive role in Cu metabolism, since it is also responsible for xylem transport of Cu but not of other metals [25,26], and contrary to Fe, Mn- and Zn-, the Cu-NA complex is very stable at mild acidic pH [27,28]. NA was suggested to participate in heavy metal stress [28]. Moreover, binding of metals by strong ligands such as NA is considered as the main detoxification strategy in non-hypertolerant plants [29].

Besides NA, the amino acid histidine (His) is another important metal-chelator. This amino acid is critical for metal coordination to metalloproteins and metal binding to the active site of metalloenzymes [30]. Free His was found to be implicated in metal hypertolerance [31,32] in addition to its decisive role in metal hyperaccumulation as shown for Ni [33]. Both NA and His form complexes with Cu with stability constants much higher than for other elements [34,35].

This study investigates for the first time how Si and/or Fe nutrition modulates the leaf micronutrient status under Cu toxicity stress, with special emphasis on the distinctive responses of metal-chelating amino acids (NA and His).

2. Results

2.1. Plant Growth and Symptoms of Copper Toxicity Stress

Total dry biomass of hydroponically grown cucumber plants was the lowest in Cu treatment (Table 1, Figure S1). Cu + Si treatment increased plant biomass by 32% as compared to Cu treatment. There was no difference in the total plant biomass between Cu + Fe and Cu + Fe + Si treatments and the control cucumber plants (Table 1). In comparison with the control, a significant decrease in the leaf DW was only recorded in Cu treatment (Table 1).

Table 1. Total plant dry weight and leaf dry weight of cucumber plants at the end of experiment. Plants were treated with excess Cu (Cu), excess Cu and Si supply (Cu + Si), excess Cu and additional Fe supply (Cu + Fe), or combined excess Cu and additional Fe and Si supply (Cu + Fe + Si); control plants (C) remained untreated.

Treatment	Dry Biomass (g Plant^{-1})	
	Total	Leaf
C	0.330 ± 0.050 c	0.223 ± 0.035 b
Cu	0.209 ± 0.047 a	0.160 ± 0.005 a
Cu + Si	0.265 ± 0.021 b	0.199 ± 0.014 b
Cu + Fe	0.291 ± 0.005 bc	0.200 ± 0.006 b
Cu + Fe + Si	0.305 ± 0.008 bc	0.217 ± 0.008 b [1]

[1] The values are means of four replicates ± SD. Different letters indicate significant differences among the treatments at $p < 0.05$.

2.2. *Concentration and Content of Microelements (Cu, Fe, Zn and Mn) in Cucumber Leaves*

Copper treatment resulted in a 2.3-fold increase in the Cu concentration in the leaves (Figure 1A). Si supply to the Cu treatment slightly but significantly ($p < 0.05$) decreased the Cu concentration (Figure 1A). However, there was no difference in leaf Cu content between Cu and Cu + Si treatments due to the changes in leaf dry weight (Table 2). Both Cu + Fe and Cu + Fe + Si treatments decreased remarkably the Cu concentration and content almost two-fold in comparison to only Cu-treated plants (Figure 1A, Table 2). The leaf concentrations of Zn, Fe and Mn were decreased about 2-fold in Cu treatment, compared to the control and it gradually increased in Cu + Si, Cu + Fe and Cu + Fe + Si treatment, respectively (Figure 1B–D). The most effective improvement in the microelement concentration was observed for Fe in Cu + Fe + Si treatment, reaching the control values (Figure 1B). The leaf content of Fe, Zn and Mn was also calculated and presented in Table 2. A remarkable decrease in the content of all examined elements, particularly for Mn (three-fold) was found in Cu treatment. Treatments: Cu + Si, Cu + Fe and Cu + Fe + Si gradually increased the content of Fe, Mn and Zn, reaching up to 87%, 71% and 63%, of the control level in Cu + Fe + Si treatment, respectively (Table 2).

Table 2. Total content of Cu, Fe, Zn and Mn in cucumber leaves (µmol per shoot). Plants were harvested after 5 days of treatment with excess Cu (Cu), excess Cu and Si supply (Cu + Si), excess Cu and additional Fe supply (Cu + Fe) or combined excess Cu and additional Fe and Si supply (Cu + Fe + Si); control plants (C) remained untreated.

Treatment	Microelement Content (µmol Per Shoot)			
	Cu	Fe	Zn	Mn
C	0.042 ± 0.001 a	0.460 ± 0.010 c	0.308 ± 0.027 d	0.199 ± 0.002 e
Cu	0.081 ± 0.003 c	0.168 ± 0.034 a	0.113 ± 0.008 a	0.068 ± 0.001 a
Cu + Si	0.079 ± 0.003 c	0.235 ± 0.021 ab	0.156 ± 0.020 b	0.090 ± 0.007 b
Cu + Fe	0.047 ± 0.004 ab	0.269 ± 0.008 b	0.159 ± 0.009 b	0.114 ± 0.009 c
Cu + Fe + Si	0.050 ± 0.001 b	0.401 ± 0.057 c	0.193 ± 0.017 c	0.142 ± 0.012 d [1]

[1] The values are means of four replicates ± SD. Different letters indicate significant differences among the treatments at $p < 0.05$.

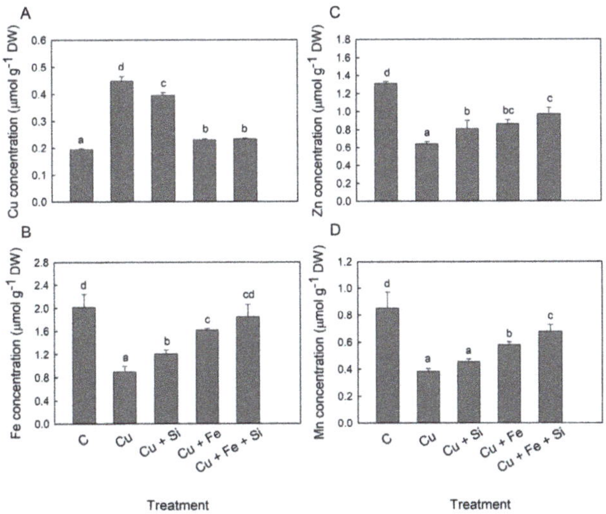

Figure 1. Concentration of Cu (**A**), Fe (**B**), Zn (**C**) and Mn (**D**) in cucumber leaves. Plants were harvested after 5 days of treatment with excess Cu (Cu), excess Cu and Si supply (Cu + Si), excess Cu and additional Fe supply (Cu + Fe) or combined excess Cu and additional Fe and Si supply (Cu + Fe + Si); control plants (C) remained untreated. Bars are means of four replicates ± SD. Different letters indicate significant differences among the treatments at $p < 0.05$.

2.3. Molar Ratio between Micronutrients

The molar ratio between elements was calculated and presented in Table 3. The lowest Fe:Cu, Zn:Cu and Mn:Cu molar ratios were noticed in Cu treatment; the Fe:Cu ratio was five-fold lower in comparison to control, whereas the Zn:Cu ratio was seven-fold decreased, from 7:1 to 1:1. The Mn:Cu molar ratio in Cu treatment was decreased to 1:1, as well. However, Cu + Si treatment increased the Fe:Cu (3:1) and Zn:Cu (2:1) ratios, while Mn:Cu ratio was the same as in Cu treatment. The Fe:Cu molar ratio was significantly improved in both Cu + Fe and Cu + Fe + Si treatments, 7:1 and 8:1, respectively (Table 3). The Zn:Cu ratio was the same in those two treatments, 4:1, while higher than in other treatments. Unlike Cu and Cu + Si, the Cu + Fe and Cu + Fe + Si treatments improved the Mn:Cu ratio, to 2:1 and 3:1, respectively (Table 3).

Table 3. Molar ratios of Fe:Cu, Zn:Cu and Mn:Cu in cucumber leaves. Plants were harvested after 5 days of treatment with excess Cu (Cu), excess Cu and Si supply (Cu + Si), excess Cu and additional Fe supply (Cu + Fe) or combined excess Cu and additional Fe and Si supply (Cu + Fe + Si); control plants (C) remained untreated.

Treatment	Fe:Cu	Zn:Cu	Mn:Cu
C	10:1	7:1	4:1
Cu	2:1	1:1	1:1
Cu + Si	3:1	2:1	1:1
Cu + Fe	7:1	4:1	2:1
Cu + Fe + Si	8:1	4:1	3:1 [1]

[1] The values are means of four replicates.

2.4. Concentration of NA and the Molar Ratio of NA to Cu, Fe, Zn and Mn

NA accumulation in leaves was induced in Cu treatment by 30% compared to control (Figure 2). Cu + Si treatment had the highest increase in NA concentration, about 300% of the control values. In

Cu + Fe plants NA concentration was decreased to 70% of the control plants values. NA concentration in Cu + Fe + Si treatment was slightly higher compared to that in the control plants though not statistically significant (Figure 2).

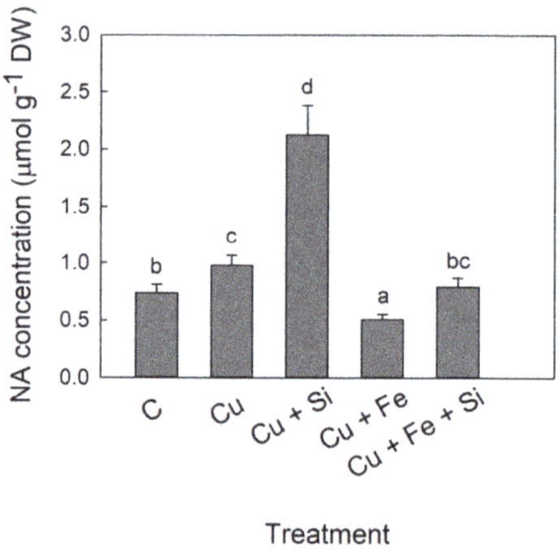

Figure 2. Concentration of nicotianamine (NA) in cucumber leaves. Plants were harvested after 5 days of treatment with excess Cu (Cu), excess Cu and Si supply (Cu + Si), excess Cu and additional Fe supply (Cu + Fe) or combined excess Cu and additional Fe and Si supply (Cu + Fe + Si); control plants (C) remained untreated. Bars are means of four replicates ± SD. Different letters indicate significant differences among the treatments at $p < 0.05$.

The molar ratio of NA to microelements in leaves of control cucumber plants was found to be as following: 3.8:1, 0.4:1, 0.6:1 and 0.9:1 for Cu, Fe, Zn and Mn, respectively. The molar ratio NA:Cu was decreased in Cu treatment while the molar ratios of NA to other elements (Fe, Zn and Mn) were significantly increased compared to control (Table 4). In Cu + Si treatment, the NA:Cu ratio was the highest among all treatments, exceeding the control value, and this treatment resulted in additionally increased the ratio between NA and the other elements compared to Cu treatment. On the contrary, the NA:Cu ratio was decreased in Cu + Fe treatment while combined Cu + Fe + Si treatment restored the NA:Cu ratio almost to the control level (Table 4). The NA:Fe, NA:Zn and NA:Mn molar ratios in Cu + Fe and Cu + Fe + Si treatments were similar to the control values (Table 4).

Table 4. The molar ratios of NA:Cu, NA:Fe, NA:Zn and NA:Mn in cucumber leaves. Plants were harvested after 5 days of treatment with excess Cu (Cu), excess Cu and Si supply (Cu + Si), excess Cu and additional Fe supply (Cu + Fe) or combined excess Cu and additional Fe and Si supply (Cu + Fe + Si); control plants (C) remained untreated.

Treatment	NA:Cu	NA:Fe	NA:Zn	NA:Mn
C	3.8:1	0.4:1	0.6:1	0.9:1
Cu	2.2:1	1.1:1	1.5:1	2.6:1
Cu + Si	5.4:1	1.8:1	2.6:1	4.7:1
Cu + Fe	2.2:1	0.3:1	0.6:1	0.9:1
Cu + Fe + Si	3.4:1	0.4:1	0.8:1	1.2:1 [1]

[1] The values are means of four replicates.

2.5. Concentration of His and the Molar Ratio of His to Cu, Fe, Zn and Mn

A continuous increase in His concentration was found in all treatments starting from Cu (1.8-fold) to Cu + Fe + Si treatment (5.4-fold) compared to control values (Figure 3).

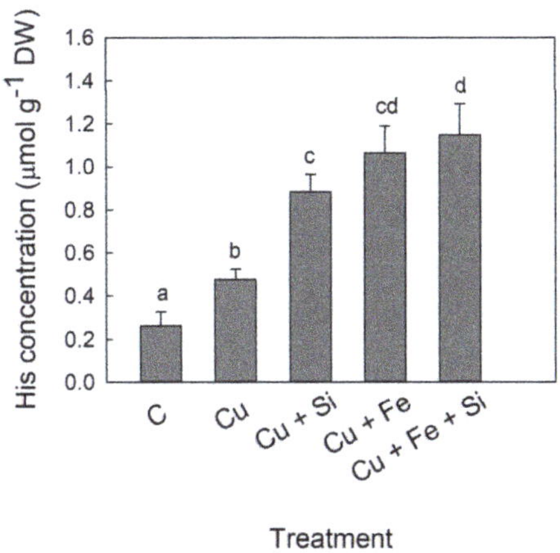

Figure 3. Concentration of histidine (His) in cucumber leaves. Plants were harvested after 5 days of treatment with excess Cu (Cu), excess Cu and Si supply (Cu + Si), excess Cu and additional Fe supply (Cu + Fe) or combined excess Cu and additional Fe and Si supply (Cu + Fe + Si); control plants (C) remained untreated. Bars are means of four replicates ± SD. Different letters indicate significant differences among the treatments at $p < 0.05$.

The changes in His and metal concentrations consequently altered the His:metal molar ratios (Table 5). In Cu treatment the His:Cu molar ratio was decreased compared to control, while Cu + Si treatment increased it above the control level. Both Cu + Fe and Cu + Fe + Si treatments had many-fold higher the His:Cu molar ratio then the control. In accordance with total His concentration, the molar ratios: His:Fe, His:Zn and His:Mn were several times greater in all examined treatments: Cu, Cu + Si, Cu + Fe and Cu + Fe + Si (Table 5).

Table 5. The molar ratios of His:Cu, His:Fe, His:Zn and His:Mn in cucumber leaves. Plants were harvested after 5 days of treatment with excess Cu (Cu), excess Cu and Si supply (Cu + Si), excess Cu and additional Fe supply (Cu + Fe) or combined excess Cu and additional Fe and Si supply (Cu + Fe + Si); control plants (C) remained untreated.

Treatment	His:Cu	His:Fe	His:Zn	NA:Mn
C	1.3:1	0.1:1	0.2:1	0.3:1
Cu	1.1:1	0.5:1	0.7:1	1.2:1
Cu + Si	2.2:1	0.7:1	1.1:1	1.9:1
Cu + Fe	4.6:1	0.7:1	1.2:1	1.8:1
Cu + Fe + Si	4.9:1	0.6:1	1.2:1	1.7:1 [1]

[1] The values are means of four replicates.

3. Discussion

The primary site of Cu toxicity is the root system; therefore many studies have been devoted to investigation of the detrimental effects of excess Cu in roots [36–38]. In our previous study, we have shown the toxic effects of 10 μM Cu treatment to cucumber plants where copper was highly accumulated in the root, reaching more than 10-fold higher values than in the leaf [18]. Apart from high Cu accumulation, excess of Cu also causes unbalanced uptake of other mineral nutrients [1,10]. Deficiency of both macro- and micronutrients was found in the leaves of different plant species exposed to Cu stress [39,40]. Herein we focused on the fine changes in molar ratios between Cu and other micronutrients and on the role of chelating amino acids in leaf under ameliorative effect of Si and/or additional Fe.

Decreased concentration and content of Fe, Zn and Mn were recorded in the leaves of Cu-treated cucumber along with higher Cu accumulation (Figure 1, Table 2). It is well documented that addition of Si may improve the nutritional status of plants despite nutrient imbalances in the growth medium [16]. Cucumber is one of the few Si-accumulating dicots with a relatively high capacity for Si accumulation, which has been shown in numerous studies [16]. In our previous study we have found that Si supply decreased Cu uptake in cucumber plants by enhancing the Cu deposition in the root cell wall of the Cu-treated cucumber [18]. However, in studies dealing with Fe, Zn or Mn deficiency the beneficial effect of Si was attributed to the metal remobilization and enhanced metal distribution rather than to a direct effect on the metal uptake [41–45]. In the present work, Si supply decreased Cu accumulation; however, there was no difference in Cu content between Cu and Cu + Si treatments due to the biomass increase of Cu + Si plants (Figure 1, Tables 1 and 2). Additionally, Si supply increased Fe, Zn and Mn concentrations and contents (Figure 1, Table 1). Similarly, in *Erica andevalensis* leaf Si supply significantly increased Zn concentration, which was lowered due to the competitive Cu–Zn interaction [19]. Si addition also restored the leaf Mn concentration in *Spartina densiflora* subjected to Cu stress [46]. Moreover, it was observed in the current study that the molar ratios of Fe:Cu and Zn:Cu were improved in the Si-supplied cucumber plants (Table 3). The optimal micronutrient ratios in plants are considered just as important as their absolute concentration particularly in terms of induced deficiency of specific nutrient [47]. Taken together, our results indicate that Si supply can alleviate to some extent the antagonistic effect of excess Cu on Fe, Zn and as well as the Mn accumulation in the leaves.

Interestingly, additional Fe supply was more effective in preventing Cu accumulation in cucumber leaves than Si supply (Figure 1A, Table 2). It has been reported that addition of Fe could ameliorate Cu toxicity symptoms due to the antagonistic interaction between Cu and Fe [7,40]. Studies in bean and maize have shown that additional Fe mitigates Cu accumulation by outcompeting Cu uptake, thus improving plant growth and decreasing oxidative stress [13,14]. We have also found that application of Fe to Cu-stressed plants resulted in a greater increase of Fe, Zn and Mn concentrations and contents than addition of Si, thus significantly improving the Fe:Cu and also Zn:Cu and Mn:Cu molar ratios, which were disturbed by the Cu treatment (Figure 1B–D, Table 3). The molar ratio between elements is in some cases the decisive indicator of stress; for example the molar ratio of Fe to other elements, rather than the total Fe concentration, has been found to be crucial for the initiation of Fe-deficiency response [48].

Simultaneous supply of Si and Fe had no additional effect regarding the Cu concentration in the leaves (Figure 1). On the other hand, this was the most effective treatment in respect to the increasing contents of Zn and Mn (Table 2). Moreover, this treatment restored the Fe content to the control level (Table 2). Consequently, the Fe:Cu, Zn:Cu and Mn:Cu molar ratios in the Cu + Fe + Si treatment were closest to the control values among all the treatments studied (Table 3). Therefore, we suggest that only when simultaneously applied with Si, the additional Fe supply could reach its full ameliorative potential regarding alleviation of the microelement deficiency induced by excess of Cu.

The accumulation of free amino acids is considered as an active response of plants to heavy metal stress [49]. Indeed, it has been shown that plants respond to excess of Cu by accumulating free amino acids [50]. However, the amount of certain amino acids rises rather than others, it refers primarily to NA and His, which have the highest binding constants for Cu compared to those for other metal ions [34,35]. Cu-NA and Cu-His complexes have been detected in plant tissues in different plant species [51–53]. In the present study, we recorded increased NA and His accumulation in the cucumber leaves under Cu toxicity stress (Figures 2 and 3); however, the NA:Cu and His:Cu molar ratios decreased compared to those in the control plants (Tables 4 and 5). Similar to our results, in the xylem sap from chicory and tomato plants both the NA and His concentrations were found elevated by Cu treatment [50]. Pich and Scholz [25] suggested that the two-fold higher NA concentration found in the leaves of Cu-treated tomato was required for the protection against phytotoxicity caused by excess Cu.

Interestingly, Si supply to Cu-stressed cucumber plants led to an additional increase in concentration of both amino acids (Figures 2 and 3). The study with Cu-treated bamboo revealed a high proportion of amino groups involved in Cu complexation, along with an increased His accumulation in Si-supplied plants [52]. This could be specific for stress conditions; however, there are no reports about Si influence on amino acids in non-stressed plants so far.

In the present study, the NA:Cu and His:Cu molar ratios were significantly improved in Cu + Si treatment so that they even exceeded the control values (Tables 4 and 5). The molar ratio of NA to metals was found constant during the plant development and as such is important for ensuring efficient loading and transport of essential micronutrients [23]. However, the significant surplus of metal chelators indicated by the molar ratio, could be decisive for achieving the tolerance to metal toxicity. The higher NA:Cu and His:Cu molar ratios, exceeding the control values, in the Si-supplied plants suggest that Si recruits the Cu-chelators to attain higher Cu tolerance. Chelation is considered as a potential mechanism that governs metal tolerance in plants by controlling the concentration of the free metal ions [54]. We have already reported that the decreased oxidative stress in the Cu + Si-treated plants as well as their improved tolerance to Cu are attributable to Cu-buffering by the increased levels of Cu-binding proteins and organic acids [18]. Taken together, the Cu-binding and Cu-chelating compounds are important mechanism of Si-alleviative effect in Cu toxicity stress.

Apart from Cu, NA is indispensable in homeostasis of Fe, but its role is ambiguous in Fe deficiency. It has been reported that the Fe deficiency did not cause biosynthesis and accumulation of NA in dicots [26,55,56]. Surprisingly, both overproduction and low NA resulted in decreased availability of Fe and affected Fe deficiency response [57,58]. Therefore, it is indicative that an optimal NA:Fe ratio is necessary for the Fe homeostasis so that an imbalance of this ratio, resulting from high or low NA, would lead to the Fe deficiency response. However, it has been found that addition of Si stimulated the NA production in Fe-deficient cucumber, leading to enhanced remobilization of Fe from the older to younger cucumber leaves [59]. Therefore, NA may have an important role in the Si-stimulated metal remobilization during metal deficiency and metal buffering in metal toxicity.

Additional Fe supply did not increase NA concentration, which is in accordance with the strong decrease of Cu accumulation observed in Cu + Fe treatment (Figure 2). However, simultaneous application of Fe and Si to Cu-stressed plants increased NA concentration, which resulted in an improved NA:Cu molar ratio, reaching the control level (Table 3). Interesting results have been obtained for His in the Cu + Fe and Cu + Fe + Si treatments. Although we recorded a lower Cu concentration in leaf compared to the concentration in Cu treatment, increased His levels were found in both Cu + Fe and Cu + Fe + Si treatments (Figure 3). Contrary to our results, concentration of His in the xylem sap of *Brassica carinata* followed the increasing trend proportional to the rising Cu concentration and had the greatest relative increase among all the amino acids analyzed [60]. Surprisingly, in the same study NA was not induced under Cu excess but rather under Cu deficiency although NA had the highest absolute concentration compared to all amino acids analyzed [60]. As suggested by Irtelli et al. [60], response to Cu stress is not only species-specific but it is also stress-specific.

4. Materials and Methods

4.1. Plant Material and Growth Conditions

Seeds of cucumber (*Cucumis sativus* L. cv. Chinese long) were surface sterilized with sodium hypochlorite, soaked in a saturated calcium sulfate solution overnight and then transferred for germination on moist filter papers for 4 days in the dark at 25 °C. The seedlings were grown for 12 days in preculture nutrient solution containing (in mM): 0.7 K_2SO_4, 0.1 KCl, 2.0 Ca $(NO_3)_2$, 0.5 $MgSO_4$ and 0.1 KH_2PO_4, and (in μM): 0.5 $MnSO_4$, 0.5 $ZnSO_4$, 0.2 $CuSO_4$, 0.01 $(NH_4)_6Mo_7O_{24}$, 10 H_3BO_3 and 20 Fe^{III}EDTA. Nutrient solution was constantly aerated and renewed completely every two days. pH was adjusted to 6 and daily checked. Plants were grown at 24/20 °C (light/dark) with a day/night regime of 16/8 h and 250 μmol photons m^{-2} s^{-1} light intensity at plant height (provided by led panels; Apollo 8,

Cidly Co., Ltd., Shenzhen, China). The experimental set up included five different groups of plants: control plants grown in standard solution containing 0.2 μM Cu and 20 μM Fe (C); plants treated with 10 μM Cu supplied as $CuSO_4$ into solution, which contained 20 μM Fe (Cu treatment); plants grown in standard solution with 20 μM Fe and treated with 10 μM Cu and 1.5 mM Si (Cu + Si treatment); plants treated with 10 μM Cu and 100 μM Fe (Cu + Fe treatment) and the last treatment contained combined 10 μM Cu, 100 μM Fe and 1.5 mM Si (Cu + Fe + Si treatment). Fe was supplied as Fe^{III}EDTA and Si was supplied as Si $(OH)_4$ freshly prepared by passing Na_2SiO_3 through a cation-exchange resin (Amberlite IR-120, H^+ form; Acros Organics, Geel, Belgium); The experiments were performed in four replicates (pots) with four plants in each pot.

4.2. Visual Observation, Plant Harvesting and Biomass Recording

The visual symptoms were recorded and the plants were harvested five days after the onset of treatments. One subsample was oven-dried at 70 °C for 72 h, weighed and subsequently ground for microelement determination. The samples for amino acids analysis were instantly frozen in liquid nitrogen after harvesting.

4.3. Determination of Microelements Cu, Fe, Zn and Mn

Dry leaf material was ground and digested in a mixture of 3 mL concentrated HNO_3 and 2 mL H_2O_2 in a microwave oven (Speedwave MWS-3+; Berghof Products + Instruments GmbH, Eningen, Germany). Cu, Fe, Zn and Mn concentrations were determined by inductively coupled plasma optical emission spectroscopy (ICP-OES, Spectro-Genesis EOP II; Spectro Analytical Instruments GmbH, Kleve, Germany). The analytical accuracy of total concentrations was evaluated using certified reference material (GBW 10015 Spinach, Institute of Geophysical and Geochemical Exploration, Langfang, China).

4.4. Extraction and Analysis of Amino Acids Nicotianamine and Histidine

For extraction of nicotianamine, leaf tissue was homogenized in liquid nitrogen and extracted in ddH_2O, incubated at 80 °C for 30 min and after centrifugation for 10 min at 16,000× *g* the supernatant was filtered and used for HPLC analysis, similarly as in Mendoza-Cózatl et al. [61]. For histidine determination, leaf tissue was homogenized in liquid nitrogen and extracted in 50% methanol, centrifuged at 4 °C and the filtered supernatant was used for analysis. Analysis of NA and His was performed according to Vasanits et al. [62] using o-phthaldialdehyde and 3-mercaptopropionic acid (OPA/MPA). Ten times diluted samples were derivatized with OPA/MPA for 5 min and immediately loaded on a reversed phase column (5.0 μm, 250 mm × 4.6 mm Luna C18 (2); Phenomenex Ltd., Torrance, CA, USA) using a Shimadzu LC-20AB Prominence liquid chromatograph (Shimadzu, Kyoto, Japan). The elution gradient was as described in Vasanits et al. [62] and the flow rate was 1.2 mL min^{-1} at 40 °C. The fluorescence intensity of OPA/MPA/AA derivative was measured at excitation and emission wavelengths of 337 and 454 nm using RF-10-AXL, fluorescence detector (Prominence, Shimadzu, Kyoto, Japan). The NA and His standards were obtained from Toronto Research Chemicals (North York, ON, Canada) and Sigma-Aldrich (St. Louis, MO, USA), respectively.

4.5. Statistical Analysis

The data were analyzed by analysis of variance (ANOVA) and tested for significance by post hoc Tukey test at a significance level of 0.05 using SPSS 21.0 software (IBM, Armonk, NY, USA).

5. Conclusions

Our results indicate that both NA and His were increased under conditions of high Cu concentration in the leaves (Cu treatment), whereas only the level of His was increased when the leaf Cu accumulation declined to a moderate level (Cu + Fe treatment). However, Si addition to either Cu or Cu + Fe treatments led to an increase in concentration of both NA and His in cucumber leaves. Therefore, the results suggest that the strategy of Si-mediated alleviation of Cu toxicity was directed primarily towards the increased Cu tolerance by Cu-chelation rather than by a marked decrease of the Cu accumulation. On the other hand, the alleviation of Cu toxicity by addition of Fe was achieved through a dramatic decrease of Cu accumulation, most probably due to an antagonism between Fe and Cu. The simultaneous addition of Si and Fe showed the best ameliorative potential regarding the Cu-induced microelement deficiency, which may be of practical importance in alleviating Cu toxicity stress in plants grown on Cu contaminated soils (e.g., post-mining and post-vineyard sites).

Supplementary Materials: The following are available online at http://www.mdpi.com/2223-7747/8/12/554/s1, Figure S1: Visual appearance of the cucumber plants

Author Contributions: Conceptualization, J.S., M.N.; methodology, D.B., P.B., and D.N.; formal analysis, D.B.; investigation, D.B., J.S.; writing—original draft preparation, D.B.; writing—review and editing, P.B., D.B., J.S., and M.N.; visualization, D.B.; funding acquisition, J.S., M.N.

Funding: This research was funded by Serbian Ministry of Education, Science and Technological Development, grant numbers ON-173005 and ON-173028.

Acknowledgments: We thank Milorad Kojic (University of Belgrade, Serbia) for English correction of the manuscript.

Conflicts of Interest: The authors declare no conflict of interest.

References

1. Marschner, H. *Mineral Nutrition of Higher Plants*; Special Publications of the Society for General Microbiology; Academic Press: London, UK, 1995.
2. Yruela, I. Copper in plants: Acquisition, transport and interactions. *Funct. Plant Biol.* **2009**, *36*, 409–430. [CrossRef]
3. Wintz, H.; Fox, T.; Wu, Y.-Y.; Feng, V.; Chen, W.; Chang, H.-S.; Zhu, T.; Vulpe, C. Expression profiles of Arabidopsis thaliana in mineral deficiencies reveal novel transporters involved in metal homeostasis. *J. Biol. Chem.* **2003**, *278*, 47644–47653. [CrossRef]
4. Cailliatte, R.; Schikora, A.; Briat, J.-F.; Mari, S.; Curie, C. High-affinity manganese uptake by the metal transporter NRAMP1 is essential for Arabidopsis growth in low manganese conditions. *Plant Cell* **2010**, *22*, 904–917. [CrossRef]
5. Vert, G.; Grotz, N.; Dédaldéchamp, F.; Gaymard, F.; Guerinot, M.L.; Briat, J.-F.; Curie, C. IRT1, an Arabidopsis transporter essential for iron uptake from the soil and for plant growth. *Plant Cell* **2002**, *14*, 1223–1233. [CrossRef]
6. Puig, S.; Andrés-Colás, N.; García-Molina, A.; Penarrubia, L. Copper and iron homeostasis in Arabidopsis: Responses to metal deficiencies, interactions and biotechnological applications. *Plant Cell Environ.* **2007**, *30*, 271–290. [CrossRef]
7. Waters, B.M.; Armbrust, L.C. Optimal copper supply is required for normal plant iron deficiency responses. *Plant Signal. Behav.* **2013**, *8*, e26611. [CrossRef]
8. Michaud, A.M.; Chappellaz, C.; Hinsinger, P. Copper phytotoxicity affects root elongation and iron nutrition in durum wheat (*Triticum turgidum durum* L.). *Plant Soil* **2008**, *310*, 151–165. [CrossRef]
9. Schmidt, W.; Bartels, M.; Tittel, J.; Fühner, C. Physiological effects of copper on iron acquisition processes in Plantago. *New Phytol.* **1997**, *135*, 659–666. [CrossRef]
10. Wallace, A.; Wallace, G.A.; Cha, J.W. Some modifications in trace metal toxicities and deficiencies in plants resulting from interactions with other elements and chelating agents—The special case of iron. *J. Plant Nutr.* **1992**, *15*, 1589–1598. [CrossRef]
11. Welch, R.M.; Norvell, W.A.; Schaefer, S.C.; Shaff, J.E.; Kochian, L.V. Induction of iron (III) and copper (II) reduction in pea (*Pisum sativum* L.) roots by Fe and Cu status: Does the root-cell plasmalemma Fe (III)-chelate reductase perform a general role in regulating cation uptake? *Planta* **1993**, *190*, 555–561. [CrossRef]
12. Alcántara, E.; Romera, F.J.; Cañete, M.; De la Guardia, M.D. Effects of heavy metals on both induction and function of root Fe (lll) reductase in Fe-deficient cucumber (*Cucumis sativus* L.) plants. *J. Exp. Bot.* **1994**, *45*, 1893–1898. [CrossRef]

13. Kumar, P.; Tewari, R.K.; Sharma, P.N. Modulation of copper toxicity-induced oxidative damage by excess supply of iron in maize plants. *Plant Cell Rep.* **2008**, *27*, 399–409. [CrossRef]
14. Pätsikkä, E.; Kairavuo, M.; Šeršen, F.; Aro, E.-M.; Tyystjärvi, E. Excess copper predisposes photosystem II to photoinhibition in vivo by outcompeting iron and causing decrease in leaf chlorophyll. *Plant Physiol.* **2002**, *129*, 1359–1367. [CrossRef]
15. Foy, C.D.; Chaney, R.L.; White, M.C. The Physiology of Metal Toxicity in Plants. *Annu. Rev. Plant Physiol.* **1978**, *29*, 511–566. [CrossRef]
16. Liang, Y.; Nikolic, M.; Bélanger, R.; Gong, H.; Song, A. Silicon-mediated tolerance to metal toxicity. In *Silicon in Agriculture*; Springer: Berlin/Heidelberg, Germany, 2015; pp. 83–122.
17. Hernandez-Apaolaza, L. Can silicon partially alleviate micronutrient deficiency in plants? A review. *Planta* **2014**, *240*, 447–458. [CrossRef]
18. Bosnić, D.; Nikolić, D.; Timotijević, G.; Pavlović, J.; Vaculík, M.; Samardžić, J.; Nikolić, M. Silicon alleviates copper (Cu) toxicity in cucumber by increased Cu-binding capacity. *Plant Soil* **2019**, *441*, 629–641. [CrossRef]
19. Oliva, S.R.; Mingorance, M.D.; Leidi, E.O. Effects of silicon on copper toxicity in Erica andevalensis Cabezudo and Rivera: A potential species to remediate contaminated soils. *J. Environ. Monit.* **2011**, *13*, 591–596. [CrossRef]
20. Frantz, J.M.; Khandekar, S.; Leisner, S. Silicon differentially influences copper toxicity response in silicon-accumulator and non-accumulator species. *J. Am. Soc. Hortic. Sci.* **2011**, *136*, 329–338. [CrossRef]
21. Keller, C.; Rizwan, M.; Davidian, J.-C.; Pokrovsky, O.S.; Bovet, N.; Chaurand, P.; Meunier, J.-D. Effect of silicon on wheat seedlings (*Triticum turgidum* L.) grown in hydroponics and exposed to 0 to 30 μM Cu. *Planta* **2015**, *241*, 847–860. [CrossRef]
22. Stephan, U.W.; Scholz, G. Nicotianamine: Mediator of transport of iron and heavy metals in the phloem? *Physiol. Plant.* **1993**, *88*, 522–529. [CrossRef]
23. Schmidke, I.; Stephan, U.W. Transport of metal micronutrients in the phloem of castor bean (Ricinus communis) seedlings. *Physiol. Plant.* **1995**, *95*, 147–153. [CrossRef]
24. Takahashi, M.; Terada, Y.; Nakai, I.; Nakanishi, H.; Yoshimura, E.; Mori, S.; Nishizawa, N.K. Role of nicotianamine in the intracellular delivery of metals and plant reproductive development. *Plant Cell* **2003**, *15*, 1263–1280. [CrossRef]
25. Pich, A.; Scholz, G. Translocation of copper and other micronutrients in tomato plants (Lycopersicon esculentum Mill.): Nicotianamine-stimulated copper transport in the xylem. *J. Exp. Bot.* **1996**, *47*, 41–47. [CrossRef]
26. Pich, A.; Scholz, G.; Stephan, U.W. Iron-dependent changes of heavy metals, nicotianamine, and citrate in different plant organs and in the xylem exudate of two tomato genotypes. Nicotianamine as possible copper translocator. *Plant Soil* **1994**, *165*, 189–196. [CrossRef]
27. Rellán-Álvarez, R.; Abadía, J.; Álvarez-Fernández, A. Formation of metal-nicotianamine complexes as affected by pH, ligand exchange with citrate and metal exchange. A study by electrospray ionization time-of-flight mass spectrometry. *Rapid Commun. Mass Spectrom.* **2008**, *22*, 1553–1562. [CrossRef]
28. Von Wirén, N.; Klair, S.; Bansal, S.; Briat, J.-F.; Khodr, H.; Shioiri, T.; Leigh, R.A.; Hider, R.C. Nicotianamine chelates both FeIII and FeII. Implications for metal transport in plants. *Plant Physiol.* **1999**, *119*, 1107–1114. [CrossRef]
29. Leitenmaier, B.; Küpper, H. Compartmentation and complexation of metals in hyperaccumulator plants. *Front. Plant Sci.* **2013**, *4*, 374. [CrossRef]
30. Sarkar, B.; Wigfield, Y. The Structure of Copper (II)-Histidine Chelate The question of the involvement of the imidazole group. *J. Biol. Chem.* **1967**, *242*, 5572–5577.
31. Nguyen, T.Q.; Hayward, A.R.; Bruce, K.E.; Hutchinson, T.C.; Emery, R.J.N. Chelator production by Deschampsia cespitosa (L.) Beauv. in adaptive Ni/Cu hyper-tolerance derived from fields in the Sudbury region and lab assessment. *Botany* **2018**, *96*, 711–721. [CrossRef]
32. Krämer, U. Metal hyperaccumulation in plants. *Annu. Rev. Plant Biol.* **2010**, *61*, 517–534. [CrossRef]
33. Callahan, D.L.; Baker, A.J.M.; Kolev, S.D.; Wedd, A.G. Metal ion ligands in hyperaccumulating plants. *JBIC J. Biol. Inorg. Chem.* **2006**, *11*, 2–12. [CrossRef] [PubMed]
34. Beneš, I.; Schreiber, K.; Ripperger, H.; Kircheiss, A. Metal complex formation by nicotianamine, a possible phytosiderophore. *Experientia* **1983**, *39*, 261–262. [CrossRef]

35. May, P.M.; Linder, P.W.; Williams, D.R. Computer simulation of metal-ion equilibria in biofluids: Models for the low-molecular-weight complex distribution of calcium (II), magnesium (II), manganese (II), iron (III), copper (II), zinc (II), and lead (II) ions in human blood plasma. *J. Chem. Soc. Dalt. Trans.* **1977**, 588–595. [CrossRef]
36. Wang, S.-H.; Zhang, H.; Zhang, Q.; Jin, G.-M.; Jiang, S.-J.; Jiang, D.; He, Q.-Y.; Li, Z.-P. Copper-Induced Oxidative Stress and Responses of the Antioxidant System in Roots of Medicago sativa. *J. Agron. Crop Sci.* **2011**, *197*, 418–429. [CrossRef]
37. Shi, J.; Yuan, X.; Chen, X.; Wu, B.; Huang, Y.; Chen, Y. Copper uptake and its effect on metal distribution in root growth zones of Commelina communis revealed by SRXRF. *Biol. Trace Elem. Res.* **2011**, *141*, 294–304. [CrossRef] [PubMed]
38. Zhang, H.; Xia, Y.; Wang, G.; Shen, Z. Excess copper induces accumulation of hydrogen peroxide and increases lipid peroxidation and total activity of copper–zinc superoxide dismutase in roots of Elsholtzia haichowensis. *Planta* **2008**, *227*, 465–475. [CrossRef]
39. Alaoui-Sossé, B.; Genet, P.; Vinit-Dunand, F.; Toussaint, M.-L.; Epron, D.; Badot, P.-M. Effect of copper on growth in cucumber plants (Cucumis sativus) and its relationships with carbohydrate accumulation and changes in ion contents. *Plant Sci.* **2004**, *166*, 1213–1218. [CrossRef]
40. Ouzounidou, G.; Ilias, I.; Tranopoulou, H.; Karataglis, S. Amelioration of copper toxicity by iron on spinach physiology. *J. Plant Nutr.* **1998**, *21*, 2089–2101. [CrossRef]
41. Nikolic, D.B.; Nesic, S.; Bosnic, D.; Kostic, L.; Nikolic, M.; Samardzic, J.T. Silicon Alleviates Iron Deficiency in Barley by Enhancing Expression of Strategy II Genes and Metal Redistribution. *Front. Plant Sci.* **2019**, *10*, 416. [CrossRef]
42. Pascual, M.B.; Echevarria, V.; Gonzalo, M.J.; Hernández-Apaolaza, L. Silicon addition to soybean (*Glycine max* L.) plants alleviate zinc deficiency. *Plant Physiol. Biochem.* **2016**, *108*, 132–138. [CrossRef]
43. Bityutskii, N.; Pavlovic, J.; Yakkonen, K.; Maksimović, V.; Nikolic, M. Contrasting effect of silicon on iron, zinc and manganese status and accumulation of metal-mobilizing compounds in micronutrient-deficient cucumber. *Plant Physiol. Biochem.* **2014**, *74*, 205–211. [CrossRef] [PubMed]
44. Pavlovic, J.; Samardzic, J.; Maksimović, V.; Timotijevic, G.; Stevic, N.; Laursen, K.H.; Hansen, T.H.; Husted, S.; Schjoerring, J.K.; Liang, Y. Silicon alleviates iron deficiency in cucumber by promoting mobilization of iron in the root apoplast. *New Phytol.* **2013**, *198*, 1096–1107. [CrossRef]
45. Jarvis, S.C.; Jones, L.H.P. The absorption and transport of manganese by perennial ryegrass and white clover as affected by silicon. *Plant Soil* **1987**, *99*, 231–240. [CrossRef]
46. Mateos-Naranjo, E.; Gallé, A.; Florez-Sarasa, I.; Perdomo, J.A.; Galmés, J.; Ribas-Carbó, M.; Flexas, J. Assessment of the role of silicon in the Cu-tolerance of the C4 grass Spartina densiflora. *J. Plant Physiol.* **2015**, *178*, 74–83. [CrossRef] [PubMed]
47. Marschner, P. *Marschner's Mineral Nutrition of Higher Plants*, 3rd ed.; Academic press: London, UK, 2012.
48. Kobayashi, T.; Itai, R.N.; Aung, M.S.; Senoura, T.; Nakanishi, H.; Nishizawa, N.K. The rice transcription factor IDEF1 directly binds to iron and other divalent metals for sensing cellular iron status. *Plant J.* **2012**, *69*, 81–91. [CrossRef]
49. Sharma, S.S.; Dietz, K.-J. The significance of amino acids and amino acid-derived molecules in plant responses and adaptation to heavy metal stress. *J. Exp. Bot.* **2006**, *57*, 711–726. [CrossRef]
50. Liao, M.T.; Hedley, M.J.; Woolley, D.J.; Brooks, R.R.; Nichols, M.A. Copper uptake and translocation in chicory (*Cichorium intybus* L. cv Grasslands Puna) and tomato (*Lycopersicon esculentum* Mill. cv. Rondy) plants grown in NFT system. II. The role of nicotianamine and histidine in xylem sap copper transport. *Plant Soil* **2000**, *223*, 243–252. [CrossRef]
51. Hazama, K.; Nagata, S.; Fujimori, T.; Yanagisawa, S.; Yoneyama, T. Concentrations of metals and potential metal-binding compounds and speciation of Cd, Zn and Cu in phloem and xylem saps from castor bean plants (Ricinus communis) treated with four levels of cadmium. *Physiol. Plant.* **2015**, *154*, 243–255. [CrossRef]
52. Collin, B.; Doelsch, E.; Keller, C.; Cazevieille, P.; Tella, M.; Chaurand, P.; Panfili, F.; Hazemann, J.-L.; Meunier, J.-D. Evidence of sulfur-bound reduced copper in bamboo exposed to high silicon and copper concentrations. *Environ. Pollut.* **2014**, *187*, 22–30. [CrossRef]

53. Mijovilovich, A.; Leitenmaier, B.; Meyer-Klaucke, W.; Kroneck, P.M.H.; Götz, B.; Küpper, H. Complexation and toxicity of copper in higher plants. II. Different mechanisms for copper versus cadmium detoxification in the copper-sensitive cadmium/zinc hyperaccumulator Thlaspi caerulescens (Ganges ecotype). *Plant Physiol.* **2009**, *151*, 715–731. [CrossRef]
54. Anjum, N.A.; Hasanuzzaman, M.; Hossain, M.A.; Thangavel, P.; Roychoudhury, A.; Gill, S.S.; Rodrigo, M.A.M.; Adam, V.; Fujita, M.; Kizek, R.; et al. Jacks of metal/metalloid chelation trade in plants—An overview. *Front. Plant Sci.* **2015**, *6*, 192. [CrossRef] [PubMed]
55. Higuchi, K.; Nishizawa, N.; Römheld, V.; Marschner, H.; Mori, S. Absence of nicotianamine synthase activity in the tomato mutant 'chloronerva'. *J. Plant Nutr.* **1996**, *19*, 1235–1239. [CrossRef]
56. Higuchi, K.; Nishizawa, N.-K.; Yamaguchi, H.; Römheld, V.; Marschner, H.; Mori, S. Response of nicotianamine synthase activity to Fe-deficiency in tobacco plants as compared with barley. *J. Exp. Bot.* **1995**, *46*, 1061–1063. [CrossRef]
57. Klatte, M.; Schuler, M.; Wirtz, M.; Fink-Straube, C.; Hell, R.; Bauer, P. The analysis of Arabidopsis nicotianamine synthase mutants reveals functions for nicotianamine in seed iron loading and iron deficiency responses. *Plant Physiol.* **2009**, *150*, 257–271. [CrossRef]
58. Cassin, G.; Mari, S.; Curie, C.; Briat, J.-F.; Czernic, P. Increased sensitivity to iron deficiency in Arabidopsis thaliana overaccumulating nicotianamine. *J. Exp. Bot.* **2009**, *60*, 1249–1259. [CrossRef]
59. Pavlovic, J.; Samardzic, J.; Kostic, L.; Laursen, K.H.; Natic, M.; Timotijevic, G.; Schjoerring, J.K.; Nikolic, M. Silicon enhances leaf remobilization of iron in cucumber under limited iron conditions. *Ann. Bot.* **2016**, *118*, 271–280. [CrossRef]
60. Irtelli, B.; Petrucci, W.A.; Navari-Izzo, F. Nicotianamine and histidine/proline are, respectively, the most important copper chelators in xylem sap of Brassica carinata under conditions of copper deficiency and excess. *J. Exp. Bot.* **2008**, *60*, 269–277. [CrossRef]
61. Mendoza-Cózatl, D.G.; Butko, E.; Springer, F.; Torpey, J.W.; Komives, E.A.; Kehr, J.; Schroeder, J.I. Identification of high levels of phytochelatins, glutathione and cadmium in the phloem sap of Brassica napus. A role for thiol-peptides in the long-distance transport of cadmium and the effect of cadmium on iron translocation. *Plant J.* **2008**, *54*, 249–259. [CrossRef] [PubMed]
62. Vasanits, A.; Kutlán, D.; Sass, P.; Molnár-Perl, I. Retention/quantitation properties of the o-phthaldialdehyde–3-mercaptopropionic acid and the o-phthaldialdehyde–N-acetyl-l-cysteine amino acid derivatives in reversed-phase high-performance liquid chromatography. *J. Chromatogr. A* **2000**, *870*, 271–287. [CrossRef]

plants

MDPI

Article

Citric Acid Assisted Phytoremediation of Chromium through Sunflower Plants Irrigated with Tannery Wastewater

Ali Imran Mallhi [1], Shahzad Ali Shahid Chatha [2], Abdullah Ijaz Hussain [2], Muhammad Rizwan [3], Syed Asad Hussain Bukhar [4], Afzal Hussain [3,5], Zahid Imran Mallhi [3], Shafaqat Ali [3,6,*], Abeer Hashem [7,8], Elsayed Fathi Abd_Allah [9], Mohammed Nasser Alyemeni [7] and Parvaiz Ahmad [7,10,*]

1 Department of Applied Chemistry, Government College University Faisalabad, Faisalabad 38000, Pakistan; imranmallhi72@gmail.com
2 Department of Chemistry, Government College University Faisalabad, Faisalabad 38000, Pakistan; Chatha222@gmail.com (S.A.S.C.); abdullahijaz@gcuf.edu.pk (A.I.H.)
3 Department of Environmental Science and Engineering, Government College University Faisalabad, Faisalabad 38000, Pakistan; mrazi1532@yahoo.com (M.R.); afzaalh345@gmail.com (A.H.); zahid.mallhi@yahoo.com (Z.I.M.)
4 Department of Agronomy, Bahauddin Zakariya University, Multan 60800, Pakistan; asadbukhari@bzu.edu.pk
5 Department of Environmental Sciences, The University of Lahore, Lahore 54000, Pakistan
6 Department of Biological Sciences and Technology, China Medical University, Taichung 40402, Taiwan
7 Botany and Microbiology Department, College of Science, King Saud University, P.O. Box. 2460, Riyadh 11451, Saudi Arabia; habeer@ksu.edu.sa (A.H.); mnalyemeni@gmail.com (M.N.A.)
8 Mycology and Plant Disease Survey Department, Plant Pathology Research Institute, ARC, Giza 12511, Egypt
9 Plant Production Department, College of Food and Agricultural Sciences, King Saud University, P.O. Box. 2460, Riyadh 11451, Saudi Arabia; eabdallah@ksu.edu.sa
10 Department of Botany, S.P. College, Srinagar, Jammu and Kashmir 190001, India
* Correspondence: shafaqataligill@yahoo.com (S.A.); parvaizbot@yahoo.com (P.A.)

Received: 19 January 2020; Accepted: 4 March 2020; Published: 19 March 2020

Abstract: Heavy metals are rapidly polluting the environment as a result of growing industrialization and urbanization. The presence of high concentrations of chromium (Cr), along with other pollutants, is widespread in tannery wastewater. In Pakistan, as a result of a severe shortage of irrigation water, farmers use tannery wastewater to grow various crops with a consequent decline in plants' yield. This experiment was performed to assess growth revival in sunflower plants irrigated with 0%, 25%, 50%, 75%, and 100% tannery wastewater, by foliar application of 0, 2.5, and 5.0 mM citric acid (CA). The wastewater treatment curtailed biomass accumulation, the growth rate, and chlorophyll contents by exacerbating the oxidative stress in sunflowers. Foliar application of CA considerably alleviated the outcomes of Cr toxicity by curbing the Cr absorption and oxidative damage, leading to improvements in plant growth, biological yield, and chlorophyll contents. It is concluded that foliar application of CA can successfully mitigate the Cr toxicity in sunflower plants irrigated with tannery wastewater.

Keywords: chromium; wastewater; sunflower; biomass; chlorophyll contents

1. Introduction

Kasur, a city in Punjab Pakistan, is famous for tanneries. According to estimates, around 144,502 to 215,036 gallons of wastewater day^{-1} is discharged into the environment by 650 registered tanneries [1].

Tannery wastewater contains several organic and inorganic pollutants [2]. Farmers mostly use tannery wastewater in periurban areas due to a scarcity of irrigation water [3,4]. There are different heavy metals in tannery wastewater, such as Cr, Cd, Mn, Pb, Fe, Ni, and Cu. Release of industrial effluent from tanneries is injurious for living organisms, including plants and animals [5]. The concentration of Cr in such water is considerably higher than other heavy metals [2]. Among heavy metals, chromium is ranked the 14th most noxious heavy metal globally; among different Cr oxidation states, Cr^{+6} is the most toxic because of the higher mobility/solubility and Cr^{+3} is the least toxic [6]. Plant growth and biomass is negatively affected through the application of industrial effluents. Outcomes of previous studies show that plant growth is negatively affected by 5 mg L^{-1} Cr in the nutrient solution [7]. Application of chromium (Cr^{+6}) contaminated wastewater significantly reduced plant growth and photosynthesis in different crops, like maize, wheat, and sunflower, by causing oxidative stress [8–10]. The physiochemical processes of plants are badly affected by chromium stress [11]. Plant growth is reduced as a result of disruption in the photosynthetic process, damage to the ultrastructure of plant cells, oxidative stress such as electrolyte leakage (EL), hydrogen peroxide (H_2O_2), malondialdehyde (MDA) and changes in miRNAs and proteins [8,12–15]. Sunflower is an oil seed crop in Pakistan and a key source of edible oil worldwide. As a hyperaccumulator plant, it can be grown under stressful conditions such as under heavy metal stress. Sunflower plants were cultivated on 82,000 hectares of land, producing 40,000 tons of oil during 2017–2018 (Pakistan Economic Survey 2017–2018). Though the sunflower plant is considered metal resistant, toxic concentrations of Cr may negatively influence its growth and development. The outcomes of the previous investigations have revealed that Cr stress induced oxidative damage to plants with a consequent decline in growth and yield [16]. As a result of limited resources and exorbitantly increasing population pressure, contamination of land and water bodies with heavy metals, especially Cr, is a potential threat to food security and safety. The current situation demands an urgent remedy to ensure the provision of quality food to the population of Pakistan.

The beneficial effects of citric acid (CA) upon the uptake of Cr as well as in growth regulation in several plant species are well documented. The CA foliar application improved the germination rate and root weight of sunflower plant by improving the activities of several antioxidants enzymes including superoxide dismutase (SOD), catalase (CAT), peroxidase (POD) and ascorbate peroxidase (APX) [17]. Foliar application of CA, combined with 5-aminolevulinic acid (ALA), considerably increased the yield of sunflower plants. Citric acid spray on radish leaves reduced the uptake of Cr and its mobility [17]. Citric acid also increased the antioxidant enzyme activities, chlorophyll fluorescence, and reduced lipid peroxidation in *Brassica napus* [18]. Citric acid alleviated the Cr (Cr^{+6}) toxicity by stimulating the antioxidant defense system in sunflower plants [19].

In the current study, it was hypothesized that CA may alleviate Cr toxicity in sunflower crop irrigated with tannery wastewater. Thus, the current study was designed to explore CA effects on morophogical, physiological, and chemical attributes of sunflower plants irrigated with tannery wastewater containing total Cr concentrations of 329 mg L^{-1}.

2. Results

2.1. Plants Growth and Biomass

Growth features of sunflower plants under various treatments of tannery wastewater, along with application of CA, are presented in Figures 1 and 2. Findings revealed that increasing concentration of wastewater treatment progressively reduced the plant height, root length, number of leaves per plant, and the leaf area (Figure 1), along with the fresh and dry weight of root, shoot, and leaves of sunflower (Figure 2). The highest decrease in the abovementioned parameters was noticed upon 100% wastewater treatments. However, CA application remarkably improved all the studied agronomic traits, both under stressed and normal growth environments. The maximum ameliorative effect of CA was observed at 5.0 mM foliar application treatment.

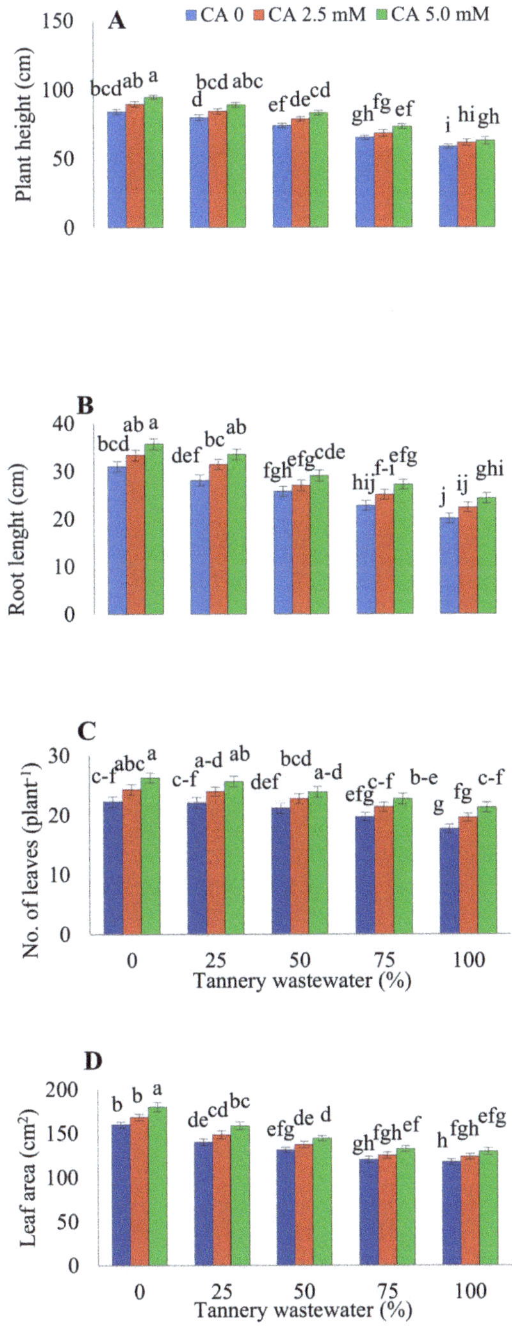

Figure 1. Impact of wastewater and citric acid (CA) on sunflower height (**A**), root lengths (**B**), number of leaves per plant (**C**), and leaf area (**D**) of sunflower plants. Data are means of 3 independent replicates, and different lettering indicates a significant difference among the values at $p < 0.05$.

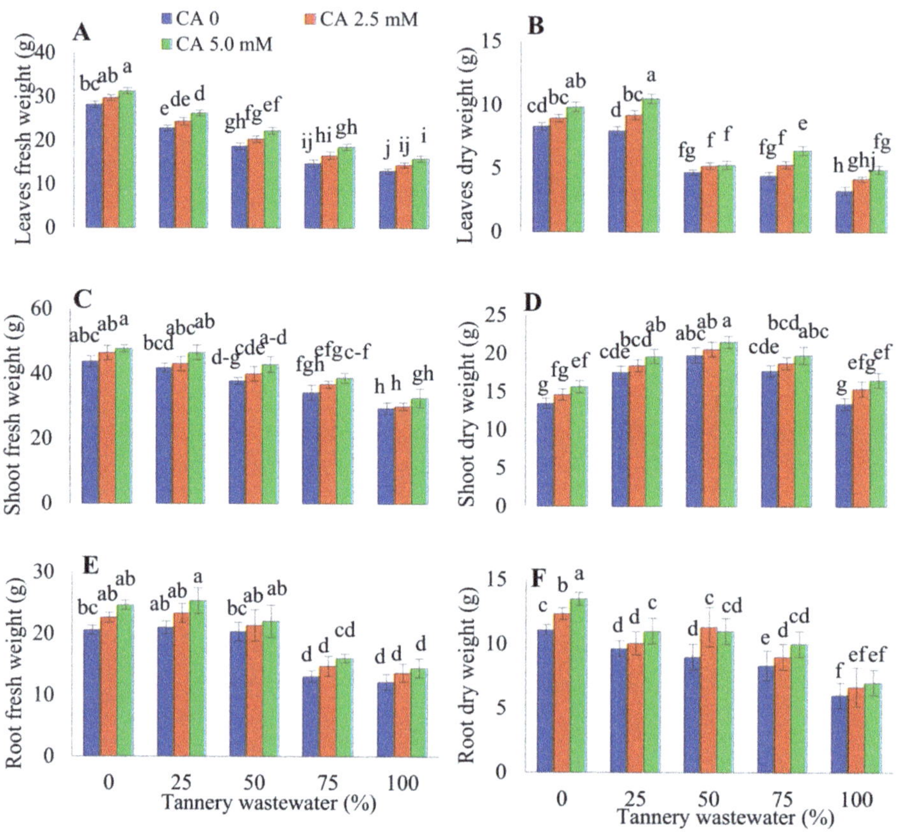

Figure 2. Impact of wastewater and citric acid (CA) treatments on leaf fresh weight (**A**), leaf dry weight (**B**), shoot fresh weight (**C**), shoot dry weight (**D**), root fresh weight (**E**) root dry weight (**F**) of sunflower plants. Data are means of 3 independent replicates, and different lettering indicates a significant difference among the values at $p < 0.05$.

2.2. Photosynthetic Pigments

The application of wastewater severely decreased Chlorophyll *a*, *b*, and the total chlorophyll (Figure 3). Citric acid treatment significantly improved the pigment contents under Cr stress, as well as normal growth conditions. However, 5.0 mM proved to be the most effective concentration of CA in mitigating the detrimental effects of wastewater on photosynthetic pigments.

2.3. Oxidative Stress Parameters

The presence of Cr in tannery effluent resulted in oxidative damage to sunflower plant (Figure 4). By increasing wastewater concentration, it sharply elevated the MDA concentration in the leaves and roots of sunflowers. The maximum value for MDA contents was recorded under 50% and 100% wastewater treatment in leaves and roots, respectively. Foliar treatment of CA significantly decreased the lipid peroxidation as revealed by reduced MDA content in plants. The most conspicuous decline in MDA content was detected in plants as compared to respective control plants under a 75% and 100% wastewater treatment along with a 5.0 mM CA application in leaves and roots, respectively.

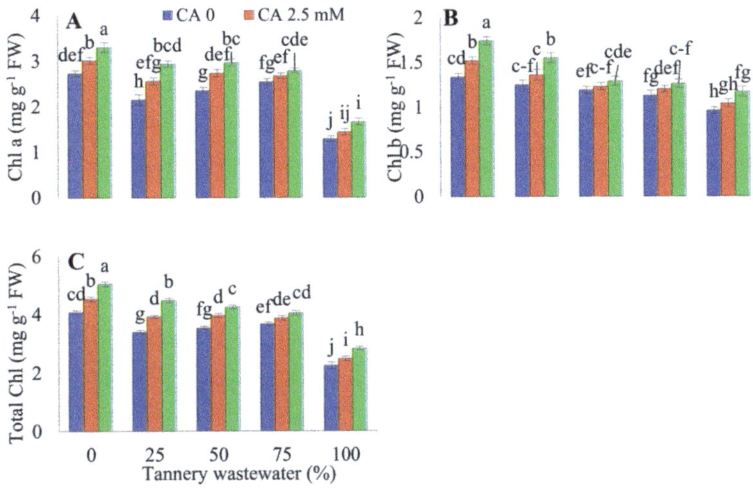

Figure 3. Impact of wastewater and citric acid (CA) treatment on chlorophyll a (**A**), chlorophyll b (**B**), and total chlorophyll (**C**) of sunflower plants. Data are means of 3 independent replicates and different lettering indicates a significant difference among values at $p < 0.05$.

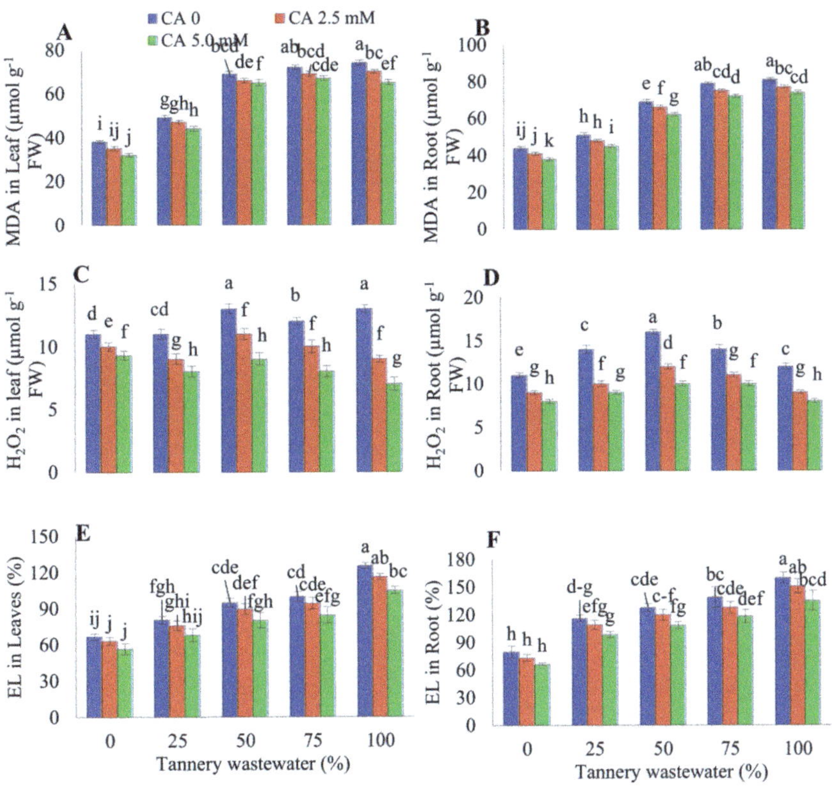

Figure 4. Impact of wastewater and citric acid (CA) treatment on leaf malondialdehyde (MDA) (**A**), roots MDA (**B**), leaves H_2O_2 (**C**), root H_2O_2 (**D**), leaf electrolyte leakage (EL) (**E**), and root EL (**F**)

of sunflower plants. Data are means of 3 independent replicates, and different lettering indicates a significant difference among the values at $p < 0.05$.

Similarly, wastewater treatment also caused oxidative damage by enhancing production of H_2O_2 in roots and leaves of plants, with the most pronounced effect at 50% wastewater application. Citric acid noticeably alleviated the oxidative damage to plants by decreasing H_2O_2 content both in the roots and leaves. The highest ameliorative effect was found in the plants irrigated with a 50% wastewater treatment in combination with a foliar application of 5.0 mM CA. Electrolyte leakage also exhibited a proportional rise with an increasing concentration of wastewater treatment, both in the leaves and roots of plants. Nonetheless, 5.0 mM CA considerably mitigated the consequences of stress as depicted by the reduction in EL in both parts of the plants.

2.4. Antioxidant Enzymes

Activities of SOD were increased with the 25% wastewater treatment and decreased thereafter with a further increase in stress level, both in the leaves and root of plants. The CA treatment enhanced SOD activities in sunflower leaves irrigated with normal tap water. Application of 2.5 mM CA enhanced the SOD activities, while 5.0 mM CA declined the SOD activities in sunflower plants irrigated with 25%–50% wastewater (Figure 5). Foliar applied CA notably improved leaf SOD activity under a 75%–100% wastewater application. In comparison with control plants, the SOD activities in roots increased considerably with a 25%–50% wastewater treatment, showing a consequent decline with a further increase in stress level. Citric acid application enhanced SOD activities in sunflower roots, with the highest improvement being observed in the control and 100% wastewater application supplemented with 2.5 mM and 5.0 mM CA treatment, respectively.

Mild to moderate levels (25%–50%) of wastewater treatment significantly decreased, while higher concentrations (75%–100%) increased the activities of POD in sunflower leaves. Application of CA significantly increased POD activities in sunflower leaves with a substantial surge at 100% wastewater irrigation, as compared with stress treatment alone. In roots, the activities POD increased with a 25% wastewater treatment and decreased with the increasing wastewater concentration. Citric acid application prominently improved the activities of POD in roots of sunflowers with the applied stress of tannery wastewater, in comparison with the stressed plants without CA treatment.

Wastewater treatment increased CAT activities at mild stress levels (25%) with a subsequent decreasing trend, both in the leaves and roots. However, CA supplementation further enhanced CAT activities in normal, as well as stressed plants in both plant parts. The effect of 5.0 mM CA concentration was the most prominent in improving CAT activities. Activities of APX were increased with a 25%–50% wastewater treatment, with a subsequent decrease under higher stress levels in both plant parts, i.e., leaves and roots. However, a 5.0 mM CA application significantly improved the APX activities of plants at all levels of wastewater treatment, as compared with those receiving no CA treatments.

2.5. Chromium Concentration

Concentration of Cr proportionally increased in sunflower root, stem, and leaves with the increasing wastewater concentration. Comparing with the control, the concentration of Cr in sunflower leaves increased by 97, 171, and 234 times, under 25%, 50%, 75%, and 100% wastewater treatments, respectively. Application of 5.0 mM CA significantly increased Cr concentration in sunflower leaves under all stress levels, with a conspicuous rise (35%) with the 100% wastewater treatment (Figure 6). The chromium concentration in the sunflower stem was increased by 92, 132, 179, and 208 times, under 25%, 50%, 75%, and 100% wastewater treatments, respectively. Exogenous CA application further increased the buildup of Cr in the sunflower stem under various concentrations of tannery wastewater. The maximum increase (36%) in stem Cr concentration was detected under a 100% tannery wastewater treatment along with a 5.0 mM CA application.

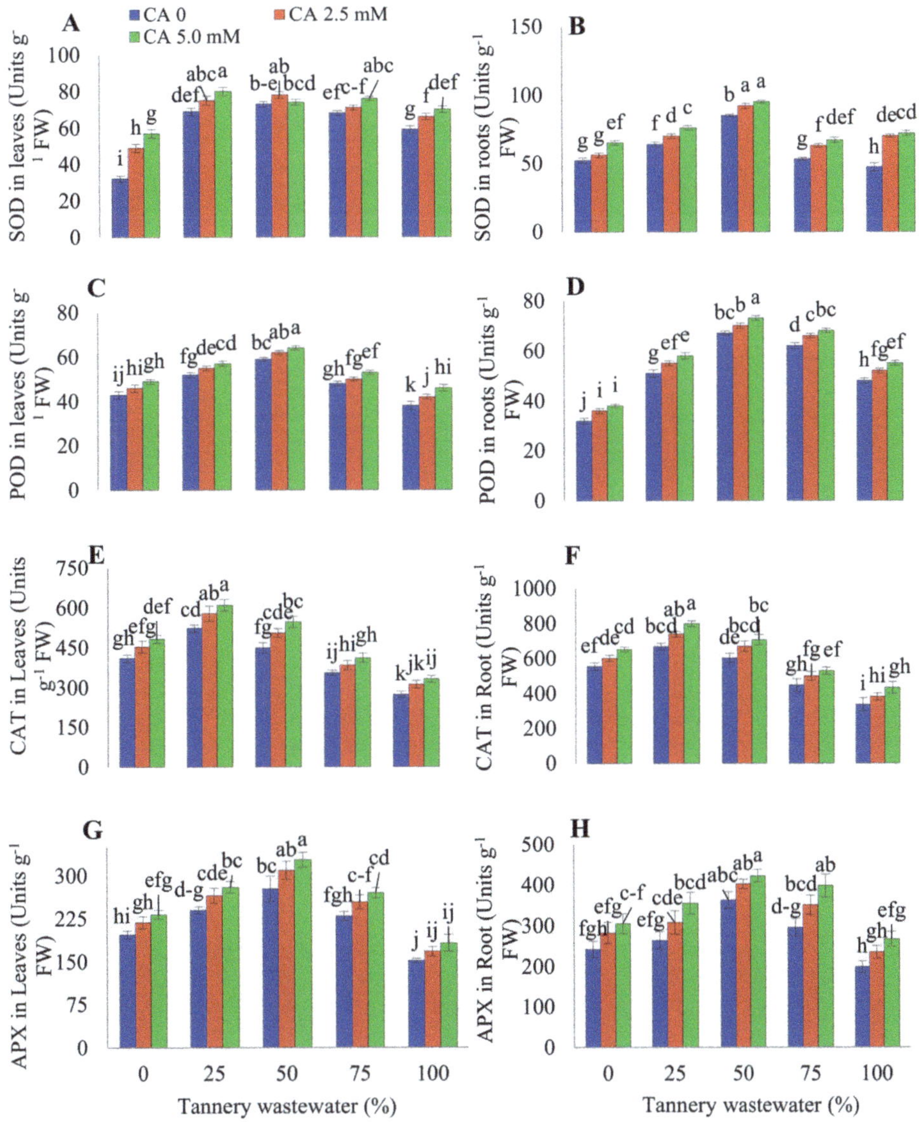

Figure 5. Impact of wastewater on and citric acid on leaf SOD (**A**), root SOD (**B**), leaf POD (**C**), root POD (**D**), leaf CAT (**E**), root CAT (**F**), leaf APX (**G**), and root APX (**H**) of sunflower plants. Data are means of 3 independent replicates, and different lettering indicates a significant difference among values at $p < 0.05$.

Heavy metals primarily affected roots of the plants under stressful conditions. In this study, the roots of stressed plants revealed comparatively higher Cr concentrations than stems and leaves. In comparison with the control, root Cr concentration was increased by 113, 211, 308, and 377 times, under 25%, 50%, 75%, and 100% wastewater treatments, respectively. Exogenously applied CA further increased the root Cr contents at all levels of wastewater application. The maximum value for Cr concentration in sunflower root was recorded with application of 5.0 mM CA with a 100% wastewater treatment.

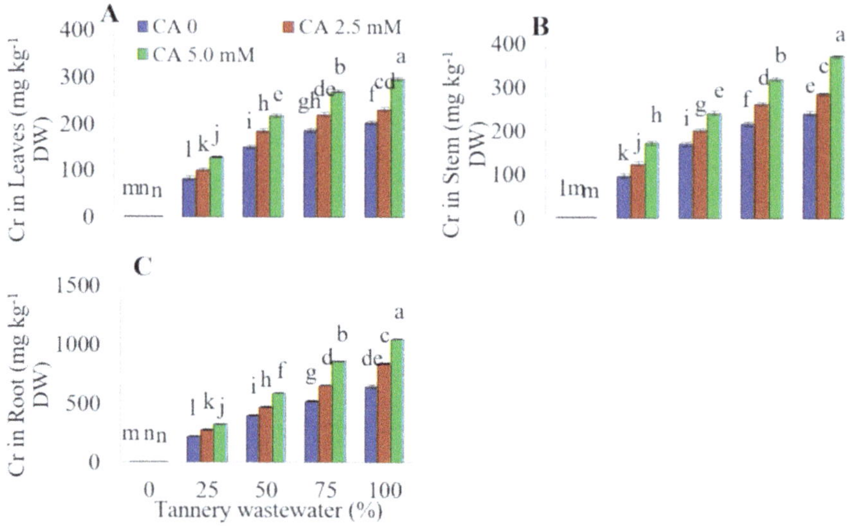

Figure 6. Impact of wastewater and citric acid on Cr concentration in leaves (**A**), stems (**B**), and roots (**C**) of sunflower plants. Data are means of 3 independent replicates and different lettering indicates a significant difference among the values at $p < 0.05$.

3. Discussion

The present study elucidated the role of CA in mitigating the Cr toxicity caused by tannery wastewater in sunflower seedlings. Besides traces of several other heavy metals, tannery wastewater holds very high concentrations of Cr [2,20]. Various morphological parameters of sunflowers, such as biomass, leaf area, number of leaves per plant, root length, and plant height considerably decreased with an increasing wastewater concentration (Figures 1 and 2), which may be attributed to the noxious amount of Cr in wastewater [20]. Previous reports have also revealed significant reductions in morphological features of various plant species under Cr toxicity [11,20]. Chromium induces morphological changes due to competition with other necessary nutrient elements, distortion of root and leaf ultrastructure, disruption in photosynthesis, and oxidative damage [12,14,21]. The findings of the current study are supported by the previous reports that Cr stress significantly decreased plant biomass, root length, and plant height in different plant species such as wheat, maize, tobacco, and spinach [11,14,19,20]. Citric acid treatment considerably improved the biomass and growth of stressed plants, in contrast to those irrigated with various wastewater concentrations without exogenous CA application (Figures 1 and 2). Improvement in the morphological characteristics of the plants might be attributed to the enhanced absorption of vital nutrients by sunflower plants [22].

It is a well-established fact that photosynthetic pigments perform a very crucial role in a plants' life due to their light harvesting function. Maqbool et al. [20] reported that Cr stress significantly affected the physiochemical attributes of the plants. In the current study, wastewater treatment prominently decreased the photosynthetic pigments in sunflower plants (Figure 3). Chlorophylls a, b, and total chlorophyll abruptly reduced with increasing tannery wastewater concentration. Reduction in plant growth may be attributed to the occurrence of high Cr concentrations in wastewater, which may have caused the structural impairment to chloroplast [11] and surplus generation of ROS under Cr induced stress [23]. Moreover, up-regulation of chlorophyllase under heavy metal stress constantly leads to an enormous reduction in chlorophyll pigments [24]. Citric acid application increased the chlorophyll contents in plants under tannery wastewater treatment. Similarly CA improved chlorophyll contents of wheat leaves under severe Cr toxicity [19]. Further, CA application also enhanced chlorophyll concentration in maize plants under drought condition [25]. Foliar application of CA caused an immense reduction in ROS generation and damage to the chloroplast, which resulted in an improvement in the pigment contents in plants [26].

Increased MDA content is an indication of membrane damage [27]. In the current study, H_2O_2, MDA contents, and EL, were considerably enhanced in plant root and leaves of sunflower plants irrigated with tannery wastewater (Figure 3). However, CA treatment markedly ameliorated the lipid peroxidation and membrane damage by scavenging the free radicals and reducing ROS production [19,25].

Antioxidant enzymes perform a very significant role in protecting the plants from oxidative stress. The present study showed enhancement in SOD, POD, CAT, and APX activities at slight to moderate stress levels. Increasing doses of wastewater drastically reduced POD and SOD activities, except POD activity in the leaves (Figure 4). Similar findings were also reported previously [28]. Mild levels of Cr stress increased antioxidant enzyme activities, whereas serious stress reduced maize antioxidant enzyme activities due to relentless oxidative injury [11]. This could be because Cr toxicity alarms the antioxidant machinery of plants into starting to scavenge the ROS. Nevertheless, higher Cr levels suppress the antioxidant system due to a continuous and increased production of ROS.

Increasing concentrations of tannery wastewater drastically increased Cr concentration in sunflower root, stems, and the leaves of sunflower plants (Figure 6). Roots exhibited a much higher amount of Cr as compared to stem and leaves of the sunflower. Similar results were observed in rice and oil-seed rapes [11,29]. This might be due to Cr immobilization by sugars as macromolecules [19], followed by compartmentalization in root cell vacuoles [30]. Exogenous application of CA promoted uptake and accumulation of Cr towards different plant parts under various levels of tannery wastewater (Figure 6). However, Ali et al. [19] observed that CA application significantly decreased the uptake of different metals and their translocations to upper parts of various plants. The CA-mediated restricted absorption of Cr, due to the protection of the membrane system of plants, and resultant increase in the uptake of essential nutrients, might be the possible reason for enhanced plant growth [31]. Organic substances like fulvic acid and humic acid are well known for their constructive effects on heavy metals' mobility and bioavailability, because they formulate composite organo-metal complexes [32–34]. Citric acid might have made complexes with Cr ions, which could be among many different possible reasons for the increase in Cr uptake by sunflower [22].

Crops may also have the ability to reduce Cr^{6+} to Cr^{3+}, which would likely to happen in roots with the help of Cr reductase enzymes, same as those present in bacteria for detoxification of heavy metals; however, such enzymes have not been identified in plants [35]. The effects of Cr on plants vary with growth medium as well as different Cr oxidation forms [36]. It is well reported that Cr (VI) has more toxic effects towards plants compared to Cr (III) [37]. Riaz et al. [36] reported that wheat growth was mainly regulated by a different oxidation form of Cr than that of the total Cr.

4. Materials and Methods

4.1. Experimental and Growth Conditions

Experiments were run at Botanical Garden of the Government College University, Faisalabad. The seeds of sunflower plants (variety Hysun-33) were carefully surface sterilized with 3% H_2O_2 for 20 min, followed by washing cautiously with distilled water. Soil was air dried and sieved through a 2 mm strainer. After that, five seeds were sown in each pot filled with 8 kg of soil. After two weeks of germination, thinning was done and only two plants were kept in each pot. Prestudy soil analyses were performed according to the established protocols, as described in Table 1. Soil texture, sodium adsorption ratio (SAR), electrical conductivity (EC), soluble ions, and different trace elements were determined by Bouyoucos [38], Page et al. [39], US Salinity Lab. Staff [40], and Soltanpour [41], respectively.

Table 1. The physicochemical characteristics of the soil used in the pot study.

Texture	Sandy Loam
Silt	15.0%
Sand	67.9%
Clay	17.10%
EC	1.96 dS m^{-1}
pH	7.61
SAR	1.89 $(mmol\ L^{-1})^{1/2}$
Available P	2.11 mg kg^{-1}
Organic matter	0.59 %
HCO_3^{-1}	2.51 mmol L^{-1}
SO_4^{-2}	11.44 mmol L^{-1}
Cl^-	5.45 mmol L^{-1}
Ca^{2+} + Mg^{2+}	13.98 mmol L^{-1}
K^+	0.04 mmol L^{-1}
Na^+	5.23 mmol L^{-1}
Available Zn $^{2+}$	0.77 mg kg^{-1}
Available Cu $^{2+}$	0.31 mg kg^{-1}
Available Cr	0.16 mg kg^{-1}

Here in Table 1, the EC stands for electrical conductivity and SAR stands for sodium adsorption ratio.

4.2. Analysis of Tannery Wastewater and Application of Treatments

Tannery wastewater was analyzed for various physicochemical characteristics through standard protocol [42] (Table 2). Two weeks old sunflower plants were irrigated every third day with five different levels (0%, 25%, 50%, 75%, and 100%) of tannery wastewater until harvesting. The number of pots were considered as one replicate per percentage of wastewater. After 1 week of the wastewater irrigation, the citric acid was sprayed at different concentrations (0, 2.5 and 5.0 mM), with a 2-day interval, throughout the experiment. The plants in the control group were cautiously sprayed with the same quantity of distilled water. Chemical fertilizers were applied as described previously [19].

4.3. Measurement of Morphological Attributes

After 8 weeks of the 1st CA treatment, the plants were harvested. After harvesting, the plants were washed with distilled water and were carefully separated into stem, leaves, and roots. Data on various agronomic traits, such as plant height, root length, fresh and dry weight of roots and shoots, number of leaves per plant, and leaf area, were measured according to the standard procedures.

4.4. Assessment of Oxidative Stress

Oxidative stress was assessed by determining EL, MDA content, and H_2O_2 concentration, according to previously described methods [43–45].

4.5. Photosynthetic Pigments and Antioxidant Enzyme Analysis

The contents of chlorophyll *a*, *b*, and total chlorophyll were assessed by following established protocol [46]. Superoxide dismutase and POD activities were examined according to Zhang et al. [47]. Activities of CAT and APX were quantified by using the already developed protocol [48].

Table 2. Characteristics of tannery wastewater used for irrigation.

Parameters	Values
COD	2897 mg L^{-1}
BOD	876 mg L^{-1}
TOC	969 mg L^{-1}
Oil & grease	11 mg L^{-1}
pH	4.13
EC	91.8 dS/m
TDS	64,968
Total Cr	329 mg L^{-1}
K^+	41 mg L^{-1}
Carbonate	ND
Ca^{2+}+Mg^{2+}	3.1 $mmol_c$ L^{-1}

Here in Table 2, COD stands for chemical oxygen demand, BOD stands for biological oxygen demand, TOC stands for total organic compounds, EC stands for electrical conductivity, TDS stands for total dissolved solids, and ND stands for not detected.

4.6. Determination of Cr Contents

Plant samples were analyzed for Cr content in leaves, stems, and roots following the protocol described by Ehsan et al. [49]. Samples (0.5 g) were ground into fine powder and burnt to ashes in a muffle furnace at 1000 °C for 12 h, followed by acid digestion overnight. Afterwards, digested samples were filtered several times to get a clear extract. Finally, samples were analyzed by atomic absorption spectrophotometer (Halo DB-20/DB-20S, Dynamica Company, London, UK) and total Cr concentration was calculated by drawing standard curve.

4.7. Statistical Analysis

Complete randomized design was applied along with 3 replicates. Analysis of variance (ANOVA) was applied using statistical software (SPSS, version 23.0 for windows; IBM Corporation, Armonk, New York, U.S). A post hoc test followed by a Duncan test was applied to see significant difference among different treatments.

5. Conclusions

This study demonstrated that wastewater application reduced the morphological and photosynthetic attributes of plants. Substantial amount of total Cr accumulated in sunflower when provided with tannery wastewater. The CA application enhanced growth and photosynthetic pigments of sunflowers by decreasing the oxidative damage. Moreover, CA increased activities of antioxidant enzymes in sunflower plants. The total Cr uptake was enhanced with enhancing the concentration of CA. Current findings suggest that citric acid application could be an easy and effective strategy to alleviate chromium toxicity. However, further investigations should be done in the future on a molecular level with detailed mechanistic approaches in this regard.

Author Contributions: Conceptualization, A.I.M., S.A.S.C., A.I.H., M.R., A.H. (Afzal Hussain), S.A., E.F.A.A. and P.A. Data curation, A.I.M., A.I.H., M.R., S.A.H.B., A.H. (Afzal Hussain), Z.I.M., S.A., A.H. (Abeer Hashem) and P.A. Formal analysis, A.I.M., M.R., A.H. (Afzal Hussain), Z.I.M., S.A. and M.N.A Funding acquisition, A.I.H., M.R., A.H. (Afzal Hussain), A.H. (Abeer Hashem), E.F.A.A. and M.N.A. Investigation, S.A.S.C., A.I.H., S.A.H.B., Z.I.M., S.A. and A.H. (Abeer Hashem). Methodology, A.I.M., M.R., A.H. (Afzal Hussain), Z.I.M., M.N.A. and P.A. Project administration, A.I.M., M.R., S.A.H.B., A.H. (Afzal Hussain), E.F.A.A. and P.A. Resources, S.A.S.C., M.R., A.H. (Afzal Hussain), A.H. (Abeer Hashem), E.F.A.A., M.N.A. and P.A. Software, A.I.H. and A.H. (Afzal Hussain). Supervision, S.A.S.C. and S.A. Validation, A.I.H. Visualization, A.I.H. and S.A.H.B. Writing-original draft, A.I.M.,

S.A.H.B., A.H. (Afzal Hussain) and M.N.A. Writing-review and editing, S.A.S.C., A.I.H., M.R., S.A.H.B., A.H. (Afzal Hussain) and E.F.A.A. All authors have read and agreed to the published version of the manuscript.

Funding: Authors are also thankful to Government College University Faisalabad, Pakistan for the research support program, Project NO. 80-ENV.Sci-2. The authors would like to extend their sincere appreciation to the Researchers Supporting Project Number (RSP-2019/134), King Saud University, Riyadh, Saudi Arabia.

Acknowledgments: Authors are also thankful to Government College University Faisalabad, Pakistan for the research support program, Project NO. 80-ENV.Sci-2. The authors would like to extend their sincere appreciation to the Researchers Supporting Project Number (RSP-2019/134), King Saud University, Riyadh, Saudi Arabia.

Conflicts of Interest: The authors declare no conflicts of interest.

References

1. Qadir, A.; Malik, R.N.; Husain, S.Z. Spatio-temporal variations in water quality of Nullah Aik tributary of the river Chenab, Pakistan. *Environ. Monit. Assess.* **2008**, *140*, 43–59. [CrossRef]
2. Ashraf, S.; Afzal, M.; Rehman, K.; Naveed, M.; Zahir, Z.A. Plant-endophyte synergism in constructed wetlands enhances the remediation of tannery effluent. *Water Sci. Technol.* **2018**, *77*, 1262–1270. [CrossRef] [PubMed]
3. Khalid, S.; Shahid, M.; Dumat, C.; Niazi, N.K.; Bibi, I.; Gul Bakhat, H.F.S.; Abbas, G.; Murtaza, B.; Javeed, H.M.R. Influence of groundwater and wastewater irrigation on lead accumulation in soil and vegetables: Implications for health risk assessment and phytoremediation. *Int. J. Phytoremediat.* **2017**, *19*, 1037–1046. [CrossRef] [PubMed]
4. Rehman, M.Z.; Khalid, H.; Akmal, F.; Ali, S.; Rizwan, M.; Qayyum, M.F.; Iqbal, M.; Khalid, M.U.; Azhar, M. Effect of limestone, lignite and biochar applied alone and combined on cadmium uptake in wheat and rice under rotation in an effluent irrigated field. *Environ. Pollut.* **2017**, *227*, 560–568. [CrossRef] [PubMed]
5. Zaheer, I.E.; Ali, S.; Rizwan, M.; Abbas, Z.; Bukhari, S.A.H.; Wijaya, L.; Alyemeni, M.N.; Ahmad, P. Zinc-lysine prevents chromium-induced morphological, photosynthetic, and oxidative alterations in spinach irrigated with tannery wastewater. *Environ. Sci. Pollut. Res.* **2019**, *26*, 28951–28961. [CrossRef] [PubMed]
6. Khalil, U.; Shakoor, M.B.; Ali, S.; Rizwan, M. Tea waste as a potential biowaste for removal of hexavalent chromium from wastewater: Equilibrium and kinetic studies. *Arab. J. Geosci.* **2018**, *11*, 573. [CrossRef]
7. Oliveira, H. Chromium as an environmental pollutant: Insights on induced plant toxicity. *J. Bot.* **2012**. [CrossRef]
8. Ali, S.; Rizwan, M.; Waqas, A.; Hussain, M.B.; Afzal, H.; Liu, S.; Alqarawi, A.A.; Hashem, A.; Abd Allah, E.F. Fulvic Acid Prevents Chromium-induced Morphological, Photosynthetic, and Oxidative Alterations in Wheat Irrigated with Tannery Waste Water. *J. Plant Growth Regul.* **2018**, *37*, 1357–1367. [CrossRef]
9. Habiba, U.; Ali, S.; Hafeez, F.; Rizwan, M.; Rehman, M.Z.; Hussain, A.; Asad, S.A. Morpho-physiological responses of maize cultivars exposed to chromium stress. *Int. J. Agric. Biol.* **2018**, *21*, 140–148.
10. Farid, M.; Ali, S.; Rizwan, M.; Ali, Q.; Saeed, R.; Nasire, T.; Abbasi, G.H.; Ishaq, M.; Rehmanig, A.; Karimh, S.T.A.; et al. Phytomanagement of chromium contaminated soils through sunflower under exogenously applied 5-aminolevulinic acid. *Ecotoxicol. Environ. Saf.* **2018**, *151*, 255–265. [CrossRef]
11. Gill, R.A.; Zang, L.; Ali, B.; Farooq, M.A.; Cui, P.; Yang, S.; Ali, S.; Zhou, W. Chromium-induced physio-chemical and ultrastructural changes in four cultivars of Brassica napus L. *Chemosphere* **2015**, *120*, 154–164. [CrossRef] [PubMed]
12. Ali, S.; Farooq, M.A.; Yasmeen, T.; Hussain, S.; Arif, M.S.; Abbas, F.; Bharwana, S.A.; Zhang, G. The influence of silicon on barley growth, photosynthesis and ultra-structure under chromium stress. *Environ. Sci. Pollut. Res.* **2013**, *22*, 10601–10609. [CrossRef] [PubMed]
13. Magenta, A.; Greco, S.; Gaetano, C.; Martelli, F. Oxidative stress and micrornas in vascular diseases. *Int. J. Mol. Sci.* **2013**, *14*, 17319–17346. [CrossRef]
14. Bukhari, S.A.H.; Wang, R.; Wang, W.; Ahmed, I.M.; Zheng, W.; Cao, F. Genotype-dependent effect of exogenous 24-epibrassinolide on chromium-induced changes in ultrastructure and physicochemical traits in tobacco seedlings. *Environ. Sci. Pollut. Res.* **2016**, *23*, 18229–18238. [CrossRef] [PubMed]
15. Schmidt, F.; Marnef, A.; Cheung, M.-K.; Wilson, I.; Hancock, J.; Staiger, D.; Ladomery, M. A proteomic analysis of oligo(dT)-bound mRNP containing oxidative stress-induced Arabidopsis thaliana RNA-binding proteins ATGRP7 and ATGRP8. *Mol. Biol. Rep.* **2010**, *37*, 839–845. [CrossRef]

16. Rizwan, M.; Ali, S.; Qayyum, M.F.; Ok, Y.S.; Zia-ur-Rehman, M.; Abbas, Z.; Hannan, F. Use of maize (Zea mays L.) for phytomanagement of Cd-contaminated soils: A critical review. Environ. *Geochem. Health* **2016**, *9*, 259–277. [CrossRef] [PubMed]
17. Ondrasek, G.H.; Bakić, M.; Zovko, L.; Filipović, C.; Meriño-Gergichevich, R.; Savić, Z.; Rengel, Z. Biogeochemistry of soil organic matter in agroecosystems and environmental implications. *Sci. Total Environ.* **2019**, *658*, 1559–1573. [CrossRef]
18. Zaheer, I.E.; Ali, S.; Rizwan, M.; Farid, M.; Shakoor, M.B.; Gill, R.A.; Najeeb, U.; Iqbal, N.; Ahmad, R. Citric acid assisted phytoremediation of copper by Brassica napus L. *Ecotoxicol. Environ. Saf.* **2015**, *120*, 310–317. [CrossRef]
19. Ali, S.; Bharwana, S.A.; Rizwan, M.; Farid, M.; Kanwal, S.; Ali, Q.; Ibrahim, M.; Gill, R.A.; Khan, M.D. Fulvic acid mediates chromium (Cr) tolerance in wheat (*Triticum aestivum* L.) through lowering of Cr uptake and improved antioxidant defense system. *Ecotoxicol. Environ. Saf.* **2015**, *89*, 66–72. [CrossRef]
20. Maqbool, A.; Ali, S.; Rizwan, M.; Ishaque, W.; Rasool, N.; Rehman, M.Z.; Bashir, A.; Abid, M.; Wu, L. Management of tannery wastewater for improving growth attributes and reducing chromium uptake in spinach through citric acid application. *Environ. Sci. Pollut. Res.* **2018**, *25*, 10848–10856. [CrossRef]
21. Shakoor, M.B.; Ali, S.; Hameed, A.; Farid, M.; Hussain, S.; Yasmeen, T.; Najeeb, U.; Bharwana, S.A.; Abbasi, G.H. Citric acid improves lead (Pb) phytoextraction in Brassica napus L. by mitigating Pb-induced morphological and biochemical damages. *Ecotoxicol. Environ. Saf.* **2014**, *109*, 38–47. [CrossRef] [PubMed]
22. Farid, M.; Ali, S.; Rizwan, M.; Ali, Q.; Abbas, F.; Bukhari, S.A.H.; Saeed, R.; Wu, L. Citric acid assisted phytoextraction of chromium by sunflower; morpho-physiological and biochemical alterations in plants. *Ecotoxicol. Environ. Saf.* **2017**, *145*, 90–102. [CrossRef] [PubMed]
23. Habiba, U.; Ali, S.; Farid, M.; Shakoor, M.B.; Rizwan, M.; Ibrahim, M.; Abbasi, G.H.; Hayat, T.; Ali, B. EDTA enhanced plant growth, antioxidant defense system, and phytoextraction of copper by *Brassica napus* L. *Environ. Sci. Pollut. Res.* **2015**, *22*, 1534–1544. [CrossRef] [PubMed]
24. Hegedüs, A.; Erdei, S.; Horváth, G. Comparative studies of H2O2 detoxifying enzymes in green and greening barley seedlings under cadmium stress. *Plant Sci.* **2001**, *160*, 1085–1093. [CrossRef]
25. Anjum, N.A.; Ahmad, I.; Mohmood, I.; Pacheco, M.; Duarte, A.C.; Pereira, E.; Umar, S.; Ahmad, A.; Khan, N.A.; Iqbal, M.; et al. Modulation of glutathione and its related enzymes in plants' responses to toxic metals and metalloids—A review. *Environ. Exp. Bot.* **2012**, *75*, 307–324. [CrossRef]
26. Shahid, M.; Dumat, C.; Silvestre, J.; Pinelli, E. Effect of fulvic acids on lead-induced oxidative stress to metal sensitive Vicia faba L. plant. *Biol. Fertil. Soils* **2012**, *48*, 689–697. [CrossRef]
27. Schmid-Siegert, E.; Stepushenko, O.; Glauser, G.; Farmer, E.E. Membranes as structural antioxidants recycling of malondialdehyde to its source in oxidation-sensitive chloroplast fatty acids. *J. Biol. Chem.* **2016**, *291*, 13005–13013. [CrossRef]
28. Meng, H.; Hua, S.; Shamsi, I.H.; Jilani, G.; Li, Y.; Jiang, L. Cadmium-induced stress on the seed germination and seedling growth of Brassica napus L., and its alleviation through exogenous plant growth regulators. *Plant Growth Regul.* **2009**, *58*, 47–59. [CrossRef]
29. Zeng, F.; Qiu, B.; Ali, S.; Zhang, G. Genotypic differences in nutrient uptake and accumulation in rice under chromium stress. *J. Plant Nutr.* **2011**, *33*, 18–528. [CrossRef]
30. Kanwal, U.; Ali, S.; Shakoor, M.B.; Farid, M.; Hussain, S.; Yasmeen, T.; Adrees, M.; Bharwana, S.A.; Abbas, F. EDTA ameliorates phytoextraction of lead and plant growth by reducing morphological and biochemical injuries in Brassica napus L. under lead stress. *Environ. Sci. Pollut. Res.* **2014**, *21*, 9899–9910. [CrossRef]
31. Nardi, S.; Pizzeghello, D.; Muscolo, A.; Vianello, A. Physiological effects of humic substances in higher plants. *Soil Biol. Biochem.* **2002**, *34*, 1527–1537. [CrossRef]
32. Kang, J.; Zhang, Z.; Wang, J.J. Influence of humic substances on bio-availability of Cu and Zn during sewage sludge composting. *Bioresour. Technol.* **2011**, *102*, 8022–8026. [CrossRef] [PubMed]
33. Kwiatkowska-Malina, J. Qualitative and quantitative soil organic matter estimation for sustainable soil management. *J. Soils Sediments* **2018**, *18*, 2801–2812. [CrossRef]
34. Raptis, S.; Gasparatos, D.; Economou-Eliopoulos, M.; Petridis, A. Chromium uptake by lettuce as affected by the application of organic matter and Cr (VI)-irrigation water: Implications to the land use and water management. *Chemosphere* **2018**, *210*, 597–606. [CrossRef] [PubMed]
35. Lytle, C.M.; Lytle, F.W.; Yang, N.; Qian, J.H.; Hansen, D.; Zayed, A.; Terry, N. Reduction of Cr (VI) to Cr (III) by wetland plants: Potential for in situ heavy metal detoxification. *Environ. Sci. Technol.* **1998**, *32*, 3087–3093. [CrossRef]

36. Riaz, M.; Yasmeen, T.; Arif, M.S.; Ashraf, M.A.; Hussain, Q.; Shahzad, S.M.; Rizwan, M.; Mehmood, M.W.; Zia, A.; Mian, I.A.; et al. Variations in morphological and physiological traits of wheat regulated by chromium species in long-term tannery effluent irrigated soils. *Chemosphere* **2019**, *222*, 891–903. [CrossRef]
37. Gomes, M.A.D.C.; Hauser-Davis, R.A.; Suzuki, M.S.; Vitória, A.P. Plant chromium uptake and transport, physiological effects and recent advances in molecular investigations. *Ecotoxicol. Environ. Saf.* **2017**, *140*, 55–64. [CrossRef]
38. Ayers, R.S.; Westcot, D.W. Water quality for agriculture, pp. 95-97 Bouyoucos GJ (1962) Hydrometer method improved for making particle size analyses of soils. *Agron. J.* **1985**, *54*, 464–465.
39. Page, A.L.; Miller, R.H.; Keeny, D.R. Methods of soil analysis (Part 2). In *Chemical and Microbiological Properties*; Soil Science Society of America: Madison, WI, USA, 1982.
40. US Salinity Laboratory Staff. Diagnosis and improvement of saline and alkali soils. In *Agriculture Handbook 60*; United States Salinity Laboratory, US Department of Agriculture: Washington, DC, USA, 1954; p. 160.
41. Soltanpour, P.N. Use of AB-DTPA soil test to evaluate elemental availability and toxicity. *Commun. Soil Sci. Plant Anal.* **1985**, *16*, 323–338. [CrossRef]
42. *Standard Methods for the Examination of Water and Wastewater*; American Public Health Association: Washington, DC, USA, 2005; pp. 1–874.
43. Dionisio-Sese, M.L.; Tobita, S. Antioxidant responses of rice seedlings to salinity stress. *Plant Sci.* **1998**, *135*, 1–9. [CrossRef]
44. Dhindsa, R.A.; Plumb-Dhindsa, P.; Thorpe, T.A. Leaf senescence: Correlated with increased permeability and lipid peroxidation, and decreased levels of superoxide dismutase and catalase. *J. Exp. Bot.* **1981**, *126*, 93–101. [CrossRef]
45. Jana, S.; Choudhuri, M.A. Glycolate metabolism of three submerged aquatic angiosperms during aging. *Aquat. Bot.* **1981**, *12*, 345–354. [CrossRef]
46. Lichtenthaler, H.K. Chlorophylls and carotenoids-the pigments of photosynthetic biomembranes. In *Methods in Enzymology*; Colowick, S.P., Kaplan, N.O., Eds.; Academic Press: New York, NY, USA, 1987; Volume 148, pp. 350–382.
47. Zhang, J.; Cai, W.W.; Lee, D.J. Occupational hazards and pregnancy outcomes. *Am. J. Ind. Med.* **1992**, *21*, 397–408. [CrossRef] [PubMed]
48. Aebi, H. Catalasein vitro. *Methods Enzymol.* **1984**, *105*, 121–126.
49. Ehsan, S.; Ali, S.; Noureen, S.; Farid, M.; Shakoor, M.B.; Aslam, A.; Bharwana, S.A.; Tauqeer, H.M. Comparative assessment of different heavy metals in urban soil and vegetables irrigated with sewage/industrial waste water. *Ecoterra* **2013**, *35*, 37–53.

plants

MDPI

Article

Citric Acid Enhances Plant Growth, Photosynthesis, and Phytoextraction of Lead by Alleviating the Oxidative Stress in Castor Beans

Zahid Imran Mallhi [1], Muhammad Rizwan [1], Asim Mansha [2,*], Qasim Ali [3], Sadia Asim [4], Shafaqat Ali [1,5,*], Afzal Hussain [1], Salman H. Alrokayan [6], Haseeb A. Khan [6], Pravej Alam [7] and Parvaiz Ahmad [8]

1 Department of Environmental Science and Engineering, Government College University, Faisalabad 38000, Pakistan; zahid.mallhi@yahoo.com (Z.I.M.); mrazi1532@yahoo.com (M.R.); afzaalh345@gmail.com (A.H.)

2 Department of Chemistry, Government College University, Faisalabad 38000, Pakistan

3 Department of Botany, Government College University, Faisalabad 38000, Pakistan; qasimbot_uaf@yahoo.com

4 Department of Chemistry, Government College Women University, Faisalabad 38000, Pakistan; sadia1asim@gmail.com

5 Department of Biological Sciences and Technology, China Medical University (CMU), Taichung City 40402, Taiwan

6 Department of Biochemistry, College of Science, King Saud University, Riyadh 11451, Saudi Arabia; salrokayan@ksu.edu.sa (S.H.A.); Haseeb_khan@yahoo.com (H.A.K.)

7 Biology Department, College of Science and Humanities, Prince Sattam bin Abdulaziz University (PSAU), Alkharj 11942, Saudi Arabia; ps.alam@psau.edu.sa

8 Botany and Microbiology Department, College of Science, King Saud University, Riyadh 11451, Saudi Arabia; parvaizbot@yahoo.com

* Correspondence: mansha.asim@gmail.com (A.M.); shafaqataligill@yahoo.com or shafaqat@mail.cmuh.org.tw (S.A.)

Received: 26 September 2019; Accepted: 15 November 2019; Published: 19 November 2019

Abstract: Lead (Pb) toxicity has a great impact in terms of toxicity towards living organisms as it severely affects crop growth, yield, and food security; thus, warranting appropriate measures for the remediation of Pb polluted soils. Phytoextraction of heavy metals (HMs) using tolerant plants along with organic chelators has gained global attention. Thus, this study examines the possible influence of citric acid (CA) on unveiling the potential phytoextraction of Pb by using castor beans. For this purpose, different levels of Pb (0, 300, 600 mg kg^{-1} of soil) and CA (0, 2.5, and 5 mM) were supplied alone and in all possible combinations. The results indicate that elevated levels of Pb (especially 600 mg kg^{-1} soil) induce oxidative stress, including hydrogen peroxide (H_2O_2) and malanodialdehyde (MDA) production in plants. The Pb stress reduces the photosynthetic traits (chlorophyll and gas exchange parameters) in the tissues of plants (leaves and roots), which ultimately lead to a reduction in growth as well as biomass. Enzyme activities such as guaiacol peroxidase, superoxide dismutase, ascorbate peroxidase, and catalase are also linearly increased in a dose-dependent manner under Pb stress. The exogenous application of CA reduced the Pb toxicity in plants by improving photosynthesis and, ultimately, plant growth. The upsurge in antioxidants against oxidative stress shows the potential of CA-treated castor beans plants to counteract stress injuries by lowering H_2O_2 and MDA levels. From the results of this study, it can be concluded that CA treatments play a promising role in increasing the uptake of Pb and reducing its phytotoxicity. These outcomes recommend that CA application could be an effective approach for the phytoextraction of Pb from polluted soils by growing castor beans.

Keywords: lead; castor beans; citric acid; phytoextraction; antioxidant enzyme

1. Introduction

Heavy metals have been of much interest to researchers and scientists regarding environmental safety, and among this, lead (Pb) has gained substantial consideration as a persuasive environmental hazard. Both natural and anthropogenic activities such as the disposal of municipal sewage sludge, fertilizer application, practices of mines, forest fires, industrial fumes, storage batteries such as lithium ion batteries, volcanic eruptions, igneous rocks, ores smelting, paints, gasoline, and explosives are the major contributors to the release of Pb in the external environment [1]. Lead has been known as an inorganic toxin (cannot be degraded) even at lower concentrations and is readily absorbed in cultivated soils, which is easily taken up by various organs of plants. The major levels of Pb exhibit in the food chain through accumulation/absorption impose severe threat towards human health [2]. In the plant environment, Pb phytotoxicity results in a stop to plant growth, depression of seed germination, disruption of cellular structures, impairment of photosynthesis, imbalance of hormones, ion homeostasis, and the over-generation of reactive oxygen species (ROS) as well as inhibition of enzyme activities [3]. As a result, crops cultivated in these Pb-polluted soils directly affect agricultural production. Therefore, it is essential to remediate Pb contamination in the soil–plant environment. The use of green-plants to remediate contaminated soils is a long term, cost-effective, and eco-friendly tool [4,5]. However, the plants' ability to uptake and translocate HMs into upper harvestable parts are dependent on soil type, plant species, and environmental conditions [6–8]. In recent research, numerous plant species have been studied which can be used to remediate the HM polluted soils, including *Brassica napus* for Pb [9], cadmium (Cd) [10], and copper (Cu) [11], while cauliflower and sunflower were used to remediate chromium (Cr) [12,13], accordingly.

Usually, many of the HMs are adsorbed in soil particles to make soil aggregates that are hard to be integrated by plants. Thus, the use of acids, which are low molecular weight organic acids like citric acid (CA), is crucial to alter the chemical activity/bioavailability of HMs and improve phytoextraction [14]. Although phytoextraction is manageable, cost-effective, and eco-friendly, it is not recommended commercially. The capacity of Pb-phytoextraction can be effectively improved by the selection of effective chelating agents like CA. Recent studies have documented the role of CA as a growth promoting agent with a chelating potential against different HMs such as Cr [15], copper (Cu) [11], Pb [16] as well as Cd [10].

Castor beans (*Ricinus communis* L.) is a tropical African plant that is well-known for its ability to grow in contaminated sites, which makes this plant a promising candidate to clean up the environment and re-cultivate polluted lands [17]. Previous scientific research has suggested that castor beans grown in specific areas of the world can accumulate Pb [18]. Castor beans is an industrial crop, is used other than in food items, and has tremendous potential in a rotation system. Therefore, it is a novel crop to cultivate for the phytoremediation of HMs because this crop has an additional economic benefit if grown in polluted sites [19].

Most of the plants used for the phytoextraction of Pb contaminated soils belong to the temperate climate, and less is known about the ability of tropical plants for the phytoextraction of HMs. By keeping in view the significance of phytoextraction to remediate polluted lands, the quest for tropical plants becomes obligatory. Therefore, this study analyzes the Pb-induced harmful effects on morph-physiological and biochemical attributes of castor beans. In addition, Pb uptake and its detoxification by the application of CA are also explored, which can be used to remediate Pb contaminated sites.

2. Results

2.1. Plant Growth and Biomass

Figure 1A–F demonstrates the negative changes in growth and biomass produced due to Pb-induced toxicity as well as the positive effect of CA application on growth and biomass characteristics of the plant. Lead application at both levels (300 and 600 mg kg^{-1}) substantially reduced plant (root and shoot) length, leaf area, the number of leaves per plant, and dry weight, and the response was dose-dependent for all parameters (Figure 1A,B). In contrast to the control treatment, the decrease in the shoot and root length was 7.21% (300 mg kg^{-1}) and 18.43% (600 mg kg^{-1}), respectively, under low and high levels of Pb stress while both doses of CA potentially improved the root and shoot lengths of Pb-exposed castor beans plants. Similarly, shoot and root dry weight, the number of leaves per plant of castor beans decreased significantly when exposed to Pb stress. Application of CA (2.5 and 5 mM) considerably improved plant growth and biomass parameters under Pb stress compared to Pb-stress-alone, as demonstrated in Figure 1A–F.

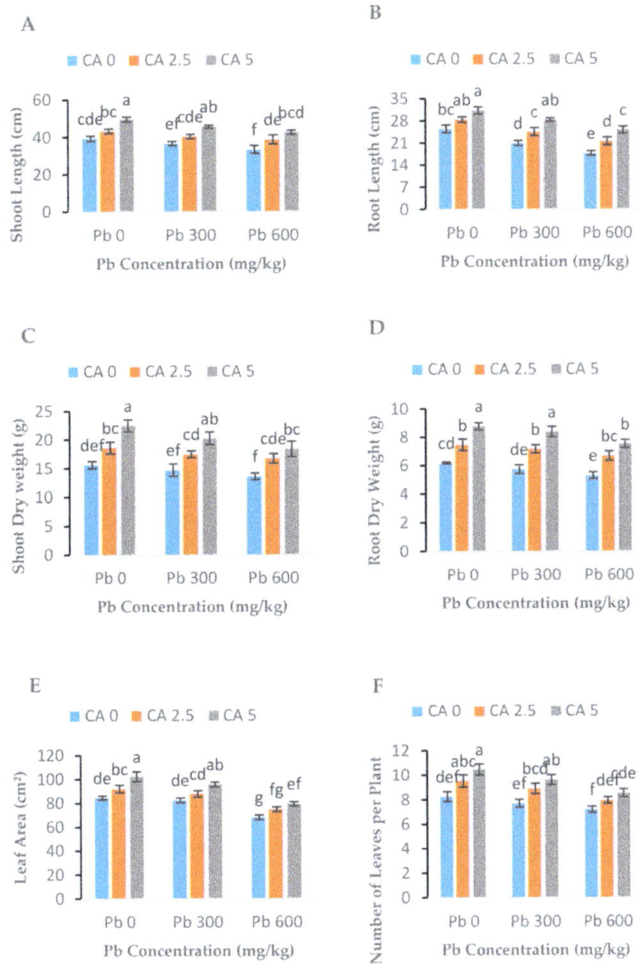

Figure 1. Effect of different Pb concentrations (0, 300, and 600 mg kg^{-1}) and CA levels (0, 2.5, and 5 mM) on shoot lengths (**A**), root lengths (**B**), shoot dry weight (**C**), roots dry weight (**D**), leaf area (**E**), numbers of leaves per plant (**F**) of castor beans plants. Values reported in figures are the mean of 3 replicates along with standard deviation. Letters show the significant difference between the treatments of $p \leq 0.05$.

2.2. Photosynthetic Pigments and Gas Exchange Parameters

Exposure to Pb stress at both levels (300 and 600 mg kg^{-1}) slightly decreased ($p < 0.05$) the photosynthetic pigments including chlorophylls (a and b), carotenoids, and total chlorophyll compared to the relevant control treatments (Figure 2A–D). When CA was applied to Pb-stressed plants, a significant improvement in all the photosynthetic parameters was observed compared to control and Pb-alone treatments (Figure 2A–D). Moreover, CA also improved these parameters where no Pb stress was applied. Citric acid (5 mM) increased the chlorophylls a and b, total chlorophyll, and carotenoids by 35, 42, 37, and 42%, respectively. Similarly, photosynthetic characteristics, including photosynthetic rate, water use efficiency, stomatal conductance, and transpiration rate, decreased significantly in Pb-exposed plants without CA (Figure 3A–D). The citric acid application significantly improved all the gas exchange characteristics, including photosynthetic rate, transpiration rate, water use efficiency, and stomatal conductance up to 33%, 37%, 43%, 47%, respectively, in Pb-exposed castor beans plants with respect to the Pb-alone treated plants (control).

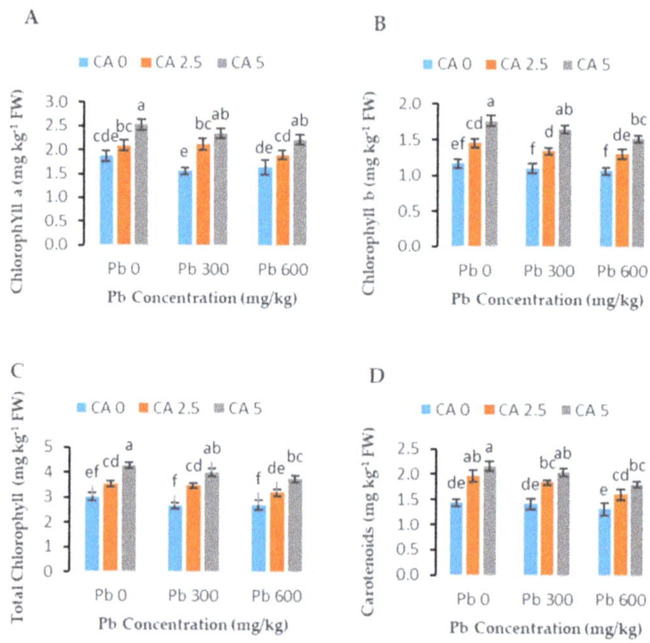

Figure 2. Effect of different Pb concentrations (0, 300, and 600 mg kg^{-1}) and CA levels (0, 2.5, and 5 mM) upon chlorophyll *a* content (**A**), chlorophyll *b* content (**B**), chlorophyll total (**C**), and the carotenoids content (**D**) of castor beans plants. Values reported in figures are the mean of 3 replicates along with standard-deviation. Letters show the significant difference between the treatments of $p \leq 0.05$.

2.3. Antioxidant Enzymes Activities

The effects of Pb stress (300 and 600 mg kg^{-1}) and CA (2.5 and 5 mM) application (alone or combined) on important antioxidant enzyme activities, i.e., POD, CAT, APX, and SOD, in roots and leaves of castor beans plants are shown in Figure 4A–H. A significant reduction in the activities of antioxidant enzymes was observed in Pb-alone treatments over the control. Although higher Pb stress (600 mg kg^{-1}) showed more reduction in enzyme activities than lower Pb stress (300 mg kg^{-1}), this decrease was significant ($p < 0.05$) in all the plants expose to Pb toxicity as compared to control plants. Under both Pb-stress levels, CA treatments significantly ($p < 0.05$) enhanced the efficiency of these enzymes compared to their respective Pb-alone treated roots and leaves of the plants.

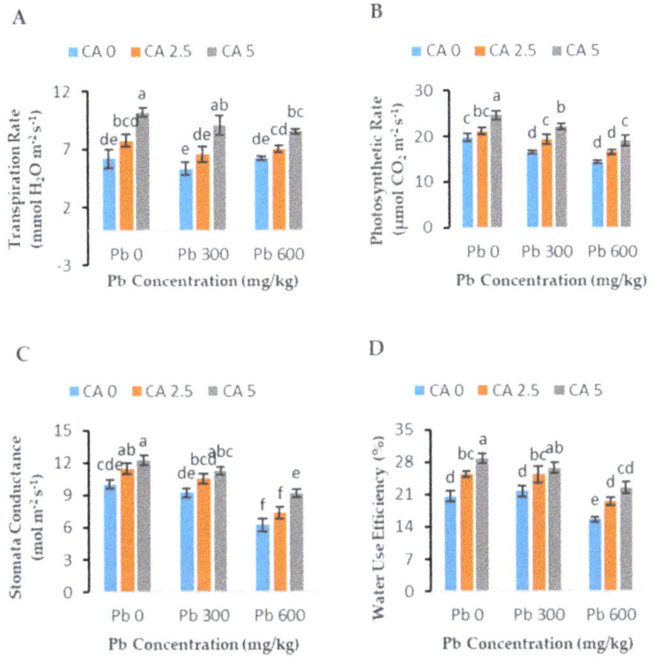

Figure 3. Effect of different Pb concentrations (0, 300, and 600 mg kg^{-1}) and CA levels (0, 2.5, and 5 mM) on transpiration rate (**A**), photosynthetic rate (**B**), stomata conductance (**C**), water use efficiency (**D**) of castor beans plants. Values reported in figures are the mean of 3 replicates along with standard deviation. Letters shows the significant difference between the treatments of $p \leq 0.05$.

2.4. *Effect of CA Application on EL, H_2O_2, and MDA*

With exposure to Pb stress, electrolyte leakage (EL), H_2O_2, and MDA concentrations were significantly enhanced in castor beans plants (leaves and roots) (Figure 5A–F). Application of CA at both levels (2.5 and 5 mM) to castor beans plants significantly ($p < 0.05$) recovered the oxidative damage caused by Pb, and the positive effects of CA were more prominent in the 5 mM CA treatment than the lower CA level. With the 5 mM CA + 600 mg kg^{-1} Pb application, the EL, H_2O_2, and MDA contents in leaves significantly ($p < 0.05$) reduced by 32%, 18%, and 21%, respectively, as compared with the respective control without CA, while in roots this reduction was 18%, 20%, and 33%, respectively.

2.5. *Lead Concentration and Uptake in Castor Beans*

The concentrations of Pb in root and shoots of castor beans were significantly enhanced with the increase in doses of Pb (Figure 6A,B). The castor beans plants stored higher Pb concentration in roots compared to shoot tissues, regardless of the applied Pb levels. The application of CA at a higher dose (5 mM) significantly increased the Pb concentration in the shoots and roots of castor beans plants. This increase in Pb concentrations was about 26% and 18% in shoots, and 25% and 28% in roots at 300 and 600 mg kg^{-1} Pb concentrations, respectively, compared to Pb-alone treated castor beans plants. Similarly, CA application at the higher dose (5 mM) significantly increased the total Pb uptake in castor beans plants; for example, the increase in total shoot Pb uptake was 46% and 39%, while in roots this increase was 48% and 49% at lower and higher levels of the Pb application, respectively (Figure 6C,D).

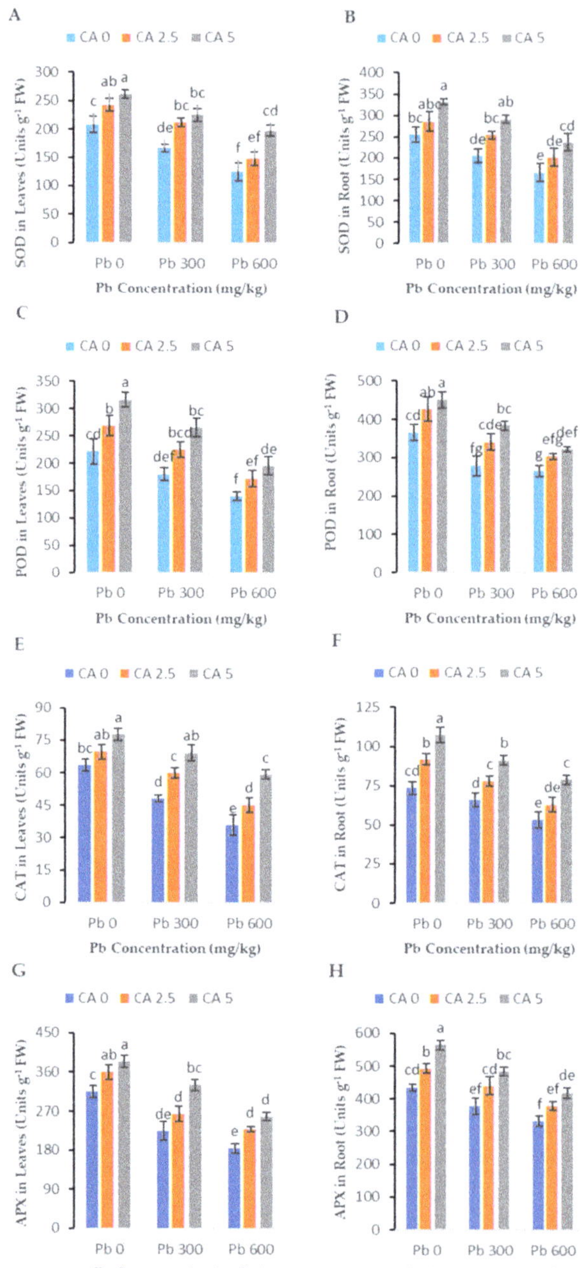

Figure 4. Effect of different Pb concentrations (0, 300, and 600 mg kg^{-1}) and CA levels (0, 2.5, and 5 mM) on leaf SOD content (**A**), roots SOD content (**B**), leaf POD content (**C**), roots POD content (**D**), leaf CAT content (**E**), roots CAT content (**F**), leaf APX content (**G**), roots APX content (**H**) of castor beans plants. Values reported in figures are the mean of 3 replicates along with standard deviation. Letters show the significant difference between the treatments of $p \leq 0.05$.

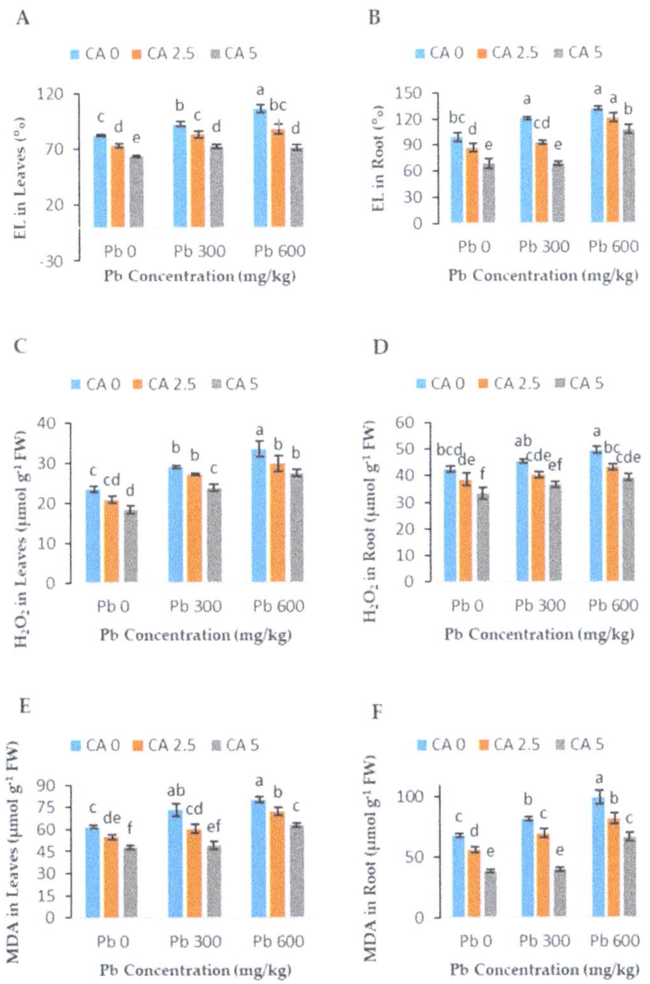

Figure 5. Effect of different Pb concentrations (0, 300, and 600 mg kg^{-1}) and CA levels (0, 2.5, and 5 mM) on EL in leaf (**A**), EL in roots (**B**), leaf H_2O_2 content (**C**), roots H_2O_2 content (**D**), leaf MDA content (**E**), roots MDA content (**F**) of castor beans plants. Values reported in figures are the mean of 3 replicates along with standard deviation. Letters show the significant difference between the treatments on $p \leq 0.05$.

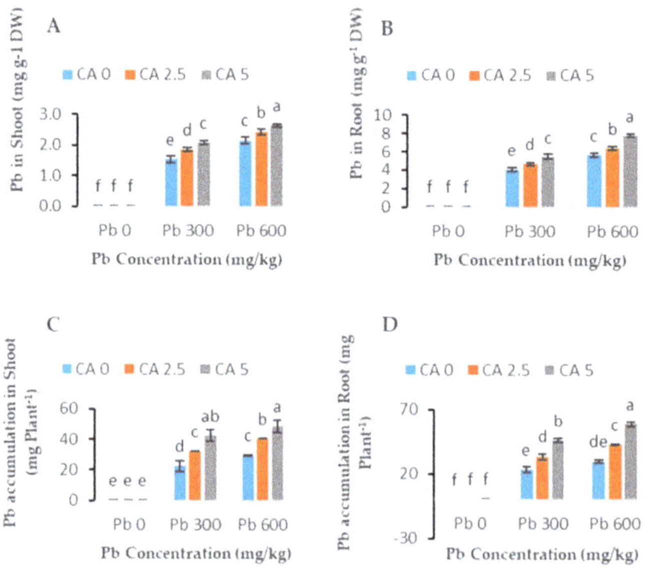

Figure 6. Effect of different Pb concentrations (0, 300, and 600 mg kg^{-1}) and CA levels (0, 2.5, and 5 mM) on Pb concentration to the plant shoot (**A**), Pb concentration to the plant root (**B**), Pb accumulation in shoot (**C**), Pb accumulation in root (**D**) of castor beans plants. Values reported in figures are the mean of 3 replicates along with standard deviation. Letters show the significant difference between the treatments of $p \leq 0.05$.

3. Discussion

3.1. Plant Growth and Biomass

This study was conducted for the determination of potential effects of Pb stress on castor beans plants, the potential of castor beans plants toward the phytoremediation of Pb, and CA's role towards the phyto-extraction of Pb by this plant. The effect of Pb on morphological attributes is prominent in our results (Figure 1), which clearly showed the retarded growth of plants under Pb stress alone. The high concentration of Pb decreased the roots and shoot lengths, plant height, and fresh and dry biomass. The greater uptake of the Pb in the plant can restrict the biomass production of plants [20,21]. The higher Pb toxicity in plants caused growth inhibition, reduced the number of leaves, and produced smaller, more brittle leaves in maize plants [22]. These detrimental effects of Pb on plant growth might be the result of a disturbance in the nutrient metabolic process in plants [23]. Lead, even at low concentrations, hampers the growth of the aerial parts and roots of plants [24], which also support the results of our study. Pb exposure considerably decreased root expansion in *Prosopis* sp. [25]. Similarly, Pb exposure in *A. Sativum* badly damaged the mitochondria, vacuolization of endoplasmic reticulum, and plasma membrane in the roots [26]. Malar et al. [27] also reported similar results, as observed in our study, and reported reduced plant growth and biomass in *H. incana* under Pb stress. Similarly, Fahr et al. [28] described the reduction in root length, plant height, fresh and dry biomass in Pb-stressed *Brassica* plants. This reduction in morphological attributes of plants might be due to the fact that Pb toxicity disturbs the intracellular biochemical reaction, cell division, and DNA synthesis in the same way as described in other various studies under HM stress [29–31].

The promotive role of CA in plants exposed to HM stress is well recognized [11,32]. The application of CA improved the growth and biomass of castor beans plants under Pb stress (Figure 1). Results of the present study were in line with the outcome described by Ehsan et al. [10], Farid et al. [13], and Afshan et al. [15] for various plants species. These studies have demonstrated that the CA application improved plant growth under various metal stresses. The application of CA plays a vital role under metal stress and promotes plant growth and biomass, which may be because of an increase in nutrient uptake by plants [33]. Improvement in biomass and plant growth might be accredited to the ability of CA to enhance the uptake of essential nutrients by the formation of complexes with nutrients [34]. The other possible reason might be that the application of CA may enhance the photosynthesis and synthesis of phytochelatins (PCs) in plants [33].

3.2. Photosynthetic Pigments

The photosynthetic pigments played a vital role in plant life as they harvest light for biochemical reactions in leaf cells. For an HM tolerance ability of a plant, chlorophyll contents may be considered one of the important indicators of HM stress [35]. Our results indicated that Pb stress significantly reduced the photosynthetic pigment in plants, which ultimately lead to the diminution in plant growth and biomass (Figure 2). A recent study had been conducted where Pb significantly affected the photosynthetic pigments in different plants [16,36]. This reduction in photosynthetic pigments might be due to the result of Pb stress, which may affect the structure of chloroplast, chloroplast membranes, plastoglobuli, promote the ion exchange in the chloroplast and inhibit the essential enzymes in the Calvin cycle [30]. All of these phenomena lead to the reduction of photosynthetic pigments, which ultimately reduce the growth of plants. The application of CA enhanced the photosynthetic pigment in stressed plants (Figure 2). A useful role of the CA towards the photosynthetic system in plants, when exposed to HMs stress, has been described in many recent research works [13,15,16]. This might be the effect of CA application, which enhanced the uptake of essential nutrients as well as the formation of photosynthetic pigments [13]. Enhancement in photosynthetic pigments in photochemical reactions may transform the light energy more effectually under the application of CA, resulting in enhancement of biomass and plant growth [30]. Citric acid significantly enhanced the contents of photosynthetic pigments of *B. napus* under Cu stress [11] and of *I. lacteal* under Pb stress [37], as consistent with our results. Similarly, the application of CA mitigated the adverse effect of Pb in *S. durmmondii* and enhanced photosynthetic activity as well as plant growth [38].

3.3. Antioxidant Enzymes

Plants develop their defensive system in the template of the antioxidant enzyme system which has a vital part in the alleviation of oxidative stress. In the current study, the activities of CAT, SOD, and POD were considerably reduced under Pb stress (Figure 3). In numerous studies, it has been reported that Pb stress reduced the activities of antioxidant enzymes in various plants [10,13,39]. Sometimes antioxidant enzymes exhibit dual behavior: mild HM stress increases the antioxidant enzyme activities while higher HM stress reduces the enzymatic activities of antioxidant enzymes. Antioxidant enzymes (SOD, POD, and CAT) activities increased in sunflower with an increase in the concentration of Cr in the soil [13]. In another study, the same response of antioxidant enzymes was observed in wheat, *B. napus*, and mung bean at a lower concentration of Cr toxicity [40]. Increased levels of CA (2.5 and 5 mM) enhanced enzymatic activities at all concentrations of Pb regarding the toxicity of Pb (Figure 3). This might be due to the growth promotive character of CA in assisting the plant to recover fast from oxidative damage [32]. In addition, the ameliorating effect of CA also helps the plant to alleviate Pb stress and to increase the activities of antioxidant enzymes, as investigated in sunflower plants [41]. Increased levels of EL and ROS also frequently increase the anti-oxidative enzyme activities as the plant tries to alleviate the metal-induced stress by activating the enzymatic defense system [42]. Citric acid application in cotton improved the antioxidant enzyme activities by mitigating Ni stress [11]. Similar kinds of improvements in antioxidant enzymes through the exogenous application of CA (2.5 and 5 mM) have been observed in *B. napus* under different metal stress [10,15,16].

3.4. H_2O_2, EL, and MDA Contents

The H_2O_2, EL, and MDA contents in plant cells indicate the degree of oxidative stress and peroxidation of membranes. Plants exposed to Pb stress confronted oxidative stress by the EL and production of ROS [12]. In this study, H_2O_2 and MDA contents and EL increased markedly with an increasing concentration of Pb in soil (Figure 5). Similar outcomes have been described in recent studies performed on wheat, barley, and *B. napus* in which the toxicity of Cr resulted in oxidative stress and a higher level of EL [40]. Similarly, sunflower showed a high production of MDA and H_2O_2 under HM stress [43]. It has already been reported that Pb stress enhanced the production of ROS, which induced severe oxidative stress in plant cells [13,39]. The ability of HMs to damage the electron transport chain and K^+ efflux may enhance the production of O^- and OH^- radicals and, ultimately, EL in plants [44]. The biotic and abiotic stress fortified by metals also increase the EL in plants [45,46]. The oxidative damage is directly correlated with a reduction of growth in plants as well as biomass [47]. In the current study, the application of CA predominantly reduced the generation of EL and ROS in castor beans plants by facilitating the activities of antioxidant enzymes. Antioxidant defensive activities play a novel role regarding the reduction of EL and ROS production [13,48]. A similar impact of CA was also detected in *B. napus* and sunflower under the stress of Cr toxicity [13,15]. In this study, the increase in castor beans growth and biomass by application of CA may be attributed to the inhibiting effect of CA on EL and ROS production.

3.5. The Accumulation of Pb in Roots and Leaves

Concentrations of Pb in roots and shoots of castor beans plants exposed to different levels of Pb along with applications of CA are shown in Figure 6. Our results clearly explained that Pb uptake increased with increasing concentrations of applied Pb in the system of soil. It is well observed that the accumulation of Pb in different parts of the plant correlates with the concentration of Pb in growth media [47]. Our results clearly showed that the concentration of Pb were significantly higher in roots as compared with the aerial parts of the plant. Similar results have been reported by many researchers [10,13,47]. That might be due to the ability of Pb to make complexes with polysaccharides such as sugar and other macromolecules.

The application of CA enhanced the uptake of Pb in roots and shoots of the castor bean. It was reported that CA application enhanced both uptake and translocation of various HMs in plants [32,49]. The chelating role of CA in soil enhances the movement and availability of HMs (Cd, Pb, Cu, and Zn) in the soil. The chelating effect of CA also made the complex with metals and enhanced the accumulation of metals by plants [50]. For example, the application of CA enhanced the uptake of Cr in sunflower [13,41]. Our results related to Pb concentrations in tissues are comparable with Kiran and Prasad [51] in which rice husk ash and biochar assisted the phytoremediation potential of castor beans plants for Pb-spiked soil. However, in our study, the Pb uptake was slighter higher as compared with Kiran and Prasad [51]. The higher Pb uptake by castor beans plants in our study might be due to the difference in plant variety, soil texture, experimental duration, and varying experimental conditions. Moreover, we applied Pb in soluble form which can be easily taken up by plants. The other potential cause of higher Pb concentrations in plants might be that the CA may decline the pH of the soil, which upgrades the mobility of metals and finally improves the uptake of HMs by plants [52,53]. Citric acid application also increased plant growth and biomass and, consequently, the accumulation and uptake of metals in plants [13,54]. Pb accumulation was found both in roots and shoots whereas maximum accumulation was found in the roots of castor beans plants. It is reported that plants have more capacity to accumulate metals in roots than in shoots. The accumulation of Pb in roots has limited the transfer of HMs to above-ground tissue as reported earlier [55]. The finding of our study showed that the application of CA improved the phytoextraction of Pb via castor beans plants and increased the biomass and growth of the plants, which might be employed for the cultivation of metal contaminated soils.

4. Materials and Methods

4.1. Soil Sampling and Analysis

Soil sampling was made from an agricultural field of the University of Agriculture Faisalabad, Pakistan. Roots and other large debris were removed from the soil and sieved by 2 mm sieve. Before conducting the experiment, basic soil analysis was performed to check the soil properties. Standard protocols were followed for various soil analysis, i.e., pH and EC by Soltanpour [56], sodium adsorption ratio and soluble ions by US Salinity Laboratory Staff [57], and Page et al. [58]. Soil particle size was determined through a protocol described by Bouyoucos [59]. The Walkley–Black method was followed for soil organic matter determination described by Jackson [60] and the calcimeter method was followed for calcium carbonate [61]. Detailed soil properties are described in Table 1.

Table 1. Soil physico-chemical properties used for the experiment.

Texture	Sandy Loam
Silt	15.0%
Sand	67.9%
Clay	17.1%
EC	1.96 dS m^{-1}
pH	7.61
Sodium adsorption ratio (SAR)	1.89 (mmol $L^{-1})^{1/2}$
Available P	2.11 mg kg^{-1}
Organic matter	0.59%
HCO_3	2.51 mmol L^{-1}
SO_4^{-2}	11.44 mmol L^{-1}
Cl^-	5.45 mmol L^{-1}
Ca^{2+} + Mg^{2+}	13.98 mmol L^{-1}
K^+	0.04 mmol L^{-1}
Na^2	5.23 mmol L^{-1}
Available Zn^2	0.77 mg kg^{-1}
Available Cu^{2+}	0.31 mg kg^{-1}
Available Cr $^{+6}$	0.16 mg kg^{-1}

4.2. Pot Experiment

A pot experiment was conducted, and for this purpose, the plastic pots, with 5 kg capacity, were filled with sieved soil. Five castor beans seeds were sown in each pot. Right after germination, plant thinning was made, and 2 plants were maintained in each pot. A complete randomized design was applied along with 3 replicates against each treatment. Plants were fertilized using NPK with a ratio of 120:50:25 kg ha^{-1}. For NPK, the salts of urea, diammonium phosphate, and potassium sulfate were used, respectively. Various concentrations of Pb (0, 300, and 600 mg kg^{-1}) were applied using $PbNO_3$ salt. First time, Pb treatments were applied after 30 days of sowing. Then, the remaining Pb solution was applied after a 7-day interval with four times of application. The total calculated amount of Pb for each treatment and replicates was added in 2 L of water, and 500 mL of water was applied each time. The final Pb concentrations in the soil were either 300 or 600 mg kg^{-1} after the four-times of soil application. This practice of Pb soil application was performed to make sure of Pb in soluble form in the soils. The higher level of Pb was selected for the identification of mechanisms of CA on the reduction of Pb toxicity in plants. Nitrogen added through the Pb salt was calculated and adjusted through N fertilizers. This was performed to avoid the effect of N in the study. The experimental plants were provided with a foliar spray of different concentrations of CA (0, 2.5, and 5 mM) at the time of Pb application in the soil, and it was provided in different intervals, similar to the Pb application. The total volume of CA used was 1 L for each treatment for all replicates.

4.3. Plants Harvesting

After 70 days of sowing, the harvesting of castor beans plants was made and their different parts were differentiated. After washing with distilled water, plants were dried in oven (70 °C) for 72 h and dry weights were recorded. Diluted HCl acid (1.0%) was used for root washing; after that roots were washed with distilled water several times until the acidic content had been fully removed. After drying at room temperature, roots were also oven-dried at 70 °C, and dry weight was noted.

4.4. Chlorophyll Contents and Gas Exchange Parameters

For estimation of chlorophyll contents, the fresh leaf samples were collected and centrifuged after the extraction in 85% acetone (*v/v*), and readings were taken at recommended wavelengths with a spectrophotometer [62]. On the same day in full sunshine (10:00–12:00 a.m.), IRGA was used for readings of photosynthetic rate, transpiration rate, stomata conductance, and water use efficiency.

4.5. Estimation of MDA, EL, H_2O_2 and Antioxidants Enzymes

The Zhang and Kirkham [63] method was followed to measure MDA contents by using thiobarbituric acid (0.1%), further described by Abbas et al. [64]. The Dionisio-Sese and Tobita [65] method was used for EL estimation. For this, the initial and final EC of the solution was noted through the extraction of samples for 2 h at 32 °C and then doing the extraction of the same samples at 121 °C for 20 min, respectively. To measure the contents of H_2O_2, the Jana and Choudhuri [66] procedure was followed. In brief, samples were homogenized with phosphate buffer (50 mM) at 6.5 pH and centrifuged for 20 min. Then H_2SO_4 (20%, *v/v*) was added and centrifuged again for 15 min, and absorbance was noted at 410 nm.

Samples were crushed in liquid nitrogen and standardized in 0.5 M phosphate buffer at a pH of 7.8 for the investigation of SOD and POD activities [67]. APX and CAT activities were estimated by the following procedures of Nakano and Asada [68] and Aebi [69], respectively. The detailed procedures are described by Abbas et al. [63].

4.6. Estimation of Pb Contents

The 1.0 g of each sample was digested at a hot plate with 4:1 of HNO_3:$HClO_4$ (*v/v*), and the concentration of Pb was estimated in digested samples by an atomic absorption spectrophotometer [70].

4.7. Statistical Analysis

At 5% probability level, ANOVA was applied for data analysis by using SPSS software (Statistics Software, Version 21.0). Tukey's HSD post hoc test was applied for multiple comparisons of the means.

5. Conclusions

The current study has shown that the promising application of CA remarkably alleviates Pb toxicity in castor beans plants at various morpho-physiological and biochemical levels. Application of CA significantly improved the content of Pb in castor beans plants and increased the growth of the plants as well as the antioxidant defense system, which further supported the plants' normal functioning and metabolism under Pb stress.

Author Contributions: Conceptualization, Z.I.M. and A.M.; Data curation, Z.I.M., S.A. and A.H.; Formal analysis, Z.I.M., M.R. and S.A.; Funding acquisition, Q.A., S.A. and S.H.A.; Investigation, M.R., A.M. and H.A.K.; Methodology, S.A. and A.H.; Project administration, M.R. and A.M.; Resources, S.A., S.H.A. and P.A.; Software, Q.A. and P.A.; Supervision, M.R.; Validation, A.M., Q.A. and P.A.; Visualization, P.A.; Writing—original draft, Z.I.M., S.A. and S.H.A.; Writing—review & editing, A.M., H.A.K. and P.A.

Funding: The authors want to say thanks to Higher Education Commission (HEC), Pakistan for financial support under HEC Project No. 20-3653/NRPU/R&D/HEC/14/437 and NRPU Project No.

5634/Punjab/NRPU/R&D/HEC/2016. The authors would like to extend their sincere appreciation to the Deanship of Scientific Research at King Saud University for funding the Research Group No. RGP-1435-066.

Conflicts of Interest: The authors declare no conflicts of interest.

References

1. Tabelin, C.B.; Igarashi, T.; Villacorte-Tabelin, M.; Park, I.; Opiso, E.M.; Ito, M.; Hiroyoshi, N. Arsenic, selenium, boron, lead, cadmium, copper, and zinc in naturally contaminated rocks: A review of their sources, modes of enrichment, mechanisms of release, and mitigation strategies. *Sci. Total Environ.* **2018**, *645*, 1522–1553. [CrossRef] [PubMed]
2. Tang, B.; Tong, P.; Xue, K.S.; Williams, P.L.; Wang, J.S.; Tang, L. High-throughput assessment of toxic effects of metal mixtures of cadmium (Cd), lead (Pb), and manganese (Mn) in nematode Caenorhabditis elegans. *Chemosphere* **2019**. [CrossRef] [PubMed]
3. Bali, S.; Jamwal, V.L.; Kaur, P.; Kohli, S.K.; Ohri, P.; Gandhi, S.G.; Bhardwaj, R.; Al-Huqail, A.A.; Siddiqui, M.H.; Ahmad, P. Role of P-type ATPase metal transporters and plant immunity induced by jasmonic acid against Lead (Pb) toxicity in tomato. *Ecotoxicol. Environ. Saf.* **2019**, *174*, 283–294. [CrossRef] [PubMed]
4. Lamhamdi, M.; El Galiou, O.; Bakrim, A.; Nóvoa-Muñoz, J.C.; Arias-Estevez, M.; Aarab, A.; Lafont, R. Effect of lead stress on mineral content and growth of wheat (*Triticum aestivum*) and spinach (*Spinacia oleracea*) seedlings. *Saudi J. Boil. Sci.* **2013**, *20*, 29–36. [CrossRef] [PubMed]
5. Kasim, W.A.; Abokassem, E.M.; Ragab, G.A.; Sewelam, N.A. Alleviation of lead stress toxicity in Vigna unguiculata by salicylic acid. *Egypt. J. Exp. Biol.* **2014**, *10*, 37–49.
6. Ali, S.; Bharwana, S.A.; Rizwan, M.; Farid, M.; Kanwal, S.; Ali, Q.; Ibrahim, M.; Gill, R.A.; Khan, M.D. Fulvic acid mediates chromium (Cr) tolerance in wheat (*Triticum aestivum* L.) through lowering of Cr uptake and improved antioxidant defense system. *Environ. Sci. Pollut. Res.* **2015**, *22*, 10601–10609. [CrossRef]
7. Ali, S.; Chaudhary, A.; Rizwan, M.; Anwar, H.T.; Adrees, M.; Farid, M.; Irshad, M.K.; Hayat, T.; Anjum, S.A. Alleviation of chromium toxicity by glycinebetaine is related to elevated antioxidant enzymes and suppressed chromium uptake and oxidative stress in wheat (*Triticum aestivum* L.). *Environ. Sci. Pollut. Res.* **2015**, *22*, 10669–10678. [CrossRef]
8. Ali, S.; Gill, R.A.; Ulhassan, Z.; Najeeb, U.; Kanwar, M.K.; Abid, M.; Mwamba, T.M.; Huang, Q.; Zhou, W. Insights on the responses of Brassica napus cultivars against the cobalt-stress as revealed by carbon assimilation, anatomical changes and secondary metabolites. *Environ. Exp. Bot.* **2018**, *156*, 183–196. [CrossRef]
9. Kanwal, U.; Ali, S.; Shakoor, M.B.; Farid, M.; Hussain, S.; Yasmeen, T.; Adrees, M.; Bharwana, S.A.; Abbas, F. EDTA ameliorates phytoextraction of lead and plant growth by reducing morphological and biochemical injuries in Brassica napus L. under lead stress. *Environ. Sci. Pollut. Res.* **2014**, *21*, 9899–9910. [CrossRef]
10. Ehsan, S.; Ali, S.; Noureen, S.; Mahmood, K.; Farid, M.; Ishaque, W.; Shakoor, M.B.; Rizwan, M. Citric acid assisted phytoremediation of cadmium by *Brassica napus* L. *Ecotoxicol. Environ. Saf.* **2014**, *106*, 164–172. [CrossRef]
11. Zaheer, I.E.; Ali, S.; Rizwan, M.; Farid, M.; Shakoor, M.B.; Gill, R.A.; Najeeb, U.; Iqbal, N.; Ahmad, R. Citric acid assisted phytoremediation of copper by *Brassica napus* L. *Ecotoxicol. Environ. Saf.* **2015**, *120*, 310–317. [CrossRef] [PubMed]
12. Hasanuzzaman, M.; Nahar, K.; Rahman, A.; Mahmud, J.A.; Alharby, H.F.; Fujita, M. Exogenous glutathione attenuates lead-induced oxidative stress in wheat by improving antioxidant defense and physiological mechanisms. *J. Plant Interact.* **2018**, *13*, 203–212. [CrossRef]
13. Farid, M.; Ali, S.; Rizwan, M.; Ali, Q.; Abbas, F.; Bukhari, S.A.H.; Saeed, R.; Wu, L. Citric acid assisted phytoextraction of chromium by sunflower; morpho-physiological and biochemical alterations in plants. *Ecotoxicol. Environ. Saf.* **2017**, *145*, 90–102. [CrossRef] [PubMed]
14. Kim, S.H.; Lee, I.S. Comparison of the ability of organic acids and EDTA to enhance the phytoextraction of metals from a multi-metal contaminated soil. *Bulletin Environ. Contamin. Toxicol.* **2010**, *84*, 255–259. [CrossRef]

15. Afshan, S.; Ali, S.; Bharwana, S.A.; Rizwan, M.; Farid, M.; Abbas, F.; Ibrahim, M.; Mehmood, M.A.; Abbasi, G.H. Citric acid enhances the phytoextraction of chromium, plant growth, and photosynthesis by alleviating the oxidative damages in *Brassica napus* L. *Environ. Sci. Pollut. Res.* **2015**, *22*, 11679–11689. [CrossRef]
16. Shakoor, M.B.; Ali, S.; Hameed, A.; Farid, M.; Hussain, S.; Yasmeen, T.; Najeeb, U.; Bharwana, S.A.; Abbasi, G.H. Citric acid improves lead (Pb) phytoextraction in *Brassica napus* L. by mitigating Pb-induced morphological and biochemical damages. *Ecotoxicol. Environ. Saf.* **2014**, *109*, 38–47. [CrossRef]
17. Olivares, A.R.; Carrillo-González, R.; González-Chávez, M.D.C.A.; Hernández, R.M.S. Potential of castor bean (*Ricinus communis* L.) for phytoremediation of mine tailings and oil production. *J. Environ. Manag.* **2013**, *114*, 316–323. [CrossRef]
18. Romeiro, S.; Lagôa, A.M.; Furlani, P.R.; Abreu, C.A.D.; Abreu, M.F.D.; Erismann, N.M. Lead uptake and tolerance of *Ricinus communis* L. *Brazilian J. Plant Physiol.* **2006**, *18*, 483–489. [CrossRef]
19. Liu, J.N.; Zhou, Q.X.; Sun, T.; Ma, L.Q.; Wang, S. Identification and chemical enhancement of two ornamental plants for phytoremediation. *Bulletin Environ. Contamin. Toxicol.* **2008**, *80*, 260–265. [CrossRef]
20. Singh, R.; Tripathi, R.D.; Dwivedi, S.; Kumar, A.; Trivedi, P.K.; Chakrabarty, D. Lead bioaccumulation potential of an aquatic macrophyte Najas indica are related to antioxidant system. *Biores. Technol.* **2010**, *101*, 3025–3032. [CrossRef]
21. Piotrowska, A.; Bajguz, A.; Godlewska-Żyłkiewicz, B.; Czerpak, R.; Kamińska, M. Jasmonic acid as modulator of lead toxicity in aquatic plant Wolffia arrhiza (*Lemnaceae*). *Environ. Exp. Bot.* **2009**, *66*, 507–513. [CrossRef]
22. Okant, M.; Kaya, C. The role of endogenous nitric oxide in melatonin-improved tolerance to lead toxicity in maize plants. *Environ. Sci. Pollut. Res.* **2019**, *26*, 11864–11874. [CrossRef] [PubMed]
23. Gopal, R.; Rizvi, A.H. Excess lead alters growth, metabolism and translocation of certain nutrients in radish. *Chemosphere* **2008**, *70*, 1539–1544. [CrossRef] [PubMed]
24. Islam, E.; Yang, X.; Li, T.; Liu, D.; Jin, X.; Meng, F. Effect of Pb toxicity on root morphology, physiology and ultrastructure in the two ecotypes of Elsholtzia argyi. *J. Hazard. Mat.* **2007**, *147*, 806–816. [CrossRef]
25. Arias, J.A.; Peralta-Videa, J.R.; Ellzey, J.T.; Ren, M.; Viveros, M.N.; Gardea-Torresdey, J.L. Effects of Glomus deserticola inoculation on Prosopis: Enhancing chromium and lead uptake and translocation as confirmed by X-ray mapping, ICP-OES and TEM techniques. *Environ. Exp. Bot.* **2010**, *68*, 139–148. [CrossRef]
26. Jiang, W.; Liu, D. Pb-induced cellular defense system in the root meristematic cells of *Allium sativum* L. *BMC Plant. Biol.* **2010**, *10*, 40. [CrossRef]
27. Malar, S.; Manikandan, R.; Favas, P.J.; Sahi, S.V.; Venkatachalam, P. Effect of lead on phytotoxicity, growth, biochemical alterations and its role on genomic template stability in Sesbania grandiflora: A potential plant for phytoremediation. *Ecotoxicol. Environ. Saf.* **2014**, *108*, 249–257. [CrossRef]
28. Fahr, M.; Laplaze, L.; El Mzibri, M.; Doumas, P.; Bendaou, N.; Hocher, V.; Bogusz, D.; Smouni, A. Assessment of lead tolerance and accumulation in metallicolous and non-metallicolous populations of Hirschfeldia incana. *Environ. Exp. Bot.* **2015**, *109*, 186–192. [CrossRef]
29. Ali, B.; Gill, R.A.; Yang, S.; Gill, M.B.; Farooq, M.A.; Liu, D.; Daud, M.K.; Ali, S.; Zhou, W. Regulation of cadmium-induced proteomic and metabolic changes by 5-aminolevulinic acid in leaves of *Brassica napus* L. *PLoS ONE* **2015**, *10*, e0123328. [CrossRef]
30. Han, Y.; Zhang, L.; Gu, J.; Zhao, J.; Fu, J. Citric acid and EDTA on the growth, photosynthetic properties and heavy metal accumulation of Iris halophila Pall. cultivated in Pb mine tailings. *Inter. Biodeter. Biodegrad.* **2018**, *128*, 15–21. [CrossRef]
31. Kumar, A.; Prasad, M.N.V.; Sytar, O. Lead toxicity, defense strategies and associated indicative biomarkers in Talinum triangulare grown hydroponically. *Chemosphere* **2012**, *89*, 1056–1065. [CrossRef] [PubMed]
32. Al Mahmud, J.; Hasanuzzaman, M.; Nahar, K.; Bhuyan, M.B.; Fujita, M. Insights into citric acid-induced cadmium tolerance and phytoremediation in *Brassica juncea* L.: coordinated functions of metal chelation, antioxidant defense and glyoxalase systems. *Ecotoxicol. Environ. Saf.* **2018**, *147*, 990–1001. [CrossRef] [PubMed]
33. Rodriguez, E.; Santos, C.; Azevedo, R.; Moutinho-Pereira, J.; Correia, C.; Dias, M.C. Chromium (VI) induces toxicity at different photosynthetic levels in pea. *Plant. Physiol. Biochem.* **2012**, *53*, 94–100. [CrossRef] [PubMed]
34. Kim, D.J.; Park, B.C.; Ahn, B.K.; Lee, J.H. Thallium uptake and translocation in barley and sunflower grown in hydroponic conditions. *Inter. J. Environ. Res.* **2016**, *10*, 575–582.

35. Huang, S.; Han, Y.; Yu, S.; Gu, J.; Zhang, L. The effect of lead and copper on the growth and physiological response of water flower *Iris pseudacorus* L. *Fresenius Environ. Bulletin.* **2011**, *20*, 2246–2250.
36. Huang, X.H.; Zhu, F.; Yan, W.D.; Chen, X.Y.; Wang, G.J.; Wang, R.J. Effects of Pb and Zn toxicity on chlorophyll fluorescence and biomass production of Koelreuteria paniculata and Zelkova schneideriana young plants. *Photosynthetica* **2019**, *57*, 688–697. [CrossRef]
37. Wang, H.Y.; Tong, H.Y.; Huang, S.Z.; Yuan, H.Y. Effects of citric acid and oxalic acid on the growth and physiology of Iris lactea var. chinensis under Pb stress. *Chin. J. Ecol.* **2010**, *29*, 1340–1346.
38. Ruley, A.T.; Sharma, N.C.; Sahi, S.V.; Singh, S.R.; Sajwan, K.S. Effects of lead and chelators on growth, photosynthetic activity and Pb uptake in Sesbania drummondii grown in soil. *Environ. Pollut.* **2006**, *144*, 11–18. [CrossRef]
39. Dias, M.C.; Mariz-Ponte, N.; Santos, C. Lead induces oxidative stress in Pisum sativum plants and changes the levels of phytohormones with antioxidant role. *Plant Physiol. Biochem.* **2019**, *137*, 121–129. [CrossRef]
40. Adrees, M.; Ali, S.; Rizwan, M.; Ibrahim, M.; Abbas, F.; Farid, M.; Zia-ur-Rehman, M.; Irshad, M.K.; Bharwana, S.A. The effect of excess copper on growth and physiology of important food crops: A review. *Environ. Sci. Pollut. Res.* **2015**, *22*, 8148–8162. [CrossRef]
41. Farid, M.; Ali, S.; Saeed, R.; Rizwan, M.; Bukhari, S.A.H.; Abbasi, G.H.; Hussain, A.; Ali, B.; Zamir, M.S.I.; Ahmad, I. Combined application of citric acid and 5-aminolevulinic acid improved biomass, photosynthesis and gas exchange attributes of sunflower (*Helianthus annuus* L.) grown on chromium contaminated soil. *Inter. J. Phytorem.* **2019**, *21*, 1–8. [CrossRef] [PubMed]
42. Demidchik, V. Reactive oxygen species and oxidative stress in plants. *Plant Stress Physiol.* **2012**, 24–58.
43. Rizwan, M.; Ali, S.; Qayyum, M.F.; Ok, Y.S.; Adrees, M.; Ibrahim, M.; Zia-ur-Rehman, M.; Farid, M.; Abbas, F. Effect of metal and metal oxide nanoparticles on growth and physiology of globally important food crops: A critical review. *J. Hazard. Mat.* **2017**, *322*, 2–16. [CrossRef] [PubMed]
44. Rizwan, M.; Ali, S.; Abbas, F.; Adrees, M.; Zia-ur-Rehman, M.; Farid, M.; Gill, R.A.; Ali, B. Role of organic and inorganic amendments in alleviating heavy metal stress in oil seed crops. *Oil Seed Crops Yield Adap. Under Environ. Stress.* **2017**, *12*, 224–235.
45. Noman, A.; Ali, S.; Naheed, F.; Ali, Q.; Farid, M.; Rizwan, M.; Irshad, M.K. Foliar application of ascorbate enhances the physiological and biochemical attributes of maize (Zea mays L.) cultivars under drought stress. *Arch. Agron. Soil Sci.* **2015**, *61*, 1659–1672. [CrossRef]
46. Arshad, M.; Ali, S.; Noman, A.; Ali, Q.; Rizwan, M.; Farid, M.; Irshad, M.K. Phosphorus amendment decreased cadmium (Cd) uptake and ameliorates chlorophyll contents, gas exchange attributes, antioxidants, and mineral nutrients in wheat (*Triticum aestivum* L.) under Cd stress. *Arch. Agron. Soil Sci.* **2016**, *62*, 533–546. [CrossRef]
47. Hu, Z.H.; Zhuo, F.; Jing, S.H.; Li, X.; Yan, T.X.; Lei, L.L.; Lu, R.R.; Zhang, X.F.; Jing, Y.X. Combined application of arbuscular mycorrhizal fungi and steel slag improves plant growth and reduces Cd, Pb accumulation in Zea mays. *Int. J. Phytorem.* **2019**, *21*, 1–9. [CrossRef]
48. Islam, F.; Yasmeen, T.; Arif, M.S.; Riaz, M.; Shahzad, S.M.; Imran, Q.; Ali, I. Combined ability of chromium (Cr) tolerant plant growth promoting bacteria (PGPB) and salicylic acid (SA) in attenuation of chromium stress in maize plants. *Plant. Physiol. Biochem.* **2016**, *108*, 456–467. [CrossRef]
49. Rathika, R.; Khalifa, A.Y.; Srinivasan, P.; Praburaman, L.; Kamala-Kannan, S.; Selvankumar, T.; Kim, W.; Govarthanan, M. Effect of citric acid and vermi-wash on growth and metal accumulation of Sorghum bicolor cultivated in lead and nickel contaminated soil. *Chemosphere* **2019**, *243*, 125327. [CrossRef]
50. Aderholt, M.; Vogelien, D.L.; Koether, M.; Greipsson, S. Phytoextraction of contaminated urban soils by *Panicum virgatum* L. enhanced with application of a plant growth regulator (BAP) and citric acid. *Chemosphere* **2017**, *175*, 85–96. [CrossRef]
51. Kiran, B.R.; Prasad, M.N.V. Biochar and rice husk assisted phytoremediation potentials of Ricinus communis L. for Lead-spiked soils. *Ecotoxicol. Env. Saf.* **2019**, *183*, 109574. [CrossRef] [PubMed]
52. Maqbool, A.; Ali, S.; Rizwan, M.; Ishaque, W.; Rasool, N.; Rehman, M.Z.; Bashir, A.; Abid, M.; Wu, L. Management of tannery wastewater for improving growth attributes and reducing chromium uptake in spinach through citric acid application. *Environ. Sci. Pollut. Res.* **2018**, *25*, 10848–10856. [CrossRef] [PubMed]
53. Rostami, S.; Azhdarpoor, A. The application of plant growth regulators to improve phytoremediation of contaminated soils: A review. *Chemosphere* **2019**, *220*, 818–827. [CrossRef] [PubMed]

54. Kaur, R.; Yadav, P.; Sharma, A.; Thukral, A.K.; Kumar, V.; Kohli, S.K.; Bhardwaj, R. Castasterone and citric acid treatment restores photosynthetic attributes in *Brassica juncea* L. under Cd (II) toxicity. *Ecotoxicol. Environ. Saf.* **2017**, *145*, 466–475. [CrossRef]
55. Karami, N.; Clemente, R.; Moreno-Jiménez, E.; Lepp, N.W.; Beesley, L. Efficiency of green waste compost and biochar soil amendments for reducing lead and copper mobility and uptake to ryegrass. *J. Hazard. Mat.* **2011**, *191*, 41–48. [CrossRef]
56. Soltanpour, P.N. Use of ammonium bicarbonate DTPA soil test to evaluate elemental availability and toxicity. *Communic. Soil Sci. Plant. Anal.* **1985**, *16*, 323–338. [CrossRef]
57. US Salinity Laboratory Staff. *Diagnosis and Improvement of Saline and Alkali Soils. Agriculture Handbook 60*; United States Salinity Laboratory, USDA: Washington, DC, USA, 1954; p. 160.
58. Page, A.L.; Miller, R.H.; Keeney, D.R. *Methods of Soil Analysis, Part 2, Chemical and Microbiological Properties*; ASA, Inc.; SSSA, Inc.: Madison, WI, USA, 1982.
59. Bouyoucos, G.J. Hydrometer method improved for making particle size analyses of soils 1. *Agron. J.* **1962**, *54*, 464–465. [CrossRef]
60. Jackson, M.L. *Soil Chem. Analysis*; Constable and Company, Ltd.: London, UK, 1962; p. 497.
61. Moodie, C.D.; Smith, H.W.; McCreery, R.A. *Laboratory Manual for Soil Fertility, (Mimeographed)*; Washington State College: Seattle, WA, USA, 1959.
62. Lichtenthaler, H.K. [34] Chlorophylls and carotenoids: Pigments of photosynthetic biomembranes. *Methods Enzymol.* **1987**, *148*, 350–382.
63. Zhang, J.; Kirkham, M.B. Drought-stress-induced changes in activities of superoxide dismutase, catalase, and peroxidase in wheat species. *Plant. Cell Physiol.* **1994**, *35*, 785–791. [CrossRef]
64. Abbas, T.; Rizwan, M.; Ali, S.; Zia-ur-Rehman, M.; Qayyum, M.F.; Abbas, F.; Hannan, F.; Rinklebe, J.; Ok, Y.S. Effect of biochar on cadmium bioavailability and uptake in wheat (*Triticum aestivum* L.) grown in a soil with aged contamination. *Ecotoxicol. Environ. Saf.* **2017**, *140*, 37–47. [CrossRef]
65. Dionisio-Sese, M.L.; Tobita, S. Antioxidant responses of rice seedlings to salinity stress. *Plant. Sci.* **1998**, *135*, 1–9. [CrossRef]
66. Jana, S.; Choudhuri, M.A. Glycolate metabolism of three submersed aquatic angiosperms: Effect of heavy metals. *Aquatic Bot.* **1981**, *11*, 67–77. [CrossRef]
67. Zhang, X.Z. *The Measurement and Mechanism of Lipid Peroxidation and SOD, POD and CAT Activities in Biological System*; Research Methodology of Crop Physiology; Agriculture Press: Beijing, China, 1992; pp. 208–211.
68. Nakano, Y.; Asada, K. Hydrogen peroxide is scavenged by ascorbate-specific peroxidase in spinach chloroplasts. *Plant Cell Physiol.* **1981**, *22*, 867–880.
69. Aebi, H. Catalase in vitro. *Methods Enzymol.* **1984**, *105*, 121–126. [PubMed]
70. Rehman, M.Z.U.; Rizwan, M.; Ghafoor, A.; Naeem, A.; Ali, S.; Sabir, M.; Qayyum, M.F. Effect of inorganic amendments for in situ stabilization of cadmium in contaminated soils and its phyto-availability to wheat and rice under rotation. *Environ. Sci. Pollut. Res.* **2015**, *22*, 16897–16906. [CrossRef] [PubMed]

plants

MDPI

Article

Exogenously-Sourced Ethylene Modulates Defense Mechanisms and Promotes Tolerance to Zinc Stress in Mustard (*Brassica juncea* L.)

M. Iqbal R. Khan [1,*], Badar Jahan [2], Mohamed F Alajmi [3], Md Tabish Rehman [3] and Nafees A. Khan [2]

1 Plant Systems Biology Laboratory, Department of Botany, School of Chemical and Life Sciences, Jamia Hamdard, New Delhi 110065, India

2 Plant Physiology and Biochemistry Laboratory, Department of Botany, Aligarh Muslim University, Aligarh 202002, India; naziabadar.2014@gmail.com (B.J.); naf9.amu@gmail.com (N.A.K.)

3 Department of Pharmacognosy, King Saud University, Riyadh 11362, Kingdom of Saudi Arabia; malajmii@ksu.edu.sa (M.F.A.); m.tabish.rehman@gmail.com (M.T.R.)

* Correspondence: iqbal.khan@jamiahamdard.ac.in; Tel.: +91-11-26059688-5536

Received: 27 October 2019; Accepted: 20 November 2019; Published: 25 November 2019

Abstract: Heavy metal (HM) contamination of agricultural soil is primarily related to anthropogenic perturbations. Exposure to high concentration of HMs causes toxicity and undesirable effects in plants. In this study, the significance of ethylene was studied in response of mustard (*Brassica juncea*) to a high level (200 mg kg^{-1} soil) of zinc (Zn) exposure. Plants with high Zn showed inhibited photosynthesis and growth with the increase in oxidative stress. Application of ethylene (as ethephon) to Zn-grown plants restored photosynthesis and growth by inhibiting oxidative stress through increased antioxidant activity, the proline metabolism glyoxalase system, and nutrient homoeostasis. The results suggested that ethylene played a role in modulating defense mechanisms for tolerance of plants to Zn stress.

Keywords: antioxidant system; ethylene; glyoxalase system; photosynthesis; proline metabolism; zinc

---

1. Introduction

The addition of heavy metals (HMs) to agricultural soil has become a leading problem worldwide due to a wide range of unrestricted and continuous anthropogenic activities, especially in developing countries. In recent times, the increasing concentration of HMs has invited attention in global research due to their non-biodegradability and their high accumulation in living things through the food web [1]. Heavy metals that serve as micronutrients, such as copper (Cu), iron (Fe), manganese (Mn), cobalt (Co), nickel (Ni), and zinc (Zn) play an essential function in plant growth and development under optimal concentrations [2,3]. Zinc, included in this category, has essential roles in binding of protein, regulation of enzyme activity, transcriptional and translational regulation, and signal transduction in being a component and co-factor of several enzymes [4,5]. The deficiency of Zn results in necrotic spots, leaf chlorosis, nutrient imbalance, altered cell division, and reduced photosynthesis and growth [6,7]. The presence of 8.0–100 μg g^{-1} DW Zn has been suggested to assist normal growth and development of plants, but at elevated concentrations of 300 μg g^{-1} it becomes a toxic pollutant [8,9] and causes overproduction of reactive oxygen species (ROS) [10,11]. Under these conditions, it significantly induces cellular damages, redox imbalance, replacement of essential functional groups, and inhibition of photosynthetic and growth processes [12].

Plants have inherent capabilities to strive under such stressful environments by way of modifying diverse defense mechanisms to scavenge/regulate the excess ROS in plants. The antioxidant enzymes, such as superoxide dismutase (SOD), ascorbate peroxidase (APX), glutathione peroxidase (GPX), glutathione S-transferase (GST), glutathione reductase (GR), monodehydroascorbate reductase (MDHAR), and dehydroascorbate reductase (DHAR), are activated for survival [13,14]. The co-substrates of ascorbate-glutathione (AsA-GSH) cycle, GSH and AsA, serve as non-enzymatic antioxidants for direct scavenging of ROS [15]. Under stress condition, increased production of a highly reactive substance, α-ketoaldehydes called methylglyoxals (MG), can damage cellular ultrastructure, causing inactivation of proteins and even cell death [16]. Consequently, efficient MG detoxification has become a necessary biochemical indicator for stress tolerance, where reduced glutathione (GSH)-dependent glyoxalase pathway efficiently performs MG detoxification via glyoxalase I (Gly I) and glyoxalase II (Gly II) enzymes [16]. Another potential mechanism for tolerance of plants to HM stress is proline accumulation [17]. Proline potentiality detoxifies excess ROS, maintains cellular osmotic environment, protects biological membranes, and stabilizes enzymes/proteins under stress conditions [18]. Thus, proline metabolism may also be considered as a strategy to increase tolerance of plants to HMs and to protect photosynthesis.

Ethylene is a simple gaseous plant hormone that interacts with nutrient uptake and potentially influences many developmental processes of plants, including photosynthesis, under optimal and stressful conditions [19,20]. Iqbal et al. [21] showed that supplementation of ethephon resulted in increased activity of nitrate reductase and ATP-sulfurylase, which accounted for enhanced assimilation of nitrogen (N) and sulfur (S) resulting in increased antioxidant activity and photosynthetic responses in mustard plants. In wheat plants under heat stress [22] and cadmium (Cd) stress [23], ethylene regulates proline production. In spite of the fact that ethylene has now been recognized as an important modulator of photosynthetic process under optimal and stressful environments, its involvement in regulation of plant tolerance to elevated Zn levels through coordination of antioxidant and glyoxalase enzyme systems and proline metabolism has not been clearly demonstrated. It is postulated that the measures that could increase N-use efficiency in plants may result in greater investment of N in cellular metabolites and induce mechanisms that could protect photosynthesis and plant dry mass, and confer tolerance to Zn stress, through involvement of enhanced activity of antioxidant and glyoxalase systems and proline biosynthesis. The exogenously applied ethylene could help in achieving these processes and in alleviation of the adverse effects of Zn-induced stress.

2. Material and Methods

2.1. Plant Material, Growth Conditions, and Treatments

Healthy seeds of mustard (*Brassica juncea* L. Czern & Coss. cv. Varuna) were sterilized using 0.01 g L^{-1} mercuric chloride solution and repeated washings with double distilled water. The seeds were sown in earthen pots filled with soil having peat and compost, 4:1 (*v/v*), and mixed with sand, 3:1 (*v/v*). In each pot, five healthy plants were maintained after seedling establishment. The experiments were conducted at the Department of Botany, Aligarh Muslim University, Aligarh, India in the naturally illuminated green house. The plants grown with 200 mg Zn kg^{-1} soil (considered on the basis of our earlier research) [24] were treated with 200 μL L^{-1} ethephon at 20 days after sowing (DAS). The source for Zn was $ZnSO_4$. In addition, a control group of plants and plants treated with 200 μL L^{-1} ethephon alone were also maintained. The application of 200 μL L^{-1} ethephon was done with a hand sprayer along with 0.5% teepol as surfactant. To maintain the effects of ethephon releasing ethylene, a high soil phosphorus (P) status was maintained as described earlier as Khan and Khan [24]. The treatments were arranged in a completely randomized block design and five replicates for each treatment were maintained (n = 3). All measurements were done at 30 DAS to record different physiological, biochemical, and growth attributes.

2.2. Measurements of Photosynthetic Traits and Plant Dry Mass Accumulation

Gas exchange parameters (stomatal conductance (gs), intercellular CO_2 concentration (Ci), and net photosynthesis (Pn) were measured in the fully expanded uppermost leaves of plants in each treatment using infrared gas analyzer (CID-340, Photosynthesis System, Bio-Science, Washington, USA). The measurements were done between 11:00 and 12:00 at light saturating intensity and at 370 ± 5 μmol mol^{-1} atmospheric CO_2 concentration.

Chlorophyll content was measured with the help of a SPAD chlorophyll meter (SPAD 502 DL PLUS, Spectrum Technologies, Aurora, IL, USA).

Activity of ribulose-1,5-bisphosphate carboxylase (Rubisco) (EC 4.1.1.39) was determined spectrophotometrically by the adopting the method of Usuda [25] by monitoring nicotinamide adenine dinucleotide (NADH) oxidation at 30 °C at 340 nm during the conversion of 3-phosphoglycerate to glycerol 3-phosphate after the addition of enzyme extract to the assay medium. Further detail is given by Khan and Khan [24].

By using a Junior-PAM chlorophyll fluorometer (Heinz Walz, Effeltrich, Germany) all the chlorophyll fluorescence parameters were studied as described earlier by Khan and Khan [24]. Calculations were done according to Khan and Khan [24] and Krall and Edwards [26].

Dry mass of plants was recorded after drying the sample in a hot air oven at 80 °C till constant weight. Leaf area was measured using a leaf area meter (LA211, Systronics, New Delhi, India).

2.3. Determination of Oxidative Stress Markers

2.3.1. Lipid Peroxidation

Lipid peroxidation in leaves was determined by estimating the content of thiobarbituric acid reactive substances (TBARS) as described by Dhindsa et al. [27]. The content of TBARS was calculated using the extinction coefficient (155 mM^{-1} cm^{-1}). The details have been given in our earlier report as Khan et al. [22].

2.3.2. Electrolyte leakage

For measuring electrolyte leakage, samples were thoroughly washed with sterile water, weighed, and then kept in closed vials with 10 mL of deionized water where they were incubated at 25 °C for 6 h using a shaker. Then, electrical conductivity (EC) was determined (C1). Samples were then again kept at 90 °C for 2 h and EC was recorded after attaining equilibrium at 25 °C (C2).

2.3.3. Methylglyoxal Content

Adopting the method of Wild et al. [28], methylglyoxal content was measured. Leaf samples were homogenized by using 5% perchloric acid followed by centrifugation at 11,000× *g*. A saturated solution of Na_2CO_3 was used to neutralize the supernatant and further mixed with Na-P and N-acetyl-L-cysteine. The product N-α-acetyl-S-(1-hydroxy-2-oxoprop-1-yl) cysteine formation was measured by spectrophotometer at 288 nm. A standard curve of MG was prepared and expressed in μmol g^{-1} FW.

2.3.4. Lipoxygenase Activity

Lipoxygenase (LOX) (EC 1.13.11.12) activity was estimated following the method of Doderer et al. [29] by monitoring the increase in absorbance at 234 nm using linoleic acid as a substrate. The LOX activity was calculated using 25 mM^{-1} cm^{-1} as an extinction coefficient.

2.4. Measurement of Ascorbate and Glutathione Content

2.4.1. Ascorbate Content

Ascorbate (AsA) content was determined following the method of Law et al. [30] with some modifications. The details have been given in our earlier report as Anjum et al. [31]. A standard curve in the range of 10–100 nmol of ascorbic acid was used for calibration. Values in both cases were corrected for the absorbance eliminating the supernatant in the blank prepared separately for AsA.

2.4.2. Reduced Glutathione Content

GSH was assayed by an enzymic recycling procedure, as detailed by Griffith [32], in which it was sequentially oxidized by 5,5-dithiobis-2-nitrobenzoic acid (DTNB) and reduced by NADPH in the presence of GR, as described earlier [24,33].

2.5. Extraction and Determination of Antioxidant Enzymes

Fresh leaf tissue (200 mg) was homogenized with an extraction buffer containing 0.05% (*v/v*) Triton X-100 and 1% (*w/v*) PVP in potassium-phosphate buffer (100 mM, pH 7.0) using chilled mortar and pestle. At 4 °C, the homogenate was centrifuged at 15,000× *g* for 20 min. The supernatant obtained after centrifugation was used to assay superoxide dismutase (SOD) (EC 1.15.1.1), GSH reductase (GR) (EC 1.6.4.2), and GSH peroxidase (GPX) (EC 1.11.1.9) enzymes, and for the assay of ascorbate peroxidase (APX) (EC 1.11.1.11). Here, 2.0 mM ascorbate was supplemented with extraction buffer. Protein was estimated according to Bradford [34] using bovine serum albumin as standard.

2.5.1. SOD

Activity of SOD was determined by monitoring the inhibition of photochemical reduction of nitro blue tetrazolium (NBT), according to the methods of Beyer and Fridovich [35] and Giannopolitis and Ries [36]. The details have been given in our earlier report as Khan and Khan [24].

2.5.2. APX

Activity of APX was determined by the method of Nakano and Asada [37] by recording the decrease in absorbance of ascorbate at 290 nm. The assay mixture contained phosphate buffer (50 mM, pH 7.0), 0.1 mM EDTA, 0.5 mM ascorbate, 0.1 mM H_2O_2, and the enzyme extract. APX activity was calculated by using the extinction coefficient 2.8 mM^{-1} cm^{-1}. One unit of the enzyme is the amount necessary to decompose 1 μmol of substrate per min at 25 °C.

2.5.3. GR

Activity of GR was determined adopting the method of Foyer and Halliwell [38] by monitoring the GSH-dependent oxidation of NADPH. The details have been given in our earlier report as Khan and Khan [24].

2.5.4. GPX

Activity of GPX was determined by adopting the method of Hasanuzzaman et al. [39]. The details have been given in our earlier report as Khan et al. [24].

2.5.5. GST

Glutathione S-transferase (GST) (EC 2.5.1.18) was determined spectrophotometrically by the method Booth et al. [40] with some modifications [41]. The reaction mixture contained 100 mM Tris–HCl buffer (pH 6.5), 1.5 mM GSH, 1 mM 1-chloro-2,4-dinitrobenzene (CDNB), and enzyme solution in a final volume of 0.7 mL. The enzyme reaction was initiated by the addition of CDNB and the increase in absorbance was measured at 340 nm for 1 min. The activity was calculated using the extinction coefficient of 9.6 mM^{-1} cm^{-1}.

2.5.6. MDHAR

Monodehydroascorbate reductase (MDHAR) (EC 1.6.5.4) activity was assayed by the method of Hossain et al. [42]. The reaction mixture contained 50 mM Tris–HCl buffer (pH 7.5), 0.2 mM NADPH,

2.5 mMAsA, and 0.5 units of AO and enzyme solution in a final volume of 0.7 mL. The reaction was started by the addition of AO. The activity was calculated from the change in ascorbate at 340 nm for 1 min using an extinction coefficient of 6.2 mM^{-1} cm^{-1}.

2.5.7. DHAR

Dehydroascorbate reductase (DHAR) (EC 1.8.5.1) activity was assayed by the method of Nakano and Asada [37]. The reaction buffer contained 50 mM K-phosphate buffer (pH 7.0), 2.5 mM GSH, and 0.1 mM DHA. The reaction was started by adding the sample solution to the reaction buffer solution. The activity was calculated from the change in absorbance at 265 nm for 1 min using an extinction coefficient of 14 mM^{-1} cm^{-1}.

2.6. Extraction and Assay of Glyoxalase Systems' Enzymes

2.6.1. Gly I Activity

Glyoxalase I (Gly I) (EC 4.4.1.5) assay was determined by the method of Hasanuzzaman et al. [43]. The assay mixture contained 100 mM K-phosphate buffer (pH 7.0), 15 mM $MgSO_4$, 1.7 mM GSH, and 3.5 mM MG in a final volume of 700 µL. The reaction was started by the addition of MG and the increase in absorbance was recorded at 240 nm for 1 min. The activity was calculated using the extinction coefficient of 3.37 mM^{-1} cm^{-1}.

2.6.2. Gly II Activity

Glyoxalase II (Gly II) (EC 3.1.2.6) activity was measured using the method of Principato et al. [44] by monitoring the formation of GSH at 412 nm for 1 min. The reaction mixture contained 100 Mm Tris–HCl buffers (pH 7.2), 0.2 mM DTNB and 1 mM S-D-lactoylglutathione (SLG) in a final volume of 1 mL. The reaction was started by the addition of SLG and the activity was calculated using the extinction coefficient of 13.6 mM^{-1} cm^{-1}.

2.7. Determination of Nutrient Content

The determination of mineral nutrients (nitrogen, N; phosphorous, P; potassium, K; and calcium, Ca) was done in acid-peroxide digested oven-dried leaf sample. K, and Ca were measured using flame photometer (Khera-391: Khera Instruments, New Delhi, India), whereas N and P were determined by using the methods of Lindner [45] and Fiske and Subba Row [46], respectively.

2.8. Determination of Proline Content and Activity of Proline Metabolizing Enzymes

2.8.1. Proline Content

Proline content was determined by adopting the ninhydrin method of Bates et al. [47]. Here, 300 mg fresh leaf samples were homogenized in 3% sulphosalicylic acid (3 mL). After this, samples were homogenate filtrated and reacted with acid ninhydrin and glacial acetic acid (1 mL each) for 1 h followed by water bath at 100 °C. The reaction mixture was extracted with toluene and the absorbance was measured at 520 nm. A standard was also prepared using L-proline.

To determine the activity of γ-glutamyl kinase (GK) (EC 2.7.2.11) and proline oxidase (PROX) (EC 1.5.99.8), enzyme extract was prepared by homogenizing 500 mg leaf sample in 0.1 M Tris–HCl buffer (pH 7.5) at 4 °C. The homogenate was centrifuged at 30,000× g for 30 min and the supernatant was used as the crude extract enzyme preparation for P5CS activity while the pellet was collected and used as extract for the assay of GK and proline oxidase.

2.8.2. GK Activity

Activity of GK was assayed by the method of Hayzer and Leisinger [48] with slight modification. The frozen sample was suspended in 10 mL of 0.1 M Tris–HCl buffer containing 1 mM 1,4-dithiothreitol

(DTT) to rupture the cell and centrifuged at 30,000× g for 30 min. The other detail has been given in our earlier report as Khan et al. [22]. Activity of GK was expressed in U mg^{-1} protein. One unit of the enzyme activity is defined as µg of glutamyl hydroxamate min^{-1} mg^{-1} protein. Glutamyl hydroxamate was used as standard.

2.8.3. POX Activity

Activity of POX was determined adopting the method of Huang and Cavalieri [49] with slight modification. The pellet was mixed with 1 mL Tricine and KOH buffer (pH 7.5) containing 6 M sucrose. This extract was used for the enzyme assay. The other detail has been given in our earlier report as Khan et al. [22]. Proline oxidase activity was expressed in U mg^{-1} protein. One unit of the enzyme activity is defined as mM DCPIP reduced min^{-1} mg^{-1} protein.

2.9. *Water Potential and Osmotic Potential*

Leaf water potential was measured on the second leaf from the top (fully expanded young leaf) of the plant by using the water potential system (Psypro, WESCOR, UT, USA). The leaf used for water potential measurement was frozen in liquid N_2 in sealed polythene bags which were thawed, and cell sap was extracted with the help of a disposable syringe. The extracted sap was used for the determination of osmotic potential using a vapor pressure osmometer (5520, WESCOR, UT, USA).

2.10. *Statistical Analysis*

Data were analyzed statistically, and standard error was calculated. Analysis of variance was performed on the data using SPSS (ver. 17.0 Inc.) to determine the significance at $p < 0.05$. Least significant difference (LSD) was calculated for the significant data to identify difference in the mean of the treatment; data are presented as mean ± SE ($n = 3$).

3. Results

3.1. *Ethephon Reverses Effects of Zn Stress on Photosynthetic and Growth Attributes*

Photosynthetic attributes were reduced in plants treated with Zn compared to the control plants. The adverse effect of Zn stress on photosynthetic parameters was reversed with ethephon application. Ethephon application to plants without Zn treatment enhanced net photosynthesis by 27.2%, stomatal conductance by 20.5%, intracellular CO_2 concentration by 17.2%, Rubisco content by 28.8%, and chlorophyll content by 25.7%, as compared to the control plants. In plants treated with Zn, ethephon supplementation restricted the adverse effects of Zn and the decrease in net photosynthesis was reduced to 18.1%, stomatal conductance to 13.1%, intracellular CO_2 concentration to 15.4%, Rubisco content to 16.4%, and chlorophyll content to 16%, in comparison to control (Table 1).

Plants grown with Zn stress exhibited a decrease in maximum PSII efficiency, intrinsic PSII efficiency, actual PSII efficiency, photochemical quenching, and electron transport rate by 26%, 23%, 29%, 24%, and 25% in comparison with the control plants. However, non-photochemical quenching increased with Zn by 59.2% as compared to the control. Ethephon application to plants without Zn stress improved all the above traits compared with control (Table 1).

Growth of plants was reduced with Zn when compared to the control plants. Ethephon application to plants without Zn treatment enhanced plant dry mass and leaf area compared to the control plants. In plants treated with Zn, ethephon supplementation restricted the adverse effects of Zn and the decrease in plant dry mass was limited to 15% as compared to control plants. Zn reduced leaf area by 30.3% compared to control (Table 1). Ethephon application to plants without Zn treatment enhanced plant dry mass and leaf area compared to the control plants. The increases in plant dry mass and leaf area were 23% and 27%, respectively, compared to the control (Table 1).

Table 1. Chlorophyll content, net photosynthesis rate, intracellular CO_2 concentration, stomatal conductance, Rubisco activity, Φ PS II, Fv/Fm, Fv′/Fm′, qP, NPQ, ETR, plant dry mass, and leaf area in the leaf of mustard (*Brassica juncea* L.) cv. Varuna at 30 days after sowing (DAS). Plants were grown with/without zinc stress and treated with 200 μL L^{-1} ethephon at 20 DAS. Data are presented as treatment mean ± SE (n = 3). Data followed by same letter are not significantly different by least significant difference (LSD) test at $p < 0.05$.

	Control	Ethephon	Zinc	Ethephon + Zinc
Chlorophyll content (SPAD value)	31.4 ± 1.3 ^{b}	39.5 ± 1.6 ^{a}	21.3 ± 0.8 ^{d}	26.4 ± 0.9 ^{c}
Net photosynthesis (μmol CO_2 $m^{-2}s^{-1}$)	19.8 ± 0.8 ^{b}	25.2 ± 0.9 ^{a}	11.2 ± 0.7 ^{d}	16.2 ± 0.6 ^{c}
Intracellular CO_2 concentration (μmol CO_2 mol^{-1})	272 ± 10.9 ^{b}	319 ± 12.7 ^{a}	187 ± 6.7 ^{d}	230 ± 8.2 ^{c}
Stomatal conductance (mmol CO_2 m^{-2} s^{-1})	380 ± 10.7 ^{b}	458 ± 12.3 ^{a}	280 ± 10.1 ^{d}	330 ± 11.8 ^{c}
Rubisco activity (μmol CO_2 m^{-2} s^{-1})	48.6 ± 1.9 ^{b}	62.6 ± 2.2 ^{a}	32.5 ± 1.17 ^{d}	40.6 ± 1.5 ^{c}
Actual PSII efficiency (Φ PS II)	0.62 ± 0.03 ^{ab}	0.68 ± 0.04 ^{a}	0.44 ± 0.02 ^{c}	0.57 ± 0.02 ^{b}
Maximum PSII efficiency (Fv/Fm)	0.77 ± 0.03 ^{a}	0.84 ± 0.03 ^{a}	0.57 ± 0.02 ^{c}	0.67 ± 0.03 ^{b}
Intrinsic PSII efficiency (Fv′/Fm′)	0.71 ± 0.03 ^{b}	0.75 ± 0.03 ^{a}	0.55 ± 0.02 ^{c}	0.65 ± 0.02 ^{b}
Photochemical quenching (qP)	0.83 ± 0.04 ^{a}	0.87 ± 0.03 ^{a}	0.63 ± 0.02 ^{c}	0.73 ± 0.03 ^{b}
Non-photochemical quenching (NPQ)	0.54 ± 0.02 ^{bc}	0.49 ± 0.03 ^{c}	0.86 ± 0.03 ^{a}	0.62 ± 0.03 ^{b}
Electron transport rate (ETR)	154 ± 6.2 ^{a}	167 ± 5.1 ^{a}	116 ± 4.2 ^{c}	135 ± 5.3 ^{b}
Plant dry mass (g $plant^{-1}$)	6.11 ± 0.6 ^{ab}	7.54 ± 0.4 ^{a}	3.58 ± 0.4 ^{c}	5.2 ± 0.5 ^{b}
Leaf area (cm^2 $plant^{-1}$)	140 ± 5.7 ^{b}	178 ± 6.1 ^{a}	97.5 ± 3.8 ^{d}	121 ± 4.6 ^{c}

3.2. Influence of Ethephon on Water Relations under Zn Stress

Application of ethephon to Zn-stressed plants increased water potential by 35% and osmotic potential by 47% as compared to the control plants. On the other hand, application of Zn alone resulted in the increase of water potential by 141% and osmotic potential by 71.4% as compared to the control plants (Figure 1).

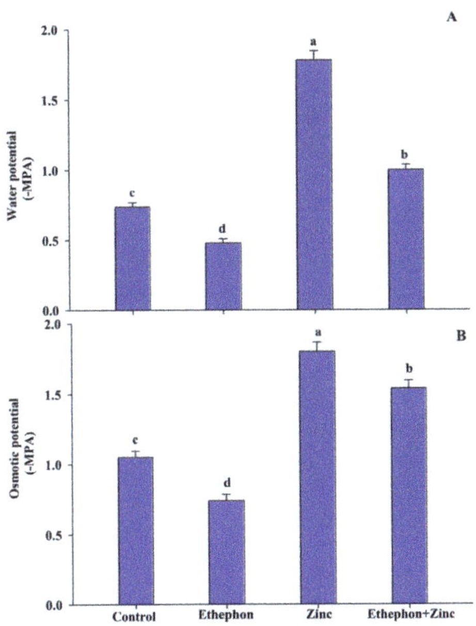

Figure 1. (**A**) Leaf water potential and (**B**) osmotic potential in the leaf of mustard (*Brassica juncea* L.) of Varuna cultivar at 30 DAS. Plants were grown with/without Zn stress and treated with 200 μL L^{-1} ethephon at 20 DAS. Data are presented as treatment mean ± SE (n = 3). Data followed by same letter are not significantly different by LSD test at $p < 0.05$.

3.3. Ethephon Reduces Zn-Induced Oxidative Stress

Plants grown with Zn showed higher oxidative stress and exhibited increased oxidative stress markers, i.e., content of TBARS and electrolyte leakage, MG content, and LOX activity (Table 2). Treatment of plants with Zn enhanced TBARS content, MG content, and LOX activity by about 95%, 61% and 51%, respectively, whereas electrolyte leakage increased by about 2.2 times in plants compared to the control. For the appraisal of the influence of ethylene in reducing Zn-induced oxidative stress, we analyzed TBARS content and electrolyte leakage after application of ethylene to these plants. Application of ethylene proved effective in lowering oxidative stress under the metal stress. Application of ethylene reduced TBARS content, MG content, and LOX activity by 28%, 13% and 12%, respectively whereas electrolyte leakage was reduced by 47% in Zn-treated plants compared to Zn-treated plants (Table 2).

Table 2. Electrolyte leakage, thiobarbituric acid reactive substances (TBARS) content, methylglyoxal (MG) content, lipoxygenase (LOX) activity in the leaf of mustard (*Brassica juncea* L.) cv. Varuna at 30 DAS. Plants were grown with/without zinc stress and treated with 200 μL L^{-1} ethephon at 20 DAS. Data are presented as treatment mean ± SE (n = 3). Data followed by same letter are not significantly different by LSD test at $p < 0.05$.

	Control	Ethephon	Zinc	Ethephon + Zinc
Electrolyte leakage (%)	1.24 ± 0.07 ^{c}	0.81 ± 0.09 ^{d}	3.93 ± 0.34 ^{a}	2.08 ± 0.15 ^{b}
TBARS content (nmol g^{-1} FW)	13.24 ± 0.6 ^{c}	8.23 ± 0.4 ^{d}	25.84 ± 0.8 ^{a}	18.56 ± 0.7 ^{b}
Methylglyoxal content (mol g^{-1} FW)	45.7 ± 1.7 ^{d}	55.9 ± 2.1 ^{c}	73.7 ± 3.0 ^{a}	64.8 ± 2.1 ^{b}
Lipoxygenase activity (μmol min^{-1} mg^{-1} protein)	10.2 ± 0.6 ^{c}	7.9 ± 0.6 ^{d}	15.4 ± 0.6 ^{a}	13.5 ± 0.6 ^{b}

3.4. Application of Ethephon Enhanced Antioxidant Systems under Zn Stress

The activities of enzymatic and non-enzymatic antioxidant systems showed modulation with ethylene treatment in both stress and non-stress conditions. Zn stress decreased AsA content as compared to the control plants. On the other hand, ethephon supplementation of Zn-exposed plants had significantly higher AsA content as compared to Zn-treated plants. Under Zn stress, GSH content was increased by 22% as compared to control plants. Application of ethephon to Zn-exposed plants had significantly higher GSH content by 65% as compared to the control plants (Figure 2).

Zinc treatment resulted in increased activity of SOD, APX, GR, and GPX by 25%, 38%, 45%, and 35%, respectively compared to the control plants. In non-stressed control plants, application of ethephon increased activity of SOD APX, GR, and GPX by 50%, 128%, 105%, and 124%, respectively, over the control. However, application of ethephon to Zn-treated plants resulted in increased activity of SOD, APX, GR, and GPX by 71%, 176%, 145%, and 165%, respectively, compared to the control (Figure 3).

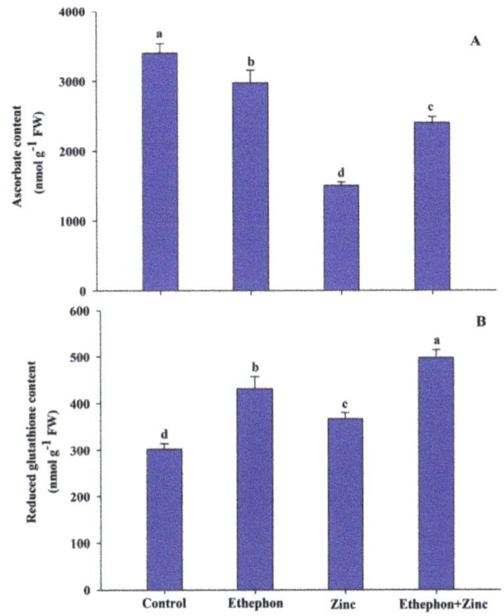

Figure 2. (**A**) Ascorbate content and (**B**) reduced glutathione content in the leaf of mustard (*Brassica juncea* L.) of Varuna cultivar at 30 DAS. Plants were grown with/without Zn stress and treated with 200 μL L^{-1} ethephon at 20 DAS. Data are presented as treatment mean ± SE (n = 3). Data followed by same letter are not significantly different by LSD test at $p < 0.05$.

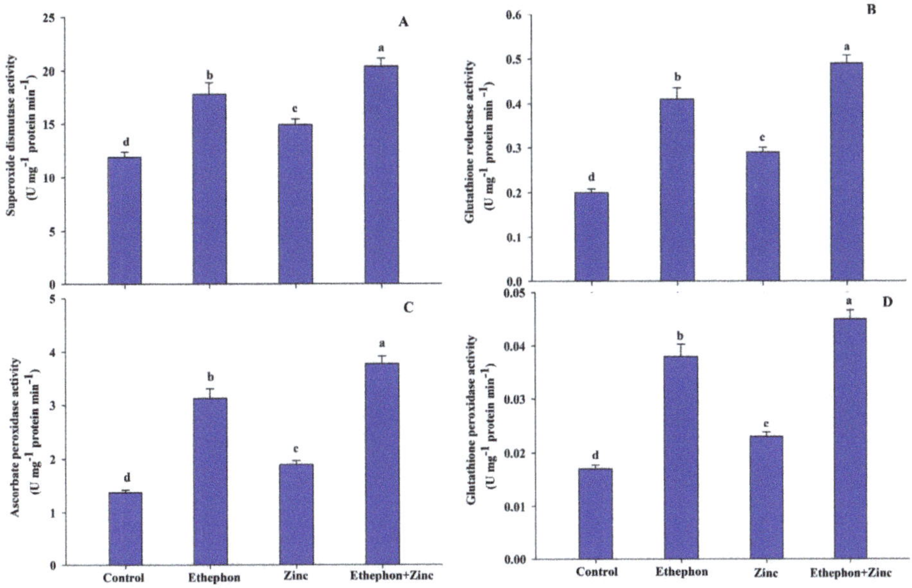

Figure 3. (**A**) Superoxide dismutase activity; (**B**) glutathione reductase activity; (**C**) ascorbate peroxidase activity; (**D**) and glutathione peroxidase activity in the leaf of mustard (*Brassica juncea* L.) of Varuna cultivar at 30 DAS. Plants were grown with/without Zn stress and treated with 200 μL L^{-1} ethephon at 20 DAS. Data are presented as treatment mean ± SE (n = 3). Data followed by same letter are not significantly different by LSD test at $p < 0.05$.

Zinc treatments also increased the activity of MDHAR, DHAR, and GST by 45%, 26%, and 67%, respectively, as compared to the control plants. Ethephon application to plants grown without Zn resulted in increased activity of MDHAR by 97%, DHAR by 53%, and GST by 219%, in comparison to control plants; in presence of Zn, ethephon application resulted in 170%, 76%, and 261% higher activity of MDHAR, DHAR, and GST, respectively, in comparison to the control plants (Figure 4).

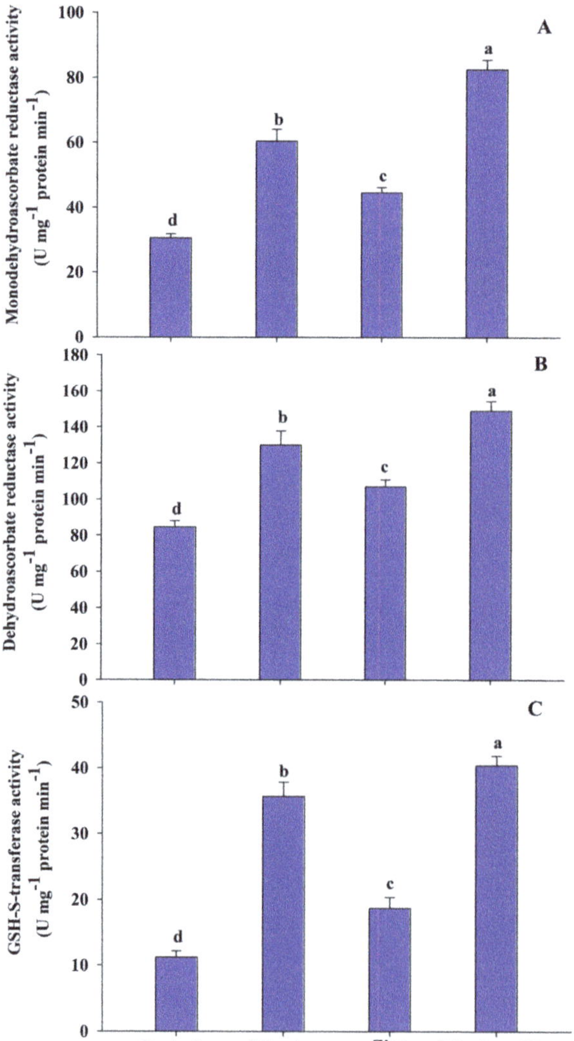

Figure 4. (**A**) Monodehydroascorbate activity; (**B**) dehydroascorbate activity; and (**C**) GSH-S-transferase activity in the leaf of mustard (*Brassica juncea* L.) of Varuna cultivar at 30 DAS. Plants were grown with/without Zn stress and treated with 200 μL L^{-1} ethephon at 20 DAS. Data are presented as treatment mean ± SE ($n = 3$). Data followed by same letter are not significantly different by LSD test at $p < 0.05$.

3.5. Ethephon Application Enhances Glyoxalase System

Gly I was increased with Zn stress by 73% over the control plants. Application of ethephon to Zn-treated plant significantly enhanced the activity of Gly I by 94% compared to the control plants

(Figure 5). The increase in Gly II with Zn stress was 110% in comparison to the control plants. However, application of ethephon to Zn-treated plants enhanced the activity of Gly II by 160% as compared to the control plants (Figure 5).

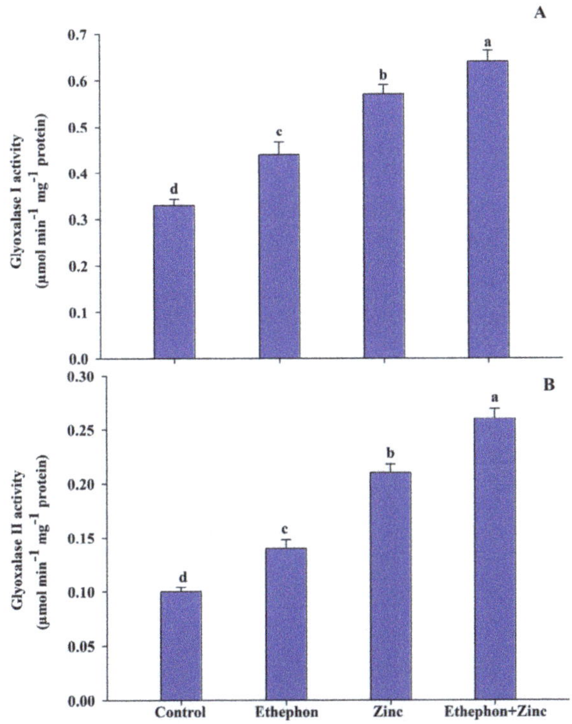

Figure 5. (**A**) Glyoxalase I activity and (**B**) glyoxalase II activity in the leaf of mustard (*Brassica juncea* L.) of Varuna cultivar at 30 DAS. Plants were grown with/without Zn stress and treated with 200 µL L^{-1} ethephon at 20 DAS. Data are presented as treatment mean ± SE (n = 3). Data followed by same letter are not significantly different by LSD test at $p < 0.05$.

3.6. Ethephon Increases Proline Metabolism under Zn Stress

In order to assess the role of proline metabolism in Zn stress tolerance, assessment of proline accumulation and activity of proline metabolizing enzymes (GK and PROX) was done. Activity of GK increased in Zn-stressed plants and also with ethephon plus Zn stress treatments. Application of ethephon increased GK activity by 170% in Zn-stressed plants compared to the control. On the other hand, activity of PROX was reduced under no-stress and Zn-stressed plants with ethephon treatment (Figure 6).

Proline accumulation increased upon ethephon application, as well as with the Zn treatments. Zinc stress induced proline biosynthesis and increased proline content by 44% in comparison to the control plants. Proline accumulation under Zn stress was further increased with ethephon application. Maximum proline accumulation resulted from ethephon application under Zn stress compared with the control (Figure 6).

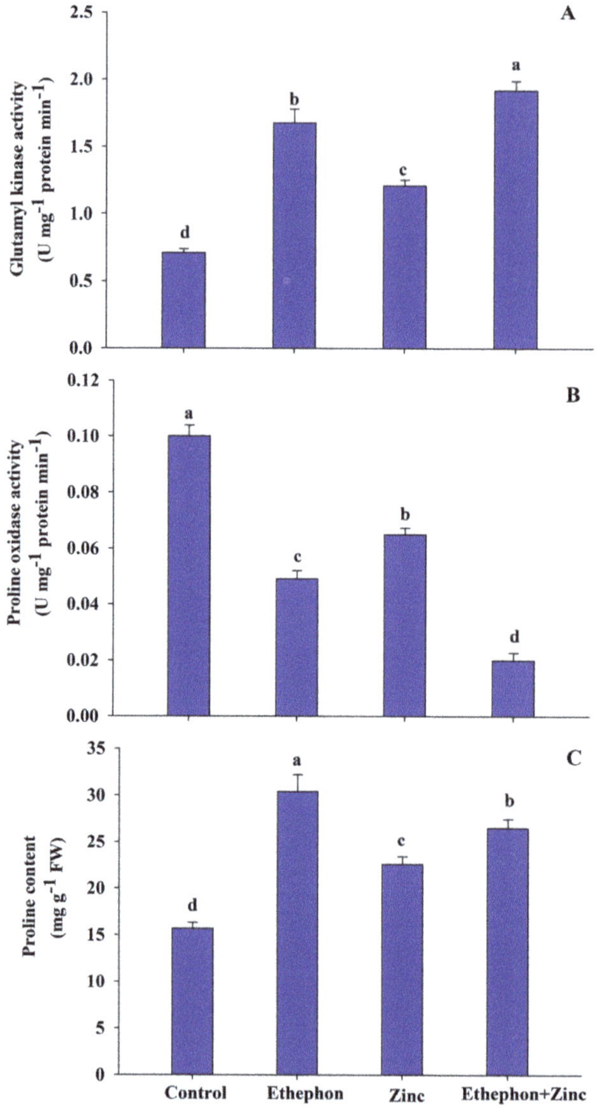

Figure 6. (**A**) Activity of glutamyl kinase; (**B**) proline oxidase; and (**C**) proline content in the leaf of mustard (*Brassica juncea* L.) of Varuna cultivar at 30 DAS. Plants were grown with/without Zn stress and treated with 200 μL L^{-1} ethephon at 20 DAS. Data are presented as treatment mean ± SE (n = 3). Data followed by same letter are not significantly different by LSD test at $p < 0.05$.

3.7. Ethephon Supplementation Maintained Nutrient Contents under Zn Stress

Treatment of ethephon alone significantly increased the content of nutrients but Zn stress decreased N, P, K, and S content in comparison to the control plants. Zn treatment decreased N by 29%, P by 23%, K by 31%, and S by 34%, as compared to control. However, application of ethephon on Zn-grown plants completely alleviated the Zn effects and increased the nutrient contents significantly in comparison to stressed plants (Figure 7).

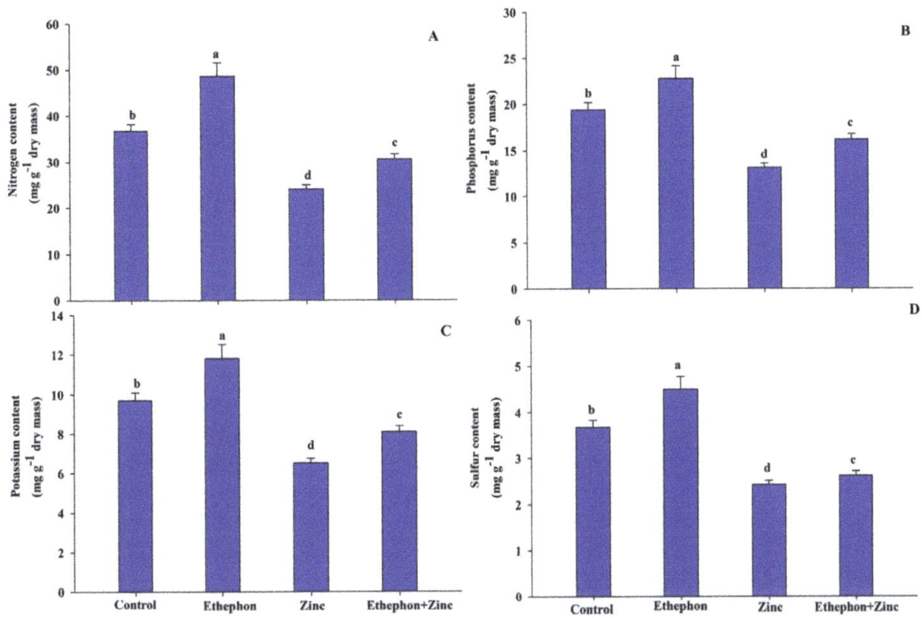

Figure 7. (**A**) Nitrogen content; (**B**) phosphorous content; (**C**) potassium content; and (**D**) sulfur content in the leaf of mustard (*Brassica juncea* L.) of Varuna cultivar at 30 DAS. Plants were grown with/without Zn stress and treated with 200 μL L^{-1} ethephon at 20 DAS. Data are presented as treatment mean ± SE (n = 3). Data followed by same letter are not significantly different by LSD test at $p < 0.05$.

4. Discussion

Zinc stress inhibited photosynthesis and plant growth in mustard plants. However, ethephon supplementation not only stimulated photosynthesis and growth under non-stress conditions but also under Zn stress. Plants show concentration-dependent requirement of Zn for optimal plant metabolism; however, excess Zn availability inhibits plant growth and development [24]. It has been reported that HMs induce lipid peroxidation in photosynthetic membranes, distort chloroplast ultrastructure, degrade photosynthetic pigments, inhibit PSII activity and the electron transport chain, and decrease both carboxylation efficiency of Rubisco and net photosynthesis [50,51]. Additionally, plants' fitness to the environment is estimated through chlorophyll fluorescence observations [52]. In the present study, the reduction in photochemical efficiency under Zn stress (Table 1) indicated the interruption in photochemical reactions that blocked electron transport system and the performance of PSII resultantly increased NPQ under Zn stress (Table 1). The results of previous study of Sirhindi et al. [53] have also reported that NPQ increases under Ni stress that was due to repressed photochemistry; this was considered as the mechanism to balance the excess absorbed light energy preventing photo inhibition.

It has been reported that ethylene potentially plays a crucial role in the adaptation of plants to HM stress [54]. Ethephon is used commercially to produce ethylene in crop plants and the ethylene released from ethephon affects several cellular, developmental, photosynthetic, and stress response processes [55]. The present study showed that exogenous ethephon supply decreased the toxic effects of Zn on photosynthetic machinery (Table 1) and enhanced the activity of PSII and gas exchange parameters. Our results are consistent with the finding of Asgher et al. [56], who reported that ethephon supplementation induced photosynthesis under Cr stress.

In the present study, Zn-stressed plants were at substantially increased oxidative stress and MG levels exhibiting increased LOX activity. These observations in mustard plants are in agreement with the finding of Molassiotis et al. [57], who found increased LOX activity which was the result of increased production of ROS due to higher oxidative stress in *Malus domestica* shoot tips under boron stress. Our results showed a possible regulatory role of exogenously applied ethylene as ethephon on ROS and MG metabolism in Zn stress tolerance in mustard plants. Indeed, ethylene acts as a vital signaling molecule for reduction of ROS [19,58]. Earlier, it has been shown that coordination and regulation of the antioxidant systems and glyoxalase systems are indispensable to attain significant tolerance against oxidative stress [59,60]. We have found that the activity of antioxidant enzymes and glyoxalase system simultaneously work to reduce excess ROS load and MG detoxification.

Antioxidant enzyme systems are part of an imperative defense system of plants against ROS caused by HMs [61]. Application of ethephon substantially induced the activity of SOD, APX, GR, GST, GPX, MDHAR, and DHAR in plants under Zn stress. Further, ethylene application enhanced the activities of Gly I and Gly II, particularly in response to Zn stress. Recently, Hassanuzzaman et al. [62] showed that the activities of Gly I and Gly II in *Oryza sativa* increased under Ni stress and Si application. Studies have shown that ethephon treatment influenced the activity of antioxidant enzymes and photosynthetic attributes, but the information on the response of plants to ethephon under Zn stress is scanty. In the present study, the effort was made to understand the role of ethylene in regulation of three major defense mechanisms (antioxidant system, glyoxalase system, and proline metabolism) along with nutrient homeostasis under high level of Zn in mustard plants. Ethylene significantly induced the defense mechanisms by up-regulating the key enzymes in the antioxidant system (SOD, APX, GR, GPX, GST), glyoxalase system (Gly I and Gly II) and proline metabolism (GK and PROX), which resulted in protection of growth and development of plants through maintaining water relations and increasing PS II activity and dry mass production.

In addition to the antioxidant and glyoxalase defense system, we also examined proline as a defense mechanism under Zn stress and to know how it was modulated with ethylene. Plants under Zn stress accumulated proline. These results are supported with the previous findings of Sirhindi et al. [53], who have shown that proline content increases in *Glycine max* under Ni toxicity. Alia et al. [63] showed that proline was involved in reducing the photodamage in the thylakoid membranes by scavenging ROS (singlet oxygen and superoxide radical anion). In the present study, ethylene-induced accumulation of proline helped in increasing osmotic potential, and thereby water potential, and protected PSII activity and photosynthesis by reducing ROS effects under Zn stress. The activity of chlorophyll fluorescence was reduced in Zn-stressed plants receiving ethylene that showed improved photosynthesis. Earlier, it has been shown that ethephon application increased the gas exchange parameters and PSII activity in mustard under HM stress [56,58]. Proline biosynthesis has been attributed in the alleviation of cytoplasmic acidosis and may maintain $NADP^+$/NADPH ratio at values compatible with plant metabolism. Rapid production of proline under stress also provides recovery from stress-induced damages to mitochondrial oxidative phosphorylation and the generation of ATP [64]. However, there has been much disagreement regarding the mechanisms by which proline reduces HM-induced oxidative stress, regulates cellular functions such as osmotic adjustment, and enhances the plant's water status [14]. In the present research, efforts were made to study the role of proline and its metabolizing enzymes in facilitating Zn detoxification and ameliorating stress in mustard plants. The increased proline accumulation after ethephon application resulted from induced GK activity and inhibited PROX activity. It has been reported that activities of GK and PROX play important roles in controlling the level of proline and environmental stress in plants [65,66]. The regulatory role of ethylene in proline metabolism and its enzymes to improve photosynthetic traits under Cd and heat stress has been shown [21,22].

HMs impact negatively on the environment and crop nutrition [67]. They disturb the uptake of essential nutrients such as N, P, K, and S which are required for normal growth of plants [68,69]. In the present study, we also focused on modulation of essential nutrients (N, P, K, and S) in mustard plants through application of ethephon under Zn stress. Ethylene has been associated with the regulation of physiological responses to nutrient homeostasis and stress tolerance responses. Efficient working of nutrient transporters and enzymes involved in nutrient assimilation enhanced nutrient uptake and this has a direct influence on the tolerance mechanisms including antioxidant system, glyoxalase system, and proline metabolism of plants under stress condition [70].

5. Conclusions and Future Prospects

Conclusively, it may be said that Zn stress adversely impacted photosynthesis and growth of plants by inhibiting metabolic pathways. Ethylene supplementation inhibited ROS production through induced defense mechanisms of antioxidant activity, proline metabolism, glyoxalase system and nutrient homoeostasis. It protected photosynthetic machinery and promoted photosynthesis and growth under Zn stress. Thus, the use of ethylene (as ethephon) may bear a prominent role in alleviation of Zn stress in mustard plants by modulating the defense mechanisms.

Author Contributions: N.A.K. and M.I.R.K. conceptualized the study and wrote the manuscript; M.I.R.K. and B.J. conducted the experiments; M.I.R.K., B.J., M.F.A. and M.T.R. analyzed the data. All authors have read the manuscript and approved without any conflicts.

Funding: This research received no external funding.

Conflicts of Interest: The authors declare that they have no conflict of interest.

References

1. Rizwan, M.; Ali, S.; Rehman, M.Z.; Maqbool, A. A critical review on the effects of zinc at toxic levels of cadmium in plants. Environ. *Sci. Pollut. Res.* **2019**, *26*, 6279–6289. [CrossRef]
2. Salla, V.; Hardaway, C.J.; Sneddon, J. Preliminary investigation of Spartina alterniflora for phytoextraction of selected heavy metals in soils from Southwest Louisiana. *Microchem. J.* **2011**, *97*, 207–212. [CrossRef]
3. Shahid, M.; Khalid, S.; Abbas, G.; Shahid, N.; Nadeem, M.; Sabir, M. Heavy Metal Stress and Crop Productivity. In *Crop Production and Global Environmental Issues*; Hakeem, K.R., Ed.; Springer: Cham, Switzerland, 2015; pp. 1–25.
4. Marschner, H. *Mineral Nutrition of Higher Plants*, 2nd ed.; Academic Press: New York, NY, USA, 1995.
5. Broadley, M.R.; White, P.J.; Hammond, J.P.; Zelko, I.; Lux, A. Zinc in plants. *New Phytol.* **2007**, *173*, 677–702. [CrossRef] [PubMed]
6. Feigl, G.; Lehotai, N.; Molnár, Á.; Ördög, A.; Rodríguez-Ruiz, M.; Palma, J.M.; Corpas, F.J.; Erdei, L.; Kolbert, Z. Zinc induces distinct changes in the metabolism of reactive oxygen and nitrogen species (ROS and RNS) in the roots of two *Brassica* species with different sensitivity to zinc stress. *Ann. Bot.* **2015**, *116*, 613–625. [CrossRef] [PubMed]
7. He, J.; Wang, Y.; Ding, H.; Ge, C. Epibrassinolide confers zinc stress tolerance by regulating antioxidant enzyme responses, osmolytes, and hormonal balance in *Solanum melongena* seedlings. *Braz. J. Bot.* **2016**, *39*, 295–303. [CrossRef]
8. Cakmak, I. Possible Roles of Zinc in Protecting Plant Cells from Damage by Reactive Oxygen Species. *New Phytol.* **2000**, *146*, 185–205. [CrossRef]
9. Zhao, H.; Zheng, W. Effects of zinc stress on growth and antioxidant enzyme responses of *Kandelia obovata* seedlings. *Toxicol. Environ. Chem.* **2015**, *97*, 1190–1201.
10. Anjum, N.A.; Singh, H.P.; Khan, M.I.R.; Masood, A.; Per, T.S.; Negi, A.; Ahmad, I. Too much is bad-an appraisal of phytotoxicity of elevated plant-beneficial heavy metal ions. *Environ. Sci. Pollut. Res.* **2015**, *22*, 3361–3382. [CrossRef]
11. Anwaar, S.A.; Ali, S.; Ali, S.; Ishaque, W.; Farid, M.; Farooq, M.A.; Najeeb, U.; Abbas, F.; Sharif, M. Silicon (Si) alleviates cotton (*Gossypium hirsutum* L.) from zinc (Zn) toxicity stress by limiting Zn uptake and oxidative damage. *Environ. Sci. Pollut. Res.* **2015**, *22*, 3441–3450. [CrossRef]
12. Khan, M.I.R.; Khan, N. *Reactive Oxygen Species and Antioxidant System in Plants: Role and Regulation under Abiotic Stress*; Springer: Berlin/Heidelberg, Germany, 2017.
13. Noctor, G.; Foyer, C.H. Ascorbate and glutathione: Keeping active oxygen under control. *Annu. Rev. Plant Physiol.* **1998**, *49*, 249–279. [CrossRef]
14. Nahar, K.; Hasanuzzaman, M.; Suzuki, T.; Fujita, M. Polyamines-induced aluminum tolerance in mung bean: A study on antioxidant defense and methylglyoxal detoxification systems. *Ecotoxicology* **2017**, *26*, 58–73. [CrossRef] [PubMed]

15. Foyer, C.H.; Noctor, G. Ascorbate and glutathione: The heart of the redox hub. *Plant Physiol.* **2011**, *155*, 2–18. [CrossRef] [PubMed]
16. Kaur, C.; Sharma, S.; Singla-Pareek, S.L.; Sopory, S.K. Methylglyoxal detoxification in plants: Role of glyoxalase pathway. *Indian J. Plant Physiol.* **2016**, *21*, 377–390. [CrossRef]
17. Per, T.S.; Khan, N.A.; Reddy, P.S.; Masood, A.; Hasanuzzaman, M.; Khan, M.I.R.; Anjum, N.A. Approaches in modulating proline metabolism in plants for salt and drought stress tolerance: Phytohormones, mineral nutrients and transgenics. *Plant Physiol. Biochem.* **2017**, *115*, 126–140. [CrossRef] [PubMed]
18. Misra, N.; Saxena, P. Effect of salicylic acid on proline metabolism in lentil grown under salinity stress. *Plant Sci.* **2009**, *177*, 181–189. [CrossRef]
19. Khan, N.A.; Khan, M.I.R.; Ferrante, A.; Poor, P. Editorial: Ethylene: A key regulatory molecule in plants. *Front. Plant Sci.* **2017**, *8*, 1782. [CrossRef]
20. Pahwa, K.; Ghai, N.; Kaur, J.; Singh, I.; Singh, S.; Dhingra, M. Influence of ethylene and cobalt chloride on photosynthetic parameters and pedicel anatomy of pigeon pea (*Cajanus cajan* L.) genotypes. *J. Environ. Biol.* **2017**, *38*, 367–374. [CrossRef]
21. Iqbal, N.; Khan, N.A.; Nazar, R.; Silva, J.A. Ethylene-stimulated photosynthesis results from increased nitrogen and sulfur assimilation in mustard types that differ in photosynthetic capacity. *Environ. Exp. Bot.* **2012**, *78*, 84–90. [CrossRef]
22. Khan, M.I.R.; Iqbal, N.; Masood, A.; Per, T.S.; Khan, N.A. Salicylic acid alleviates adverse effects of heat stress on photosynthesis through changes in proline production and ethylene formation. *Plant Signal. Behav.* **2013**, *8*, e26374. [CrossRef]
23. Khan, M.I.R.; Nazir, F.; Asgher, M.; Per, T.S.; Khan, N.A. Selenium and sulfur influence ethylene formation and alleviate cadmium-induced oxidative stress by improving proline and glutathione production in wheat. *J. Plant Physiol.* **2015**, *173*, 9–18. [CrossRef]
24. Khan, M.I.R.; Khan, N.A. Ethylene reverses photosynthetic inhibition by nickel and zinc in mustard through changes in PS II activity, photosynthetic nitrogen use efficiency, and antioxidant metabolism. *Protoplasma* **2014**, *251*, 1007–1019. [CrossRef] [PubMed]
25. Usuda, H. The activation state of ribulose 1, 5-bisphosphate carboxylase in maize leaves in dark and light. *Plant Cell Physiol.* **1985**, *26*, 1455–1463.
26. Krall, J.P.; Edwards, G.E. Relationship between photosystem II activity and CO_2 fixation in leaves. *Physiol. Plant.* **1992**, *86*, 180–187. [CrossRef]
27. Dhindsa, R.H.; Plumb-Dhindsa, P.; Thorpe, T.A. Leaf senescence correlated with increased level of membrane permeability, lipid peroxidation and decreased level of SOD and CAT. *J. Exp. Bot.* **1981**, *32*, 93–101. [CrossRef]
28. Wild, R.; Ooi, L.; Srikanth, V.; Münch, G. A quick, convenient and economical method for the reliable determination of methylglyoxal in millimolar concentrations: The N-acetyl-L-cysteine assay. *Anal. Bioanal. Chem.* **2012**, *403*, 2577–2581. [CrossRef]
29. Doderer, A.; Kokkelink, I.; van der Veen, S.; Valk, B.; Schram, A.; Douma, A. Purification and characterization of two Lipoxygenase isoenzymes from germinating barley. *Biochim. Biophys. Acta* **1992**, *112*, 97–104. [CrossRef]
30. Law, M.E.; Charles, S.A.; Halliwell, B. Glutathione and ascorbic acid in spinach (*Spinacia oleracea*) chloroplasts: The effect of hydrogen peroxide and of paraquat. *Biochem. J.* **1983**, *210*, 899–903. [CrossRef]
31. Anjum, N.A.; Umar, S.; Iqbal, M.; Khan, N.A. Cadmium causes oxidative stress in mung bean by affecting the antioxidant enzyme system and ascorbate-glutathione cycle metabolism. *Russ. J. Plant Physiol.* **2011**, *58*, 92–99. [CrossRef]
32. Griffith, O.W. Determination of GSH disulphide using GSH reductase and 2-vinylpyridine. *Anal. Biochem.* **1980**, *106*, 207–211. [CrossRef]
33. Masood, A.; Khan, M.I.R.; Fatma, M.; Asgher, M.; Per, T.S.; Khan, N.A. Involvement of ethylene in gibberellic acid-induced sulfur assimilation, photosynthetic responses, and alleviation of cadmium stress in mustard. *Plant Physiol. Biochem.* **2016**, *104*, 1–10. [CrossRef]
34. Bradford, M.M. A rapid and sensitive method for the Quantitation of micro-gram quantities of proteins utilising the principle of protein–dye binding. *Anal. Biochem.* **1976**, *72*, 248–254. [CrossRef]
35. Beyer, W.F.; Fridovich, I. Assaying for superoxide dismutase activity: Some large consequences of minor changes in conditions. *Anal. Biochem.* **1987**, *161*, 559–566. [CrossRef]

36. Giannopolitis, C.N.; Ries, S.K. Superoxide dismutase. I. Occurrence in higher plants. *Plant Physiol.* **1977**, *59*, 309–314. [CrossRef] [PubMed]
37. Nakano, Y.; Asada, K. Hydrogen peroxide is scavenged by ascorbate specific peroxidase in spinach chloroplasts. *Plant Cell Physiol.* **1981**, *22*, 867–880.
38. Foyer, C.H.; Halliwell, B. The presence of glutathione and glutathione reductase in chloroplasts: A proposed role in ascorbic acid metabolism. *Planta* **1976**, *133*, 21–25. [CrossRef]
39. Hasanuzzaman, M.; Hossain, M.A.; Fujita, M. Exogenous selenium pretreatment protects rapeseed seedlings from cadmium-induced oxidative stress by up regulating antioxidant defense and methylglyoxal detoxification systems. *Biol. Trace Elem. Res.* **2012**, *149*, 248–261. [CrossRef]
40. Booth, J.; Boyland, E.; Sims, P. An enzyme from rat liver catalyzing conjugation. *Biochem. J.* **1961**, *79*, 516–524.
41. Hossain, M.A.; Hossain, M.Z.; Fujita, M. Stress induced changes of methylglyoxal level and glyoxalase I activity in pumpkin seedlings and cDNA cloning of glyoxalase I gene. *Aust. J. Crop Sci.* **2009**, *3*, 53–64.
42. Hossain, M.A.; Nakano, Y.; Asada, K. Monodehydroascorbate reductase in spinach chloroplasts and its participation in the regeneration of ascorbate for scavenging hydrogen peroxide. *Plant Cell Physiol.* **1984**, *25*, 385–395.
43. Hasanuzzaman, M.; Nahar, K.; Anee, T.I.; Khan, M.I.R.; Fujita, M. Silicon-mediated regulation of antioxidant defense and glyoxalase systems confers drought stress tolerance in *Brassica napus* L. *S. Afr. J. Bot.* **2018**, *115*, 50–57. [CrossRef]
44. Principato, G.B.; Rosi, G.; Talesa, V.; Giovanni, E.; Uotila, L. Purification and characterization of two forms of glyoxalase II from the liver and brain of wistar rats. *Biochim. Biophys. Acta* **1987**, *911*, 349–355. [CrossRef]
45. Lindner, R.C. Rapid analytical methods for some of the more common inorganic constituents of plant tissues. *Plant Physiol.* **1944**, *19*, 70–89. [CrossRef] [PubMed]
46. Fiske, C.H.; Subba Row, Y. The colorimetric determination of phosphorus. *J. Biol. Chem.* **1925**, *66*, 375–400.
47. Bates, L.E.; Waldren, R.P.; Teare, I.D. Rapid determination of free proline for water stress studies. *Plant Soil* **1973**, *39*, 205–207. [CrossRef]
48. Hayzer, D.J.; Leisinger, T.H. The gene-enzyme relationships of proline biosynthesis in *Escherichia coli*. *J. Gen. Microbial.* **1980**, *118*, 287–293. [CrossRef]
49. Huang, A.H.C.; Cavalieri, A.J. Proline oxidase and water stress induced proline accumulation in spinach leaves. *Plant Physiol.* **1979**, *63*, 531–535. [CrossRef]
50. Mishra, S.; Dubey, R. Heavy Metal Toxicity induced Alterations in Photosynthetic Metabolism in Plants. In *Handbook of Photosynthesis*; Pessarakli, M., Ed.; CRC Press: Boca Raton, FL, USA, 2005; pp. 827–844.
51. Chen, Y.; Wu, N.; Zhang, Z.W.; Yuan, M.; Yuan, S. Perspective of Monitoring Heavy Metals by Moss Visible Chlorophyll Fluorescence Parameters. *Front. Plant Sci.* **2019**, *10*, 35. [CrossRef]
52. Bączek-Kwinta, R.; Antonkiewicz, J.; Łopata-Stasiak, A.; Kępka, W. Smoke compounds aggravate stress inflicted on Brassica seedlings by unfavorable soil conditions. *Photosynthetica* **2019**, *57*, 1–8. [CrossRef]
53. Sirhindi, G.; Mir, M.A.; Abd-Allah, E.F.; Ahmad, P.; Gucel, S. Jasmonic acid modulates the physio-biochemical attributes, antioxidant enzyme activity, and gene expression in Glycine max under nickel toxicity. *Front. Plant Sci.* **2016**, *7*, 591. [CrossRef]
54. Thao, N.P.; Khan, M.I.R.; Thu, N.B.A.; Hoang, X.L.T.; Asgher, M.; Khan, N.A.; Tran, L.S.P. Role of ethylene and its cross talk with other signaling molecules in plant responses to heavy metal stress. *Plant Physiol.* **2015**, *169*, 73–84. [CrossRef]
55. Khan, N.A.; Mir, M.R.; Nazar, R.; Singh, S. The application of ethephon (an ethylene releaser) increases growth, photosynthesis and nitrogen accumulation in mustard (*Brassica juncea* L.) under high nitrogen levels. *Plant Biol.* **2008**, *10*, 534–538. [CrossRef] [PubMed]
56. Asgher, M.; Per, T.S.; Verma, S.; Pandith, S.A.; Masood, A.; Khan, N.A. Ethylene supplementation increases PSII Efficiency and alleviates chromium-inhibited photosynthesis through increased nitrogen and sulfur assimilation in mustard. *J. Plant Growth Regul.* **2018**, *37*, 1300–1317. [CrossRef]
57. Molassiotis, A.; Sotiropoulos, T.; Tanou, G.; Diamantidis, G.; Therios, I. Boron induced oxidative damage and antioxidant and nucleolytic responses in shoot tips culture of the apple rootstock EM9 (*Malus domestica* Borkh). *Environ. Exp. Bot.* **2006**, *56*, 54–62. [CrossRef]
58. Khan, N.A.; Asgher, M.; Per, T.S.; Masood, A.; Fatma, M.; Khan, M.I.R. Ethylene potentiates sulfur-mediated reversal of cadmium inhibited photosynthetic responses in mustard. *Front. Plant Sci.* **2016**, *7*, 1628. [CrossRef] [PubMed]

59. Álvarez Viveros, M.F.; Inostroza-Blancheteau, C.; Timmermann, T.; González, M.; Arce-Johnson, P. Overexpression of Gly I and Gly II genes in transgenic tomato (*Solanum ycopersicum Mill.*) plants confers salt tolerance by decreasing oxidative stress. *Mol. Biol. Rep.* **2013**, *40*, 3281–3290. [CrossRef]
60. Mostofa, M.G.; Hossain, M.A.; Fujita, M. Trehalose pretreatment induces salt tolerance in rice (*Oryza sativa* L.) seedlings: Oxidative damage and co-induction of antioxidant defense and glyoxalase systems. *Protoplasma* **2015**, *252*, 461–475. [CrossRef] [PubMed]
61. Halliwell, B.; Gutteridge, J. Oxygen toxicity, oxygen radicals, transition metals and disease. *Biochem. J.* **1984**, *219*, 1–14. [CrossRef]
62. Hasanuzzaman, M.; Alam, M.M.; Nahar, K.; Mohsin, S.M.; Bhuyan, M.B.; Parvin, K.; Hawrylak-Nowak, B.; Fujita, M. Silicon-induced antioxidant defense and methylglyoxal detoxification works coordinately in alleviating nickel toxicity in *Oryza sativa* L. *Ecotoxicology* **2019**, *28*, 261–276. [CrossRef]
63. Alia; Saradhi, P.P.; Mohanty, P. Involvement of proline in protecting thylakoid membranes against free radical-induced photodamage. *J. Photochem. Photobiol.* **1997**, *38*, 253–257. [CrossRef]
64. Hare, P.D.; Cress, W.A. Metabolism implications of stress-induced proline accumulation in plants. *Plant Growth Regul.* **1997**, *21*, 79–102. [CrossRef]
65. Szabados, L.; Savoure, A. Proline: Multifunctional amino acid. *Trends Plant Sci.* **2010**, *15*, 89–97. [CrossRef] [PubMed]
66. Iqbal, N.; Umar, S.; Khan, N.A. Nitrogen availability regulates proline and ethylene production and alleviates salinity stress in mustard (*Brassica juncea*). *J. Plant Physiol.* **2015**, *178*, 84–91. [CrossRef] [PubMed]
67. Antonkiewicz, J.; Czesława, J.; Małgorzata, K.; Renata, S. Nickel bioaccumulation by the chosen plant species. *Acta Physiol. Plant.* **2016**, *38*, 40. [CrossRef]
68. Munns, R.; Tester, M. Mechanisms of salinity tolerance. *Annu. Rev. Plant Biol.* **2008**, *59*, 651–681. [CrossRef] [PubMed]
69. Hussain, S.; Khan, A.M.; Rengel, Z. Zinc-biofortified wheat accumulates more cadmium in grains than standard wheat when grown on cadmium-contaminated soil regardless of soil and foliar zinc application. *Sci. Total Environ.* **2019**, *654*, 402–408. [CrossRef] [PubMed]
70. Ahanger, M.A.; Agarwal, R.M. Salinity stress induced alterations in antioxidant metabolism and nitrogen assimilation in wheat (*Triticum aestivum* L) as influenced by potassium supplementation. *Plant Physiol. Biochem.* **2017**, *115*, 449–460. [CrossRef]

plants

MDPI

Article

Silver Can Induce Oxidative Stress in Parallel to Other Chemical Elicitors to Modulate the Ripening of Chili Cultivars

Arijit Ghosh, Indraneel Saha, Debabrata Dolui, Arnab Kumar De, Bipul Sarkar and Malay Kumar Adak *

Plant Physiology and Plant Molecular Biology Research Unit, Department of Botany, University of Kalyani, Kalyani, Nadia-741235, India; arijitgo625@gmail.com (A.G.); indraneelsaha92@gmail.com (I.S.); debabratabotany@gmail.com (D.D.); akde26@gmail.com (A.K.D.); bipulsarkar007@gmail.com (B.S.)

* Correspondence: mkadak09@gmail.com; Tel.: +91-9432418218

Received: 5 January 2020; Accepted: 5 February 2020; Published: 12 February 2020

Abstract: Two chili cultivars, i.e., cv. Bullet and cv. Tejaswini, were evaluated on postharvest related ripening characteristics with varying durations under hydrogen peroxide, putrescine and silver treatments. The reducing sugar was inversely related to the maximum values at 7 days of ripening. Silver and putrescine were the most regulatory in terms of changing of the total carbohydrate content as compared to hydrolysis of the total reducing sugar. Regarding pectin methylesterase activity, both chilies were consistent, regardless of the number of days of incubation. Still, putrescine and silver were significant contributors to variations in cv. Bullet and cv. Tejaswani. For the pigment content, lycopene and chlorophyll increased in a linear manner, although these treatments significantly varied over time. Hydrogen peroxide and putrescine were responsible for the maximum accumulation of lycopene for both the cultivars, whereas, only cv. Tejaswani displayed maximum carotenoid for putrescine. Silver for both chili varieties was the most inhibitory for lycopene and carotenoid content. Superoxide had a good impact on the accumulation of lipid peroxides, irrespective of the chili variety. The maximum accumulation of lipid peroxide was recorded at seven days of treatment. Phenolics and flavonoids were in decreasing order for both the chili varieties, progressing through the days of the study period in a similar manner. Silver was the main contributor to variations in the phenolics and flavonoid contents in cv. Tejaswani. The solubilization of total carbohydrate into reducing sugar was in an inverse relationship, with the maximum values being reached at 7 days of ripening.

Keywords: oxidative stress; ripening physiology; silver; chemical elicitors; chili

1. Introduction

Fruit ripening is established as a collective physiological process, with turnover as well the synthesis of bio-molecules being major events during fruit maturation. Changes of color, texture, and the solubilization of stored carbohydrates and other complex bio-molecules are facilitated by some gene expressions. This is more attributed to the elevation of CO_2 concentration through increased respiratory rates [1]; however, this is the case mostly for climacteric fruits. For non-climacteric ones, it is not essentially accompanied with any changes in CO_2 efflux. From a biochemical point of view, the total soluble solids and their conversion into sugars, the conversion of organic acids to keto-sugars, the turnover of flavon compounds are the most important. Leto-sugars, dextrans, and other disaccharides are responsible for the major variations in the characteristics of fruit types [2]. Fruit ripening and abscission are facilitated by programmed cell death, and are regulated by some signaling mechanisms [3]. Both endogenous compounds, like growth substances and secondary metabolites, are also involved in the ripening processes. Likewise, abiotic factors including the accumulation of minerals in soil, temperature fluctuations, UV intensity, and hypoxic or anoxic exposure, can cause variations in the contents of flavonoids and phenolic compounds. Incidentally, all these physical and chemical elicitations are targeted on a cellular level, with a characteristic change in the redox of ripening tissues. Consequently, ripening fruit are highly exposed to a burst of ROS depending upon the fruit type [4]. The exogenous application of chemical residues as elicitors has been observed to modulate fruit ripening with a precise ratio of ROS generation, as well its degradation. Moreover, a precise ratio of the concentration and function on different biochemical reactions is also proportionate to ROS metabolism [5]. In postharvest physiology, the signaling through some elicitors, which may be either endogenous or even exogenously applied synthetic moieties, is quite relevant. Besides silver, other chemical moieties have been observed to modulate the ripening characteristics, including fruit texture, color, softness etc. On a cellular level, changing the bonding pattern of cell wall residues is related to an extension of the shelf life of fruits. A certain level of ROS accumulation with wall bound NOX activity may represent a characteristic biomarker for fruit ripening phenomena. The maturation of fruit varieties through changes in specific characteristics is also regarded as being related to biomarkers under conditions of chemical elicitation, with silver being the main contributor. This biomarker would be more aligned to the ROS metabolism of ripening fruit tissues, as well as ROS related gene expression [6]. Therefore, the analysis and correlation of those cellular responses as biomarkers would comprise unraveling the key factors in the ripening processes. This may be worthy of closer examination with a suitable fruit sample. *Capsicum* has been under investigation for both its qualitative and quantitative nutritional traits due to the acquisition of secondary metabolites [7]. *Capsicum*, a non-climacteric fruit, has also been important as a nutritional supplement with antioxidation properties. This could be important in horticultural practices, with induced or enhanced changes in ripening patterns during the postharvest period. This study sets out to monitor oxidative exposure and its impact on the ripening process in two distinctly different *Capsicum* varieties. Additionally, it seeks to determine the efficacy of a few chemical elicitors, including silver, during postharvest storage.

2. Results

In our observations, distinguished phenotypic changes were recorded as plants responded to different treatments. Besides this, the cellular and physiological responses of two chili varieties (Bullet and Tejaswani) as a function of three elicitors and their interactions during ongoing postharvest storage were validated statistically by Analysis of Variance (ANOVA) following determination of the Least Significant Difference (LSD) (Tables S1–S10).

2.1. *Effect of Chemical Treatments on Lycopene and Carotenoid Content*

On the basis of chemical treatments, the two varieties of *Capsicum* fruit varied in their responses in a distinct manner. Monitoring of the delayed changes related to ripening was undertaken with consideration of varietal differences. In the initial observation, a detachment or loosening of the fruit stalk was observed, which is an indication of senescence. However, this varied significantly with duration during the treatments, compared to the control. As compared to the control, regardless of the chemical treatments, changes in fruit color were not prominent, even up to 3 days, but became distinct after 5 and 7 days (Figure 1). At 7 days, the contents of pigments such as lycopene and carotenoids in the fruit were significantly varied compared to the control throughout the treatments. As the fruit progressed through the days of incubation, the lycopene content was maximally affected at 7 days, regardless of the variety (Figure 2a). However, significant variation was observed with treatment with putrescine, but not to the same extent for *cv.* Bullet and *cv.* Tejaswani, with a 60.74% decrease and a 0.14-fold increase, respectively compared to the control. Still, in both chili varieties, $AgNO_3$ also showed significant effects by the down regulation of lycopene content throughout the days of treatment compared to the control. The maximum levels of regulation were 90.48% and 96.42% fall for *cv.* Tejaswani and *cv.* Bullet compared to the control at 7 days. For carotenoids, the activities for both chili varieties were also significantly varied ($p < 0.01$, 0.05) under each treatment (Figure 2b). Thus, the

$AgNO_3$ appeared to be down regulatory for carotenoid, by 88.02% and 93.75% compared to the control. In contrast, putrescine treatment was discriminatory for carotenoid content for the two chili variety. At 7 days, putrescine significantly ($p \leq 0.001$) reduced the pigment content by 69.01% for *cv.* Bullet, whereas for *cv.* Tejaswani, it was only 10.45%.

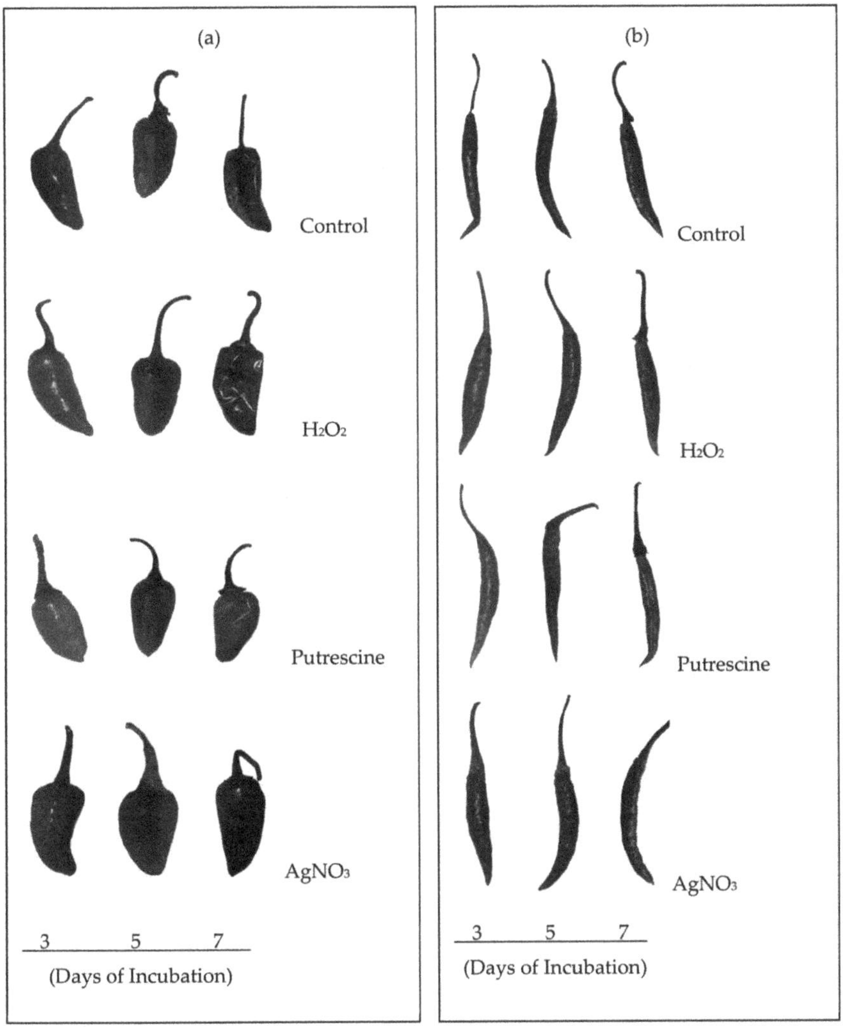

Figure 1. Morphological changes of chili fruits: *cv.* Bullet (**a**) and *cv.* Tejaswani (**b**) under various treatments, i.e., H_2O_2, putrescine, and $AgNO_3$ after 3, 5, and 7 days.

2.2. Changes of Reducing Sugar and Total Sugar Content

The most significant turnover of metabolites in the fruit ripening process is from sugar profiles. Significant changes in the total sugar and reducing sugar contents of the two chili varieties are presented in Figure 3a,b respectively. In both varieties, over the course of the treatment period, reducing sugar and total sugar contents with all treatments were quite discriminatory with respect to control. Compared to the control at 7 days, the reducing sugar contents in all the treatments were maximally changed for both chili varieties. Treatment with H_2O_2 was shown to change the sugar content, yielding a 0.35-fold increase for *cv.* Tejaswani, but a 30.34% decrease in the case of *cv.* Bullet chilies at 7 days. Similar trends were also observed on days 3 and 5 under H_2O_2 treatment. For *cv.* Tejaswani, the changes of total soluble sugar content were almost the opposite, showing a 21.74% decrease and 0.09-fold increase under putrescine and $AgNO_3$ at 7 days, compared to the bullet chili variety. This change in total sugar content was brought forward from days 3 and 5, but was not necessarily always significant ($p \leq 0.01$, 0.001). Two varieties were significantly distinct in their trend of total soluble sugar accumulation at 5 and 7 days. This was attributed to the H_2O_2, putrescine, and $AgNO_3$ treatments for both varieties. At 5 days, with the exception of the putrescine treatment, an increase of total carbohydrate was recorded in *cv.* Tejaswani: 0.09-fold and 0.09-fold increases respectively for H_2O_2 and $AgNO_3$. *cv.* Bullet chilies, in contrast, showed a reduced carbohydrate content, i.e., 22.15%, 55.69%, and 77.85% for the H_2O_2, putrescine and $AgNO_3$ treatments, compared to the control. On the 7th day, the Bullet chili variety accumulated sugar maximally under H_2O_2, putrescine, and $AgNO_3$. For the tejaswani variety, a significant promotional effect, i.e., a 0.21-fold increase was recorded only with H_2O_2 treatment at 7 days. However, in the Tejaswani variety, the putrescine and $AgNO_3$ treatments were quite inhibitory, i.e., 42.86% and 75% reductions of total sugar contents at 7 days.

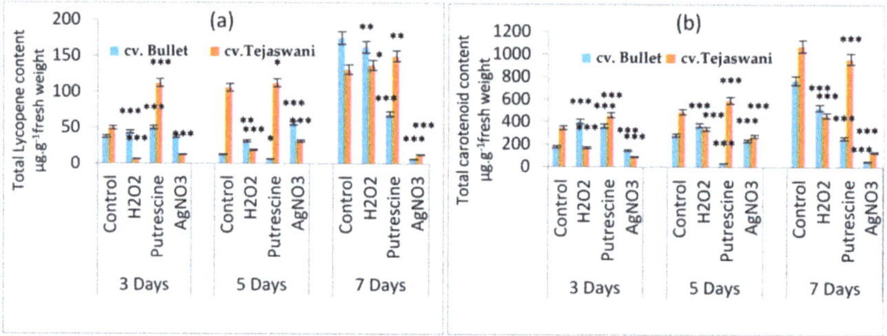

Figure 2. Changes of total lycopene (**a**) and total carotenoid content (**b**) of two chili types, i.e., *cv.* Bullet and *cv.* Tejaswani, under different treatments (Control, H_2O_2, putrescine and $AgNO_3$) and durations (3, 5, and 7 days). The vertical bars represent the data with the mean of the three replicates with ± *SE* ($n = 3$) from an individual set of experiments, and significant differences between treatments, as calculated by a student's t-test marked as * ($p \leq 0.05$) ** ($p \leq 0.01$) *** ($p \leq 0.001$).

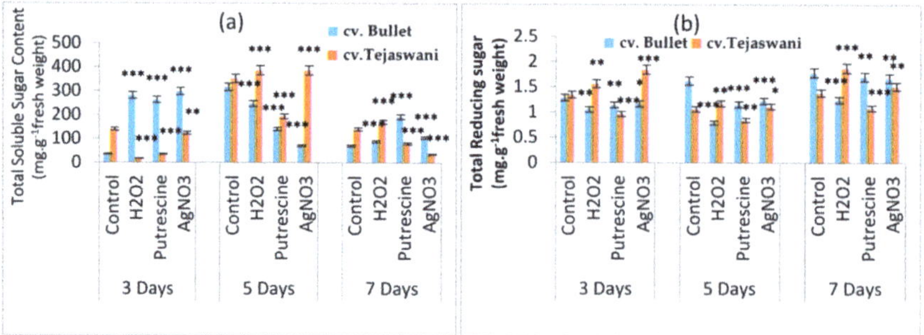

Figure 3. Changes of total soluble sugar (**a**) and total reducing sugars (**b**) of two chili types, i.e., *cv.* Bullet and *cv.* Tejaswani, under different treatments (Control, H_2O_2, putrescine, and $AgNO_3$) and durations (3, 5, and 7 days). The vertical bars represent the data with means of three replicates with ± *SE* ($n = 3$) from an individual set of experiments, and significant differences between treatments, as calculated by a student's t-test marked as *($p \leq 0.05$) **($p \leq 0.01$) ***($p \leq 0.001$).

2.3. Influence of Treatments on H_2O_2 and Lipid Peroxide Production

The ripening event through physiological intervention was also facilitated by the induction of reactive oxygen species (ROS). This idea also materialized in the present experiment through the accumulation of O_2^- in both chili varieties during incubation under all treatments for 3, 5, and 7 days. Both the varieties were maximally affected at 7 days with significant ($p \leq 0.01$, 0.001) decreases in O_2^- contents compared to days 3 and 5 with respect to the control (Figure 4a). Therefore, the most striking effects were the gradual loss of O_2^- by 60.86%, 60.86%, and 30.43% for *cv.* Bullet under H_2O_2, putrescine, and $AgNO_3$ respectively. In contrast, the Tejaswani chili variety could not modulate its O_2^- content under H_2O_2 and putrescine, but reduced it significantly under $AgNO_3$, i.e., by 42.86% at 7 days compared to the control. In contrast, at 5 days, it was more promising and consistent in terms of the up regulation of O_2^- for all treatments on the Tejaswani chili variety. As a consequence of O_2^- generation, the changes of membrane permeability were also quite expected, with the generation of lipid peroxide content in both varieties. The trend in MDA concentration throughout treatments showed an inconsistency with a rise and fall which was observed in both chili varieties (Figure 4b). As in the example, at 3 days, the Bullet variety became more prone to oxidative damage by increased MDA content compared to the Tejaswani variety. In contrast, on day 5, the *cv.* Tejaswani was shown to produce a higher MDA content then *cv.* Bullet, irrespective of treatments. However, the maximum variability was observed at 7 days, where the H_2O_2 treatment yielded a 6.78% and 32.31% decrease in the MDA content compared to the control for the Bullet and Tejaswani varieties, respectively. In the Tejasani variety, putrescine was less effective under the same conditions, whereas silver yielded a reduced MDA content, with a significant ($p \leq 0.001$) variation. Thus, *cv.* Bullet and *cv.* Tejaswani showed reduced levels of MDA by 22.03% and 83.08% as compared to the control.

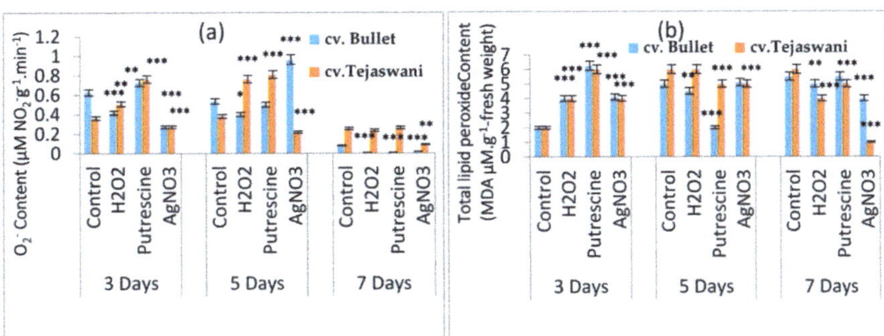

Figure 4. Changes of total O_2^-content (**a**)and total lipid peroxide content (**b**) of two chili types, i.e., *cv.* Bullet and *cv.* Tejaswani, under different treatments (control, H_2O_2, putrescine, and $AgNO_3$) and durations (3, 5, and 7 days). The vertical bars represent the data with means of three replicates with ± SE ($n = 3$) from the individual set of experiments, and significant differences between treatments as calculated by student's t-test marked as *($p \leq 0.05$) ** ($p \leq 0.01$) *** ($p \leq 0.001$).

2.4. Changes of Phenolic Content and total Flavonoid Content

Important constituents for fruit ripening are the phenolic residues and their corresponding derivatives. A turnover of phenolic content was observed in *cv.* Bullet and *cv.* Tejaswani throughout the experiment period (Figure 5a); the observed maximum at 7 days indicated the involvement of phenolic residues up to the final stages of ripening. The maximum changes at 7 days were recorded in all the treatments for both varieties with up and down regulation. Thus, the *cv.* Tejaswani variety had a linear increase of phenol content compared to the control, i.e., increases by 0.1 fold, 0.4 fold, and 0.48 fold under H_2O_2, putrescine, and $AgNO_3$ treatments, respectively. In contrast, the Bullet chili variety was also sensitive those treatments, showing 37.48%, 25.65%, and 46.85% increases under

H_2O_2, putrescine, and $AgNO_3$ treatments, respectively. Flavonoids and other important residues for total phenolic content were more involved in the initial days of the treatments. A gradual decline in flavonoid content undoubtedly indicated depletion through the fruit ripening process (Figure 5b). In the present experiment, the Tejaswani variety was inconsistent in the up and down regulation of flavonoid content. At 3 days, however, H_2O_2 was shown to have up regulated the content by 0.76 fold compared to the control. *cv.* Bullet did not show any significant variation with the treatments; however, the peak in the flavonoid content recorded with putrescine increased by 0.138 fold at 3 days.

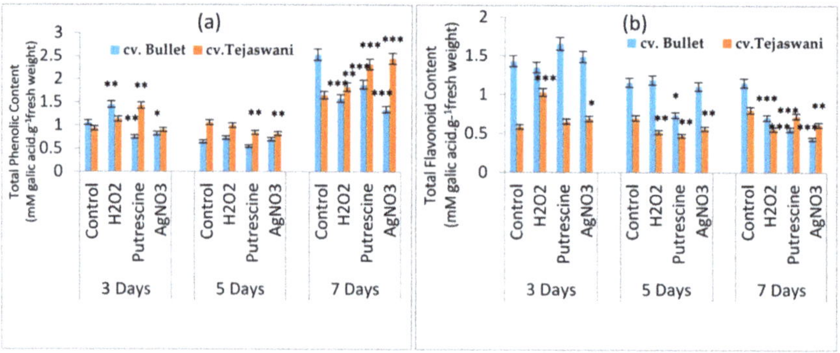

Figure 5. Changes of total phenolic content(**a**)and total flavonoid content (**b**) of two chili types, i.e., *cv.* Bullet and *cv.* Tejaswani, under different treatments (Control, H_2O_2, putrescine, and $AgNO_3$) and durations (3, 5, and 7 days). The vertical bars represent the data with the means of three replicates with ± SE (n = 3) from an individual set of experiments, and significant differences between treatments as calculated by a student's t-test marked as *($p \leq 0.05$) **($p \leq 0.01$) ***($p \leq 0.001$).

2.5. Activity of Pectin Methylesterase (PME) under Different Treatments

PME is a cell wall lysing enzyme that was shown to be the most discriminatory (Figure 6). However, chemical treatments were varied, with differential patterns of enzyme activities at 3,5, and 7 days for both varieties. An up regulation of PME activity compared to the control was maximized for the Bullet variety, then the Tejaswani variety at 3 days under all the treatments. This trend reversed for the *cv.* Bullet, with maximum activity through H_2O_2, putrescine, and $AgNO_3$ at 5 days. At 7 days, an admixture of heterogeneity in terms of the activity between *cv.* Bullet and *cv.* Tejaswani was recorded. Thus, H_2O_2 did not yield any significant changes for the two chili varieties, but for putrescine and $AgNO_3$, the activities were distinct between two varieties. With putrescine, the activities were 50% subdued for *cv.* Tejaswani, which increased by 0.25 fold for *cv.* Bullet compared to the control. On the other hand, $AgNO_3$ remained non-significant for both chili varieties against the control at 7 days.

2.6. Bioaccumulation of Silver in the Tissue

As expected from the silver nitrate treatment, the chili varieties were compatible in metal accumulation. Figure 7 illustrates the accumulation of silver in the pulp, albeit one that varied in concentration in the two varieties at 7 days. As compared to the control, a significant ($p \leq 0.05$) accumulation was recorded, i.e., 10.71 and 9.67 fold respectively, in cv. Bullet and cv. Tejaswani. Therefore, the tissue flexibility is up taken with metals. The responses of silver as an inducer are likely to be variable for the two chili varieties.

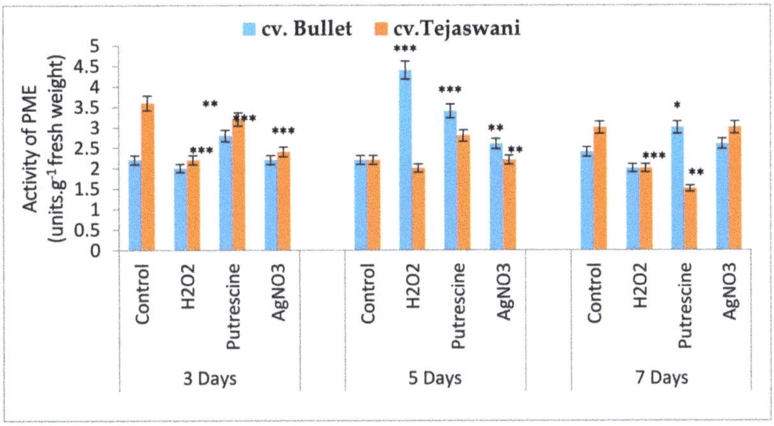

Figure 6. Assay of PME activity of two chili types, *cv.* Bullet and *cv.* Tejaswani, under different treatments (Control, H_2O_2, putrescine, and $AgNO_3$) and durations (3, 5, and 7 days). The vertical bars represent the data with the mean of three replicates with ± SE (n = 3) from an individual set of experiments, and significant differences between the treatments as calculated by a student's t-test marked as *($p \leq 0.05$) **($p \leq 0.01$) ***($p \leq 0.001$).

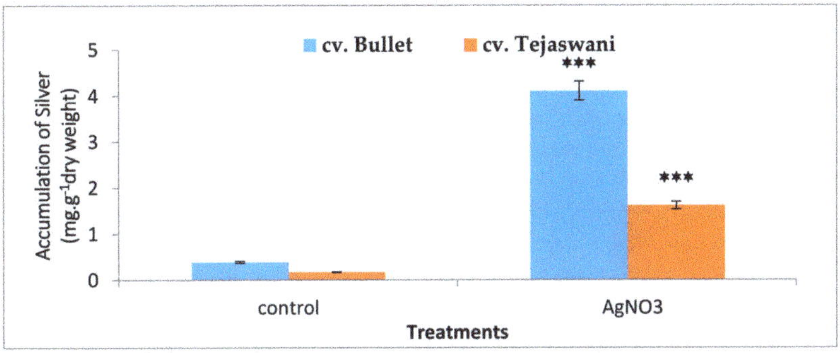

Figure 7. Changes in accumulation of silver of two chilitypes, *cv.* Bullet and *cv.* Tejaswani, under treatments (control and $AgNO_3$) after 7 days. The vertical bars represent the data with the mean of three replicates with ± *SE* (n = 3) from an individual set of experiments, and significant differences between the treatments as calculated by a student's t-test marked as *($p \leq 0.05$) **($p \leq 0.01$) ***($p \leq 0.001$).

3. Discussion

The ripening of fruit is a developmental process where physiological cellular events occur in plants. This research examined the delayed ripening of fruits in postharvest storage. It also proposes a physical/chemical system for the prolonged storage of fruits. In the present experiment, a chemical system employing H_2O_2, putrescine, and silver was used to extend the maximum shelf life of chili fruits under ambient laboratory conditions. From the aforementioned results, the efficacy of these chemicals is discussed in relation to the modulation of the most significantly contributing elements over the course of the treatments. The pigment system, including lycopene and carotenoids, the conversion of reserve carbohydrates into simpler forms, the acquisition of ROS and its impacts, the accumulation of secondary metabolites, the fruit skin's softening enzymes, and finally, the absorption of metals like silver into fruit tissue, were important. These contributed to the overall ripening phenomenon, either directly and indirectly [8]. Silver, a toxic metal, remains effective in threshold concentrations in terms of influencing ripening [9]. Of the chemical elicitors used in earlier studies, silver has been shown to be able to induce some degenerative effects following concomitant disorders on human health, albeit beyond threshold concentrations in tissues [10]. In contrast, this metal has a unique ability, i.e., controlling ethylene functionales by blocking one of their receptors. Silver has been in demand, regardless of fruit type (climacteric or non-climacteric) in postharvest storage [11]. Therefore, the question of whether and how such a metal in minimal concentrations might be a modulator along with H_2O_2 and putrescine is worthy of exploration. The present study is a description in which two chili varieties were evaluated in terms of their cellular and physiological responses when interacting with silver ion-mediated ROS. This was elucidated with two chemical elicitors i.e., H_2O_2 and putrescine, which had modulated responses in their own way [12,13]. The discriminatory responses of fruits of those cultivars under the influence of the aforementioned chemical elicitors made it more complex to decipher the ripening process throughout the postharvest period. Likewise, the readily observable characteristic of fruit color, as measured herein, regarding lycopene and carotenoid contents, were interesting to study. The regulation of pigment content in the fruit throughout the ripening process indicates the metabolic status of tissues due to other interfering reactions. Thus, throughout the various stages of ripening (at 3, 5, and 7 days), an almost consistent trend of pigment content (lycopene and carotenoid) was recorded. The retaining of lycopene, a carotenoid, is important for two basic processes: quenching of indigenously produced ROS, and as a ripening index on the fruit's skin [14]. Plant responses to oxidative stress vary according to genotypic specificity. Thus, increases in the total soluble sugar content at 5 days for the Tejaswani chili variety might be due to some changes in carbohydrate metabolism after oxidative stress. ROS often induces some sort of anomaly in tissue hydration/turgidity vis-à-vis water deficit [15]. It would be logical to examine the increase in sugar content linked to osmotic adjustment under such an ROS-induced water deficit. Both chili varieties might overcome the effect of H_2O_2 and thereby sustain the fruit skin integrity, leading to delayed ripening. This is also true for both chili varieties but may not be the case with $AgNO_3$. Silver has an intrinsic ability to modulate the ripening process, either by the down regulation of ethylene biosynthesis or by other, related phenomenon [16]. Thus, for both varieties, silver had down regulated the lycopene and carotenoid contents. In the context to polyamine, (putrescine used herein) the situation is more or less the same as that with H_2O_2. Polyamine, with its auto oxidation, may generate a ROS like H_2O_2, which might interfere with ripening related activities [17]. Therefore, in comparison, silver would not be a better choice to delay ripening as far as the lycopene and carotenoid contents are concerned. As discussed, ROS is an important moiety in the developmental processes of plants, including fruit ripening [18]. In the present experiment, regardless of the chili variety, a fair accumulation of O_2^- supported the sensitivity in different chemical treatments to the ripening process. The maximum activity of O_2^- was recorded at the ripening period, along with other factors like H_2O_2, putrescine, and $AgNO_3$. For *cv.* Tejaswani chilies, the role of putrescine in the metabolism of O_2^- may be due to the auto-oxidation of poly amine in the generation of H_2O_2. The latter may also be converted into other free radicals like OH^-. Polyamines, though relevant to stress tolerance, would are ROS producers, on account of their auto-oxidation abilities [19]. The activity of ROS in other gene induction related to cell wall bound peroxidase has been a feature for climacteric fruit ripening [20]. As a consequence of ROS accumulation, oxidized bio-metabolites are hazardous for tissues, as they initiate more free radical generation. In the present experiment, regardless of the duration, the fruit experienced a significant accumulation of lipid peroxides, variably with H_2O_2, putrescine, and $AgNO_3$ treatments. Upon observation of the nonlinearity between O_2^- and lipid peroxides, OH^- assumed the other pathways for free radical generation. ROS generation and its products are common consequences of plant developmental phases, including fruits maturation [21]. In the latter periods of ripening, *cv.* Bullet and *cv.* Tejaswani experienced a significant accumulation of lipid peroxide, even without any chemical treatments. These indicate that the ripening process is ROS dependent, irrespective of any chemical elicitors related to oxidative exposure. Among the few indigenous moieties which inhibit or alter the ripening process, phenolics and their derivatives are the most important. If ROS accelerates or induces processed, then it must be reversed by some other things which quench the intrinsic energy of ROS [22]. Flanonoids, with their various glycosidic assemblages, have anti-oxidation abilities in macromolecules [23]. For total phenolics, putrescine was able to maintain the phenolic content for both cultivars; silver was the next most effective elicitor for the Tejaswani chili variety. This is a genotypic variation which depends upon an elicitor for the same metabolic bio-accumulation. Nevertheless, polyamine is universal for its compatibility with phenolic metabolism through any of its biosynthetic pathways. Therefore, apart from H_2O_2 and silver, polyamine could be used for the delayed ripening of fruit. This has been illustrated in fruits where polyamines moderated ripening-specific degenerative processes [24]. One of the most important determinants of fruit ripening is the activity of cell wall hydrolyzing enzymes [25,26]. Thus, we see the expression of cell wall softening enzymes like PME recording the most variability in the early ripening period, but this later becomes inconsistent. Therefore, fruits are expected to receive impulses from H_2O_2, putrescine and $AgNO_3$ as elicitors. It is interesting to note that H_2O_2 suppressed the activity of PME, and that $AgNO_3$ sustained the activity, regardless of the fruit type (*cv.* Bullet and *cv.* Tejaswani). This assumes the ability of H_2O_2 as a ROS to down regulate PME expression, whereas $AgNO_3$ remains insignificant. Briefly, the pigment content of the fruit skin, secondary metabolites (as phenolics) and cell wall softening-related enzyme activity were significantly affected by chemical elicitation. This statement is most aligned with the findings on day 7 under all the treatments. This is because any fruit which sustains its metabolites after 7 days of storage by suppressing any loss or turn over causing decay is preferred. The behavior of ripening is under genotypic control for any chemical elicitors or abiotic factors. So, the selection of such elicitation may not be precise, but subjected to chili variety, in which flexibility in response to elicitors and maximum utilization in modulation would be the selection criteria.

4. Materials and Methods

4.1. Plant Materials and Treatments

The present experiment was conducted in the Laboratory of Plant Physiology and Biochemistry, Department of Botany, University of Kalyani. Two varieties of chili, i.e., short-round (*cv.* Bullet) and long-slender (*cv.* Tejaswani), were collected (plant age: 90 days) from horticultural farms at the Agricultural University after maturation into uniform sizes. The collected fruits were checked for any disease or defects, disinfected with a sodium hypochlorite solution, and thoroughly washed with deionized water. The sanitized fruits were divided into four sets according to the chemical treatments used: control (distilled water), H_2O_2 (40 mM), putrescine (1 mM), and $AgNO_3$ (40 μM). Those concentrations caused the minimum decay and loss of freshness of the fruits throughout the maximum duration of the experiment for both the chili varieties. All the treatment sets were transferred to growth chambers which were kept at 35–37 °C, 90–98% relative humidity, and under a light intensity of 5000 Lux. Under these conditions, the fruits were kept for 3, 5, and 7 days; the solutions were changed every other day. Samplings were made after treatment on days 3, 5, and 7 day of incubation. Three replicates (30 fruits each) per treatment were used. All samples were frozen in liquid nitrogen and stored in –80 °C until use in further biochemical assays.

4.2. Estimation of Carbohydrate Content

Fruits from different treatments were crushed into fine powders under cold conditions. The reducing sugar was extracted in 70% hot ethanol following repeated boiling, concentrated in a water bath and to reduce the sugar, and 2,4-dinitrosalicylic acid (DNS) reagent was added with crystalline phenol in the ratio of 5:1 dissolved in a 1% NaOH solution. The color was reduced by adding 40% Rochelle salt (Potassium sodium tartrate), and the absorbance was read at 520 nm [27].

For total carbohydrate detection, 500 mg of fruit tissue was taken in different test tubes containing 5.0 mL of 2.5 N HCl, and hydrolyzed for 3 h in a boiling water bath. After cooling, the solution was made neutral by the addition of sodium carbonate powder. The extract was centrifuged at 10,000× *g* for 10 min and the supernatant was collected. The reaction mixture was made by mixing the aliquot with 0.2% anthrone reagent in an acidic state under cold conditions. The reaction mixture was kept in a boiling water bath for 30 min and the color was read at 630 nm [28]. The concentration of sugars was calculated using the standards with dextrose, and expressed on a fresh weight basis.

4.3. Estimation of Total Lycopene and Carotenoid Content

First, 0.5 g of material from each sample was crushed into a fine powder and saponified by the addition of 2.5 mL 1.2% alcoholic potassium hydroxide. The saponified extract was mixed with 10.0 mL petroleum ether (AR grade) in a separating funnel and mixed. Upon separation, the aqueous layer was discarded and the upper pigment containing layer was saved. The process continued until the lower layer became transparent, at which time the final volumes for each sample were pulled. The absorbance was read at 450 nm and 503 nm against petroleum ether as the blank. The amount of lycopene and carotenoids were measured using the technique outlined in [29].

$$\text{Amount of total lyopene} = (3.12 \times A_{503} \times \text{volume of the samplesolution} \times 100)/\text{weight of the sample} \quad (1)$$

$$\text{Amount of total carotenoid} = (3.12 \times A_{450} \times \textit{volume of the samplesolution} \times 100 \times 4)/\textit{weight of the sample} \quad (2)$$

4.4. Estimation of Total Phenolic and Flavonoid Content

The extraction made in an alcoholic solution and an aliquot of 0.1 mL were mixed thoroughly with Folin-Ciocalteu reagent, followed by the addition of 20% (w/v) Na_2CO_3. After incubation of the mixture at 37 °C, the absorbance was recorded at 750 nm. Gallic acid was used and the total content of phenolics was expressed as equivalence of gallic acid. Flavonoid content was estimated; 1.0 g fresh sample was thoroughly crushed and extracted in 50% aqueous methanol. The mixture was concentrated in a water bath, and centrifuged at 8000× *g* at 4 °C for 10 min. To the solution, 10% $AlCl_3$ was added. The reaction was checked by adding 1.0 mL 1 M NaOH and further diluted with ethanol. The absorbance was read at 510 nm. The flavonoid content was estimated from the calibration curve with gallic acid.

4.5. Detection of H_2O_2, O_2^-, and Lipid Peroxide

To determine the H_2O_2 content,1 g of sample from each treatment was crushed in a 0.01 M phosphate buffer (pH 7.0), and the homogenate was centrifuged at 12,000× *g* for 15 min. Then, the supernatant was mixed with a chromate reagent (5% chromate: glacial acetic acid = 3:1) and kept in a boiling water bath for 10 min. After the development of a green color, the absorbance of the solution was measured at 570 nm. The result was expressed in μM H2O2 g^{-1} fresh weight. For the detection of O_2^-, fresh samples were thoroughly crushed in a 65 mM phosphate buffer with a pH of 6.5. The homogenate was centrifuged at 12,000× *g* for 15 min at 4 °C. The supernatant was rediluted with 65 mM phosphate buffer with a pH of 7.8, to which 10 mM hydroxylamine hydrochloride was added. The mixture was then incubated at 25 °C for 30 min. To the mixture, 10 mM sulfanilamide and 7 Mm alfa-napthylamine were added; this was kept at 37 °C for 20 min. The change of absorbance was recorded taking a reagent blank at 530 nm. The content of O_2^- was estimated from the $NaNO_2$ calibration curve, and the extraction coefficient of O_2^- as 12.8 mM^{-1} cm^{-1}. Then, 1.0 g tissue was thoroughly crushed in 65 mM phosphate buffer (pH 6.5) followed by deprotenization with 0.1% cold TCA. The supernatant was collected by centrifugation at 12,000× *g* for 10 min at 4 °C. A reaction mixture of 0.5% thiobarbituric acid (TBA) in 20% (w/v) TCA was added. After incubation at 95 °C for 30 min followed by cooling for 1 h, the precipitation was discarded by centrifugation at 10,000× *g* for 5 min. The supernatant was saved, read at 532 nm and 600 nm, and expressed malondialdehyde (MDA) content/g-fresh weight [30].

4.6. Assay of Pectin Methylesterase (PME)

First, 1.0 g of fresh sample was thoroughly crushed into a fine powder, added to liquid N_2, and homogenized with a buffer of 2.25 gm NaCl-EDTA salt. It was centrifuged 10,000× *g* for 15 min at 4 °C. The supernatant was collected and the residues were re-extracted with the same buffer for 15 min. The supernatant was collected and dissolved in 100 mL cold ethanol with 1.5 mL NaCl. For the reaction, a 1% pectin solution in 5.0 mL 1 M NaCl was used to adjust the pH to 7.5. Then, 0.5 mL enzyme extract was added, followed by the continuous addition of 0.1 N NaOH for 1 h. From a blank of heat-killed enzyme, the alkali consumption was recorded over the time. The activity of the enzymes, expressed as units/mL, was recorded following the method described in [31].

4.7. Quantification of Silver Content

One gram of sample was digested in 5 ml triacid mixture (i.e., 98% Sulfuric acid: 69% Nitric acid: 70% Perchloric acid = 3:3:1) at 200 °C for 1 min until the solution becomes clear. Digested samples were diluted 10 times, and the quantification of silver was then done by atomic absorption spectroscopy (PerkinElmer Optima 3300XL, Midland, Canada) [32]. The silver content was expressed as mg/g of plant dry weight.

4.8. Statistical Analysis

A student's t- test was computed using Microsoft Excel to determine significant differences among treatments at $p \leq 0.001$, $p \leq 0.01$ and $p \leq 0.05$ levels of significance. To detect the variations between the two chili varieties under the three treatments (i.e., H_2O_2, putrescine, $AgNO_3$) as compared to the control, Analysis of Variance (ANOVA), followed by LSD, was calculated using the IBM-SPSS software (Armonk, New York, USA). The Tables for ANOVA may be found in the Supplementary section.

5. Conclusions

Fruit ripening is a collective degenerative process. Significant physiological responses in fruit were examined under the application of H_2O_2, putrescine, and silver. During postharvest storage, fruits received a significant amount of ROS with a differential bioaccumulation of silver. The variability of silver accumulation would be a selection pressure for cultivar sensitivity to metals. On the other hand, a high accumulation of ROS in pulp tissues might have assisted in the regulation of ripening under the various treatments. For oxidative damages, fruits were more relieved with silver, although this was not the case for all the cultivars. Therefore, the choice of treatment in terms of the antioxidation of metabolites might determine possible storage ability. The retention of metabolites like complex carbohydrates and pigments appeared to be sensitive to all the treatments. The secondary metabolites may be other criteria which could increase the shelf life of fruit when treated with polyamines as well as H_2O_2.Interestingly, cell wall loosening activity and the maintenance of redox would meet the regulations of postharvest storage, more so than with silver. Silver apparently showed some advantages over other treatments for its retention of the visual characteristics of chili fruits undergoing delayed ripening. Additionally, putrescine would be next in line in extending shelf life in postharvest storage. Apart from laboratory-based experiments, the use of silver and other chemicals needs to be examined on plants in natural/field conditions.

Supplementary Materials: The following are available online at http://www.mdpi.com/2223-7747/9/2/238/s1, **Table S1.** Lycopene content. **Table S2.** Total Carotenoid content. **Table S3.** Total Sugar Content. **Table S4:** Total Reducing Sugar content. **Table S5.** Changes of total O^{2-} content. **Table S6.** Changes of total MDA content. **Table S7.** Changes of total phenolic content. **Table S8.** Changes of total Flavonoid Content. Table **S9.** Assay of Pectin Methylesterase. **Table S10.** Silver Content.

Author Contributions: Conceptualization, A.G. and M.K.A.; methodology, A.K.D.; validation, A.G., I.S. and D.D.; formal analysis, A.G. and I.S.; investigation, A.G.; data curation, A.G. and D.D.; writing—original draft preparation, A.G. and M.K.A.; supervision, M.K.A.; writing—review and editing, M.K.A. All authors have read and agreed to the published version of the manuscript.

Funding: This research received no external funding

Acknowledgments: This work is financially supported by DST-PURSE II program on University of Kalyani, DST, Govt. of INDIA, New Delhi and University Grant Commission providing fellowship as UGC-JRF.

Conflicts of Interest: The authors declare that there is no conflict of interest.

References

1. Martínez-Lüscher, J.; Morales, F.; Sánchez-Díaz, M.; Delrot, S.; Aguirreolea, J.; Gomes, E.; Pascual, I. Climate change conditions (elevated CO_2 and temperature) and UV-B radiation affect grapevine (*Vitisvinifera* cv. Tempranillo) leaf carbon assimilation, altering fruit ripening rates. *Plant Sci.* **2015**, *236*, 168–176. [CrossRef]
2. Basu, S.; Roychoudhury, A.; Saha, P.P.; Sengupta, D.N. Differential antioxidative responses of indica rice cultivars to drought stress. *Plant Growth Regul.* **2009**, *60*, 51–59. [CrossRef]
3. Wang, T.-D.; Zhang, H.-F.; Wu, Z.-C.; Li, J.-G.; Huang, X.-M. Sugar Uptake in the Aril of Litchi Fruit Depends on the Apoplasmic Post-Phloem Transport and the Activity of Proton Pumps and the Putative Transporter LcSUT4. *Plant Cell Physiol.* **2015**, *56*, 377–387. [CrossRef] [PubMed]
4. Decros, G.; Baldet, P.; Beauvoit, B.; Stevens, R.; Flandin, A.; Colombié, S.; Gibon, Y.; Pétriacq, P. Get the Balance Right: ROS Homeostasis and Redox Signalling in Fruit. *Front. Plant Sci.* **2019**, *10*, 1091. [CrossRef] [PubMed]
5. Del Río, L.A.; López-Huertas, E. ROS generation in peroxisomes and its role in cell signaling. *Plant Cell Physiol.* **2016**, *57*, 1364–1376. [CrossRef] [PubMed]
6. Ramos, P.; Parra-Palma, C.; Figueroa, C.R.; Zuñiga, P.E.; Valenzuela-Riffo, F.; Gonzàlez, J.; Gaete-Eastman, C.; Morales-Quintana, L. Cell wall-related enzymatic activities and transcriptional profiles in four strawberry (*Fragaria x ananassa*) cultivars during fruit development and ripening. *Sci. Hortic.* **2018**, *238*, 325–332. [CrossRef]
7. Bortolin, R.C.; Caregnato, F.F.; Junior, A.M.D.; Zanotto-Filho, A.; Moresco, K.S.; Rios, A.D.O.; Salvi, A.D.O.; Ortmann, C.F.; De Carvalho, P.; Reginatto, F.H.; et al. Chronic ozone exposure alters the secondary metabolite profile, antioxidant potential, anti-inflammatory property, and quality of red pepper fruit from Capsicum baccatum. *Ecotoxicol. Environ. Saf.* **2016**, *129*, 16–24. [CrossRef]
8. Brandes, N.; Zude-Sasse, M. Respiratory patterns of European pear (*Pyruscommunis* L. 'Conference') throughout pre- and post-harvest fruit development. *Heliyon* **2019**, *5*, e01160. [CrossRef]
9. Kumar, V.; Parvatam, G.; Ravishankar, G.A. AgNO3—A potential regulator of ethylene activity and plant growth modulator. *Electron. J. Biotechnol.* **2009**, *12*, 8–9. [CrossRef]
10. Hussain, M.; Raja, N.I.; Iqbal, M.; Aslam, S. Applications of plant flavonoids in the green synthesis of colloidal silver nanoparticles and impacts on human health. *Iran. J. Sci. Technol. Trans. A Sci.* **2019**, *43*, 1381–1392. [CrossRef]
11. Tavakoli, S.A.; Mirzaei, S.; Rahimi, M.; Tavassolian, I. Assessment of peppermint, clove, cumin essential oils and silver nano particles on biochemical and shelf life of *Citrus limon*(L.). *Indian J. Biochem. Biophys.* **2019**, *56*, 269–275.
12. Al-Saikhan, M.S.; Shalaby, T.A. Effect of hydrogen peroxide (H_2O_2) treatment on physicochemical characteristics of tomato fruits during post-harvest storage. *Aust. J. Crop Sci.* **2019**, *13*, 798. [CrossRef]
13. Hosseini, M.S.; Babalar, M.; Askari, M.A.; Zahedi, S.M. Comparison the effect of putrescine application on postharvest quality of *Pyruscommunis* cv. "Shah-Miveh" and "Spadona.". *Food Sci. Nutr.* **2018**, *7*, 14–21. [CrossRef] [PubMed]
14. Müller, L.; Caris-Veyrat, C.; Lowe, G.; Böhm, V. Lycopene and Its Antioxidant Role in the Prevention of Cardiovascular Diseases—A Critical Review. *Crit. Rev. Food Sci. Nutr.* **2015**, *56*, 1868–1879. [CrossRef]
15. Ahmed, B.; Hashmi, A.; Khan, M.S.; Musarrat, J. ROS mediated destruction of cell membrane, growth and biofilms of human bacterial pathogens by stable metallic AgNPs functionalized from bell pepper extract and quercetin. *Adv. Powder Technol.* **2018**, *29*, 1601–1616. [CrossRef]
16. An, J.; Almasaud, R.A.; Bouzayen, M.; Zouine, M.; Chervin, C. Auxin and ethylene regulation of fruit set. *Plant Sci.* **2020**, *292*, 110381. [CrossRef]
17. Seifi, H.S.; Shelp, B.J. Spermine Differentially Refines Plant Defense Responses against Biotic and Abiotic Stresses. *Front. Plant Sci.* **2019**, *10*, 117. [CrossRef]
18. Zhang, Y.; Gong, Y.; Chen, L.; Peng, Y.; Wang, Q.; Shi, J. Hypotaurine delays senescence of peach fruit by regulating reactive oxygen species metabolism. *Sci. Hortic.* **2019**, *253*, 295–302. [CrossRef]
19. Podlešáková, K.; Ugena, L.; Spíchal, L.; Doležal, K.; De Diego, N. Phytohormones and polyamines regulate plant stress responses by altering GABA pathway. *New Biotechnol.* **2019**, *48*, 53–65. [CrossRef]
20. Kärkönen, A.; Kuchitsu, K. Reactive oxygen species in cell wall metabolism and development in plants. *Phytochemistry* **2015**, *112*, 22–32. [CrossRef]
21. De Oliveira, A.B.; de Almeida Lopes, M.M.; Moura, C.F.; de Siqueira Oliveira, L.; de Souza, K.O.; Gomes Filho, E.; Urban, L.; de Miranda, M.R. Effects of organic vs. conventional farming systems on quality and antioxidant metabolism of passion fruit during maturation. *Sci. Hortic.* **2017**, *222*, 84–89. [CrossRef]

22. Berni, R.; Luyckx, M.; Xu, X.; Legay, S.; Sergeant, K.; Hausman, J.-F.; Lutts, S.; Cai, G.; Guerriero, G. Reactive oxygen species and heavy metal stress in plants: Impact on the cell wall and secondary metabolism. *Environ. Exp. Bot.* **2019**, *161*, 98–106. [CrossRef]
23. Zhang, L.; Liu, P.; Li, L.; Huang, Y.; Pu, Y.; Hou, X.; Song, L. Identification and Antioxidant Activity of Flavonoids Extracted from Xinjiang Jujube (Ziziphus jujube Mill.) Leaves with Ultra-High Pressure Extraction Technology. *Molecules* **2018**, *24*, 122. [CrossRef] [PubMed]
24. Fortes, A.M.; Agudelo-Romero, P.; Pimentel, D.; Alkan, N. Transcriptional Modulation of Polyamine Metabolism in Fruit Species under Abiotic and Biotic Stress. *Front. Plant Sci.* **2019**, *10*, 816. [CrossRef]
25. Jain, V.; Chawla, S.; Choudhary, P.; Jain, S. Post-harvest calcium chloride treatments influence fruit firmness, cell wall components and cell wall hydrolyzing enzymes of Ber(*Ziziphusmauritiana*Lamk.) fruits during storage. *J. Food Sci. Technol.* **2019**, *56*, 4535–4542. [CrossRef]
26. Zhang, L.; Wang, P.; Chen, F.; Lai, S.; Yu, H.; Yang, H. Effects of calcium and pectin methylesterase on quality attributes and pectin morphology of jujube fruit under vacuum impregnation during storage. *Food Chem.* **2019**, *289*, 40–48. [CrossRef]
27. Hodge, J.E. Determination of reducing sugars and carbohydrates. *Methods Carbohydr. Chem.* **1962**, *1*, 380–394.
28. Miller, G.L. Estimation of reducing sugar by dinitrosalicylic acid method. *Anal. Chem.* **1972**, *31*, 426–428. [CrossRef]
29. Zakaria, M.; Simpson, K.; Brown, P.R.; Krstulović, A. Use of reversed-phase high-performance liquid chromatographic analysis for the determination of provitaminA carotenes in tomatoes. *J. Chromatogr. A* **1979**, *176*, 109–117. [CrossRef]
30. Heath, R.L.; Packer, L. Photoperoxidation in isolated chloroplasts: I. Kinetics and stoichiometry of fatty acid peroxidation. *Arch. Biochem. Biophys.* **1968**, *125*, 189–198. [CrossRef]
31. Hobson, G.; Norris, F.W.; Schryver, S.B.; Resch, C.E.; Buston, H.W.; Nanji, H.R. Polygalacturonase in normal and abnormal tomato fruit. *Biochem. J.* **1964**, *92*, 324–332. [CrossRef] [PubMed]
32. Choleva, T.G.; Tsogas, G.Z.; Giokas, D.L. Determination of silver nanoparticles by atomic absorption spectrometry after dispersive suspended microextraction followed by oxidative dissolution back-extraction. *Talanta* **2019**, *196*, 255–261. [CrossRef] [PubMed]

plants

MDPI

Article

Metabolic Changes Induced by Silver Ions in *Carlina acaulis*

Sławomir Dresler [1,*], Barbara Hawrylak-Nowak [2,*], Maciej Strzemski [3], Magdalena Wójciak-Kosior [3], Ireneusz Sowa [3], Agnieszka Hanaka [1], Iwona Gołoś [1], Agnieszka Skalska-Kamińska [3], Małgorzata Cieślak [4] and Jozef Kováčik [5]

1 Department of Plant Physiology and Biophysics, Institute of Biological Science, Maria Curie-Skłodowska University, 20-033 Lublin, Poland; agnieszka.hanaka@poczta.umcs.lublin.pl (A.H.); fantan8@wp.pl (I.G.)

2 Department of Botany and Plant Physiology, Faculty of Environmental Biology, University of Life Sciences in Lublin, Akademicka 15, 20-950 Lublin, Poland

3 Department of Analytical Chemistry, Medical University of Lublin, 20-093 Lublin, Poland; maciej.strzemski@poczta.onet.pl (M.S.); kosiorma@wp.pl (M.W.-K.); i.sowa@umlub.pl (I.S.); agnieszka.skalska-kaminska@umlub.pl (A.S.-K.)

4 Łukasiewicz—Textile Research Institute, Scientific Department of Unconventional Technologies and Textiles, Brzezińska 5/15, 92-103 Łódź, Poland; cieslakm@iw.lodz.pl

5 Department of Biology, University of Trnava, Priemyselná 4, 918 43 Trnava, Slovak Republic; jozkovacik@yahoo.com

* Correspondence: dresler.slawomir@gmail.com (S.D.); barbara.nowak@up.lublin.pl (B.H.-N.); Tel.: +48-81-537-5078 (S.D.); +48-81-445-60-96 (B.H.-N.)

Received: 5 November 2019; Accepted: 15 November 2019; Published: 17 November 2019

Abstract: Silver is one of the most toxic heavy metals for plants, inducing various toxic symptoms and metabolic changes. Here, the impact of Ag(I) on *Carlina acaulis* physiology and selected metabolites was studied using two Ag concentrations (1 or 10 μM) after 14 days of exposure. The higher concentration of Ag(I) evoked reduction of growth, while 1 μM Ag had a growth-promoting effect on root biomass. The translocation factor (<0.04) showed that Ag was mainly retained in the roots. The 1 μM Ag concentration increased the level of low-molecular-weight organic acids (LMWOAs), while 10 μM Ag depleted these compounds in the roots. The increased concentration of Ag(I) elevated the accumulation of phytochelatins (PCs) in the roots and reduced glutathione (GSH) in the shoots (but not in the roots). At 1 μM, Ag(I) elevated the level of phenolic and triterpene acids, while the 10 μM Ag treatment increased the carlina oxide content in the roots. The obtained results indicate an alteration of metabolic pathways of *C. acaulis* to cope with different levels of Ag(I) stress. Our data imply that the intracellular binding of Ag(I) and nonenzymatic antioxidants contribute to the protection against low concentrations of Ag ions.

Keywords: heavy metals; thiols; phenolic metabolites; organic acids

1. Introduction

The exposure of plants to toxic levels of heavy metals (HMs) induces many changes in their metabolism [1–3]. In response to metal stress, plants activate several HM tolerance mechanisms, including enzymatic and nonenzymatic antioxidant systems or HM-binding compounds such as phytochelatins or organic acids [4]. Moreover, HM excess has a typically significant impact on the synthesis and accumulation of secondary metabolites (SMs) [5,6]. The role of SMs in HM detoxification may be related to the presence of hydroxyl groups in phenolic compounds, while thiols or ascorbate are generally protective compounds in plants [7–9].

Among HMs, Ag ions are considered as one of the most toxic agents for plants [9]. Their presence in the environment leads to several negative and toxic symptoms, including growth reduction, disruption of macro- and micronutrient uptake and homeostasis, or depletion of photosynthetic pigments [9,10]. As the environmental pollution with this metal has increased over the past few decades owing to human activities [11], the influence of Ag on living organisms has received increasing attention [12,13].

Carlina acaulis L. from the Asteraceae family is a monocarpic perennial herb. Xerothermic and calcareous grasslands are natural habitats of these plants [14]. However, *Carlina* genus plants are also found in HM-contaminated areas, for example, waste deposits from metal mining and smelting [15]. As a pseudometallophyte, *C. acaulis* can be an attractive object to study the HM tolerance and detoxification mechanisms. Various biochemical pathways leading to the production of a spectrum of metabolites such as triterpenes (oleanolic acid and ursolic acid), essential oil compounds (carlina oxide), or phenolic acids (chlorogenic acid and 3,5-dicaffeoylquinic acid) make this species ideal for the research of SM–HM interactions.

The aim of this study was to investigate the metabolic response of *C. acaulis* exposed to two concentrations of Ag(I) (1 or 10 μM). The objectives of the study included (i) estimation of *C. acaulis* tolerance to Ag(I) stress, (ii) assessment of Ag accumulation and translocation as well as the content of mineral nutrients, (iii) evaluation of nonenzymatic antioxidants and ligands, and (iv) assessment of the impact of Ag(I) on the production of secondary metabolites.

2. Results

2.1. Impact of Ag(I) on Plant Growth and H_2O_2 Accumulation

Both the shoot and root biomass of *C. acaulis* was variously affected by the two concentrations of Ag(I) applied to the cultivation solution (Figure 1). It was found that, contrary to 1 μM Ag, 10 μM Ag caused a reduction of shoot fresh weight (FW) by more than 52% in comparison with the control. Moreover, the higher Ag(I) concentration resulted in visible necrotic symptoms on the leaves, while the presence of 1 μM Ag did not negatively affect the phenotype (Figure 2). Similar to shoots, the higher concentration of Ag(I) reduced root biomass (50% of the control). Interestingly, plants exposed to 1 μM Ag showed significantly higher root biomass (126% of the control). Although there was no negative effect of 1 μM Ag on the shoot FW, this treatment slightly increased the H_2O_2 accumulation in the leaves (Figure 3). Contrary to the control leaves, both Ag(I) concentrations induced appearance of brown spots after 3′,3-diaminobenzidine (DAB) staining, which indicated induction of H_2O_2 accumulation.

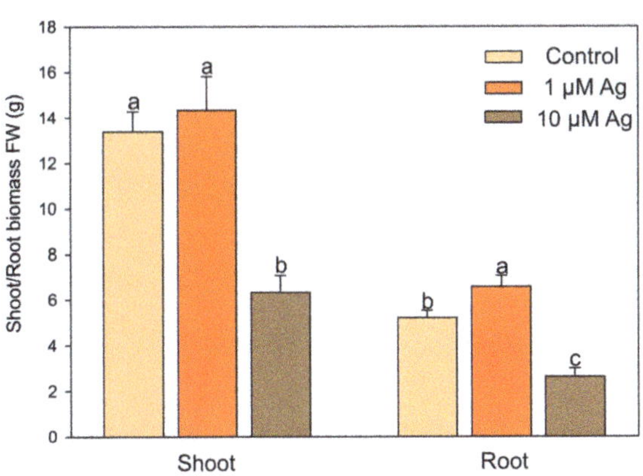

Figure 1. Effect of Ag(I) ions on the fresh weight of *Carlina acaulis* shoots and roots after 14 days of exposure. Data are mean ± SE (n = 25); values followed by the same letter are not significantly different ($p < 0.05$, Tukey's test).

Figure 2. Phenotype of *C. acaulis* plants cultivated for 14 days at various Ag(I) concentrations.

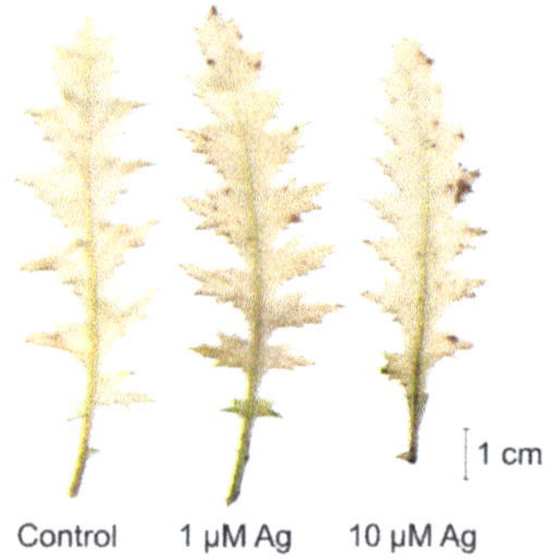

Figure 3. Histochemical detection of H_2O_2 in the leaves of *C. acaulis* growing for 14 days in the control conditions and at 1 or 10 μM Ag.

2.2. Bioconcentrations of Ag and Selected Mineral Nutrients

The accumulation of Ag in *C. acaulis* was significantly affected by its concentration in the nutrient medium (Figure 4). In the 10 μM treatment, the concentration of Ag in the shoots and roots significantly increased, reaching 6.1- and 13.6-fold higher levels in comparison with 1 μM Ag. It was also found that Ag was poorly transferred to the aboveground parts of the plants. The values of the Ag translocation factor (TF) calculated as the ratio of the Ag content in the shoots and roots (TF = 0.04 for 1 μM Ag; TF = 0.02 for 10 μM Ag) indicate that most Ag(I) ions taken up by roots are retained in these organs.

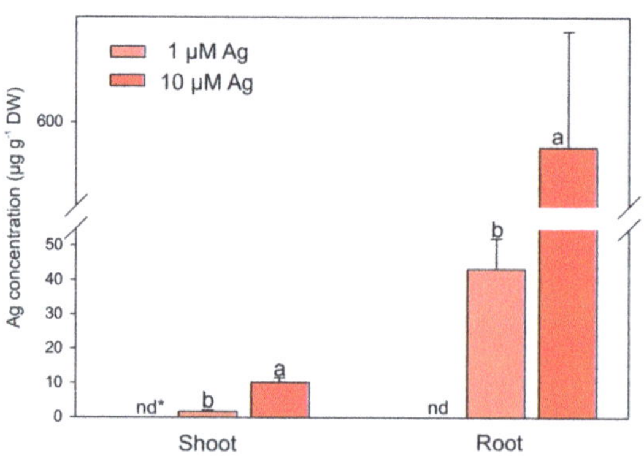

Figure 4. Ag concentrations in the roots and shoots of *C. acaulis* grown under two Ag(I) treatments for 14 days. Data are mean ± SE (n = 4); values followed by a different letter are significantly different within a particular organ ($p < 0.05$, Tukey's test); * nd—not detected.

The concentrations of both K and P in the roots and Ca in the shoots decreased at 10 μM Ag (Table 1). In turn, the higher concentration of Ag caused an increase in the level of Ca in the roots, while plants exposed to 1 μM Ag contained more Ca in their shoots. Moreover, it was observed that the lower concentration of Ag ions significantly decreased the Cu level in the shoots. However, the accumulation of K and P in the shoots, Mg and Zn in both the shoots and roots, and Cu in the roots did not differ significantly between the treatments.

Table 1. Concentration of selected mineral nutrients in the shoot and roots of *C. acaulis* grown under two Ag(I) treatments for 14 days. Data are mean ± SE (n = 4); values in the columns followed by the same letter are not significantly different between the treatments ($p < 0.05$, Tukey's test).

Treatment	K (mg g^{-1} DW)		P (mg g^{-1} DW)	
	Shoot	Root	Shoot	Root
Control	55.45 ± 3.83 a	41.93 ± 2.35 a	3.48 ± 0.35 a	5.69 ± 0.37 a
1 μM Ag	55.01 ± 4.52 a	40.17 ± 5.22 a	3.46 ± 0.65 a	5.79 ± 0.48 a
10 μM Ag	52.02 ± 1.53 a	25.19 ± 2.19 b	2.75 ± 0.23 a	3.84 ± 0.19 b
	Ca (mg g^{-1} DW)		**Mg (mg g^{-1} DW)**	
	Shoot	Root	Shoot	Root
Control	12.22 ± 0.73 b	5.09 ± 0.42 b	4.26 ± 0.31 a	2.32 ± 0.09 a
1 μM Ag	14.37 ± 0.47 a	5.49 ± 0.47 b	4.20 ± 0.56 a	2.74 ± 0.03 a
10 μM Ag	9.26 ± 0.21 c	7.67 ± 036 a	3.75 ± 0.21 a	2.52 ± 0.04 a
	Zn (μg g^{-1} DW)		**Cu (μg g^{-1} DW)**	
	Shoot	Root	Shoot	Root
Control	21.57 ± 4.02 a	58.76 ± 7.82 a	5.32 ± 0.17 a	14.06 ± 1.40 a
1 μM Ag	24.89 ± 2.90 a	61.48 ± 12.02 a	3.73 ± 0.41 b	11.54 ± 1.14 a
10 μM Ag	20.64 ± 2.86 a	56.08 ± 9.68 a	4.54 ± 0.11 ab	14.92 ± 2.17 a

2.3. Changes in Low-Molecular-Weight Organic Acids (LMWOAs), Thiols, and Ascorbic Acid (AsA) Accumulation under Ag(I) Exposure

In the 1 μM Ag treatment, the contents of LMWOAs (malic and citric acids) in the shoots and roots were elevated (Figure 5): malate and citrate represented 304% and 227% of the control value in the shoots and 230% and 320% of the control value in the roots, respectively. Similarly, the higher concentration of Ag(I) also elevated the malic acid amount in the shoots. On the contrary, plants grown at 10 μM Ag showed a decrease in the malate and citrate content by 70% and 77%, respectively, compared with the control.

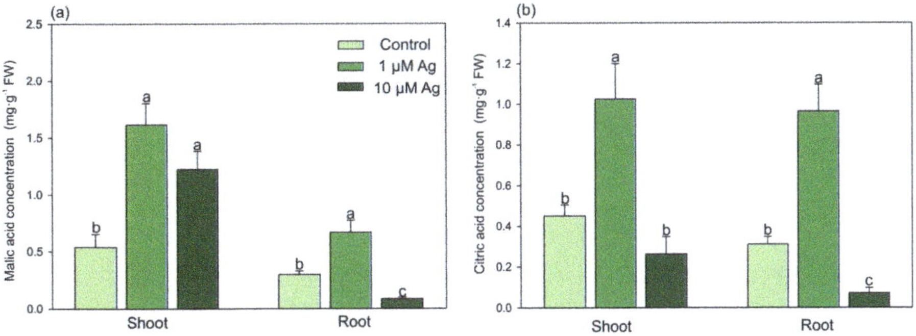

Figure 5. Effect of Ag(I) on (**a**) malic acid and (**b**) citric acid concentrations in the shoots and roots of *C. acaulis* after 14 days of exposure. Data are mean ± SE (n = 5); values for individual organs followed by the same letter are not significantly different ($p < 0.05$, Tukey's test).

The exposure to Ag increased the level of reduced glutathione (GSH) in the shoots (Figure 6). The highest concentration of this thiol-peptide (approx. 3-fold higher than in the control) was found at 10 μM Ag. In turn, the higher Ag(I) concentration significantly depleted the level of GSH in the roots. The accumulation of phytochelatins (PCs) in the shoots was below the detection limit, while their accumulation in the roots was dependent on the Ag(I) dose. It was found that Ag ions stimulated the PC accumulation from 0.05 to 7.70 and 21.42 nmol -SH g^{-1} FW in the control, 1 μM Ag, and 10 μM Ag treatments, respectively.

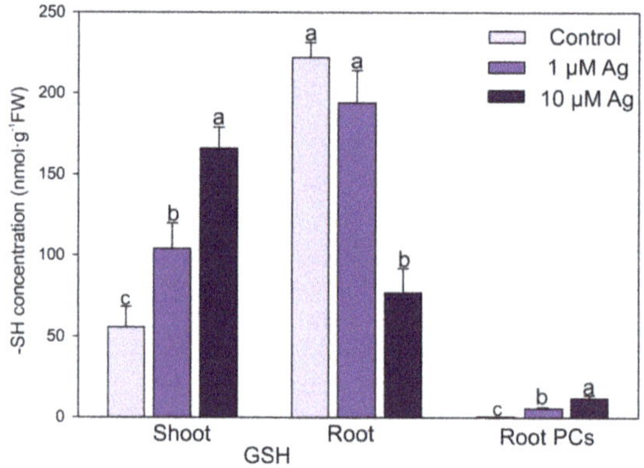

Figure 6. Effect of Ag(I) on reduced glutathione (GSH) concentrations in the shoots and roots and the total content of phytochelatins (PCs) in the roots of *C. acaulis* after 14 days of exposure. Data are mean ± SE (n = 5); values for individual organs followed by the same letter are not significantly different ($p < 0.05$, Tukey's test).

The level of AsA in the shoots did not differ between the Ag(I) treatments; however, significant depletion of the AsA content in the roots was found at 10 μM Ag (Figure 7).

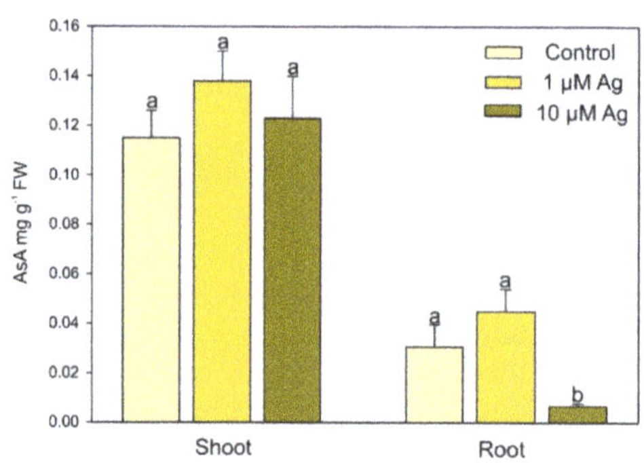

Figure 7. Effect of Ag(I) on ascorbic acid (AsA) concentrations in the shoots and roots of *C. acalulis* after 14 days of exposure. Data are mean ± SE (n = 5); values for individual organs followed by the same letter are not significantly different ($p < 0.05$, Tukey's test).

2.4. Changes in the Level of Selected Secondary Metabolites under Ag(I) Exposure

Two major phenolic acids, namely, chlorogenic acid (Figure 8a) and 3,5-dicaffeoylquinic acid (Figure 8b), were quantified in the shoots and roots of *C. acaulis*. Contrary to the roots, where no changes were found, the level of chlorogenic acid was significantly elevated in the shoots in response to the Ag(I) exposure and its concentrations increased by 2.9- and 1.5-fold at 1 and 10 μM Ag compared with the control. Only the lower Ag(I) dose had a significant positive effect on the concentration of 3,5-dicaffeoylquinic acid both in the shoots and roots.

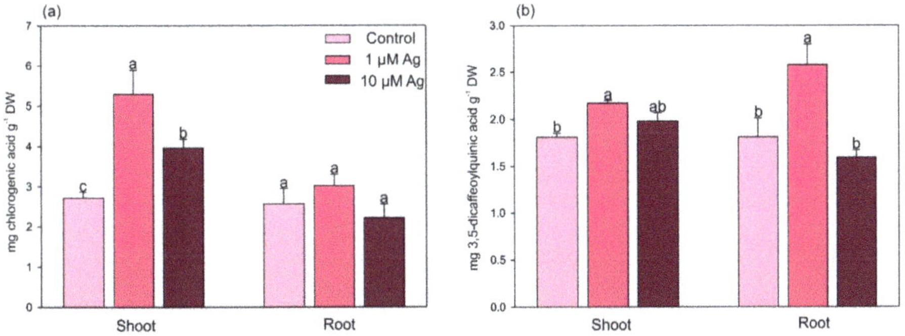

Figure 8. Effect of Ag(I) on the concentrations of (**a**) chlorogenic acid and (**b**) 3,5-dicaffeoylquinic acid in the shoots and roots of *C. acalulis* after 14 days of exposure. Data are mean ± SE (n = 5); values for individual organs followed by the same letter(s) are not significantly different ($p < 0.05$, Tukey's test).

The triterpene (oleanolic and ursolic) acids were detected in the shoots only, while carlina oxide was present in the root tissues only (Figure 9). The content of oleanolic and ursolic acids significantly increased under 1 μM Ag only. On the contrary, only the higher concentration of Ag(I) elevated the accumulation of carlina oxide in the roots.

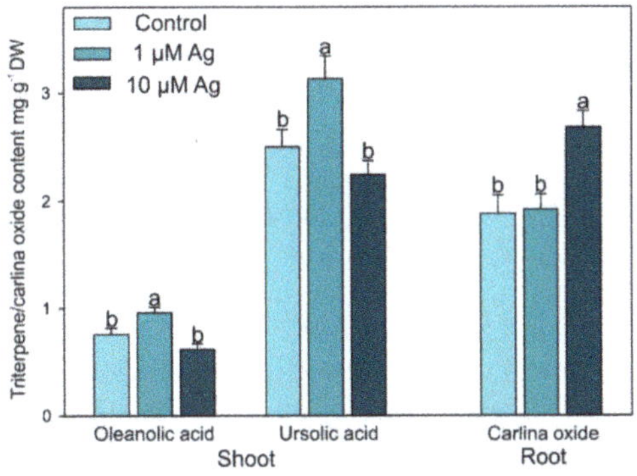

Figure 9. Effect of Ag(I) on the concentration of triterpene acids in the shoots and carlina oxide level in the roots of *C. acalulis* after 14 days of exposure. Data are mean ± SE (n = 5); values for individual organs followed by the same letter are not significantly different ($p < 0.05$, Tukey's test).

3. Discussion

Reduction of growth is one of the most visible toxicity symptoms of HM excess in plants, and the negative impact of toxic concentrations of Cd, Pb, Cu, Zn, Ni, or Mn on plant biomass has been reported repeatedly [1,5]. Since Ag ions are highly toxic, exposure of plants to Ag evokes many negative effects, including reduction of biomass or decreased chlorophyll accumulation [16]. In agreement with these findings, 10 μM Ag(I) depleted the shoot and root biomass and produced visible necrotic symptoms on the leaves. On the contrary, the lower Ag(I) concentration slightly but significantly stimulated root biomass, indicating the hormetic effect of this dose. Similarly, the stimulatory effect of low Ag doses on plant growth has been previously reported [10,17–20]. Yang et al. [17] indicated that the exposure of rice plants to 50 μg L^{-1} Ag increased the root biomass by 52% and the root length by 72%. A beneficial effect of sublethal concentrations of Ag (<0.02 mg L^{-1}) was also observed in *Arabidopsis thaliana* [18]. Also, our preliminary study showed that long-time exposure of *C. acaulis* to 0.1 μM Cd positively influenced the shoot and root biomass (unpublished data). The phytostimulative impact of a low concentration of HMs, the so-called hormetic effect, is probably an adaptive process [17,21]. The impact of Ag on ethylene biosynthesis (antagonist of its biosynthesis) may also play a role in the observed growth changes [22]. Moreover, it cannot be excluded that the hormetic effect was the result of the positive impact of the low Ag(I) concentration on the uptake and contents of some nutrients, for example, Ca in the leaves.

In this study, Ag was accumulated mostly in the roots. This finding is in agreement with previous reports, which showed that the translocation of this element to the aboveground parts of plants is very low in various species [9,18,23]. As reported by Wang et al. [18], Ag ions are quickly taken up by roots from the nutrient solution. They found distribution of Ag in hydroponically cultured *A. thaliana* mainly in roots (68%), followed by 30% in media, and only 2% of Ag accumulated in aboveground parts.

It is well known that exposure of plants to HM stress affects the uptake and accumulation of essential nutrients [1,3,24,25]. This phenomenon is a result of several mechanisms, including competition for transporters or a direct impact of Ag ions on plasma membrane calcium channels [26]. Interestingly, 1 μM Ag evoked an increase in the Ca concentration in the shoots, while 10 μM Ag stimulated accumulation of Ca in the roots. The obtained results indicate a significant impact of Ag(I) ions on the Ca bioconcentration and translocation. Given the role of Ca in HM detoxification [26], the increasing content of Ca under Ag(I) exposure may indicate some role of this element in plant Ag resistance. The higher Ag(I) dose decreased the root K and P accumulation, and similar depletion of P has also been observed in *Spirodela polyrhiza* [16]. In contrast to our data, the exposure of *A. thaliana* to 2.0 mg L^{-1} of Ag(I) (ca. 18.5 μM) did not change the content of some mineral nutrients in the leaves; however, 2.0 mg L^{-1} of Ag nanoparticles significantly reduced the accumulation of K, Fe, and Zn in the leaves by 71%, 50%, and 49%, respectively [19].

Quantitative changes in LMWOAs are frequently observed in various species under HM stress due to their chelating ability. This can lead to reduction of HM phytoavailability (chelation of HMs with exudates) or their intercellular detoxification (chelation of HMs in the cytosol) [2,27,28]. Moreover, it has been pointed out that LMWOAs participate in long-distance translocation of HMs to parts of plants with low metabolic activity such as the cell wall or trichomes [27]. Here, we observed that the low concentration of Ag ions significantly increased the concentrations of malic and citric acids in both the shoots and roots. The phenomenon of elevated accumulation of LMWOAs under HM stress has been observed in higher plants [2]. In the present study, a dose-dependent impact of the Ag(I) treatments on the content of LMWOAs was observed. A similar effect was found in previous research, where a low Cd concentration increased the level of malic and citric acids in the shoots and roots of *C. acaulis*, but increased accumulation of these acids was detected only in the shoots at a high Cd dose (acute stress, data not published). The reduced content of LMWOAs in the roots of plants treated with 10 μM Ag may be a result of increased exudation limiting the metal availability in the medium. On the contrary, the low Ag dose stimulated the accumulation of LMWOAs in the organs, indicating intercellular metal chelation as a detoxification mechanism.

Phytochelatins are essential intercellular ligands of HMs, and their content significantly increased in response to both Ag(I) doses. Ag belongs to a group of metals that induce PC synthesis in plants and yeasts [29]. It was also shown that Ag is well bound by PCs [30]. However, we found low ability of Ag(I) to stimulate PC biosynthesis in the roots and no induction of PC synthesis in the shoots. The accumulation of these thiols in the roots of *C. acaulis* treated with Ag(I) was lower in comparison to their accumulation after Cd exposure (unpublished) or in other plant species stressed by Cd or Pb ([4] and references therein). This fact can be explained by the higher strength of Cd, Pb, or Zn ions in PC induction compared with Ag [31]. Moreover, it was noted that the PC content in *Ricinus communis* collected from an area of silver mines, contrary to Cd or Pb, was not correlated with the Ag(I) content in the plants [32]. Lack of PC induction by Ag ions was also observed earlier in *Scenedesmus vacuolatus* [33]. The level GSH (i.e., a precursor of PCs) increased in the shoots with the increasing

Ag(I) concentration in the nutrient medium. At the same time, accumulation of GSH in the roots was suppressed by the higher Ag dose. In agreement with our data, an Ag dose of 20–60 mg L^{-1} (as $AgNO_3$) strongly elevated the GSH content in wheat callus cultures after 24 h of exposure [34]. GSH together with AsA are part of the nonenzymatic scavenging system [35]. It has been shown that these compounds protect plant cells against HM-induced oxidative damage [36,37]. The elevated level of GSH (but not AsA) in the leaves under the Ag(I) excess may indicate a protective role of this compound in the leaf cells under Ag(I) exposure. In turn, the higher Ag(I) concentration significantly decreased both GSH and AsA accumulation in the roots, which was in agreement with earlier data [38]. This depletion of antioxidant systems may have been evoked by the higher Ag amounts in the root tissue with consequent reduction of growth and vitality.

Since they typically contain one or more free hydroxyl groups, phenolic compounds play a significant role in reactive oxygen species (ROS) scavenging [39]. We found that the Ag(I) stress stimulated the accumulation of both major phenolic acids (chlorogenic and 3,5-dicaffeoylquinic). This fact indicates that an increase in the level of phenolic compounds may participate in Ag resistance mechanisms. The chlorogenic acid level in chamomile tissues (Asteraceae family) has been observed to increase under various doses of different metals, which indicates a more general contribution to the antioxidative potential of plants [5]. Although there is evidence that HM ions can change the synthesis and accumulation of phenolic compounds in plants, the influence of HMs on this process is highly dependent on the level of the stress factor [2,15]. Khan et al. [40] showed that *Pennisetum glaucum* exposed to increasing concentrations of Ag (2, 4, and 6 mM) for 24 h exhibited approximately 2-fold higher total phenolic and flavonoid contents compared with control plants. Similarly, an increasing effect of Ag ions on the accumulation of phenolic compounds was observed in *Bacopa monnieri* [41] and *Hordeum vulgare* [42]. In turn, *Carlina vulgaris* collected from a highly HM-polluted area accumulated lower amounts of phenolic compounds and flavonoids than plants inhabiting nonpolluted areas [15].

The lower dose of Ag(I) enhanced accumulation of both triterpene acids (oleanolic and ursolic), while an elevated level of carlina oxide in the roots was found at the higher dose of Ag(I). The role of triterpenes in protection against HM ions has been considered previously, as these compounds have antioxidant properties and can protect plant cells against ROS damage [43]. Previous studies showed that HM stress elevated the ursolic acid level in *Prunella vulgaris* [44], while Cd stress increased accumulation of oleanolic acid in *Achrynthes bidentate* cell cultures [45].

4. Materials and Methods

4.1. Plant Materials and Growth Conditions

C. acaulis L. achenes collected from the Botanical Garden of Maria Curie-Skłodowska University in Lublin (voucher specimen no. 2005A) were germinated on the surface of garden soil. The young seedlings (10-days old) were transferred into polyethylene pots filled with garden soil. The plants (one per pot) were cultivated over 28 days to achieve an approximately 12 cm diameter of the leaf rosette. Afterwards, the plants were carefully washed with distilled water and transferred into pots filled with 0.5 L of half-strength Hoagland's solution. The plants were acclimated to hydroponic conditions for 5 days and thereafter divided into three groups (25 plants per treatment): (i) control plants cultivated without addition of Ag, (ii) plants cultivated with 1 μM of Ag, and (iii) plants cultivated with 10 μM of Ag. Silver was added in the form of nitrate ($AgNO_3$, Sigma-Aldrich, St. Louis, MO, USA). The plants were grown in controlled conditions in a growth chamber at 18/25 °C (night/day) under light-emitting diodes at a photosynthetic photon flux density of 150 μmol m^{-2} s^{-1} and relative humidity of 60%–65%. The plants were harvested 14 days after the addition of silver, separated into shoots and roots (the roots were washed with deionized water), and weighed to determine fresh biomass. For the determination of LMWOAs, PCs, GSH, and AsA, the samples (biological repeats n = 5) were frozen in liquid nitrogen and stored at −80 °C. Similar aliquots were dried at room temperature for determination of secondary metabolites and at 70 °C to constant weight for measurement of Ag and mineral nutrients (biological repeats n = 4).

4.2. Determination of Silver and Mineral Nutrients

Aliquots of powdered samples were digested in a 5 mL mixture of HNO_3:H_2O (2:8 *v/v*) in a microwave digestion apparatus (TOPwave, Analytick Jena AG, Jena, Germany). The content of elements was measured using an ICP-OES PlasmaQuant PQ 9000 Elite (Analityk Jena AG, Jena, Germany). The effective plasma power was 1300 W, and the argon flow rate of the plasma, auxiliary, and nebulizer were 12, 0.5, and 0.6 L/min, respectively.

4.3. Analysis of LMWOAs, Ascorbic Acid, and Thiols

An Agilent 7100 Capillary Electrophoresis (Agilent Technologies, Santa Clara, CA, USA) was applied for analysis of LMWOAs, AsA, GSH, and PCs according to protocols published previously [37,46,47].

4.4. Quantification of Secondary Metabolites

Triterpenes (ursolic and oleanolic acid), chlorogenic acid, 3,5-dicaffeoylquinic acid, and carlina oxide were analyzed in 100% methanolic extracts using high-performance liquid chromatography (VWR Hitachi Chromaster 600 chromatograph, Merck, Darmstadt, Germany). The triterpenes were analyzed using the RP18e LiChrosper 100 column (Merck, Darmstadt, Germany) (25 cm × 4.9 mm i.d., 5 μm particle size) according to the method reported previously [48]. The C18 reversed-phase column Kinetex (Phenomenex, Torrance, CA, USA) was used to separate phenolic acids as described in the previous work [49]. The carlina oxide content was measured according to the method reported previously [50].

4.5. Visualization of H_2O_2

Hydrogen peroxide (H_2O_2) in the *C. acaulis* leaves was visualized using the 3′,3-diaminobenzidine method [51].

4.6. Statistical Analysis and Experimental Design

The completely randomized design of the experiment involved three treatments (control, 1 μM Ag, and 10 μM Ag) with 25 plants per treatment. The whole experiment was performed two times in the same growth conditions. The data were statistically analyzed using one-way analysis of variance (ANOVA). The differences between the treatments were determined with Tukey's test at the 0.05 probability level. Statistic ver. 13.3 software (TIBCO Software Inc. 2017, Palo Alto, CA, USA) was used to carry out all statistical analysis.

5. Conclusions

Our data confirm that the responses of *C. acaulis* to Ag(I) are dose dependent. Although Ag(I) slightly increased the H_2O_2 concentrations in the leaves, its accumulation in the shoots was low. On the contrary, Ag was mainly retained in the roots, but the low Ag(I) dose even stimulated root growth. Only the higher Ag(I) concentration disturbed the K, P, and Ca balance, while the low Ag(I) stress affected Ca and Cu accumulation. The stimulation of the content of LMWOAs in plants exposed to 1 μM Ag together with the significantly higher content of phenolic and triterpene acids observed in this treatment may indicate both chelation of Ag(I) and induction of antioxidative mechanisms against Ag excess. In contrast, the higher Ag dose generally reduced the content of organic (malic and citric) acids, AsA, or GSH in the roots and, in combination with the relatively high accumulation of Ag in the roots, may be the cause of growth depression.

Author Contributions: S.D., M.S., M.C., and J.K. conceived the experiments and the experimental design; S.D. did project administration; S.D. and I.G. cultivated the plants; S.D., I.G., and A.H. performed LMWOA, GSH, AsA, and PC analysis; M.S. and M.W.-K. analyzed the secondary metabolites; I.S. and A.S.-K. analyzed the elements; S.D. conducted statistical analysis and data visualization; S.D. prepared the original manuscript draft; J.K., and B.H.-N. were involved in the experimental ideas and critically revised the manuscript.

Funding: This research received no external funding.

Acknowledgments: The authors thank Anna Zoń for the English language editing.

Conflicts of Interest: The authors declare no conflict of interest.

References

1. Kováčik, J.; Dresler, S.; Wójciak-Kosior, M.; Hladký, J.; Babula, P. Metabolic changes induced by manganese in chamomile. *Plant Physiol. Biochem.* **2018**, *133*, 127–133. [CrossRef] [PubMed]
2. Dresler, S.; Wójciak-Kosior, M.; Sowa, I.; Stanisławski, G.; Bany, I.; Wójcik, M. Effect of short-term Zn/Pb or long-term multi-metal stress on physiological and morphological parameters of metallicolous and nonmetallicolous *Echium vulgare* L. populations. *Plant Physiol. Bioch.* **2017**, *115*, 380–389. [CrossRef] [PubMed]
3. Dresler, S.; Rutkowska, E.; Bednarek, W.; Stanisławski, G.; Kubrak, T.; Bogucka-Kocka, A.; Wójcik, M. Selected secondary metabolites in *Echium vulgare* L. populations from nonmetalliferous and metalliferous areas. *Phytochemistry* **2017**, *133*, 4–14. [CrossRef] [PubMed]
4. Clemens, S. Molecular mechanisms of plant metal tolerance and homeostasis. *Planta* **2001**, *212*, 475–486. [CrossRef] [PubMed]
5. Kováčik, J.; Klejdus, B. Dynamics of phenolic acids and lignin accumulation in metal-treated *Matricaria chamomilla* roots. *Plant Cell Reports* **2008**, *27*, 605–615. [CrossRef]
6. Sytar, O.; Kumar, A.; Latowski, D.; Kuczynska, P.; Strzałka, K.; Prasad, M.N.V. Heavy metal-induced oxidative damage, defense reactions, and detoxification mechanisms in plants. *Acta Physiol. Plant.* **2013**, *35*, 985–999. [CrossRef]
7. Niggeweg, R.; Michael, A.J.; Martin, C. Engineering plants with increased levels of the antioxidant chlorogenic acid. *Nat. Biotechnol.* **2004**, *22*, 746–754. [CrossRef]
8. Michalak, A. Phenolic compounds and their antioxidant activity in plants growing under heavy metal stress. *Pol. J. Environ. Stud.* **2006**, *15*, 523–530.
9. Ratte, H.T. Bioaccumulation and toxicity of silver compounds: A review. *Environ. Toxicol. Chem.* **1999**, *18*, 89–108. [CrossRef]
10. Blamey, F.P.C.; Kopittke, P.M.; Wehr, J.B.; Kinraide, T.B.; Menzies, N.W. Rhizotoxic effects of silver in cowpea seedlings. *Environ. Toxicol. Chem.* **2010**, *29*, 2072–2078. [CrossRef] [PubMed]
11. Purcell, T.W.; Peters, J.J. Sources of silver in the environment. *Environ. Toxicol. Chem.* **1998**, *17*, 539–546. [CrossRef]
12. Das, D.; Lakshmi, V.; Das, N.; Vimala, R. Studies on toxicity of ag (I) on plants and microbes. *Res. J. Pharm. Biol. Chem. Sci.* **2013**, *4*, 176–188.
13. Wallace, A.; Mueller, R.T. Recovery from acute silver toxicity by bush beans grown in solution culture. *J. Plant Nutr.* **1980**, *2*, 93–95. [CrossRef]
14. Strzemski, M.; Wójciak-Kosior, M.; Sowa, I.; Załuski, D.; Verpoorte, R. Historical and traditional medical applications of *Carlina acaulis* L.—A critical ethnopharmacological review. *J. of Ethnopharmacol.* **2019**, *239*, 111842. [CrossRef]
15. Strzemski, M.; Wójciak-Kosior, M.; Sowa, I.; Załuski, D.; Szwerc, W.; Sawicki, J.; Kocjan, R.; Feldo, M.; Dresler, S. Carlina vulgaris L as a source of phytochemicals with antioxidant activity. *Oxid. Med. Cell. Longev.* **2017**, *2017*, 1891849. Available online: https://www.hindawi.com/journals/omcl/2017/1891849/ (accessed on 15 April 2019). [CrossRef]
16. Jiang, H.-S.; Li, M.; Chang, F.-Y.; Li, W.; Yin, L.-Y. Physiological analysis of silver nanoparticles and $AgNO_3$ toxicity to *Spirodela polyrhiza*. *Environmen. Toxicol. Chem.* **2012**, *31*, 1880–1886. [CrossRef]
17. Yang, Q.; Shan, W.; Hu, L.; Zhao, Y.; Hou, Y.; Yin, Y.; Liang, Y.; Wang, F.; Cai, Y.; Liu, J.; et al. Uptake and transformation of silver nanoparticles and ions by rice plants revealed by dual stable isotope tracing. *Environ. Sci. Technol.* **2019**, *53*, 625–633. [CrossRef]

18. Wang, J.; Koo, Y.; Alexander, A.; Yang, Y.; Westerhof, S.; Zhang, Q.; Schnoor, J.L.; Colvin, V.L.; Braam, J.; Alvarez, P.J.J. Phytostimulation of poplars and Arabidopsis exposed to silver nanoparticles and Ag^+ at sublethal concentrations. *Environ. Sci. Technol.* **2013**, *47*, 5442–5449. [CrossRef]
19. Qian, H.; Peng, X.; Han, X.; Ren, J.; Sun, L.; Fu, Z. Comparison of the toxicity of silver nanoparticles and silver ions on the growth of terrestrial plant model Arabidopsis thaliana. *J. Environ. Sci.* **2013**, *25*, 1947–1956. [CrossRef]
20. Cvjetko, P.; Milošić, A.; Domijan, A.-M.; Vinković Vrček, I.; Tolić, S.; Peharec Štefanić, P.; Letofsky-Papst, I.; Tkalec, M.; Balen, B. Toxicity of silver ions and differently coated silver nanoparticles in *Allium cepa* roots. *Ecotoxicol. Environ. Saf.* **2017**, *137*, 18–28. [CrossRef]
21. Calabrese, E.J. Evidence that hormesis represents an "overcompensation" response to a disruption in homeostasis. *Ecotoxicol. Environ. Saf.* **1999**, *42*, 135–137. [CrossRef] [PubMed]
22. Ma, J.-H.; Yao, J.-L.; Cohen, D.; Morris, B. Ethylene inhibitors enhance in vitro root formation from apple shoot cultures. *Plant Cell Rep.* **1998**, *17*, 211–214. [CrossRef]
23. Hirsch, M.P. Availability of sludge-borne silver to agricultural crops. *Environ. Toxicol. Chem.* **1998**, *17*, 610–616. [CrossRef]
24. Kováčik, J.; Dresler, S. Calcium availability but not its content modulates metal toxicity in *Scenedesmus quadricauda*. *Ecotoxicol. Environ. Saf.* **2018**, *147*, 664–669. [CrossRef] [PubMed]
25. Nedjimi, B. Heavy metal tolerance in two Algerian saltbushes: A review on plant responses to cadmium and role of calcium in its mitigation. In *Plant Nutrients and Abiotic Stress Tolerance*; Hasanuzzaman, M., Fujita, M., Oku, H., Nahar, K., Hawrylak-Nowak, B., Eds.; Springer Singapore: Singapore, 2018; pp. 205–220, ISBN 978-981-10-9044-8.
26. Klíma, P.; Laňková, M.; Vandenbussche, F.; Van Der Straeten, D.; Petrášek, J. Silver ions increase plasma membrane permeability through modulation of intracellular calcium levels in tobacco BY-2 cells. *Plant Cell Rep.* **2018**, *37*, 809–818. [CrossRef] [PubMed]
27. Kutrowska, A.; Szelag, M. Low-molecular weight organic acids and peptides involved in the long-distance transport of trace metals. *Acta Physiol. Plant.* **2014**, *36*, 1957–1968. [CrossRef]
28. Hawrylak-Nowak, B.; Dresler, S.; Matraszek, R. Exogenous malic and acetic acids reduce cadmium phytotoxicity and enhance cadmium accumulation in roots of sunflower plants. *Plant Physiol. Bioch.* **2015**, *94*, 225–234. [CrossRef]
29. Zenk, M.H. Heavy metal detoxification in higher plants—A review. *Gene* **1996**, *179*, 21–30. [CrossRef]
30. Mehra, R.K.; Tran, K.; Scott, G.W.; Mulchandani, P.; Saini, S.S. Ag(I)-binding to phytochelatins. *J. Inorg. Biochem.* **1996**, *61*, 125–142. [CrossRef]
31. Grill, E.; Winnacker, E.-L.; Zenk, M.H. Phytochelatins, a class of heavy-metal-binding peptides from plants, are functionally analogous to metallothioneins. *Proc. Natl. Acad. Sci. USA* **1987**, *84*, 439–443. [CrossRef]
32. Figueroa, J.A.L.; Wrobel, K.; Afton, S.; Caruso, J.A.; Felix Gutierrez Corona, J.; Wrobel, K. Effect of some heavy metals and soil humic substances on the phytochelatin production in wild plants from silver mine areas of Guanajuato, Mexico. *Chemosphere* **2008**, *70*, 2084–2091. [CrossRef] [PubMed]
33. Le Faucheur, S.; Schildknecht, F.; Behra, R.; Sigg, L. Thiols in Scenedesmus vacuolatus upon exposure to metals and metalloids. *Aquat. Toxicol.* **2006**, *80*, 355–361. [CrossRef] [PubMed]
34. Barbasz, A.; Kreczmer, B.; Oćwieja, M. Effects of exposure of callus cells of two wheat varieties to silver nanoparticles and silver salt ($AgNO_3$). *Acta Physiol. Plant.* **2016**, *38*, 1–11. [CrossRef]
35. Jozefczak, M.; Remans, T.; Vangronsveld, J.; Cuypers, A. Glutathione is a key player in metal-induced oxidative stress defenses. *Int. J. Mol. Sci.* **2012**, *13*, 3145–3175. [CrossRef] [PubMed]
36. Kováčik, J.; Klejdus, B.; Babula, P.; Hedbavny, J. Ascorbic acid affects short-term response of Scenedesmus quadricauda to cadmium excess. *Algal Res.* **2017**, *24*, 354–359. [CrossRef]
37. Dresler, S.; Maksymiec, W. Capillary zone electrophoresis for determination of reduced and oxidised ascorbate and glutathione in roots and leaf segments of Zea mays plants exposed to Cd and Cu. *Acta Sci. Pol. Hortorum Cultus* **2013**, *12*, 143–155.
38. Bagherzadeh Homaee, M.; Ehsanpour, A.A. Silver nanoparticles and silver ions: Oxidative stress responses and toxicity in potato (*Solanum tuberosum* L.) grown in vitro. *Hortic. Environ. Biotechnol.* **2016**, *57*, 544–553. [CrossRef]
39. Sakihama, Y.; Cohen, M.F.; Grace, S.C.; Yamasaki, H. Plant phenolic antioxidant and prooxidant activities: Phenolics-induced oxidative damage mediated by metals in plants. *Toxicology* **2002**, *177*, 67–80. [CrossRef]

40. Khan, I.; Raza, M.A.; Bin Khalid, M.H.; Awan, S.A.; Raja, N.I.; Zhang, X.; Min, S.; Wu, B.C.; Hassan, M.J.; Huang, L. Physiological and biochemical responses of pearl millet (*Pennisetum glaucum* L.) seedlings exposed to silver nitrate ($AgNO_3$) and silver nanoparticles (AgNPs). *Int. J. Environ. Res. Public Health* **2019**, *16*, 2261. [CrossRef]
41. Krishnaraj, C.; Jagan, E.G.; Ramachandran, R.; Abirami, S.M.; Mohan, N.; Kalaichelvan, P.T. Effect of biologically synthesized silver nanoparticles on *Bacopa monnieri* (Linn.) Wettst. plant growth metabolism. *Process Biochem.* **2012**, *47*, 651–658. [CrossRef]
42. Fayez, K.A.; El-Deeb, B.A.; Mostafa, N.Y. Toxicity of biosynthetic silver nanoparticles on the growth, cell ultrastructure and physiological activities of barley plant. *Acta Physiol. Plant.* **2017**, *39*, 155. [CrossRef]
43. Do Nascimento, P.G.G.; Lemos, T.L.G.; Bizerra, A.M.C.; Arriaga, Â.M.C.; Ferreira, D.A.; Santiago, G.M.P.; Braz-Filho, R.; Costa, J.G.M. Antibacterial and antioxidant activities of ursolic acid and derivatives. *Molecules* **2014**, *19*, 1317–1327. [CrossRef] [PubMed]
44. Wu, Z.; Guo, Q.; Wang, Q.; Zhou, L.; Zhang, Z.; Zhang, L.; Huang, T. Effects of lead, copper and cadmium stresses on growth and inherent quality of *Prunalla vulgaris*. *Zhongguo Zhongyao Zazhi* **2010**, *35*, 263–267. (In Chinese) [PubMed]
45. Wang, Q.J.; Lei, X.Y.; Zheng, L.P.; Wang, J.W. Molecular characterization of an elicitor-responsive 3-hydroxy-3-methylglutaryl coenzyme A reductase gene involved in oleanolic acid production in cell cultures of *Achyranthes bidentata*. *Plant Growth Regul.* **2017**, *81*, 335–343. [CrossRef]
46. Dresler, S.; Hanaka, A.; Bednarek, W.; Maksymiec, W. Accumulation of low-molecular-weight organic acids in roots and leaf segments of Zea mays plants treated with cadmium and copper. *Acta Physiol. Plant.* **2014**, *36*, 1565–1575. [CrossRef]
47. Dresler, S.; Hawrylak-Nowak, B.; Kováčik, J.; Pochwatka, M.; Hanaka, A.; Strzemski, M.; Sowa, I.; Wójciak-Kosior, M. Allantoin attenuates cadmium-induced toxicity in cucumber plants. *Ecotoxicol. Environ. Saf.* **2019**, *170*, 120–126. [CrossRef]
48. Strzemski, M.; Wójciak-Kosior, M.; Sowa, I.; Rutkowska, E.; Szwerc, W.; Kocjan, R.; Latalski, M. Carlina species as a new source of bioactive pentacyclic triterpenes. *Ind. Crop. Prod.* **2016**, *94*, 498–504. [CrossRef]
49. Sowa, I.; Paduch, R.; Strzemski, M.; Zielińska, S.; Rydzik-Strzemska, E.; Sawicki, J.; Kocjan, R.; Polkowski, J.; Matkowski, A.; Latalski, M.; et al. Proliferative and antioxidant activity of *Symphytum officinale* root extract. *Nat. Prod. Res.* **2018**, *32*, 605–609. [CrossRef]
50. Strzemski, M.; Wójciak-Kosior, M.; Sowa, I.; Kocjan, R.; Tyszczuk-Rotko, K. Methodological approach to determine carlina oxide—A main volatile constituent of *Carlina acaulis* L. essential oil. *Talanta* **2019**, *191*, 504–508. [CrossRef]
51. Hawrylak-Nowak, B.; Dresler, S.; Rubinowska, K.; Matraszek-Gawron, R.; Woch, W.; Hasanuzzaman, M. Selenium biofortification enhances the growth and alters the physiological response of lamb's lettuce grown under high temperature stress. *Plant Physiol. Bioch.* **2018**, *127*, 446–456. [CrossRef]

plants

Article

Influence of PGPB Inoculation on *HSP70* and *HMA3* Gene Expression in Switchgrass under Cadmium Stress

Nahmina Begum *, Zhaoyang Hu, Qingsheng Cai * and Laiqing Lou *

College of Life Sciences, Nanjing Agricultural University, Nanjing 210095, China; 2016216002@njau.edu.cn
* Correspondence: nahmina20007@yahoo.com (N.B.); qscai@njau.edu.cn (Q.C.); loulq@njau.edu.cn (L.L.); Tel.: +86-25-84395187 (Q.C.); Fax: +86-25-84396542 (Q.C.)

Received: 13 October 2019; Accepted: 13 November 2019; Published: 14 November 2019

Abstract: This study aimed to evaluate the gene expression of *HSP70* and *HMA3* in the switchgrass inoculated with plant-growth-promoting-bacteria (PGPB) under cadmium (Cd) stress and to observe the benefit of PGPB in plant growth and development. Plants were grown in hydroponic culture and treated with PGPB inoculants: *Pseudomonas grimontii*, *Pantoea vagans*, *Pseudomonas veronii*, and *Pseudomonas fluorescens* with the strains Bc09, So23, E02, and Oj24, respectively. The experimental results revealed that *HSP70* and *HMA3* genes expressed highly in the PGPB-inoculated plants under Cd stress. In addition, the expression of *HSP70* and *HMA3* genes was considerably higher in the first two days after successive four-day exposure of Cd in plants compared to the last two days of exposure. Increased biomass and indole-3-acetic-acid production with reduced Cd accumulation were observed in the PGPB-inoculated plants under Cd stress compared to the Cd-control plants. These PGPB, with their beneficial mechanisms, protect plants by modifying the gene expression profile that arises during Cd-toxic conditions and increased the healthy biomass of switchgrass. This demonstrates there is a correlation among the growth parameters under Cd stress. The PGPB in this study may help to intensify agriculture by triggering mechanisms to encourage plant growth and development under heavy metal stress.

Keywords: agriculture; Cd stress; environmental; gene expression; PGPB; switchgrass

1. Introduction

Currently, biotic and abiotic stresses frequently limit plant production. Heavy metal pollution is a topic of great interest among all abiotic stresses [1,2]. Heavy metal pollution endangers plants, animals, and humans and it is now an alarming global problem. Heavy metal exposure has been shown to cause changes in plant protein expression [1]. In particular, cadmium (Cd) is a threat to living organisms because it accumulates in plants in contaminated environment and can enter the food chain [3]. Cd disturbs normal plant metabolism and results in poor growth and development in host plants [4]. Plants have developed diverse mechanisms for the alleviation of stress and re-establishment of cellular homeostasis, to include expression of metal transporters for metal uptake and vacuolar transport, chelators for tolerance and detoxification of heavy metals, and chaperones that deliver and traffic metal ions [5,6]. Heat-shock proteins (*HSPs*) are well established chaperones involved in "housekeeping" the cell processes [7]. *HSPs* with molecular weight from 10 to 200 kDa participate in cell signaling under stress conditions [8]. The principal *HSPs* are split into five conserved classes according to their approximate molecular weight: *HSP100*, *HSP90*, *HSP70*, *HSP60*, and small heat shock proteins (*sHSPs* of 15 to 42 kDa in molecular weight) [9]. Exposure to Cd resulted in an increased accumulation of stress-related proteins such as *HSPs*, chaperones, foldases, proteases and pathogenesis-related (PR) proteins in the leaf and root tissues of poplar plants, but root tissues exhibited decreased proteins from the primary metabolism pathways [10]. The root tissues of different grass plants of the Yellowstone National Park exhibited higher s*HSPs* compared to shoot tissues under extreme heat exposure [11]. Under metal toxic conditions, toxic metal ions interfere with the cellular protein homeostasis through the folding process and enhance the production of improperly folded proteins followed by endoplasmic reticulum (ER) stress, resulting in decreased cell viability. Under these circumstances, cells initiate a different quality control system to fine-tune the protein homeostasis [8,12]. The *HSP70* family members have been studied extensively in a wide range of plant species, as they accumulate during the environmental stresses [13], primarily localizing to the cytosol, ER, and mitochondria [8]. Numerous studies have shown a correlation between *HSP70* overexpression and heavy metal tolerance, but the cellular function of *HSP70* under stress has not been completely revealed [14]. Together, the *HSP70* chaperones and co-chaperones prevent the accumulation of nascent proteins as aggregates and confirm the appropriate folding of proteins while transferring the proteins to their destination [15]. One of our research groups has highlighted the inevitable role of the HSP network to normalize the protein function under Cd stress in the switchgrass using deep sequencing [16].

HMA3 is a member of the heavy metal ATPase (HMA) family [3]. *OsHMA3* is a tonoplast-localized transporter for Cd in the roots of rice (*Oryza sativa*) and when it overexpressed, Cd accumulation in grain is reduced [17]. The overexpression of *OsHMA3* causes higher Cd concentration in roots by enhancing the vacuolar sequestration of Cd in the roots. As a consequence, *OsHMA3* can reduce Cd accumulation in grain efficiently and enhance plant tolerance to Cd [3,18]. Notably, overexpression of *OsHMA3* reduced Cd accumulation in grain without affecting Fe and Zn [19].

Plant-growth-promoting-bacteria (PGPB) can enhance plant growth by different direct and/or indirect mechanisms [20]. PGPB of agronomic importance can diminish the detrimental environmental impacts on plants and help to enhance agriculture [21,22]. Switchgrass (*Panicum virgatum* L.) is a model perennial bioenergy crop. It has a high growth rate, good biomass yield, and minimal nutrient requirements, is highly adapted to grow on marginal lands, and has a wide range of habitats [20,23]. This bioenergy crop is largely dependent on natural arbuscular mycorrhizal fungi (AMF), but little is known about the interaction between switchgrass and AMF [24]. In addition, some exotic endophytic PGPB can also colonize switchgrass, for example, *Paraburkholderia phytofirmans* PsJN, *Bacillus*, *Pseudomonas*, *Pantoea*, and *Enterobacter*, etc. [25–27]. Recently, the term induced systemic tolerance (IST) has been proposed to describe PGPB-induced physical and chemical changes that occur in host plants due to the tolerance triggered by PGPB-plants interactions during various abiotic stresses [28]. In our previous report, we explained the details of the four PGPB strains, that is, *Pseudomonas grimontii* strain Bc09, *Pantoea vagans* strain So23, *Pseudomonas veronii* strain E02, and *Pseudomonas fluorescens* strain Oj24. These four strains are highly Cd tolerant, produce indole-3-acetic acid (IAA), have 1-aminocyclopropane-1-carboxylic-acid deaminase (ACCD) activities, can solubilize insoluble phosphate, and can colonize inside switchgrass [27]. These PGPB mechanisms are recognized as the IST that mainly evolved during abiotic stresses as described earlier. In the present study, we were eager to evaluate IAA production by these four PGPB in inoculated-switchgrass under Cd stress. IAA production by many PGPB plays a key role in the growth and development of the plants. IAA can induce tissue differentiation, cell division, and elongation, lateral-root formation, etc. [29]. It is reported that a number of bacteria can produce IAA, for example, *Azospirillum brasilense*, *Enterobacter cloacae* UW5, *Pantoea agglomerans*, and *Pseudomonas putida* [30–34].

Currently, we are unaware of any reports regarding IAA production and *HSP70* and *HMA3* gene expression in switchgrass-microbe interactions under Cd stress. Therefore, this experiment focused on *HSP70* and *HMA3* gene expression as well as IAA production and growth promotion due to the switchgrass-PGPB interaction under Cd stress. Additionally, we observed the correlation between these parameters by performing principal component analysis (PCA).

2. Results

2.1. Biomass of the PGPB-Inoculated Plants with and without Cd Stress

The plant biomass (fresh and dry weight) of the PGPB-inoculated (single strain and Mixed-combination of the four strain) group under both Cd-amended and Cd-nonamended conditions was measured and is presented in Figure 1A–F. Among the treatments both under Cd stress and without stress, E02-inoculated plants showed significantly higher ($p < 0.05$) root fresh weight (0.9 g) and the next highest root fresh weight (0.6 g) was observed in the Oj24-inoculated plants (Figure 1A). Although the PGPB-inoculated plants with Cd-exposure achieved higher root fresh weight ranging from 3% (3 g in So23 + Cd) to 140% (31 g in Mixed + Cd) compared to the noninoculated Cd-control plants but possessed no significant difference among them (Figure 1A). In the case of root dry weight, among all the treatments E02-inoculated plant (0.1 g) achieved a significantly higher ($p < 0.05$) weight, gradually second and third higher group was found as in Mixed + Cd-inoculated plants (0.08 g) and Bc09 + Cd-inoculated plants (0.07 g) (Figure 1D). The noninoculated Cd-control plants showed lower dry weight (0.02 g) among all the treatments. Under Cd stress, the PGPB-inoculated plants obtained a higher root dry weight than the Cd-control plants that ranged from 46% (0.04 g in So23 + Cd) to 217% (0.08 g in Mixed + Cd) (Figure 1D).

Among all the treatments, Oj24-inoculated plants (1.9 g) exhibited significantly higher ($p < 0.05$) shoot fresh weight and noninoculated Cd-control plants (0.5 g) possessed significantly lower ($p < 0.05$) shoot fresh weight (Figure 1B). Under Cd stress, the PGPB-inoculated plants and noninoculated Cd-control plants exhibited reduced shoot fresh weight compared to the plants that grew normally. However, in the Cd-groups, PGPB-inoculated plants exhibited greater shoot fresh weight, which ranged from 22% (0.6 g in So23 + Cd) to 112% (1 g in Oj24 + Cd) compared to the noninoculated Cd-control plants (Figure 1B), but there was no significant difference among those results. In consideration of shoot dry weight, without Cd stress E02-inoculated (0.3 g) and Oj24-inoculated plants (0.27 g) gained significantly higher ($p < 0.05$) shoot dry weight. Under Cd stress, Mixed + Cd-inoculated plants (0.29 g) exhibited significantly higher ($p < 0.05$) shoot dry weight (Figure 1E). However, all the PGPB-inoculated plants under Cd stress gained from 0.1% (0.1 g in So23 + Cd) to 196% (0.3 g in Mixed + Cd) more shoot dry weight compared to the noninoculated Cd-control plants (Figure 1E). Furthermore, in the Cd-group, Bc09-, E02-, Oj24-, and Mixed-inoculated plants were significantly ($p < 0.05$) different (Figure 1E).

The E02-inoculated plants (2.6 g) and Oj24-inoculated plants (2.4 g) were observed to have significantly higher ($p < 0.05$) total fresh weight, but these two treatments were not significantly different from each other (Figure 1C). A significantly lower ($p < 0.05$) total fresh weight was observed in the noninoculated Cd-control plants (0.7 g) (Figure 1C). Under Cd stress, the plants experienced a reduced total fresh weight compared to under normal conditions. The PGPB-inoculated plants under Cd stress exhibited from 17% (0.8 g in So23 + Cd) to 115% (1 g in Mixed + Cd) more total fresh weight compared to the noninoculated Cd-control plants, although there was no significant difference among them (Figure 1C). The E02-inoculated plants (0.4 g) possessed significantly higher ($p < 0.05$) total dry weight among all the treatments. The next significant higher total dry biomass was observed in the Mixed + Cd-inoculated plants (0.37 g) and Oj24-inoculated plants (0.34 g) (Figure 1F). If we consider only the Cd-group, then PGPB-inoculated plants gained from 9% (0.1 g in So23 + Cd) to 200% (0.4 g in Mixed + Cd) more total dry weight ranged compared to the noninoculated Cd-control plants (Figure 1F). Moreover, plants grown in normal conditions differed significantly with each other, except the control and So23-inoculated plants, while in the Cd-group all the treatments differed significantly with each other except the Cd-control and So23 + Cd treatment.

In this study, all the PGPB-inoculated (single and mixed) and noninoculated plants with and without Cd-exposure were grown under the same environmental conditions at the same time. We observed that Cd-exposure imparts a negative pressure on plants, as the PGPB-inoculated plants under Cd stress showed less biomass in some cases compared to the PGPB-inoculated plants without Cd amendment, but the biomass was greater in the PGPB-inoculated plants under Cd stress compared to the noninoculated Cd-control plant (Figure 1C,F). Among all the treatments, E02-inoculated plants grown without Cd-amended conditions gained significantly ($p < 0.05$) more root and total fresh biomass and root, shoot, and total dry biomass (Figure 1A,C–F). In addition, Oj24-inoculated plants grown without Cd-exposure also achieved significantly more shoot fresh and total biomass ($p < 0.05$) (Figure 1B,C). Furthermore, mixed-inoculated plants grown under Cd stress exhibited significantly ($p < 0.05$) more root, shoot, and total dry biomass (Figure 1D–F). On the other hand, noninoculated Cd-controls possessed significantly ($p < 0.05$) less shoot fresh weight and total fresh weight among all the treatments (Figure 1B,C).

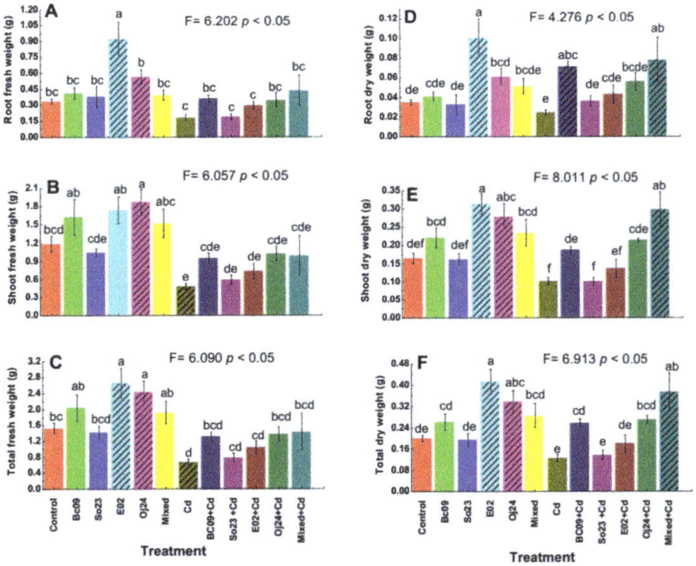

Figure 1. Biomass of switchgrass cv Alamo. (**A**) Root fresh weight, (**B**) shoot fresh weight, (**C**) total fresh weight, (**D**) root dry weight, (**E**) shoot dry weight, and (**F**) total dry weight. Control = noninoculated control plant without Cd stress; Cd = noninoculated Cd-control plant with 20 μ mol L^{-1} $CdCl_2$ stress; Bc09, So23, E02, Oj24, and Mixed (combination of the four strain) = plant inoculated with endophytic PGPB without Cd stress; and Bc09 + Cd, So23 + Cd, E02 + Cd, Oj24 + Cd, and Mixed + Cd (combination of the four strain) = plant inoculated with endophytic PGPB with 20 μ mol L^{-1} $CdCl_2$ stress. The bar represents ±SE (n = 5). Columns not sharing a significance letter are significantly different according to Duncan's Multiple Range Test at $p < 0.05$. The dash pattern indicates the biomass of that treatment is significantly different among the treatments.

2.2. Determination of Cd Concentrations inside Plant Tissues

Cd concentrations in the plant (root, shoot, and whole plant) and the translocation factor (TF) were determined and are presented in Figure 2A–D. The concentrations of Cd in the roots and shoots of the PGPB-inoculated plants were lower compared to those in the noninoculated Cd-control plants, except for the So23-inoculated plants. The So23-inoculated plants exhibited significantly ($p < 0.05$) more Cd content in the roots and shoots (42% and 58%, respectively) compared to the control plants. We observed in the biomass study that the fresh and dry biomass of the So23-inoculated plant was lower than the noninoculated control plant. The increased Cd concentration in the plants might be a cause behind the lower biomass of So23-inoculated plants under Cd stress. Without Cd stress, the biomass shows no difference with control plants. However, Bc09-, E02-, Oj24- and mixed-inoculated plants had 53%, 21%, 54%, and 43% lower Cd contents, respectively, compared to the noninoculated Cd-control plants. The roots contained more Cd than the shoots (Figure 2A–C).

The TF of plants is the measurement of the ability to transfer metals from the roots to shoots [35]. This translocation depends largely on the types of metals and substrates [36]. The Bc09-inoculated plants had significantly ($p < 0.05$) higher TF (13%), whereas the mixed-inoculated plants had significantly ($p < 0.05$) (54%) lower TFs compared to the noninoculated Cd-control plants (Figure 2D). It is known that if the Cd concentration of at least 100 mg Kg^{-1} is accumulated [37] and the TF is greater than 1, then a plant is considered as hyper-accumulator [38]. In our study, the TF was less than 1 in both PGPB-inoculated and noninoculated Cd-control plants. Furthermore, the shoots had a Cd concentration of less than 100 mg Kg^{-1}, while the roots contained more than 100 mg Kg^{-1} Cd. Therefore, this plant could be introduced as a Cd-accumulator rather than a Cd-hyperaccumulator.

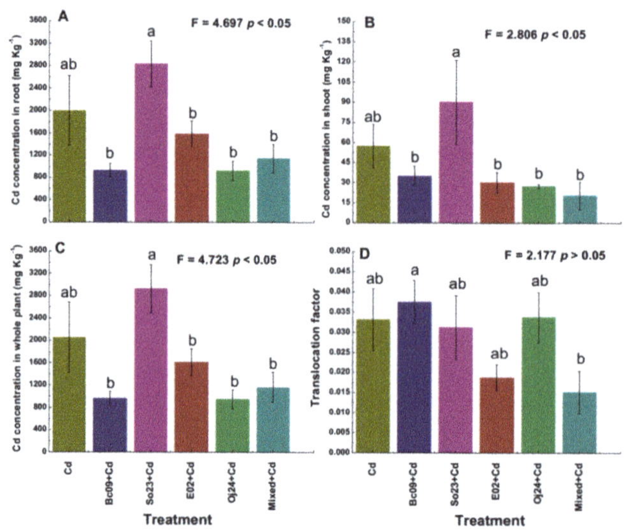

Figure 2. Uptake of Cd in switchgrass cv Alamo under 20 μ mol L^{-1} $CdCl_2$ stress. (**A**) Cd concentration in root, (**B**) Cd concentration in shoot, and (**C**) Cd concentration in whole plant, and (**D**) translocation factor (TF). TF represents metal concentration in shoot/root and denotes metal translocation. Cd = noninoculated Cd-control plant with 20 μ mol L^{-1} $CdCl_2$ stress and Bc09 + Cd, So23 + Cd, E02 + Cd, Oj24 + Cd, and Mixed + Cd (combination of the four strain) = plant inoculated with endophytic PGPB with 20 μ mol L^{-1} $CdCl_2$ stress. The bar represents ± SE (n = 5). Columns not sharing a significance letter are significantly different according to Duncan's Multiple Range Test at $p < 0.05$.

2.3. *Concentrations of IAA in Plants Determined through HPLC*

The concentration of IAA was determined through high-performance liquid chromatography (HPLC), where the concentration of IAA was detected in the roots but not in the shoots in our triplicate experiments. The IAA concentration was higher in the PGPB-inoculated plants without Cd-amendment, with a difference ranging from 3.6% to 82.2%, except for the Bc09-inoculated plants. Under Cd-amended conditions, the noninoculated Cd-control and E02-inoculated plants contained lower IAA concentrations compared to the noninoculated control plants. The Bc09-, So23-, Oj24-, and mixed-inoculated plants contained 69.4%, 57.8%, 22%, and 29.8% higher IAA concentrations compared to those of the control plants, respectively (Figure 3). Among all the treatments, the E02-inoculated plants grown without Cd-exposure and Bc09-inoculated plants grown with Cd-exposure contained significantly ($p < 0.05$) higher IAA concentrations than the noninoculated control plants. Other treatments showed no significant differences among them.

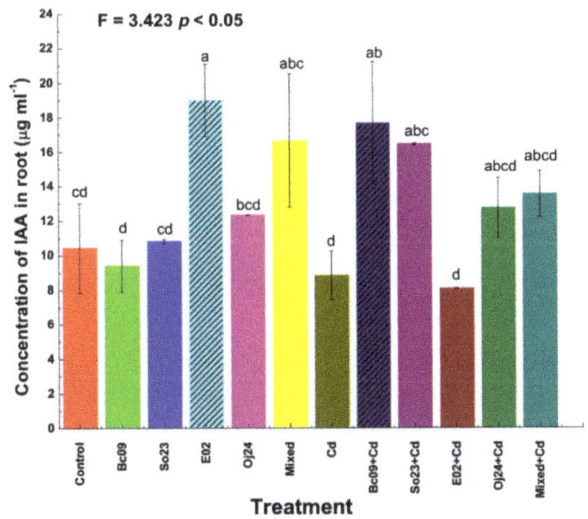

Figure 3. The IAA concentration in the root of switchgrass cv Alamo. Control = noninoculated control plant without Cd stress; Cd = noninoculated Cd-control plant with 20 μ mol L^{-1} $CdCl_2$ stress; Bc09, So23, E02, Oj24, and Mixed (combination of the four strain) = plant inoculated with endophytic PGPB without Cd stress; and Bc09 + Cd, So23 + Cd, E02 + Cd, Oj24 + Cd, and Mixed + Cd (combination of the four strain) = plant inoculated with endophytic PGPB with 20 μ mol L^{-1} $CdCl_2$ stress. The bar represents ±SE (n = 5). Columns not sharing a significance letter are significantly different according to Duncan's Multiple Range Test at $p < 0.05$. The dash pattern indicates the biomass of that treatment is significantly different among the treatments.

2.4. Expression of the HSP70 Gene in PGPB-Inoculated and Noninoculated Switchgrass

The relative gene expression level of three genes of the *HSP70* gene family, that is, Pavir.5KG619900.1 (*HSP70A*), Pavir.9KG488300.1 (*HSP70B*), and Pavir.5KG300400.1 (*HSP70C*), were analyzed by the qRT-PCR method and found to express in all of the plants after four consecutive days with and without Cd stress (Figure 4A–C). It should be noted that both with and without Cd stress, *HSP70* gene expression in the PGPB-inoculated plants was upregulated compared with that in the control plant. On the first day of Cd exposure, the *HSP70A* gene expression level was 10.9–14.5 fold higher in the noninoculated Cd-control and PGPB-inoculated plants compared to the noninoculated control plants. This gradually decreased after the second, third, and fourth day of exposure. In the second day, the gene expression level was 7.4–13 fold higher compared to the noninoculated control plants, which became 6.8–12.3 fold higher on the third day for the same treatment groups. Finally, on the fourth day the gene expression level was 2.7–5.2 fold higher compared to the noninoculated control plants (Figure 4A). Plants inoculated with So23 (on the first to third day), Bc09 (on the second day), E02 (on the first day), Oj24 (on the fourth day), and mixed (on the first day) had significantly ($p < 0.05$) higher levels of gene expression compared to the Cd-control.

In the case of *HSP70*B gene, on the first day of Cd exposure, the expression level was 9.3–15.7 fold higher in noninoculated Cd-control, PGPB-inoculated plants compared to the noninoculated control plants. The gene expression level was also observed to gradually decreased after second, third and fourth day of exposure to Cd. The expression was 8.4–13.8 fold higher; 6.6–12.6 fold higher; and 3.4–10.5 fold higher in noninoculated Cd-control and PGPB-inoculated plants compared to the noninoculated control plants (Figure 4B). The *HSP70*B gene expressed significantly more ($p < 0.05$) in the So23-inoculated plants on all four days. In addition, the gene also expressed significantly more ($p < 0.05$) in the E02-inoculated plant on the first three consecutive days, compared to the Cd-control.

In the case of the *HSP70*C gene, the expression level was 10–16.2 fold higher, 9.8–13.2 fold higher, 8–12.5 fold higher, and 5.2–11.7 fold higher in the noninoculated Cd-control and PGPB-inoculated plants after one day, two days, three days, and four days of Cd-exposure, respectively (Figure 4C), where a gradual decrease in gene expression was also observed in the last two days of Cd exposure compared to the first two days of exposure. The *HSP70*C gene expressed significantly more ($p < 0.05$) in the So23-inoculated plant on all the four consecutive days, in the Bc09-inoculated plant on the second and third day, in E02-inoculated plant on the first day, and in the mixed-inoculated plants on the fourth day.

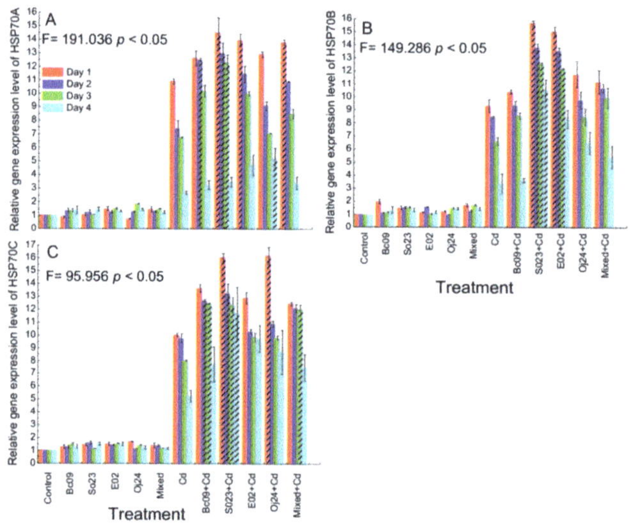

Figure 4. Expression of *HSP70* gene in the switchgrass cv Alamo. Relative gene expression level of (**A**) *HSP70*A = Pavir.5KG619900.1; (**B**) *HSP70*B = Pavir.9KG488300.1; (**C**) *HSP70*C = Pavir.5KG300400.1. Control = noninoculated control plant without Cd stress; Cd = noninoculated Cd-control plant with 20 μ mol L^{-1} $CdCl_2$ stress; Bc09, So23, E02, Oj24, and Mixed (combination of the four strain) = plant inoculated with endophytic PGPB without Cd stress; and Bc09 + Cd, So23 + Cd, E02 + Cd, Oj24 + Cd, and Mixed + Cd (combination of the four strain) = plant inoculated with endophytic PGPB with 20 μ mol L^{-1} $CdCl_2$ stress. The dash pattern indicates the biomass of that treatment is significantly different among the treatments. The expression of *HSP70* genes was determined by quantitative real-time PCR. FTSH protease 4 (FTSH4) was used as internal control. Each column represents the relative gene expression levels (mean value ± SD, with three biological repeats each with 3 technical repeats) calculated using $2^{-\Delta\Delta CT}$ method.

2.5. Expression of HMA3 Gene in PGPB-Inoculated and Noninoculated Switchgrass

The relative gene expression level of three genes of the *HMA3*, that is, AP13CTG05330TIGR01512 (*HMA3A*), AP13CTG10982TIGR01512 (*HMA3B*), and Pavir.Ba03387.1 (*HMA3C*), was analyzed by the qRT-PCR method. Gene expressions were evaluated during four consecutive days after Cd exposure and were found to express in the PGPB-inoculated plants under both Cd-amended and Cd-nonamended conditions (Figure 5A–C).

In the case of the *HMA3*A gene, on the first, second, third and fourth day of exposure to Cd stress, the expression level was 9.6–14.6 fold higher, 9.3–14.3 fold higher, 8.8–12.8 fold higher, and 8–11.9 fold higher in the noninoculated Cd-control and PGPB-inoculated plants compared to the noninoculated-control plants (Figure 5A). The gene expressed significantly more in the So23-inoculated plant on all four consecutive days. Moreover, plants inoculated with Bc09 and E02 strain showed significant ($p < 0.05$) higher *HMA3*A gene expression on the first two days compared to the noninoculated Cd-control plant.

In case of *HMA3*B gene, the expression level was 11.5–17.8 fold higher, 10.6–17.1 fold higher, 8–14.3 fold higher, and 7–13.8 fold higher in noninoculated Cd-control and PGPB-inoculated plants compared to the noninoculated control plants on the first day, second day, third day, and fourth day of exposure to Cd stress (Figure 5B). After Cd exposure on all four consecutive days, the So23-inoculated plant showed significantly ($p < 0.05$) higher gene expression among all the treatments. Furthermore, E02-inoculated plant showed significantly ($p < 0.05$) higher gene expression on the first two days, whereas the Bc09-inoculated plants showed significantly ($p < 0.05$) higher gene expression on the first days among the four days of Cd-exposure.

In the case of the *HMA3*C gene, the expression level was 10.4–17.5 fold, 8.8–16.1 fold, 7.6–13.7 fold, and 5.2–13.2 fold higher in the noninoculated Cd-control and PGPB-inoculated plants after one day, two days, three days, and four days of Cd exposure (Figure 5C). The So23-inoculated plant showed significantly ($p < 0.05$) higher gene expression among all the treatments on all four consecutive days. Furthermore, Bc09-inoculated plant showed significantly ($p < 0.05$) higher gene expression on the first two days and the E02-inoculated plants showed significantly ($p < 0.05$) higher gene expression on the third day.

It should be noted that the gene expression level in the PGPB-inoculated plant was higher on the first day of Cd exposure and gradually decreased on the second, third and fourth day of exposure in case of all three *HMA3* genes.

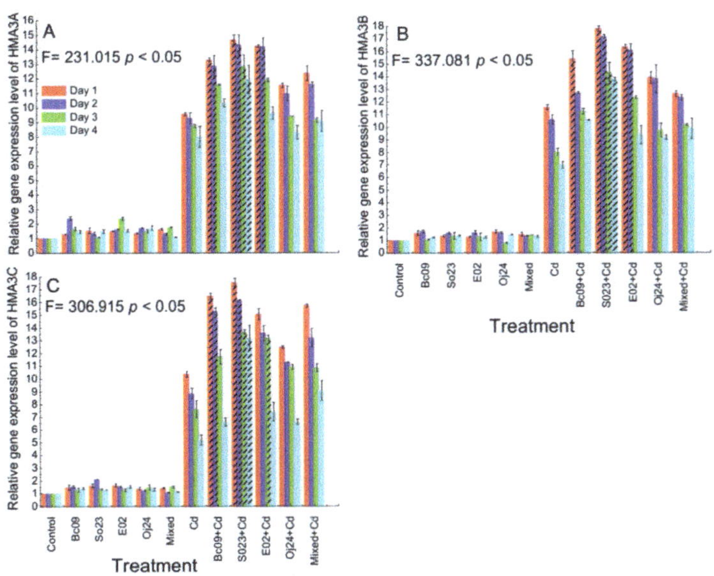

Figure 5. Expression of *HMA3* genes in the roots and shoots of the switchgrass cv Alamo. Relative gene expression level of (**A**) *HMA3*A = AP13CTG05330TIGR01512; (**B**) *HMA3*B = AP13CTG10982TIGR01512; (**C**) *HMA3*C = Pavir.Ba03387.1. Control = noninoculated control plant without Cd stress; Cd = noninoculated Cd-control plant with 20 μ mol L^{-1} $CdCl_2$ stress; Bc09, So23, E02, Oj24, and Mixed (combination of the four strain) = plant inoculated with endophytic PGPB without Cd stress; and Bc09 + Cd, So23 + Cd, E02 + Cd, Oj24 + Cd, and Mixed + Cd (combination of the four strain) = plant inoculated with endophytic PGPB with 20 μ mol L^{-1} $CdCl_2$ stress. The dash pattern indicates the biomass of that treatment is significantly different among the treatments. The expression of *HMA3* genes was determined by quantitative real-time PCR. FTSH protease 4 (FTSH4) was used as internal control. Each column represents the relative gene expression levels (mean value ± SD, with three biological repeats each with 3 technical repeats) calculated using $2^{-\Delta\Delta CT}$ method.

2.6. PCA to Evaluate the Correlation between Plant Growth Parameters, Cd Concentrations, TF, IAA Concentrations, and HSP70 and HMA3 Gene Expression Level

There is more than one response variable in most research studies and variables need to be systematically analyzed to take advantage of the information about the relationships between them. PCA is used to simplify a number of related variables simultaneously. In the combined data set, PCA provided two principal factors, F1 (X-axis) and F2 (Y-axis), where the eigenvalue was > 1 and able to explain approximately 87.72% of the variability of the total data (PC 1 variance of 61.18% and PC 2 variance of 26.54%) (Figure 6). Figure 6 represents a biplot analysis of data into PCs where the concentration of Cd in the roots (RCd), the concentration of Cd in the shoots (SCd), TF, expression level of *HSP70* gene and *HMA3* in the plants corresponded to PC 1 and root dry weight (RDW), shoot dry weight (SDW), and concentration of IAA in the roots (IAA-R) corresponded to PC 2. The correlation matrix among these parameters is shown in Table 1. From Figure 6 and Table 1, it was observed that RDW and SDW were significantly positively correlated. IAA-R and RDW were also significantly positively correlated. SDW showed a significant negative correlation with RCd and SCd. However, RCd and SCd were significantly positively correlated. In addition, TF was also significantly positively correlated with RCd and SCd. Again, TF was significantly positively correlated with *HSP70* and *HMA3*. However, *HSP70* and *HMA3* exhibited highly significant positive correlations between them. Furthermore, *HSP70* and *HMA3* possessed a significant positive correlation with RCd and SCd. Guo et al. [39] also reported a significant positive correlation between *FIHMA3* (a gene homologous to *OsHMA3* in rice) gene expression and Cd accumulation in the roots of *Festulolium loliaceum* (Huds.), which might be responsible for reducing Cd toxicity through vacuolar sequestration of Cd into roots. From the PCA biplot (Figure 6), it was confirmed that the PGPB-inoculated plants and the noninoculated Cd-control plants under Cd stress were far from the same group of plants without Cd stress. These findings provide evidence that both *HSP70* and *HMA3* gene expression in the PGPB-inoculated plants, together with the beneficial activities of PGPB, such as the production of IAA, ACCD activities, and phosphate solubilization, etc. [27], under Cd stress played an important role in limiting the entry of or detoxifying the Cd in the cytoplasm to reduce the Cd toxicity.

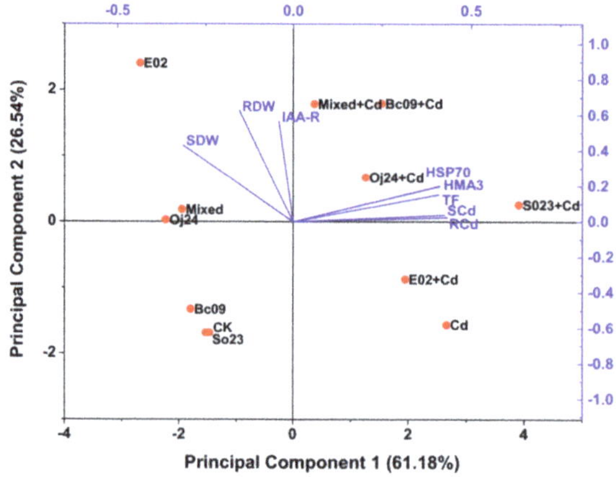

Figure 6. Principal Component Analysis (PCA) biplot based on correlation matrix. CK = noninoculated control with no Cd stress; Bc09, So23, E02, Oj24 and Mixed = PGPB (Bc09, So23, E02, Oj24 and combination of four)-inoculated Alamo with no Cd stress. Cd = noninoculated Cd-control under 20 μ mol L^{-1} $CdCl_2$ stress; Bc09 + Cd, So23 + Cd, E02 + Cd, Oj24 + Cd and Mixed + Cd = PGPB (Bc09, So23, E02, Oj24 and combination of four)-inoculated Alamo under 20 μ mol L^{-1} $CdCl_2$ stress. RDW = root dry weight, SDW = shoot dry weight, RCd = Cd concentration in root, SCd = Cd concentration in shoot, TF = translocation factor, IAA-R = Concentration of IAA in root, HSP70 = HSP70 gene expression level, HMA3 = HMA3 gene expression level.

Table 1. Correlation matrix among growth parameters, concentrations of Cd, TF, concentrations of IAA, *HSP70*, and *HMA3* gene expression in switchgrass. RDW = root dry weight, SDW = shoot dry weight, RCd = Cd concentration in root, SCd = Cd concentration in shoot, TF = translocation factor, IAA-R = Concentration of IAA in root, *HSP70* = *HSP70* gene expression level, *HMA3* = *HMA3* gene expression level. The correlation matrix was constructed using Pearson bivariate correlation analysis.

Parameter	RDW	SDW	RCd	SCd	TF	IAA-R	*HSP70*	*HMA3*
RDW	1							
SDW	0.830 **	1						
RCd	−0.293	−0.627 *	1					
SCd	−0.304	−0.655 *	0.970 **	1				
TF	−0.109	−0.495	0.797 **	0.821 **	1			
IAA-R	0.685 *	0.417	−0.039	0.083	0.083	1		
HSP70	−0.013	−0.379	0.857 **	0.788 **	0.888 **	0.02	1	
HMA3	−0.017	−0.404	0.858 **	0.796 **	0.883 **	0.043	0.993 **	1

* Correlation is significant at the 0.05 level (2-tailed). ** Correlation is significant at the 0.01 level (2-tailed).

3. Discussion

Heavy metals, for example, Cd, impose an imbalance in the cellular homeostasis by silencing necessary enzymes and hindering the functions of proteins [8]. In order to balance the homeostasis, plants respond to abiotic stressors by activating the stress response genes. The *HSP*s gene is one of the well-studied genes that is activated during intracellular stress conditions and maintained cellular homeostasis [16,40–42]. This finding has a good agreement with our current study, where the *HSP70* gene was expressed both under Cd stress and without Cd stress in our experimental results. The reason behind this interesting phenomenon might be in response to the *HSP*s gene in various biological systems, such as stabilization, appropriate folding, and assembly of the proteins under both favorable and unfavorable environmental conditions [6,43]. Notably, the *HSP70* gene was highly expressed under Cd stress, especially in the PGPB-inoculated plants, compared to the noninoculated plants. In addition, on the first day and second day, *HSP70* gene expression increased considerably after Cd stress in PGPB-inoculated and noninoculated Cd-control plants, and then its expression gradually reduced on the third and fourth days relative to the first two days. The reason may be the plant responded rapidly after exposure to Cd and, thus, may have increased its gene expression. Wang et al. [44] investigated *HSP70* gene expression in creeping bentgrass under prolonged heat stress and observed the presence of *HSP70* in all treatments before and after heat stress, although the expression was higher after stress. Recently, one of our research groups discovered that 22 *HSP* genes were highly expressed under Cd stress, and in *Arabidopsis*, one *HSP*-encoding gene was overexpressed and provided plants with Cd tolerance. These findings support the evidence that the *HSP* network re-establishes normal protein function and cellular homeostasis in switchgrass under Cd stress [16]. Several articles have described the adverse effects of high or low levels of Cd toxicity on plants by changing the protein profiles [45–47]. Under Cd stress, *HSP70* was differentially expressed in tomato roots [48] and young poplar leaves [49], with significantly increased expression in germinating rice seedlings [50] and *Arabidopsis thaliana* [51]. Interestingly, in our study, the gene expression level was significantly higher in the PGPB-inoculated plants under Cd stress. Therefore, our finding demonstrated evidence that PGPB might have a protective capability to combat against Cd stress. However, few reports exist in the case of *HSP70* gene expression in plant–microbe interactions, especially endophytic PGPB interactions, under Cd toxic conditions, whereas several reports have described that the overexpression of *HSP70* has a positive correlation with the acquisition of tolerance under heavy metal stress [8]. To our knowledge, this report is the first attempt to discover *HSP70* gene expression in the endophytic PGPB-inoculated switchgrass under Cd stress.

PGPB can enhance plant tolerance under abiotic stresses through a mechanism known as induced systemic tolerance (IST) [41,52], in which PGPB have several notable activities, including the synthesis of phytohormones, such as IAA [53], ACCD activities that allow the degradation of ethylene precursor to relieve plant stress, solubilization of insoluble phosphate and *HSP* gene expression [54]. In our study, we also observed similar results, as the production of IAA and *HSP70* gene expression increased in the plants under Cd stress compared to the control with no stress and, notably, the PGPB-inoculated plants under Cd stress had better results than the noninoculated Cd-control plants.

There is a substantial difference between switchgrass cv Alamo and rice. Alamo has genome sizes of 1600 Mb, is tetraploid, is an inbreeder and is a perennial. The genome size of rice is 373 Mb, and rice is a diploid, has an outcrossing breeding system and is an annual. Although genetic organization and habit are dissimilar, they possess significant synteny and collinearity [55,56]. In this study, we evaluated the expression level of the three rice homologous gene *HMA3* in switchgrass cv Alamo. To the best of our knowledge, this is the first time the expression of *HMA3* gene expression in switchgrass has been examined. We observed an upregulated gene expression level under Cd stress, which was similar to the *HMA3* gene expression results in rice [18]. It is important to mention that this gene was expressed both under Cd-amended and Cd-nonamended conditions in the switchgrass, while under Cd stress the expression was enhanced, especially in the PGPB-inoculated switchgrass. In addition, in the Cd-control and PGPB-inoculated plants, an enhanced level of *HMA3* gene expression was also noted on the first two days of Cd-exposure compared to the last two days of Cd-exposure, similar to *HSP70* genes. The mineral analysis of the Cd group showed that the Cd concentration were significantly lower in PGPB-inoculated Alamo compared to the noninoculated Cd-control plants. However, the exception was that the So23-inoculated plant showed significantly higher *HSP70* and *HMA3* gene expression but displayed higher Cd concentrations both in roots and shoots compared to the noninoculated Cd-control plants with less biomass achievement. In our previous study, we reported that So23 required a longer time than the other three bacterial strains to increase its growth rate under highly Cd-toxic conditions. Furthermore, uptake of Cd from high Cd containing sources was higher in the So23 strain [27]. Thus, we conclude that this characteristic of this bacterium might be a reason for the higher Cd content in plants. The *OsHMA2* promoter could control *OsHMA3* gene expression and, thus, reduce Cd accumulation in rice grains by sequestering more Cd into the vacuoles of different tissues after uptake by the root cells [3,17]. As a result of the enhanced sequestration of Cd in the vacuoles in the root cells, the Cd concentration decreased in the shoots. If the tonoplast localized transporter for Cd *OsHMA3* became functionless, then it would cause a reduced accumulation of Cd in the roots but a higher accumulation in the shoots [19]. However, functional *OsHMA3* induced vacuolar sequestration of Cd in the roots and accumulated more Cd, resulting in enhanced tolerances to Cd. In natural conditions, without stress this gene is usually expressed at a lower level [18]. In the present study, the PGPB-inoculated switchgrass gained more biomass and had reduced Cd concentrations compared to the noninoculated-Cd control. One of the underlying mechanisms behind this might be the expression of the *HMA3* gene. Extensive research is needed on this topic. Similar results were observed in the case of *AtHMA3* in *Arabidopsism*, a homolog of *OsHMA3*, which showed transport activity for Pb and Cd when expressed in a yeast mutant [57]. The underlying mechanism behind this transport-substrate affinity in switchgrass species needs to be investigated in the future. Previously, we reported bacterial plant growth-promoting activities, such as phytohormone IAA production, ACCD activities, phosphate solubilization, Cd tolerances and toxicity tests [27]. Currently, we were interested in observing the IAA production by PGPB-inoculated and noninoculated plants with and without Cd-amended conditions. Interestingly, we detected IAA in roots but not in the shoots. The IAA production was higher in the PGPB-inoculated plants than in the Cd-control plants under Cd stress.

In the PCA biplot and correlation matrix, we observed that the growth, Cd concentration in plant tissues, TF, IAA production, *HSP70,* and *HMA3* gene expression were correlated with each other. In addition, we observed differences in gene expression patterns between plants that were grown normally and under Cd stress; interestingly, the expression was higher in the PGPB-inoculated plants under Cd stress. From previous reports, it is known that a higher TF indicates increased Cd transportation in the shoots and roots. However, a lower TF is associated with reduced Cd accumulation in the shoots and roots [38]. This was reflected in the correlation analysis, where a significant positive correlation between TF, RCd, and SCd was observed. The gene expression of *HSP70* and *HMA3* in the plants were also positively correlated with TF. Finally, the results of this study demonstrated enhanced plant growth and biomass under Cd stress conditions, especially in the PGPB-inoculated plants compared to the noninoculated plants in this experiment. The PCA biplot also revealed that the PGPB-inoculated plants and noninoculated-control plants under Cd stress were far from the same group of plants without Cd stress. Based on their gene expression levels, *HSP70* and *HMA3* were more highly expressed in the single PGPB-inoculated plants than in the co-inoculated (mixed) plants under Cd stress (Figures 4 and 5). The same and/or different PGPB in independently and/or in groups can trigger beneficial mechanisms differently under the same environmental situation and gene expression level, and proteomic profiles could also be modified in different manners [21]. However, the plant-PGPB interaction is dependent on plant genotype and bacterial strain [58]. Regardless of the mechanisms used, PGPB helps plants to enhance growth and yields; therefore, we obtained higher biomass in all the PGPB-inoculated plants compared to the Cd control plants under Cd stress.

This study concludes that PGPB has a profound influence on the expression of *HSP70* and *HMA3* gene in switchgrass under Cd stress. They are environmentally friendly agents to help plants during Cd toxic conditions and could assist in increase biomass production of switchgrass. Moreover, this will increase the success of unfavorable land use and intensify agricultural production.

4. Materials and Methods

4.1. Plant Cultivation, Harvest, Measurement of Biomass and Determination Cd Concentrations

Plant cultivation and harvest and the determination of Cd concentrations in plant tissues and TFs were carried out according to Begum et al. [27]. The PGPB strains were Bc09, So23, E02, Oj24, and a mixed culture of these four strains. The Cd stress that with the application of 20 μ mol L^{-1} $CdCl_2$ for 4 days. The plants were grown in hydroponic culture in a greenhouse for 30 days from germination to harvest. There were two groups of plants: one group was under Cd stress conditions, and another group was grown without Cd stress. Each plant treatment had 6 replicate pots with 7 plants in each pot. A sampling of plants for the assessment of gene expression analysis was performed on the first day (instantly after exposure to Cd), the second day, the third day, and fourth day of exposure to Cd.

The concentrations of Cd and TFs inside plant tissues were calculated as follows [27]:

$$\text{Cd concentration in plant tissues} = \frac{\text{Cd root or shoot } \times \text{Volume}}{\text{Dry weight root or shoot}} \tag{1}$$

where "Cd root or shoot" is the inductively coupled plasma optical emission spectrometer (ICP-OES, Optima 2100 DV, Perkin Elmer, Gaithersburg, MD, USA) machine reading of the Cd concentration in the shoots or roots in mg L^{-1} and "Dry weight root or shoot" is the dry weight of roots or shoots in kg;

$$\text{TF} = \frac{\text{Cd shoot}}{\text{Cd root}} \tag{2}$$

where "Cd shoot" is the concentration of Cd in the shoots (mg kg $^{-1}$) and "Cd root" is the concentration of Cd in the roots (mg kg^{-1}).

4.2. Determination of IAA through HPLC

IAA was determined in the plant samples through HPLC (Agilent Technologies, Waldbronn, Karlsruhe, Germany) performed following the method described by Cui et al. [59]. One gram of root and shoot samples were measured and immediately frozen in liquid nitrogen for rapid cooling to stop the metabolic activities and then stored at –80 °C. Then, the samples were ground in 5 mL of 50% precooled methanol. The ground samples were then incubated at 4 °C for 12 hr followed by centrifugation (Dynamic Dynamica Scientific Ltd., Suzhou, China) at 10,000 *g* for 10 min at 4 °C. The supernatant was transferred to a fresh tube and stored at 4 °C. To the sample, 3 mL of 50% precooled methanol was added and incubated at 4 °C for 12 hr. After centrifugation at 10,000 *g* for 10 min at 4 °C, the supernatant was collected and stored with the previously collected supernatant. To absorb the phenolic substances and pigments 0.2 g polyvinyl-poly pyrrolidone (PVPP) was added to the extract and the extract was incubated in a shaker incubator (Crystal IS-RDV1, Dallas, USA) at 4 °C for 60 min. Again the samples were centrifuged at 10,000 *g* for 10 min at 4 °C. The supernatant was collected and then passed through a C18 column (Sep-Pak Cartridges, Waters Corporation, Milford, MA, USA). The samples were then freeze-dried with a continuous supply of N_2 gas. Then, the lyophilized samples were dissolved by adding 2.5 mL 50% precooled methanol. The dissolved samples were then filtered through a 0.22 μm ultrafiltration membrane. An aliquot was then injected into the high-performance liquid chromatography (HPLC) instrument. The chromatographic conditions for the Agilent 1290 Infinity system (Agilent Technologies, Waldbronn, Karlsruhe, Germany) were as follows: The mobile phase was 0.6% acetic acid (volume fraction) and chromatographic grade methanol gradient elution; the column temperature was 35 °C; the injection volume was 2 μL; the flow rate was 0.3 mL min^{-1}; the detection wavelength was 254 nm.

4.3. Gene Expression of HSP70 and HMA3 in PGPB-Inoculated and Noninoculated Switchgrass

4.3.1. Total RNA Extraction

Plant tissue (1 g fresh seedlings) was ground into a fine powder using liquid nitrogen. Total RNA was isolated and purified separately according to the manufacturer's protocol using the Spin Column Plant Total RNA Purification Kit (Sangon Biotech, Co. Ltd. Shanghai, China). Extracted RNA was quantified using a Nanodrop ND-2000 spectrophotometer (Nanodrop Technologies, Inc., Wilmington, DE, USA). The 260/280 nm ratio of samples ranged from 1.9 to 2.1, and the average RNA integrity number (RIN) was over 8. These results indicated that the extracted RNA was of high quality without any apparent degradation and was suitable for further cDNA synthesis.

4.3.2. Reverse Transcriptional Polymerase Reaction (RT-PCR)

A HiScript^{TM} Q RT SuperMix for qPCR (Vazyme, Nanjing, China) was used to synthesize the cDNAs. The synthesized cDNA was diluted 20 times with distilled water for use in qRT-PCR.

4.3.3. Quantitative Real-time PCR (qRT-PCR)

The qRT-PCR was applied to analyze the relative expression level of a candidate gene in diluted cDNA samples. The primer sequences for the *HSP70*, *HMA3*, and the internal control FTSH protease 4 (FTSH4) genes are given in Table 2. All reactions were conducted in three biological replicates, and the reactions without template were used as negative controls. A Roche-480 system (Roche, Rotkreuz, Switzerland) was used to perform qRT-PCR using a final volume of 20 μL reaction mixture. The qRT-PCR program began with a denaturation step first (95 °C, 1 min), and then continued with 40 amplification cycles, which were programmed as 95 °C for 5 sec, 57 °C for 30 sec, and 72 °C for 30 sec. The CT values for the target and standard control genes were retrieved from the qRT-PCR system, and the comparative threshold $2^{-\Delta\Delta CT}$ method was applied to quantify the relative expression of the given gene [60,61].

Table 2. The primer sequences.

Gene Name	Gene ID	Primer Sequences (5′ to 3′)
HSP70	Pavir.5KG619900.1 (*HSP70A*)	GAGCTGTGCAAGAGCATCAA TTCTTGGTTGGGATGGTGGT
	Pavir.9KG488300.1 (*HSP70B*)	ATCGACTTCTACGCGACCAT CTGCGACTTGTCCATCTTGG
	Pavir.5KG300400.1 (*HSP70C*)	AAGATCACCATCACCAGCGA CGTACGTCTCCAGCTTGTTG
HMA3	AP13CTG05330TIGR01512 (*HMA3A*)	GTGACCAAGTCATGGGAGGA GTGCAACAGCCAAAGAAAGC
	AP13CTG10982TIGR01512 (*HMA3B*)	GGAAGACTGCACGAACCATC CACAGCCCTTGTTGCTAGTC
	Pavir.Ba03387.1 (*HMA3C*)	GTTCTGGGAGCACAGGACAT AGTTCCCGTCTTGTCGAATG
FTSH4 (Internal control)		TGGATGGCTTTAAGCAGAATGA CAAAACGCCCAGGTCTGACT

4.4. PCA to Evaluate the Correlation between Plant Growth Parameters, Cd Concentrations, TFs, IAA Concentration, and HSP70 and HMA3 Gene Expression Levels

A PCA was performed to determine the correlation between plant growth parameters, Cd concentrations, TFs, IAA concentration, and *HSP70* and *HMA3* gene expression levels. The original variables in a linear combination are called the principal component to illustrate the variation in a particular orthogonal dimension. To determine the correlation of each principal component with each of the original variables, the Pearson bivariate correlation was performed [62].

4.5. Statistical Analyses

All data were statistically analyzed using IBM SPSS software version 23 (SPSS Inc.). The data are presented as the mean values with standard errors. The analysis was performed using the Duncan Multiple Range Test (DMRT). Significant differences were analyzed by one-way ANOVA at $p < 0.05$. The correlation matrix was constructed using Pearson bivariate correlation analysis using IBM SPSS software version 23 (SPSS Inc.). Graph preparation and principal component analysis (PCA) were performed using OriginPro 9.0 software (Origin Lab, Corporation, Northampton, USA).

5. Conclusions

Application of beneficial endophytic PGPB in agriculture seems to be a good approach in the future to ensure the high yield potential of plants and food safety. To the best of our knowledge, this is the first report on *HSP70* and *HMA3* gene expression in the PGPB-inoculated switchgrass under Cd stress. These results will provide information for future research on heavy metal resistance in plants with the assistance of PGPB.

Author Contributions: N.B. contributed to conceptualization, data curation, formal analysis, investigation, methodology, software, validation, visualization, writing—original draft, review, and editing. Z.H. contributed to investigation and resources. Q.C. contributed to conceptualization, funding acquisition, project administration, supervision and writing—review & editing. L.L. contributed to resources, supervision, writing—review & editing.

Funding: This research received no external funding.

Acknowledgments: Financial support from the Natural Science Foundation of China (NSFC, 31372359) and PAPD (A Project Funded by the Priority Academic Program Development of Jiangsu Higher Education Institutions) are gratefully acknowledged. The authors thank Mahmuda Akter, Hokkaido University, Japan, for her sincere help regarding the principal component analysis (PCA). The authors also express gratitude to Ripa Akter Sharmin, GM Al Amin, and Mashiur Rahman Bhuiyan, College of Agriculture, Nanjing Agricultural University and Muhammad Arshad, The State Key Laboratory of Crop Genetics and Germplasm Enhancement, Nanjing Agricultural University for their help during the experimental procedure.

Conflicts of Interest: The authors declare no conflict of interest.

References

1. Cvjetko, P.; Zovko, M.; Balen, B. Proteomics of heavy metal toxicity in plants. *Arh. Hig. Rada. Toksikol.* **2014**, *65*, 1–18. [CrossRef] [PubMed]
2. Zhang, J.; Martinoia, E.; Lee, Y. Vacuolar Transporters for Cadmium and Arsenic in Plants and their Applications in Phytoremediation and Crop Development. *Plant Cell Physiol.* **2018**, *59*, 1317–1325. [CrossRef] [PubMed]
3. Wang, C.; Guo, W.; Cai, X.; Li, R.; Ow, D.W. Engineering low-cadmium rice through stress-inducible expression of OXS3-family member genes. *New Biotechnol.* **2019**, *48*, 29–34. [CrossRef] [PubMed]
4. Al Mahmud, J.; Hasanuzzaman, M.; Nahar, K.; Bhuyan, M.B.; Fujita, M. Insights into citric acid-induced cadmium tolerance and phytoremediation in *Brassica juncea* L.: Coordinated functions of metal chelation, antioxidant defense and glyoxalase systems. *Ecotox. Environ. Safe.* **2018**, *147*, 990–1001. [CrossRef] [PubMed]
5. Clemens, S. Molecular mechanisms of plant metal tolerance and homeostasis. *Planta* **2001**, *212*, 475–486. [CrossRef] [PubMed]
6. Hossain, Z.; Komatsu, S. Contribution of proteomic studies towards understanding plant heavy metal stress response. *Front. Plant Sci.* **2013**, *3*, 310. [CrossRef]
7. Sørensen, J.G.; Kristensen, T.N.; Loeschcke, V. The evolutionary and ecological role of heat shock proteins. *Ecol. Lett.* **2003**, *6*, 1025–1037. [CrossRef]
8. Hasan, M.K.; Cheng, Y.; Kanwar, M.K.; Chu, X.-Y.; Ahammed, G.J.; Qi, Z.-Y. Responses of Plant Proteins to Heavy Metal Stress—A Review. *Front. Plant Sci.* **2017**, *8*. [CrossRef]
9. Chen, B.; Feder, M.E.; Kang, L. Evolution of heat-shock protein expression underlying adaptive responses to environmental stress. *Mol. Ecol.* **2018**, *27*, 3040–3054. [CrossRef]
10. Kieffer, P.; Planchon, S.; Oufir, M.; Ziebel, J.; Dommes, J.; Hoffmann, L.; Hausman, J.-F.; Renaut, J. Combining proteomics and metabolite analyses to unravel cadmium stress-response in poplar leaves. *J. Proteome Res.* **2008**, *8*, 400–417. [CrossRef]
11. Stout, R.G.; Al-Niemi, T.S. Heat-tolerant Flowering Plants of Active Geothermal Areas in Yellowstone National Park. *Ann. Bot.* **2002**, *90*, 259–267. [CrossRef] [PubMed]
12. Amm, I.; Sommer, T.; Wolf, D.H. Protein quality control and elimination of protein waste: The role of the ubiquitin–proteasome system. *BBA Mol. Cell Res.* **2014**, *1843*, 182–196. [CrossRef] [PubMed]
13. Gupta, S.C.; Sharma, A.; Mishra, M.; Mishra, R.K.; Chowdhuri, D.K. Heat shock proteins in toxicology: How close and how far? *Life Sci.* **2010**, *86*, 377–384. [CrossRef] [PubMed]
14. Wang, W.; Vinocur, B.; Shoseyov, O.; Altman, A. Role of plant heat-shock proteins and molecular chaperones in the abiotic stress response. *Trends Plant Sci.* **2004**, *9*, 244–252. [CrossRef] [PubMed]
15. Park, C.J.; Seo, Y.S. Heat shock proteins: A review of the molecular chaperones for plant immunity. *Plant Pathol. J.* **2015**, *31*, 323. [CrossRef]
16. Song, G.; Yuan, S.; Wen, X.; Xie, Z.; Lou, L.; Hu, B.; Cai, Q.; Xu, B. Transcriptome analysis of Cd-treated switchgrass root revealed novel transcripts and the importance of HSF/HSP network in switchgrass Cd tolerance. *Plant Cell Rep.* **2018**, *37*, 1485–1497. [CrossRef]
17. Shao, J.F.; Xia, J.; Yamaji, N.; Shen, R.F.; Ma, J.F. Effective reduction of cadmium accumulation in rice grain by expressing OsHMA3 under the control of the OsHMA2 promoter. *J. Exp. Bot.* **2018**, *69*, 2743–2752. [CrossRef]
18. Sasaki, A.; Yamaji, N.; Ma, J.F. Overexpression of OsHMA3 enhances Cd tolerance and expression of Zn transporter genes in rice. *J. Exp. Bot.* **2014**, *65*, 6013–6021. [CrossRef]
19. Ueno, D.; Yamaji, N.; Kono, I.; Huang, C.F.; Ando, T.; Yano, M.; Ma, J.F. Gene limiting cadmium accumulation in rice. *Proc. Natl. Acad. Sci. USA* **2010**, *107*, 16500–16505. [CrossRef]
20. Santoyo, G.; Moreno-Hagelsieb, G.; del Carmen Orozco-Mosqueda, M.; Glick, B.R. Plant growth-promoting bacterial endophytes. *Microbiol. Res.* **2016**, *183*, 92–99. [CrossRef]
21. Ibort, P.; Imai, H.; Uemura, M.; Aroca, R. Proteomic analysis reveals that tomato interaction with plant growth promoting bacteria is highly determined by ethylene perception. *J. Plant Physiol.* **2018**, *220*, 43–59. [CrossRef]
22. Hussain, S.S.; Mehnaz, S.; Siddique, K.H.M. Harnessing the Plant Microbiome for Improved Abiotic Stress Tolerance. In *Plant Microbiome: Stress Response*; Egamberdieva, D., Ahmad, P., Eds.; Springer: Singapore, 2018.

23. Zanetti, F.; Zegada-Lizarazu, W.; Lambertini, C.; Monti, A. Salinity effects on germination, seedlings and full-grown plants of upland and lowland switchgrass cultivars. *Biomass Bioenerg.* **2019**, *120*, 273–280. [CrossRef]
24. Emery, S.M.; Kinnetz, E.R.; Bell-Dereske, L.; Stahlheber, K.A.; Gross, K.L.; Pennington, D. Low variation in arbuscular mycorrhizal fungal associations and effects on biomass among switchgrass cultivars. *Biomass Bioenerg.* **2018**, *119*, 503–508. [CrossRef]
25. Esmaeel, Q.; Miotto, L.; Rondeau, M.; Leclère, V.; Clément, C.; Jacquard, C.; Sanchez, L.; Barka, E.A. *Paraburkholderia phytofirmans* PsJN-Plants Interaction: From Perception to the Induced Mechanisms. *Front. Microbiol.* **2018**, *9*, 2093. [CrossRef] [PubMed]
26. Afzal, S.; Begum, N.; Zhao, H.; Fang, Z.; Lou, L.; Cai, Q. Influence of endophytic root bacteria on the growth cadmium tolerance and uptake of switchgrass (*Panicum virgatum* L.). *J. Appl. Microbiol.* **2017**, *123*, 498–510. [CrossRef] [PubMed]
27. Begum, N.; Afzal, S.; Zhao, H.; Lou, L.; Cai, Q. Shoot endophytic plant growth-promoting bacteria reduce cadmium toxicity and enhance switchgrass (*Panicum virgatum* L.) biomass. *Acta Physiol. Plant.* **2018**, *40*, 170. [CrossRef]
28. Govindasamy, V.; George, P.; Raina, S.K.; Kumar, M.; Rane, J.; Annapurna, K. Plant-Associated Microbial Interactions in the Soil Environment: Role of Endophytes in Imparting Abiotic Stress Tolerance to Crops. In *Advances in Crop Environment Interaction*; Bal, S.K., Mukherjee, J., Choudhury, B.U., Dhawan, A.K., Eds.; Springer Singapore: Singapore, 2018.
29. Estenson, K.; Hurst, G.B.; Standaert, R.F.; Bible, A.N.; Garcia, D.; Chourey, K.; Doktycz, M.J.; Morrell-Falvey, J.L. Characterization of Indole-3-acetic Acid Biosynthesis and the Effects of This Phytohormone on the Proteome of the Plant-Associated Microbe *Pantoea* sp. YR343. *J. Proteome Res.* **2018**, *17*, 1361–1374. [CrossRef] [PubMed]
30. Broek, A.V.; Gysegom, P.; Ona, O.; Hendrickx, N.; Prinsen, E.; Van Impe, J.; Vanderleyden, J. Transcriptional analysis of the *Azospirillum brasilense* indole-3-pyruvate decarboxylase gene and identification of a cis-acting sequence involved in auxin responsive expression. *Mol. Plant Microbe Interact.* **2005**, *18*, 311–323. [CrossRef]
31. Kulkarni, G.; Nayak, A.; Sajjan, S.; Oblesha, A.; Karegoudar, T. Indole-3-acetic acid biosynthetic pathway and aromatic amino acid aminotransferase activities in *Pantoea dispersa* strain GPK. *Lett. Appl. Microbiol.* **2013**, *56*, 340–347. [CrossRef]
32. Patten, C.L.; Glick, B.R. Role of *Pseudomonas putida* indoleacetic acid in development of the host plant root system. *Appl. Environ. Microbiol.* **2002**, *68*, 3795–3801. [CrossRef]
33. Ryu, R.J.; Patten, C.L. Aromatic amino acid-dependent expression of indole-3-pyruvate decarboxylase is regulated by TyrR in *Enterobacter cloacae* UW5. *J. Bacteriol.* **2008**, *190*, 7200–7208. [CrossRef] [PubMed]
34. Sergeeva, E.; Hirkala, D.L.M.; Nelson, L.M. Production of indole-3-acetic acid, aromatic amino acid aminotransferase activities and plant growth promotion by *Pantoea agglomerans* rhizosphere isolates. *Plant Soil* **2007**, *297*, 1–13. [CrossRef]
35. Liu, Y.; Yang, Y.; Li, C.; Ni, X.; Ma, W.; Wei, H. Assessing Soil Metal Levels in an Industrial Environment of Northwestern China and the Phytoremediation Potential of Its Native Plants. *Sustainability* **2018**, *10*, 2686. [CrossRef]
36. Pachura, P.; Ociepa-Kubicka, A.; Skowron-Grabowska, B. Assessment of the availability of heavy metals to plants based on the translocation index and the bioaccumulation factor. *Desalin. Water Treat.* **2016**, *57*, 1469–1477. [CrossRef]
37. Liu, C.; Lou, L.; Deng, J.; Li, D.; Yuan, S.; Cai, Q. Morph-physiological responses of two switchgrass (*Panicum virgatum* L.) cultivars to cadmium stress. *Grassl. Sci.* **2016**, *62*, 92–101. [CrossRef]
38. Wang, Z.; Liu, X.; Qin, H. Bioconcentration and translocation of heavy metals in the soil-plants system in Machangqing copper mine, Yunnan Province, China. *J. Geochemic. Explor.* **2019**, *200*, 159–166. [CrossRef]
39. Guo, Q.; Meng, L.; Humphreys, M.W.; Scullion, J.; Mur, L.A.J. Expression of FlHMA3, a P1B2-ATPase from *Festulolium loliaceum*, correlates with response to cadmium stress. *Plant Physiol. Biochem.* **2017**, *112*, 270–277. [CrossRef]
40. Luo, H.; Sun, X.; Li, Z. Negative Regulator of the Abiotic Stress Response. Google Patents US2015/0337327A1, 26 November 2015.

41. Fukami, J.; de la Osa, C.; Ollero, F.J.; Megías, M.; Hungria, M. Co-inoculation of maize with *Azospirillum brasilense* and *Rhizobium tropici* as a strategy to mitigate salinity stress. *Funct. Plant Biol.* **2018**, *45*, 328–339. [CrossRef]
42. Song, J.; Weng, Q.; Ma, H.; Yuan, J.; Wang, L.; Liu, Y. Cloning and expression analysis of the Hsp70 gene ZmERD2 in *Zea mays*. *Biotechnol. Biotech. Equip.* **2016**, *30*, 219–226. [CrossRef]
43. Liu, J.; Pang, X.; Cheng, Y.; Yin, Y.; Zhang, Q.; Su, W.; Hu, B.; Guo, Q.; Ha, S.; Zhang, J.; et al. The Hsp70 Gene Family in *Solanum tuberosum*: Genome-Wide Identification, Phylogeny, and Expression Patterns. *Sci. Rep.* **2018**, *8*, 16628. [CrossRef]
44. Wang, K.; Zhang, X.; Goatley, M.; Ervin, E. Heat shock proteins in relation to heat stress tolerance of creeping bentgrass at different N levels. *PLoS ONE* **2014**, *9*, e102914. [CrossRef] [PubMed]
45. Asea, A.A.; Kaur, P.; Calderwood, S.K. (Eds.) *Heat Shock Proteins and Plants*; Springer: Berlin/Heidelberg, Germany, 2016; p. 341.
46. Esposito, S.; Loppi, S.; Monaci, F.; Paoli, L.; Vannini, A.; Sorbo, S.; Maresca, V.; Fusaro, L.; Lentini, M.; De Lillo, A. In-field and in-vitro study of the moss *Leptodictyum riparium* as bioindicator of toxic metal pollution in the aquatic environment: Ultrastructural damage, oxidative stress and HSP70 induction. *PLoS ONE* **2018**, *13*, e0195717. [CrossRef] [PubMed]
47. Yu, X.Z.; Fan, W.J.; Lin, Y.J. Analysis of gene expression profiles for metal tolerance protein in rice seedlings exposed to both the toxic hexavalent chromium and trivalent chromium. *Int. Biodeterior. Biodegrad.* **2018**, *129*, 102–108. [CrossRef]
48. Rodríguez-Celma, J.; Rellán-Álvarez, R.; Abadía, A.; Abadía, J.; López-Millán, A.F. Changes induced by two levels of cadmium toxicity in the 2-DE protein profile of tomato roots. *J. Proteom.* **2010**, *73*, 1694–1706. [CrossRef]
49. Kieffer, P.; Dommes, J.; Hoffmann, L.; Hausman, J.F.; Renaut, J. Quantitative changes in protein expression of cadmium-exposed poplar plants. *Proteomics* **2008**, *8*, 2514–2530. [CrossRef]
50. Ahsan, N.; Lee, S.H.; Lee, D.G.; Lee, H.; Lee, S.W.; Bahk, J.D.; Lee, B.H. Physiological and protein profiles alternation of germinating rice seedlings exposed to acute cadmium toxicity. *CR Biol.* **2007**, *330*, 735–746. [CrossRef]
51. Sarry, J.E.; Kuhn, L.; Ducruix, C.; Lafaye, A.; Junot, C.; Hugouvieux, V.; Jourdain, A.; Bastien, O.; Fievet, J.B.; Vailhen, D. The early responses of *Arabidopsis thaliana* cells to cadmium exposure explored by protein and metabolite profiling analyses. *Proteomics* **2006**, *6*, 2180–2198. [CrossRef]
52. Kaushal, M.; Wani, S.P. Rhizobacterial-plant interactions: Strategies ensuring plant growth promotion under drought and salinity stress. *Agric. Ecosyst. Environ.* **2016**, *231*, 68–78. [CrossRef]
53. Spaepen, S.; Vanderleyden, J. Auxin signaling in *Azospirillum brasilense*: A proteome analysis. In *Biological Nitrogen Fixation*; Bruijn, F.J.D., Ed.; John Wiley & Sons Inc.: Hoboken, NJ, USA, 2015; pp. 937–940.
54. Lim, J.H.; Kim, S.D. Induction of drought stress resistance by multi-functional PGPR *Bacillus licheniformis* K11 in pepper. *Plant Pathol. J.* **2013**, *29*, 201–208. [CrossRef]
55. Sharma, M.K.; Sharma, R.; Cao, P.; Harkenrider, M.; Jenkins, J.; Grimwood, J.; Zhang, J.; Udvardi, M.K.; Schmutz, J.; Ronald, P.C. Targeted switchgrass BAC library screening and sequence analysis identifies predicted biomass and stress response-related genes. *BioEnergy Res.* **2016**, *9*, 109–122. [CrossRef]
56. Sharma, M.K.; Sharma, R.; Cao, P.; Jenkins, J.; Bartley, L.E.; Qualls, M.; Grimwood, J.; Schmutz, J.; Rokhsar, D.; Ronald, P.C. A genome-wide survey of switchgrass genome structure and organization. *PLoS ONE* **2012**, *7*, e33892. [CrossRef] [PubMed]
57. Gravot, A.; Lieutaud, A.; Verret, F.; Auroy, P.; Vavasseur, A.; Richaud, P. AtHMA3, a plant P1B-ATPase, functions as a Cd/Pb transporter in yeast. *FEBS Lett.* **2004**, *561*, 22–28. [CrossRef]
58. Benizri, E.; Kidd, P.S. The Role of the Rhizosphere and Microbes Associated with Hyperaccumulator Plants in Metal Accumulation. In *Agromining: Farming for Metals. Mineral Resource Reviews*; Van der Ent, A., Echevarria, G., Baker, A., Morel, J., Eds.; Springer: Cham, Switzerland, 2018; pp. 157–188.
59. Cui, H.; Cao, X.; Wang, J.; Xiong, A.; Hou, X.; Li, Y. Effects of exogenous GR24 on the growth of axillary bud of non-heading Chinese cabbage. *J. Nanjing Agric. Univ.* **2016**, *39*, 366–372.
60. Livak, K.J.; Schmittgen, T.D. Analysis of relative gene expression data using real-time quantitative PCR and the $2^{-\Delta\Delta CT}$ method. *Methods* **2001**, *25*, 402–408. [CrossRef] [PubMed]

61. Schmittgen, T.D.; Livak, K.J. Analyzing real-time PCR data by the comparative CT method. *Nat. Protoc.* **2008**, *3*, 1101. [CrossRef]
62. Jackson, J.E. Principal Components and Factor Analysis: Part I—Principal Components. *J. Qual. Technol.* **1980**, *12*, 201–213. [CrossRef]

plants

MDPI

Article

Effect of *Rhododendron arboreum* Leaf Extract on the Antioxidant Defense System against Chromium (VI) Stress in *Vigna radiata* Plants

Vandana Gautam [1], Pooja Sharma [1], Palak Bakshi [1], Saroj Arora [1], Renu Bhardwaj [1], Bilal Ahmad Paray [2], Mohammed Nasser Alyemeni [3] and Parvaiz Ahmad [3,4,*]

1 Department of Botanical and Environmental Sciences, Guru Nanak Dev University, Amritsar 143005, India; vndu.gndu@gmail.com (V.G.); pooja.sharma@gmail.com (P.S.); p.bakshi@gmail.com (P.B.); sarojarora.gndu@gmail.com (S.A.); renubhardwaj82@gmail.com (R.B.)

2 Zoology Department, College of Science, King Saudi University, Riyadh 11451, Saudi Arabia; bparay@ksu.edu.sa

3 Botany and Microbiology Department, College of Science, King Saudi University, Riyadh 11451, Saudi Arabia; mnyemeni@ksu.edu.sa

4 Department of Botany, S.P. College, Srinagar 190001, India

* Correspondence: parvaizbot@yahoo.com; Tel.: +98-58376669

Received: 25 November 2019; Accepted: 21 January 2020; Published: 29 January 2020

Abstract: In the current investigation, we studied role of *Rhododendron* leaf extract in *Vigna radiata* grown under chromium metal stress. We observed that seed treatment with *Rhododendron* leaf extract resulted in the recuperation of seedling growth under chromium toxicity. Seed treatment with *Rhododendron* leaf extract significantly improved the contents of anthocyanin and xanthophyll pigments under stress. The antioxidative defense system triggered after *Rhododendron* extract treatment, resulting in the increased actions of antioxidant enzymes. Oxidative stress induced by the assembly of reactive oxygen species was reduced after *Rhododendron* extract treatment under chromium toxicity as indicated by the enhanced contents of non-enzymatic antioxidants, namely ascorbic acid, tocopherol, and glutathione. Furthermore, *Rhododendron* leaf extract treatment under chromium metal stress also encouraged the biosynthesis of organic acids, polyphenols, as well as amino acids in *Vigna radiata*. Statistical analysis of the data with multiple linear regression also supported that *Rhododendron* leaf extract can effectively ease chromium metal-induced phytotoxicity in *Vigna radiata*.

Keywords: *Rhododendron arboreum*; *Vigna radiata*; enzymes activity; chromium (Cr); polyphenols

1. Introduction

Heavy metals, put forth by the modern urbanized and industrialized environment, are major chemical agents causing hindrances in normal growth, functioning, productivity, and survival of plants as well as animals [1]. In recent times, heavy metal toxicity in farming systems along with edible crops has become a major concernnot only from an environmental but also from a health point of view. Chromium (Cr) is the prevalent element, whichexists in different oxidation states in nature, amongst which, the hexavalent chromium [Cr(VI)] is the most pernicious [2]. Volcanic dust and igneous rocks are the natural sources of chromium; however, agricultural land is severely contaminated with chromium metal by inappropriate waste dumping or seepage from tanneries and burning of coal or oil [3,4]. Usually, chromium is not persistent in the atmosphere as it is deftly deposited in soil or dissolved in water. Therefore, it is readily available to the roots of plants budding in the infected region. Even in low concentrations, Cr is reported to enhance the production of reactive oxygen species (ROS), including hydrogen peroxide (H_2O_2) and superoxide anion,which lead to oxidative stress [5]. Oxidative stress hinders seed germination, growth, and metabolism in plants along with cell injury. Living organisms are equipped with an antioxidant defense system against ROS, which consists of enzymatic and non-enzymatic antioxidants [6]. Antioxidant enzymes include superoxide dismutase (SOD), catalase (CAT), ascorbate peroxidase (APOX), glutathione peroxidase (GPOX), guaiacol peroxidase (POD), polyphenol oxidase (PPO), dehydroascorbate reductase (DHAR), glutathione-S-transferase (GST), and glutathione reductase (GR) [7]. Non-enzymatic antioxidants comprise of ascorbic acid, tocopherol, glutathione (GSH), and polyphenols. Tocopherol repairs peroxyl radicals and terminates the chain of lipid peroxidation [8,9]. Thus, the antioxidative defense system helps to overcome oxidative stress. However, this system does not work efficiently when living organisms encounter severe metal stress, resulting in a need for external supplements to beat the oxidative stress.

A literature survey reveals the oxidative stress protective and plant growth regulatory effects of several plant extracts. For example, tea seed extract has been reported to regulate the growth of *Brassica* sp., *Avena sativa*, and *Hordeum vulgare* [10]. *Rhododendron arboreum* of the Ericaceae family is native to Bhutan, China, India, and Nepal. The trees grow in North and North-East India. The name 'rhododendron' is derived from the Greek word 'rhodo' meaning rose and 'dendron' meaning tree. *Rhododendron* is the national flower of Nepal and the state tree of Uttarakhand (India). This evergreen tree is important from an economical as well as a horticultural perspective. Also, it is widely used by the tribal people of North India for gastronomic as well as traditional restorative purposes. In our previous study [11], its antioxidant and antimutagenic properties were evaluated along with the active principals responsible for those activities. It has also been reported to be a resource of a number of phytoconstituents of therapeutic value by numerous authors [12–14]. So, keeping in view the eco-friendly and remunerative nature of plant extracts, the present study was designed to evaluate the Cr metal stress-ameliorating activity of *Rhododendron arboreum* leaves using *Vigna radiata*.

Due to its high protein content and low price, *Vigna radiata* has become the most widespread food item in every household all over the globe. In India, it is frequently used as food, especially for babies, old, and sick people, because of its easy digestibleness and great nutritional value. Its fiber is also an outstanding feed for farm animals. In addition, the capability of *Vigna radiata* to fix atmospheric nitrogen and add organic matter to the soil are major characteristics to maintain soil fertility. It has been previously reported that Cr exposure has harmful effects on the physiological and biochemical processes of *Vigna radiata* L [15]. Furthermore, metal stresses decrease the productivity of crops and go through the food chain, resulting in terrible health problems to the consumers, including human beings and other herbivores [16]. The *Vigna radiata* crop was selected for the present research since it fits well as a model because of its short life cycle and low maintenance nature.

The current analysis was carried out to investigate the effect of methanol extract of *Rhododendron arboreum* leaves (MEL) on Cr-treated *Vigna radiata* plants. The analyzed biochemical parameters included anthocyanin and xanthophyll pigment contents, antioxidant enzymes estimation, non-enzymatic antioxidants (ascorbic acid, tocopherol, and glutathione) analysis, and organic acids, polyphenols, and amino acids analysis.

2. Materials and Methods

The stress protective effect of methanol extract of *Rhododendron arboreum* leaves (MEL) on Cr(VI)-treated *Vigna radiata* plants was observed by analyzing the pigment contents, protein content, enzymatic antioxidants, non-enzymatic antioxidants, polyphenols, amino acids, and organic acids.

2.1. Raising of Plant Material

The certified seeds of *Vigna radiata* var. SML-668 were procured from Punjab Agricultural University, Ludhiana (India). The healthy seeds of *Vigna radiata* were surface sterilized for one minute with 0.01% mercuric chloride ($HgCl_2$) and rinsed with distilled water several times. The seeds were given a pre-sowing treatment with different concentrations (125, 250, and 500 ppm) of *Rhododendron* leaf extract for six hours. Petriplates lined with whatman filter paper 1 were supplemented with 250 μM chromium (IV). These solutions were prepared in distilled water. The plants were grown in petriplates kept in a seed germinator for 7 days under controlled conditions of 25±0.5 °C temperature, 16-h photoperiod, 8-h dark period, 175 μmol/m^2/s light intensity, and 60% relative humidity. After 7 days of growth, the plants were supplemented with half-strength Hoagland medium and the petriplates were monitored for 7 days, after which the plants were harvested and independent experiments were performed. Each petriplate contained 20 seeds and all the treatments were given in triplicates in order to ensure valid statistical analysis.

2.2. *Rhododendron arboreum Methanol Leaf Extract(MEL)*

The fresh leaves of *Rhododendron arboreum* leaves were washed with double-distilled water, dried in the shade, and ground to a fine powder in a mixergrinder. Them, 1 kg of powder was extracted with 80% methanol and dried in a vacuum rotary evaporator to get 68.38 g (6.83%) of methanol leaf extract. The nontoxic and maximum yield producing (in terms of the root length, shoot length, fresh weight, and dry weight) concentrations of the MEL (methanol extract of *Rhododendron arboreum* leaves) were identified by carrying out different test experiments on *Vigna radiata* seeds. Three concentrations, i.e., 125, 250, and 500 ppm, were selected for the present study. A 1000 ppm mother stock of MEL was prepared and working concentrations of 125, 250, and 500 ppm were prepared from the mother stock by serial dilutions. The mother stock and working concentrations were stored in a refrigerator at 4 °C.

2.3. *Cr(VI)Metal*

Heavy metal Cr(VI) was given in the form of potassium chromate (K_2CrO_4) (Qualigens Fine Chemicals Pvt. Ltd., Mumbai, India). The IC_{50} value (250 μM) of the metal was determined and used for the treatments. The working 250 μM concentration of Cr(VI) was prepared from 10 mM mother stock.

2.4. *Treatments*

All the treatments given in the present study are shown in Table 1.

Table 1. Different treatments given to *Vigna radiata* in the present investigation.

Sr. No.	Name of the Treatment	MEL (ppm)	Cr(VI) (μM)
1.	Control	0	0
2.	MEL-1	125	0
3.	MEL-2	250	0
4.	MEL-3	500	0
5.	Cr(VI)	0	250
6.	Cr(VI) + MEL-1	125	250
7.	Cr(VI) + MEL-2	250	250
8.	Cr(VI) + MEL-3	500	250

2.5. *Phenolic Pigments:*

2.5.1. Xanthophyll Content

The plant tissue was tested for xanthophyll content using the method of Lawrence [17]. Hexane, acetone, absolute alcohol, and toluene were mixed in the ratio of 10:7:6:7 to prepare the extractant. Then, 50 mg of dried plant sample was converted into a fine powder using a pestle and mortar. The powdered plant sample was mixed with 30 mL of extractant solution in a volumetric flask and shaken for 10–15 min. The hot saponification process was started by adding 2 mL of methanolic KOH (40%). The contents were shaken for 1–2 min, then the flask was kept at 56 °C in a water bath followed by incubation for one hour in cool dark conditions. Then, 30 mL of hexane was added to the reaction mixture with continued vigorous shaking and the final volume of 100 mL was made with the addition of 10% Na_2SO_3. The flask was again incubated for one hour in dark conditions. The absorbance (A) of the upper phase was noted at 474 nm.

2.5.2. Anthocyanin Content

The anthocyanin content of the fresh plant sample was estimated by the method given by Mancinelli [18]. Under chilled conditions, 1g of fresh plant sample was crushed using a pestle and mortar, in an extraction mixture, which consisted of methanol, distilled water, and hydrochloric acid in a ratio of 79:20:1. The crushed tissue was centrifuged for 20 min at 13,000 rpm and 4 °C. The A of the supernatant was recorded at 530 and 657 nm.

2.6. *Protein Content*

The protein content of the samples was determined by following the method given by Lowry et al. [19]. Onegram of fresh plant samples was weighed and crushed in 3 mL of 50 mM potassium phosphate buffer with pH 7.0 using a pestle and mortar under ice cold conditions. The homogenates were centrifuged for 20 min at 13,000 rpm and 4 °C. In total, 100 μL of supernatant was mixed with 900 μL of distilled water followed by the addition of 5 mL of reagent C. The test tube with 1 mL of distilled water served as a blank. This mixture was mixed well and allowed to stand for 10 min followed by the addition of 500 μL of Folin–Ciocalteau reagent. The reaction mixture was mixed thoroughly and incubated at room temperature for 30 min in dark conditions. The optical density of the blue color was measured at 660 nm. A standard curve of the protein solution representing concentration vs. absorbance was plotted and a linear regression equation was obtained, which was used to calculate the protein content of the samples, which was expressed as mg g^{-1}fresh weight.

2.7. *Enzymatic Antioxidants*

2.7.1. Preparation of Plant Extracts

Onegram of fresh plant sample was weighed and crushed in 3 mL of extractant buffer using a pestle and mortar under ice cold conditions. The homogenates were centrifuged for 20 min at 13,000 rpm and 4°C. The supernatant was collected for analysis of the biochemical parameters. The extractant buffer for the estimation of the activities of (POD, CAT, GR, APOX, DHAR, and GPOX) enzymes was 50 mM potassium phosphate buffer with pH 7.0. For estimation of the SOD activity, samples were homogenized in 50 mM sodium carbonate buffer at pH 10.2. The homogenates for GST were prepared in 0.2 M potassium phosphate buffer with pH 7.4 whereas for PPO, 0.1 M potassium phosphate buffer at pH 7.0 was used.

2.7.2. Guaiacol Peroxidase (POD, EC. 1.11.1.7)

The guaiacol peroxidase activity was estimated by method given by Pütter [20]. To prepare 3 mL of the reaction mixture 50 mM phospahte buffer, 20.1 mM of guaiacol solution, 12.3 mM of H_2O_2, and 60μL of enzyme extract were added to a cuvette. The rate of the oxidation of guaiacol and formation of guaiacol dehydrogenation product (GDHP) was recorded at 436 nm for 1 min at 3-s intervals at 25 °C. One unit of enzyme activity is defined as the amount of enzyme necessary to catalyze the oxidation of guaiacol and the development of 1 μM of GDHP $min^{-1}g^{-1}$fresh weight. The enzyme activity was expressed as UA mg g^{-1} protein.

2.7.3. Catalase (CAT, EC 1.11.1.6)

The estimation of the catalase activity was performed according to the method of Aebi [21]. First, 3 mL of the reaction mixture consisted of 50 mM phosphate buffer, 15 mM H_2O_2, and 60 μL of enzyme extract. The breakdown of hydrogen peroxide caused a decline in the optical density, which was recorded for one minute at 240 nm and 25 °C. The quantity of enzyme necessary to release half peroxide oxygen from hydrogen peroxide is referred to as one unit of the enzyme activity and was expressed as UA mg g^{-1} protein

2.7.4. Superoxide Dismutase (SOD, EC 1.15.1.1)

The superoxide dismutase activity of the fresh plant samples was estimated by the method proposed by Kono [22]. To estimate the SOD activity, a reaction mixture containing 50 mM sodium carbonate buffer, 24 μM Nitro Blue Tetrazolium (NBT), 0.1 mM ethylenediaminetetraacetic acid (EDTA), and 0.03% Triton X-100 was put in a cuvette and the reaction was started by the addition of 1 mM hydroxylamine hydrochloride. Then, 70 μL of enzyme extract were added to the reaction mixture and the increase in the absorbance was observed at 560 nm for 2 min at 6-s intervals at 25 °C. The amount of enzyme required to inhibit NBT reduction up to 50% is defined as one unit of enzyme and was expressed as UA mg g^{-1} protein.

2.7.5. Ascorbate Peroxidase (APOX, EC. 1.11.1.11)

The APOX activity of plant samples was estimated by following the method given by Nakano and Asada [23]. First, 50 mM phosphate buffer, 0.5 mM ascorbate, 1 mM H_2O_2, and 70 μLof enzyme extract comprised the 3 mL reaction mixture. The decrease in absorbance was recorded at 290 nm. One unit of ascorbate peroxidase was estimated by measuring the amount of enzyme required to oxidize 1 μM of ascorbate min^{-1} g^{-1}fresh weight. The APOX activity was expressed as UA mg g^{-1} protein

2.7.6. Glutathione Reductase (GR, EC 1.6.4.2)

The glutathione reductase activity of fresh plant samples was assayed by using the method proposed by Carlberg and Mannervik [24]. The 3-mL reaction mixture in the cuvette consisted of 50 mM phosphate buffer, 1 mM EDTA, 0.1 mM reduced nicotinamide adenine dinucleotide phosphate (NADPH), 1 mM glutathione disulphide, and 75 μL of enzyme extract. The decrease in absorbance per minute was observed at 340 nm at intervals of 3 s at 25 °C. One unit of enzyme activity is defined as the amount of enzyme required to oxidize 1.0 μM of NADPH min^{-1} g^{-1} fresh weight. The activity was expressed as UA mg g^{-1} protein.

2.7.7. Dehydroascorbate Reductase (DHAR, EC. 1.8.5.1)

The dehydroascorbate reductase activity of plant samples was estimated using the method given by Dalton et al. [25]. The 3-mL reaction mixture contained 50 mM phosphate buffer, 0.1 mM EDTA, 1.5 mM reduced glutathione and 0.2 mM dehydroascorbate, and 75 μL of enzyme extract. An increase in the absorbance per minute was noted at 265 nm at 3-s intervals and 25 °C. One unit of dehydroascorbate reductase is defined as the quantity of enzyme needed to catalyze the development of 1 μM of ascorbate min^{-1} g^{-1} fresh weight of plant tissue. The DHAR activity was indicated by UA mg g^{-1} protein.

2.7.8. Polyphenol Oxidase (PPO, EC 1.10.3.1)

The polyphenol oxidase activity was estimated according to the method proposed by Kumar and Khan [26]. For the estimation of the enzyme, 1 mL of phosphate buffer, 0.5 mL of catechol, and 0.25 mL of enzyme extract were added to the cuvette and it was incubated for 2 min. The reaction was initiated by the addition of 0.5 mL of H_2SO_4 and any change in absorbance was noted at 495 nm for 1 min. One unit of enzyme activity is defined as the quantity of enzyme necessary to oxidize 1 μM of catechol. The activities of the enzyme were expressed as UA mg g^{-1} protein.

2.7.9. Glutathione-S-Transferase (GST, EC 2.5.1.13)

The glutathione-S-transferase enzyme activity was estimated according to the method given by Habig and Jakoby [27]. The 3-mL reaction mixture in the cuvette contained 0.2 M phosphate buffer, 1 mM each of GSH and 1-Chloro-2,4-dinitrobenzene (CDNB), and the reaction was set off by adding 70 μL of enzyme extract. The increase in absorbance at 340 nm was recorded for 1 min at 3-s intervals. One unit of enzyme activity is defined as the quantity of enzyme catalyzing the development of 1 μM of conjugated GSH-CDNB min^{-1} g^{-1} plant tissue at 25 °C. The activity was revealed as UA mg g^{-1} protein.

2.7.10. Glutathione Peroxidase (GPOX, EC 1.11.1.7)

The glutathione peroxidase activity of fresh plant samples was analyzed according to the method given by Flohé and Günzler [28]. The composition of the 3-mL reaction mixture in the cuvette consisted of 50 mM phosphate buffer, 0.5 mM EDTA, 1 mM of GSH, 1 mM of sodium azide, 0.15 mM NADPH and 0.15 mM H_2O_2, and 75 μL of enzyme extract. A decrease in the absorbance due to oxidation of NADPH was measured for 1 min at 340 nm at 3-s intervals. One unit of enzyme activity is defined as the amount of enzyme required to oxidize 1.0 μM of NADPH $min^{-1}g^{-1}$ fresh tissue at 37 °C and the activities were revealed as UA mg g^{-1} protein.

2.8. Non-Enzymatic Antioxidants

2.8.1. Ascorbic Acid Content

The ascorbic acid content was estimated by the method of Roe and Kuether [29]. First, 1 g of fresh plant sample was crushed in 3 mL of 50 mM Tris buffer (pH 10) using a pestle and mortar under chilled conditions. The homogenate was centrifuged at 13,000 rpm and 4 °C for 20 min. Then, 500 μL of supernatant were mixed with 4 mLof distilled water, 0.5 mL 50% TCA, and 100 mg of activated charcoal in test tubes. The mixture was thoroughly mixed and filtered through Whatman filter paper No. 1. To the filtrate, 0.4 mL of 2,4-Dinitrophenylhydrazine (DNPH) reagent was added and incubation was given at 37 °C for 3 h. Cooling was done using an ice bath followed by the addition of 1.6 mL of cold H_2SO_4 (65%). Again, incubation was given for 30 min at room temperature and the absorbance was taken at 520 nm. Ascorbic acid in the concentration of 1 mg/100 mL was used as a standard and its content in the sample was expressed as mg ascorbic acid g^{-1} fresh weight.

2.8.2. Tocopherol Content

Tocopherol (vitamin E) was estimated by the method given by Martinek [30]. First, 1 g of fresh plant sample was crushed in 3 mL of 50 mM Tris buffer (pH 10) using a pestle and mortar under chilled conditions. The homogenate was centrifuged at 13,000 rpm and 4 °C for 20 min. Then, 500 μL of supernatant were mixed with 500 μL distilled water and 500 μL $FeCl_3$ (0.12% in absolute ethanol) in test tubes followed by vigorous shaking until precipitates formed. The reaction mixture was mixed with 500 μL xylene and vortexed for 30 s. The solution was centrifuged for 10 min at 3000 rpm. The upper xylene layer was mixed with an equal volume of 2,4,6-Tripyridyl-S-triazine (TPTZ) (12% in n-propanol) and the absorbance was read at 600 nm. A 1 mg 100 mL^{-1} concentration of tocopherol was used as the standard tocopherol content in sample and was expressed as mg tocopherol g^{-1} fresh weight.

2.8.3. Glutathione Content

The GSH content was estimated by the method given by Sedlak and Lindsay [31]. First, 1 g of fresh plant sample was crushed in 3 mL of 0.2 M Tris buffer (pH 8.2) using a pestle and mortar under chilled conditions. The homogenate was centrifuged at 13,000 rpm and 4 °C for 20 min. Then, 100 μL of supernatant was mixed with 1 mL of 0.2 M Tris buffer (pH 8.2), 50 μL DTNB (5,5′-dithiobis-2-nitrobenzoic acid), and 4 mL of absolute methanol in test tubes followed by an incubation of 15 min at room temperature. The reaction mixture was centrifuged for 15 min at 3000 rpm and the absorbance of the supernatant was noted at 412 nm. GSH at a concentration of 1mg 100^{-1} mL was used as a standard for determining the GSH content in the sample. The value was expressed as mg GSH g^{-1} fresh weight.

2.8.4. Glutathione Imaging

Tagging of glutathione in the roots of *Vigna radiata* was done with MCB (monochlorobimane) dye according to the method of Hartmann et al. [32]. MCB produces a fluorescent adduct with GSH. In the meanwhile, sodium azide depletes adenosine triphosphate (ATP) from the cells in order to avoid vacuolar seizure of the MCB-GSH adduct caused by ATP. The root samples were dipped in the 25 μM MCB dye containing 5 mM sodium azide for about 15 to 20 min. Samples were then mounted in distilled water on a glass slide and covered with a cover slip. The slides were then observed under a confocal microscope at an excitation wavelength of 351–364 nm and emission wavelength of 477 nm. The intensity of the blue color indicates the measure of glutathione in the roots.

2.9. *Polyphenol Estimation*

Qualitative as well as quantitative analysis of the plant samples for polyphenolic compounds was carried out using ultra-high-performance liquid chromatography (UHPLC). First, 500 mg of fresh plant material were crushed in 2 mL of high-performance liquid chromatography (HPLC)-grade methanol using a pestle and mortar. The solution was centrifuged at 13,000 rpm for 20 min and filtered using 0.2-micron filter paper. The portrayal of phenolic compounds was executed using a 130 MPa Shimadzu UHPLC (Nexera) system, equipped with a DGU-20As prominence degasser, LC-30AD liquid chromatograph, SIL-30AC autosampler, CTO-10AS VP column oven, CBM-20A communications bus module, and SPD-M20A photodiode array detector (PDA). The chromatography was carried out at room temperature with a flow rate of 1 mL/min at λ 280 nm. A 150 × 4.6 mm C-18 column with a pore size of 5 μm was used. The injection volume was 5 μL. The solvent system comprised of solvent A (0.01% acetic acid in water) and solvent B (methanol). The gradient runs were 70% A and 30% B, reaching 45% B at 12 min, 75% B at 13.5 min, 75% B at 15 min, 50% B at 16.6 min, 25% B at 18 min, 25% B at 20 min, 30% B at 21 min, and stopped at 22 min, with an elution of 4 min. The mixture of 11 polyphenolic standards, namely gallic acid ($C_7H_6O_5$), catechin ($C_{15}H_{14}O_6$), chlorogenic acid ($C_{16}H_{18}O_9$), epicatechin ($C_{15}H_{14}O_6$), caffeic acid ($C_9H_8O_4$), umbelliferone ($C_9H_6O_3$), coumaric acid ($C_9H_8O_3$), rutin ($C_{27}H_{30}O_{16}$), ellagic acid ($C_{14}H_6O_8$), quercetin ($C_{15}H_{10}O_7$), and kaempferol ($C_{15}H_{10}O_6$), was diluted with methanol at different concentrations by serial dilution for quantitative analysis. The calibration curves were generated by plotting the concentrations versus peak areas. The detection of every compound was based on a combination of the retention time and spectral similarity. The detection and quantification limit for all the detected compounds were calculated on the basis of signal-to-noise ratio (S/N) of 3 and 10 with the corresponding standard solution, respectively.

2.10. *Amino Acid Profiling*

The amino acid analysis was done using an amino acid analyzer by making slight modifications to the method given by Iriti et al. [33]. First, 0.25 mg of fresh plant sample were crushed in 1.25 mL of 80% methanol using a pestle and mortar in chilled conditions. The homogenized sample was centrifuged for 20 min at 13,000 rpm. The supernatant was mixed with 6% sulfosalicylic acid in a ratio of 1:1 and the solution was further diluted with 0.1 N HCl in a ratio of 1:4. The solution was filtered using 0.2-micron filter paper. Using an auto sampler, 1 μL of sample was injected into the C-18 silica-bonded, 150-mm-long amino acid column with a pore size of 120 Å and particle size of 5 μm. The amino acid analysis was carried out using a Shimadzu, Nexera X_2 amino acid analyzer. A 150-mm-long C-18 silica-bonded column with a 120 Å pore size and 5 micrometer particle size was used for the analysis of amino acids. The injection volume was 1 μL, run time was 32 min, oven temperature was 40 °C, and pump flow rate was 1 mL min^{-1}. Mobile phase "A" consisted of 5.6 pH phosphate buffer; mobile phase "B" comprised of ultra-pure water, methanol, and acetonitrile in the ratio of 15:40:45. The detection of every amino acid was based on a combination of the retention time and spectral similarity. The detection and quantification limits for all the detected amino acids were calculated on the basis of the signal-to-noise ratio (S/N) of 3 and 10 with the corresponding standard solution, respectively.

2.11. Organic Acid Profiling

Organic acids in the test samples of 7-day-old *Vigna radiata* plants were estimated using GC-MS according to the method given by Chen et al. [34]. The 7-day-old *Vigna radiata* plants were washed with distilled water, dried at room temperature, and crushed to make powder with the help of a pestle and mortar. Then, 50 mg of powder of the dried plant sample was mixed with 500 μL of 0.5 N HCl and 500 μL of methanol. The mixture was shaken for 3 h followed by centrifugation for 10 min at 10,000× *g*. To the supernatant, 300 μL of methanol and 100 μL of 50% H_2SO_4 were added and overnight incubation was given at 60 °C using a water bath. The solution was cooled to 25 °C followed by the addition of 800 μL of chloroform and 400 μL of distilled water. The solution was subjected to vortexing for one minute and the lower chloroform layer was used for the assessment of organic acids. To estimate the content of organic acids, 2 μL of the chloroform layer were injected into the GC-MS system (Shimadzu GC-MS-QP2010 Plus) in split mode. A DB-5 ms analytical column was used and the initial column temperature for one minute was 50 °C, which was raised to 125 °C at a rate of 25 °C min^{-1} followed by a further increase to 300 °C at a rate of 10 °C/min and held for 15 min. The injection temperature was 250 °C, the carrier gas used was helium, and the gas flow in the column was 1.7 mL min^{-1}. The ion source temperature and interface temperature was set at 200 °C and 280 °C, respectively. The quantification of organic acids was done using a standard curve.

2.12. Statistical Analysis

All experimental measurements were carried out in triplicate. The mean values and standard error were calculated. The data were analyzed statistically by two-way analysis of variance (ANOVA), as described by Bailey [35]. Tukey's multiple comparison test was used to compare the difference among means by using the honest significant difference HSD values. Comparisons with P-values ≤ 0.05 were considered as significantly different. Multiple regression analysis with interaction (MLR) was carried out to identify the nature of the effect brought about by independent variables Cr, MEL, and their interaction (Cr(VI) × MEL). β-regression values provided the relative effects of the independent variables:

$$X_1 = \mathrm{Cr(VI)},\ X_2 = \mathrm{MEL}\ \text{and}\ X_1X_2 = \mathrm{Cr(VI)} \times \mathrm{MEL}.$$

3. Results

3.1. Phenolic Pigments

The data obtained on alterations in the contents of anthocyanin and xanthophylls pigments due to the effect of different treatments of Cr(VI), MEL, and their interaction in *Vigna radiata* plants are presented in Table 2. It was noted that in relation to 1.2 μg anthocyanin g^{-1} FW in control plants, the anthocyanin content in 7-day-old plants of *Vigna radiata* increased to 13.75 μg g^{-1} FW with the Cr(VI) metal treatment. MEL application to Cr(VI)-treated plants further enhanced the content and the plants grown in combination with Cr(VI) and 500 ppm MEL showed a maximum increase of 36.61 μg g^{-1} FW in the anthocyanin content. Various treatments had a significant role in modulating the increase of the xanthophylls pigment. The xanthophyll content in 7-day-old Cr(VI) metal-stressed plants was elevated to 28.81 μg g^{-1} FW in contrast to 16.1 μg g^{-1} FW in control plants. The exogenous supplementation of MEL at the 125, 250, and 500 ppm concentrations to Cr(VI)-stressed plants further enhanced the xanthophyll content to 34.95, 42.37, and 48.94 μg g^{-1} FW, respectively. Anthocyanin (HSD 5.92) and xanthophylls (HSD 5.67) showed a statistically significant difference in the two-way ANOVA analysis. Positive β values in the MLR analysis indicated the positive effect of Cr(VI), and the interaction of Cr(VI) × MEL on the anthocyanin and xanthophyll contents.

Table 2. Effect of Cr(VI), different concentrations of methanol extract of *Rhododendron arboreum* leaves (MEL), and their combinations on anthocyanin and xanthophyll content in 7-day-old *Vigna radiata* seedlings. The values are the means of three replicates ± S.D. (standard deviation). Tukeys test performed and significance checked at *$p \leq 0.05$. the F ratio is the term in which ANOVA is represented; Cr(VI) = Dose, MEL treatments = Treatment and combination of Cr(VI) and MEL = Dose × treatment.

Concentrations		Anthocyanin (μg g^{-1} FW)	Xanthophyll (μg g^{-1} FW)
Cr(VI) (μM)	MEL (ppm)		
0	0	1.20 ± 0.58	16.10 ± 0.97
0	125	2.78 ± 1.11	18.85 ± 1.32
0	250	5.39 ± 1.12	22.24 ± 0.63
0	500	11.89 ± 1.37	25.21 ± 1.94
250	0	13.75 ± 0.42	28.81 ± 0.97
250	125	17.47 ± 3.23	34.95 ± 2.54
250	250	24.34 ± 3.20	42.37 ± 3.19
250	500	36.61 ± 3.07	48.94 ± 2.77
F-ratio Treatment $_{(1,16)}$		429.96*	492.25*
F-ratio Dose $_{(3,16)}$		74.46*	60.38*
F-ratio Treatment × Dose $_{(3,16)}$		9.86*	8.55*
HSD		5.92	5.67

Parameter	MLR Equation	β-Regression Coefficient			Multiple Correlation
		Cr(VI)	(MEL)	(Cr(VI) × MEL)	
Anthocyanin	Y = 0.53 + 0.04 Cr(VI) + 0.02 (MEL) + 0.0001 (Cr(VI) × MEL)	0.54	0.36	0.38	0.99*
Xanthophyll	Y = 16.61 + 0.05 Cr(VI) + 0.01 (MEL) + 9 × 10^{-5} Cr(VI) × MEL)	0.61	0.31	0.34	0.99*

3.2. Protein Content

Table 3 represents the variations in the protein content of 7-day-old *Vigna radiata* plants due to Cr(VI) treatment and MEL supplementation. The protein content in untreated plants was observed to be 18.70 mg g^{-1} FW, which lowered to 9.60 mg g^{-1} FW in the Cr(VI)-treated plants. The supplementation of 125, 250, and 500 ppm MEL concentrations to the Cr(VI)-treated plants resulted in the elevation of the protein content to 12.60, 15.73, and 16.83 mg g^{-1} FW, respectively. The ANOVA analysis depicted a statistically significant difference in the protein content, with an HSD value of 1.57 (Table 4). The MLR analysis revealed the negative impact of Cr(VI) treatment (negative β value) and positive impact of the Cr(VI) × MEL application (positive β value) on the protein content.

Table 3. Effect of Cr(VI), different concentrations of methanol extract of *Rhododendron arboreum* leaves (MEL), and their combinations on the protein content and activities of superoxide dismutase (SOD) and guaiacol peroxidase (POD) enzymes in 7-day-old *Vigna radiata* seedlings. The values are the means of three replicates ± S.D. (standard deviation). Tukeys test performed and significance checked at *$p \leq 0.05$. The F ratio is the term in which ANOVA is represented; Cr(VI) = Dose, MEL treatments = Treatment and combination of Cr(VI) and MEL = Dose × treatment.

Concentrations		Protein Content (μg/g FW)	SOD (UA mg g^{-1} Protein)	POD (UA mg g^{-1} Protein)
Cr(VI) (μM)	MEL (ppm)			
0	0	18.70 ± 0.45	1.37 ± 0.11	132.08 ± 2.59
0	125	14.23 ± 0.45	1.70 ± 0.47	132.35 ± 2.09
0	250	15.90 ± 0.30	1.98 ± 1.24	133.38 ± 4.68
0	500	16.50 ± 0.90	2.09 ± 0.05	146.67 ± 1.27
250	0	9.60 ± 0.30	6.08 ± 0.66	160.58 ± 8.76
250	125	12.60 ± 0.90	7.71 ± 0.08	184.77 ± 9.37
250	250	15.73 ± 0.45	8.51 ± 0.17	196.39 ± 1.86
250	500	16.83 ± 0.25	9.22 ± 0.09	241.91 ± 36.27
F-ratio Treatment $_{(1,16)}$		135.06*	779.27*	113.04*
F-ratio Dose $_{(3,16)}$		43.05*	14.63*	13.10*
F-ratio Treatment × Dose $_{(3,16)}$		93.06*	5.56*	6.05*
HSD		1.57	1.51	38.97

Parameter	MLR Equation	β-Regression Coefficient			Multiple Correlation
		Cr(VI)	(MEL)	(Cr(VI) × MEL)	
Protein Content	Y= 16.79 + −0.02 Cr(VI) + −0.002 (MEL) + 7 × 10^{-6} (Cr(VI) × MEL)	−1.17	−0.14	1.05	0.86*
SOD	Y= 1.48 + 0.02 Cr(VI) + 0.001 (MEL) +2 × 10^{-6} (Cr(VI) × MEL)	0.80	0.08	0.24	0.99*
POD	Y= 129.53 + 0.12 Cr(VI) + 0.03 (MEL) + 0.0005 (Cr(VI) × MEL)	0.43	0.15	0.59	0.99*

Table 4. Effect of Cr(VI), different concentrations of methanol extract of *Rhododendron arboreum* leaves (MEL), and their combinations on the activities of catalase (CAT), ascorbate peroxidase (APOX), and glutathione reductase (GR) enzymes in 7-day-old *Vigna radiata* seedlings. The values are the means of three replicates ± SD. (standard deviation). Tukeys test performed and significance checked at $^{*}p \leq 0.05$. The F ratio is the term in which ANOVA is represented; Cr(VI) = Dose, MEL treatments = Treatment and combination of Cr(VI) and MEL = Dose × treatment.

Concentrations		CAT (UA mg g^{-1} Protein)	APOX (UA mg g^{-1} Protein)	GR (UA mg g^{-1} Protein)
Cr(VI) (μM)	MEL (ppm)			
0	0	7.27 ± 1.19	14.28 ± 1.78	27.01 ± 3.62
0	125	7.56 ± 0.65	14.76 ± 4.32	25.96 ± 3.10
0	250	7.82 ± 1.86	15.02 ± 3.47	28.20 ± 2.26
0	500	8.40 ± 1.50	16.61 ± 2.82	28.44 ± 1.78
250	0	16.40 ±1.92	41.66 ± 6.27	47.16 ± 3.71
250	125	17.54 ± 2.19	53.04 ± 2.61	57.16 ± 1.24
250	250	18.58 ± 2.65	61.30 ± 5.74	61.44 ± 3.56
250	500	23.22 ± 2.06	80.94 ± 2.72	74.78 ± 3.92
F-ratio Treatment $_{(1,16)}$		217.87*	725.75*	689.60*
F-ratio Dose $_{(3,16)}$		5.24*	28.78*	24.07*
F-ratio Treatment × Dose $_{(3,16)}$		2.76	22.64*	18.55*
HSD		5.24	11.33	8.63

Parameter	MLR Equation	β-Regression Coefficient			Multiple Correlation
		Cr(VI)	(MEL)	(Cr(VI) × MEL)	
CAT	Y= 7.27 + 0.03 Cr(VI) + 0.002 (MEL) + 5 × 10^{-6} (Cr(VI) × MEL)	0.73	0.07	0.33	0.99*
APOX	Y= 14.15 + 0.11 Cr(VI) + 0.004 (MEL) + 0.0003 (Cr(VI) × MEL)	0.57	0.03	0.51	0.99*
GR	Y= 26.54 + 0.08 Cr(VI) + 0.003 (MEL) + 0.0002 (Cr(VI) × MEL)	0.61	0.04	0.47	0.99*

3.3. Enzymatic Antioxidants

The antioxidant enzymes were significantly altered by Cr(VI) exposure and supplementation with various concentrations of MEL. The superoxide dismutase (SOD) activity of 7-day-old *Vigna radiata* plants was enhanced from 1.37 (control) to 6.08 UA mg g^{-1} protein under Cr(VI) treatment (Table 3). The combination of Cr(VI) and MEL further increased the SOD activity, and maximum enhancement to 9.22 UA mg g^{-1} protein was noted in the case of Cr(VI) × 500 ppm MEL. Similar trends were observed in POD (Table 3), CAT, APOX, GR (Table 4), DHAR, PPO, GST, and GPOX (Table 5) as Cr(VI) increased the enzyme activity, and co-application of MEL further enhanced the enzyme activity.

Two-way ANOVA analysis revealed that all the enzymes, SOD, POD, CAT, APOX, GR, DHAR, PPO, GST, and GPOX, had a statistically significant difference except in the case of the F ratio of the Cr(VI) × MEL interaction in CAT activity. The HSD values for SOD, POD, CAT, APOX, GR, DHAR, PPO, GST, and GPOX were noted to be 1.51, 38.97, 5.24, 11.33, 8.63, 7.59, 1.29, 6.79, and 1.63, respectively. The β values in the MLR analysis for all the enzymes were positive for all the treatments, indicating the positive correlation.

3.4. Non-Enzymatic Antioxidants

The effect of individual and binary treatments of Cr(VI) and MEL on the contents of non-enzymatic antioxidants in 7-day-old *Vigna radiata* plants is presented in Table 6. The ascorbic acid content was observed to be enhanced to 22.6 μg g^{-1} FW in Cr(VI)-treated plants in response to 11.97 μg g^{-1} FW in the control. Supplementation with 500 ppm MEL to metal-stressed plants further elevated the ascorbic acid content to 49.05 μg g^{-1} FW. The tocopherol content in the plants was increased with Cr(VI) (47.88 μg g^{-1} FW) as compared to the control (21.88 μg g^{-1} FW). The content was further enhanced by the application of a combination of Cr(VI) and 125, 250 and 500 ppm MEL (62.52, 82.07, and 98.99 μg g^{-1} FW). The glutathione content was also observed to increase with Cr(VI) exposure to the plants (481.71 μg g^{-1} FW) as compared to the control (299.79 μg g^{-1} FW). Priming of metal-stressed plants with MEL elevated the glutathione content even higher (635.45 μg g^{-1} FW at Cr(VI) × 500 ppm MEL). Ascorbic acid, tocopherol, and glutathione contents in 7-day-old *Vigna radiata* plants showed a statistically significant difference at $p \leq 0.05$ in the two-way ANOVA analysis, with HSD values of 3.42, 10.001, and 50.8, respectively. MLR analysis revealed that ascorbic acid, tocopherol, and glutathione increased (positive β values) with Cr(VI) treatment and Cr(VI) × MEL interactions.

Table 5. Effect of Cr(VI), different concentrations of methanol extract of *Rhododendron arboreum* leaves (MEL), and their combinations on the activities of dehydroascorbate reductase (DHAR), polyphenol oxydase (PPO), glutathione S-transferase (GST), and glutathione peroxidase (GPOX) enzymes in 7-day-old *Vigna radiata* seedlings. The values are the means of three replicates ± SD. (standard deviation). Tukeys test performed and significance checked at *$p \leq 0.05$. The F ratio is the term in which ANOVA is represented; Cr(VI) = Dose, MEL treatments = Treatment and combination of Cr(VI) and MEL = Dose × treatment.

Concentrations		DHAR (UA mg g^{-1} Protein)	PPO (UA mg g^{-1} Protein)	GST (UA mg g^{-1} Protein)	GPOX (UA mg g^{-1} Protein)
Cr(VI) (μM)	MEL (ppm)				
0	0	26.95 ± 2.57	2.54 ± 0.42	16.80 ± 1.47	16.41 ± 0.34
0	125	25.71 ± 2.85	2.75 ± 0.27	16.91 ± 0.18	17.75 ± 0.39
0	250	27.80 ± 2.65	3.13 ± 0.13	17.90 ± 0.82	17.82 ± 0.13
0	500	30.57 ± 0.75	3.74 ± 0.37	18.56 ± 0.29	17.98 ± 0.34
250	0	54.28 ± 2.85	8.96 ± 0.86	34.64 ± 4.46	26.09 ± 0.67
250	125	69.62 ± 1.08	10.83 ± 0.43	45.82 ± 2.23	27.65 ± 0.79
250	250	78.92 ± 4.32	12.47 ± 0.43	56.23 ± 3.94	26.49 ± 0.21
250	500	92.65 ± 2.67	13.58 ± 0.37	64.51 ± 1.65	28.67 ± 1.05
F-ratio Treatment $_{(1,16)}$		1770.58*	2035.90*	1115.51*	1708.56*
F-ratio Dose $_{(3,16)}$		65.78*	45.55*	49.10*	13.93*
F-ratio Treatment × Dose $_{(3,16)}$		44.24*	16.64*	38.32*	3.09*
HSD		7.59	1.29	6.79	1.63

Parameter	MLR Equation	β-Regression Coefficient			Multiple Correlation
		Cr(VI)	(MEL)	(Cr(VI) x MEL)	
DHAR	Y= 25.9 + 0.12 Cr(VI) + 0.008 (MEL) + 0.0003 (Cr(VI) × MEL)	0.63	0.06	0.44	0.99*
PPO	Y= 2.5 + 0.02Cr(VI) + 0.002 (MEL) +3 × 10^{-6} Cr(VI) × MEL)	0.79	0.10	0.25	0.99*
GST	Y= 16.71 + 0.08 Cr(VI) + 0.003 (MEL) + 0.0002 (Cr(VI) × MEL)	0.56	0.03	0.51	0.99*
GPOX	Y= 16.92 + 0.03 Cr(VI) + 0.002 (MEL) + 7 × 10^{-7} (Cr(VI) × MEL)	0.94	0.09	0.05	0.99*

Table 6. Effect of Cr(VI), different concentrations of methanol extract of *Rhododendron arboreum* leaves (MEL), and their combinations on ascorbic acid, tocopherol, and glutathione content in 7-day-old *Vigna radiata* seedlings. The values are the means of three replicates ± SD. (standard deviation). Tukeys test performed and significance checked at *$p \leq 0.05$. The F ratio is the term in which ANOVA is represented; Cr(VI) = Dose, MEL treatments = Treatment and combination of Cr(VI) and MEL = Dose × treatment.

Concentrations		Ascorbic Acid (μg g^{-1} FW)	Tocopherol (μg g^{-1} FW)	Glutathione (μg g^{-1} FW)
Cr(VI) (μM)	MEL (ppm)			
0	0	11.97 ± 0.36	21.88 ± 1.90	299.79 ± 7.68
0	125	15.80 ± 0.72	30.42 ± 1.02	333.10 ± 8.87
0	250	19.70 ± 1.88	32.22 ± 0.62	384.35 ± 20.33
0	500	26.45 ± 0.49	40.76 ± 4.12	443.28 ± 16.01
250	0	22.60 ± 0.41	47.88 ± 0.69	481.71 ± 11.74
250	125	28.45 ± 1.27	62.52 ± 0.04	407.41 ± 20.33
250	250	39.30 ± 0.65	82.07 ± 4.85	520.15 ± 29.10
250	500	49.05 ± 2.23	98.99 ± 7.33	635.45 ± 19.34
F-ratio Treatment $_{(1,16)}$		1097.11*	828.52*	396.82*
F-ratio Dose $_{(3,16)}$		326.84*	107.15*	107.27*
F-ratio Treatment × Dose $_{(3,16)}$		32.78*	27.12*	13.43*
HSD		3.42	10.001	50.8

Parameter	MLR Equation	β-Regression Coefficient			Multiple Correlation
		Cr(VI)	(MEL)	(Cr(VI) × MEL)	
Ascorbic Acid	Y= 12.15 + 0.04 Cr(VI) + 0.02 (MEL) + 0.0001 (Cr(VI) × MEL)	0.47	0.46	0.37	0.99*
Tocopherol	Y= 23.59 + 0.1 Cr(VI) + 0.03 (MEL) + 0.0003 (Cr(VI) × MEL)	0.52	0.25	0.45	0.99*
Glutathione	Y= 301.33 + 0.51 Cr(VI) + 0.29 (MEL) + 0.0003 (Cr(VI) × MEL)	0.63	0.53	0.14	0.94*

3.5. Glutathione Imaging

Tagging of glutathione in the roots of *Vigna radiata* was done by incubating the root sections in monochlorobimane dye (MCB) according to the method of Hartmann et al. [32] and the imaging was visualized using confocal microscopy. MCB produces a fluorescent adduct with glutathione. The intensity of the blue color indicates the measure of glutathione in roots. The intensity of the blue color indicates that the glutathione content increased with Cr(VI) stress as compared to the control plants. The amplification in the amount of blue color proves that co-application of MEL (500 ppm) along with Cr(VI) stress further increased the production of glutathione (Figure 1).

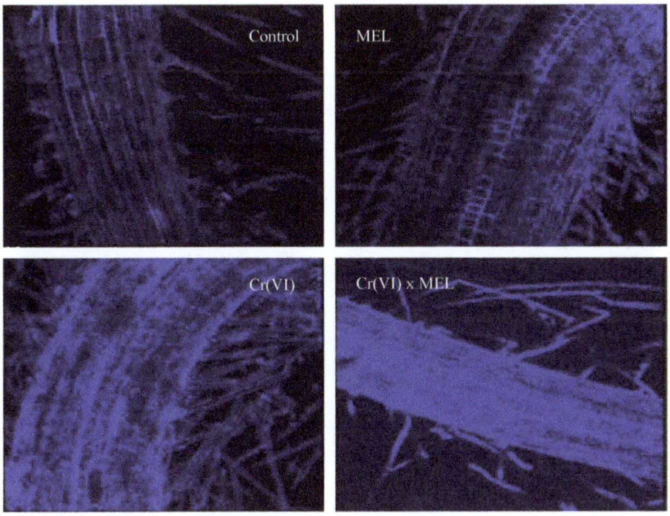

Figure 1. Effect of seed soaking treatment of methanol extract of *Rhododendron arboreum* leaves (MEL) on glutathione accumulation in *Vigna radiata* seedlings under Cr(VI) (250 μM) stress.

3.6. Polyphenol Content

Qualitative as well as quantitative analysis of the plant samples for polyphenolic compounds was carried out using ultra high-performance liquid chromatography (UHPLC). Five polyphenols, namely gallic acid, chlorogenic acid, caffeic acid, catechin, and coumaric acid, were found to be present in the plants. A comparative change in the polyphenolic contents of *Vigna radiata* plants grown under Cr(VI) exposure and the combination of Cr(VI) × MEL treatments is represented in Table 7. The contents of all the polyphenols increased with Cr(VI) stress. The supplementation of MEL to Cr(VI)-exposed plants further enhanced the polyphenol content and the maximum increase was observed in the case of the combination of Cr(VI) × 500 ppm MEL. Two-way ANOVA analysis of the polyphenols revealed a statistically significant difference. However, in case of catechin, the interaction of Cr(VI) × MEL treatment was noted to be non-significant. The β values from MLR analysis depict that gallic acid was positively affected by Cr(VI) treatment and negatively affected by co-application of Cr(VI) × MEL, whereas chlorogenic acid, caffeic acid, catechin, and coumaric acid were positively affected by the Cr(VI) and Cr(VI) × MEL treatments.

3.7. Amino Acid Content

The total amino acids content decreased to 591.47 μg g^{-1} FW with Cr(VI) stress from 2504.69 μg G^{-1} FW in control plants. MEL recovered the amino acid content in the Cr(VI)-stressed plants, and the maximum recovery was observed to be 2208.89 μg g^{-1} FW in plants given the combination of 500 ppm MEL with Cr(VI). In total, 17 amino acids, namely aspartic acid, glutamine, β-alanine, lysine, glutamic acid, asparagines, serine, isoleucine, glycine, threonine, citrulline, arginine, GABA (gamma-aminobutyric acid), cystine, ornithine, proline, and methionine, were found to be present in the plant samples. The variations in the contents of these amino acids due to treatment with Cr(VI) and MEL alone or in combination with each other are presented in Tables 8–11.

Table 7. Effect of Cr(VI), different concentrations of methanol extract of *Rhododendron arboreum* leaves (MEL), and their combinations on polyphenol contents in 7-day-old *Vigna radiata* seedlings. The values are the means of three replicates ± S.D. (standard deviation). Tukeys test performed and significance checked at $^*p \leq 0.05$. The F ratio is the term in which ANOVA is represented; Cr(VI) = Dose, MEL treatments = Treatment and combination of Cr(VI) and MEL = Dose × treatment.

Concentrations		Gallic Acid ($\mu g\ g^{-1}$ FW)	ChlorogenicAcid ($\mu g\ g^{-1}$ FW)	Caffeic Acid ($\mu g\ g^{-1}$ FW)	Catechin ($\mu g\ g^{-1}$ FW)	Coumaric Acid ($\mu g\ g^{-1}$ FW)
Cr(VI) (μM)	MEL (ppm)					
0	0	9.60 ± 1.40	1.74 ± 0.05	0.10 ± 0.002	263.80 ± 16.15	0.18 ± 0.004
0	125	9.87 ± 0.60	1.64 ± 0.01	0.15 ± 0.008	268.03 ± 49.93	0.16 ± 0.006
0	250	10.46 ± 1.06	1.74 ± 0.03	0.16±0.003	282.89 ± 55.04	0.18 ± 0.006
0	500	13.57 ± 0.91	1.81 ± 0.03	0.18 ± 0.004	287.84 ± 9.55	0.21 ± 0.006
250	0	16.05 ± 0.43	4.60 ± 0.11	0.94 ± 0.009	331.08 ± 53.67	0.44 ± 0.020
250	125	13.22 ± 0.71	3.21 ± 0.64	0.84 ± 0.01	422.96 ± 10.10	0.35 ± 0.010
250	250	15.18 ± 0.61	4.22 ± 0.25	0.96 ± 0.04	430.51 ± 20.83	0.59 ± 0.020
250	500	17.51 ± 0.50	5.26 ± 0.12	1.24 ± 0.16	447.50 ± 35.40	0.74 ± 0.050
F-ratio Treatment $_{(1,16)}$		185.09*	618.21*	124.54*	79.75*	1207.79*
F-ratio Dose $_{(3,16)}$		24.66*	19.73*	16.50*	4.35*	87.41*
F-ratio Treatment × Dose $_{(3,16)}$		3.93*	14.41*	10.08*	2.17	59.07*
HSD		2.35	0.72	0.16	102.71	0.06

Parameter	MLR Equation	β-Regression Coefficient			Multiple Correlation
		Cr(VI)	(MEL)	(Cr(VI) × MEL)	
Gallic Acid	Y= 9.08 + 0.02 Cr(VI) + 0.008 (MEL) + 1 × 10^{-6} (Cr(VI) × MEL)	0.96	0.54	−0.2	0.93*
Chlorogenic Acid	Y= 1.69 + 0.008 Cr(VI) + 0.0002 (MEL) + 8 × 10^{-7} (Cr(VI) × MEL)	0.76	0.02	0.24	0.94*
Caffeic Acid	Y= 0.12 + 0.002 Cr(VI) + 0.0001 (MEL) + 2× 10^{-7} (Cr(VI) × MEL)	0.83	0.06	0.21	0.99*
Catechin	Y= 264.5 + 0.39 Cr(VI) + 0.05 (MEL) + 0.0006 (Cr(VI) × MEL)	0.67	0.12	0.34	0.96*
Coumaric Acid	Y= 0.17 + 0.0008 Cr(VI) + 7 × 10^{-6} (MEL) + 3 × 10^{-7} (Cr(VI) × MEL)	0.51	0.05	0.53	0.96*

Table 8. Effect of Cr(VI), different concentrations of methanol extract of *Rhododendron arboreum* leaves (MEL), and their combinations on aspartic acid, glutamine, β-alanine, and lysine contents in 7-day-old *Vigna radiata* seedlings. The values are the means of three replicates ± SD. (standard deviation). Tukeys test performed and significance checked at $^*p \leq 0.05$. The F ratio is the term in which ANOVA is represented; Cr(VI) = Dose, MEL treatments = Treatment and combination of Cr(VI) and MEL = Dose × treatment.

Concentrations		Aspartic Acid ($\mu g\ g^{-1}$ FW)	Glutamine ($\mu g\ g^{-1}$ FW)	β-Alanine ($\mu g\ g^{-1}$ FW)	Lysine ($\mu g\ g^{-1}$ FW)
Cr(VI) (μM)	MEL (ppm)				
0	0	79.00 ± 0.85	40.61 ± 0.35	168.83 ± 0.86	35.11 ± 0.48
0	125	45.71 ± 2.43	37.30 ± 0.46	90.79 ± 0.34	17.81 ± 0.62
0	250	63.18 ± 0.39	38.45 ± 0.50	101.34 ± 0.51	18.85 ± 0.85
0	500	74.18 ± 0.68	39.72 ± 0.98	114.37 ± 0.48	20.73 ± 0.81
250	0	32.16 ± 0.60	8.82 ± 1.12	35.96 ± 0.98	18.54 ± 1.44
250	125	73.39 ± 0.57	30.08 ± 0.71	115.62 ± 0.22	23.70 ± 1.03
250	250	75.90 ± 0.13	36.40 ± 0.51	124.54 ± 1.40	29.54 ± 0.78
250	500	78.04 ± 0.51	39.60 ± 0.57	164.09 ± 0.67	30.84 ± 1.14
F-ratio Treatment $_{(1,16)}$		2.43	1295.79*	769.44*	43.96*
F-ratio Dose $_{(3,16)}$		509.11*	529.81*	2956.89*	48.24*
F-ratio Treatment × Dose $_{(3,16)}$		1520.74*	655.19*	17456.18*	285.57*
HSD		2.87	1.98	2.19	2.64

Parameter	MLR equation	β-Regression Coefficient			Multiple correlation
		Cr(VI)	(MEL)	(Cr(VI) × MEL)	
Aspartic Acid	Y= 63.48 + −0.06 Cr(VI) + 0.009 (MEL) + 0.0003 (Cr(VI) × MEL)	−0.17	0.1	0.7	0.62
Glutamine	Y= 39.03 + −0.08 Cr(VI) + 5 × 10^{-6} (MEL) +0.0002 (Cr(VI) × MEL)	−1.12	−0.0009	0.94	0.88*
β-Alanine	Y= 135.01 + −0.3 Cr(VI) + −0.074 (MEL) + 0.0012 (Cr(VI) × MEL)	−0.95	−0.34	1.31	0.8*
Lysine	Y= 27.81 + −0.03 Cr(VI) + −0.02 (MEL) + 0.0002 (Cr(VI) × MEL)	−0.6	−0.64	1.26	0.71*

Table 9. Effect of Cr(VI), different concentrations of methanol extract of *Rhododendron arboreum* leaves (MEL), and their combinations on glutamic acid, asparagine, serine, and isoleucine contents in 7-day-old *Vigna radiata* seedlings. The values are the means of three replicates ± S.D. (standard deviation). Tukeys test performed and significance checked at $*p \leq 0.05$. The F ratio is the term in which ANOVA is represented; Cr(VI) = Dose, MEL treatments = Treatment and combination of Cr(VI) and MEL = Dose × treatment.

Concentrations		Glutamic Acid (μg g^{-1} FW)	Asparagine (μg g^{-1} FW)	Serine (μg g^{-1} FW)	Isoleucine (μg g^{-1} FW)
Cr(VI) (μM)	MEL (ppm)				
0	0	29.83 ± 0.76	1134.65 ± 48.39	10.01 ± 0.38	174.95 ± 1.75
0	125	15.84 ± 0.79	443.42 ± 9.80	1.23 ± 0.09	65.96 ± 0.77
0	250	18.80 ± 0.87	652.13 ± 14.26	1.94 ± 0.13	68.31 ± 0.60
0	500	22.84 ± 0.70	686.02 ± 12.23	3.68 ± 0.56	86.11 ± 1.10
250	0	8.65 ± 1.31	355.18 ± 34.14	0.58 ± 0.15	13.75 ± 3.03
250	125	25.35 ± 0.71	824.64 ± 26.16	6.15 ± 0.06	87.03 ± 1.25
250	250	27.43 ± 0.52	875.57 ± 16.85	6.98 ± 0.12	104.83 ± 0.99
250	500	28.47 ± 0.50	974.98 ± 22.29	7.84 ± 0.89	146.25 ± 16.09
F-ratio Treatment $_{(1,16)}$		3.83	7.20*	49.47*	20.52*
F-ratio Dose $_{(3,16)}$		73.33*	58.91*	30.23*	49.36*
F-ratio Treatment × Dose $_{(3,16)}$		489.49*	651.33*	451.69*	447.64*
HSD		2.29	73.65	1.153	16.62

Parameter	MLR equation	β-Regression Coefficient			Multiple correlation
		Cr(VI)	(MEL)	(Cr(VI) × MEL)	
Glutamic Acid	Y= 23.43 + −0.03 Cr(VI) + −0.007 (MEL) + 0.0002 (Cr(VI) × MEL)	−0.61	−0.19	1.02	0.66
Asparagine	Y= 851.39 + −1.31 Cr(VI) + −0.55 (MEL) + 0.0065 (Cr(VI) × MEL)	−0.66	−0.42	1.13	0.64
Serine	Y= 6.15 + −0.01 Cr(VI) + −0.009 (MEL) + 9 × 10^{-6} (Cr(VI) × MEL)	−0.54	−0.5	1.13	0.64
Isoleucine	Y= 127.8 + −0.37 Cr(VI) + −0.13 (MEL) + 0.001 (Cr(VI) × MEL)	−0.99	−0.52	1.37	0.78*

Table 10. Effect of Cr(VI), different concentrations of methanol extract of *Rhododendron arboreum* leaves (MEL), and their combinations on glycine, threonine, citrulline, arginine, and gamma-aminobutyric acid (GABA) contents in 7-day-old *Vigna radiata* seedlings. The values are the means of three replicates ± SD. (standard deviation). Tukeys test performed and significance checked at $*p \leq 0.05$. The F ratio is the term in which ANOVA is represented; Cr(VI) = Dose, MEL treatments = Treatment and combination of Cr(VI) and MEL = Dose × treatment.

Concentrations		Glycine (μg g^{-1} FW)	Threonine (μg g^{-1} FW)	Citrulline (μg g^{-1} FW)	Arginine (μg g^{-1} FW)	GABA (μg g^{-1} FW)
Cr(VI)(μM)	MEL (ppm)					
0	0	9.47 ± 0.56	6.39 ± 0.68	22.55 ± 1.18	647.02 ± 12.48	5.009 ± 0.26
0	125	4.64 ± 0.40	2.72 ± 0.22	10.97 ± 0.81	97.11 ± 5.87	1.89 ± 0.04
0	250	5.48 ± 0.46	3.38 ± 0.72	12.60 ± 0.50	164.57 ± 6.72	2.47 ± 0.40
0	500	6.07 ± 0.76	3.81 ± 0.06	15.61 ± 1.07	181.85 ± 6.55	3.06 ± 0.67
250	0	3.58 ± 0.32	1.37 ± 0.42	4.91 ± 0.75	78.43 ± 7.27	1.18 ± 0.37
250	125	6.77 ± 0.37	3.72 ± 0.09	15.93 ± 0.82	311.96 ± 49.42	2.89 ± 0.12
250	250	7.25 ± 0.29	4.22 ± 0.27	19.47 ± 0.62	466.28 ± 18.01	3.49 ± 0.43
250	500	8.14 ± 0.96	4.86 ± 0.76	20.94 ± 0.52	575.68 ± 11.60	4.16 ± 0.83
F-ratio Treatment $_{(1,16)}$		0.006	7.06*	0.12	108.4*	0.84
F-ratio Dose $_{(3,16)}$		6.24*	5.26*	45.23*	91.77*	6.97*
F-ratio Treatment × Dose $_{(3,16)}$		73.37*	56.22*	305.2*	725.5*	41.19*
HSD		1.59	1.38	2.32	56.87	1.31

Parameter	MLR Equation	β-Regression Coefficient			Multiple Correlation
		Cr(VI)	(MEL)	(Cr(VI) × MEL)	
Glycine	Y= 7.42 + −0.01 Cr(VI) + −0.005 (MEL) + 5 × 10^{-6} (Cr(VI) × MEL)	−0.77	−0.47	1.2	0.67
Threonine	Y= 4.84 + −0.01 Cr(VI) + −0.003 (MEL) + 4 × 10^{-6} (Cr(VI) × MEL)	−0.96	−0.46	1.19	0.7
Citrulline	Y= 17.32 + −0.03 Cr(VI) + −0.009 (MEL) + 0.0001 (Cr(VI) × MEL)	−0.76	−0.29	1.17	0.72*
Arginine	Y= 423.61 + −1.09 Cr(VI) + −0.69 (MEL) + 0.006 (Cr(VI) × MEL)	−0.65	−0.61	1.34	0.76*
GABA	Y= 3.64 + −0.008 Cr(VI) + −0.002 (MEL) + 3 × 10^{-6} (Cr(VI) × MEL)	−0.83	−0.39	1.18	0.69

Table 11. Effect of Cr(VI), different concentrations of methanol extract of *Rhododendron arboreum* leaves (MEL), and their combinations on cystine, ornithine, proline, methionine, and total amino acid contents in 7-day-old *Vigna radiata* seedlings. The values are the means of three replicates ± SD. (standard deviation). Tukeys test performed and significance checked at $^{*}p \leq 0.05$. The F ratio is the term in which ANOVA is represented; Cr(VI) = Dose, MEL treatments = Treatment and combination of Cr(VI) and MEL = Dose × treatment.

Concentrations		Cystine (μg g^{-1} FW)	Ornithine (μg g^{-1} FW)	Proline (μg g^{-1} FW)	Methionine (μg g^{-1} FW)	Total Amino Acids (μg g^{-1} FW)
Cr(VI) (μM)	MEL (ppm)					
0	0	28.42 ± 2.48	9.91 ± 1.03	44.49 ± 9.005	51.98 ± 0.79	2504.69 ± 54.9
0	125	8.56 ± 2.02	5.82 ± 0.92	14.39 ± 2.14	26.21 ± 0.55	893.15 ± 24.40
0	250	11.94 ± 1.48	6.58 ± 0.44	16.94 ± 0.90	33.15 ± 1.47	1223.58 ± 13.29
0	500	14.78 ± 0.85	7.38 ± 0.51	19.38 ± 0.59	43.61 ± 0.67	1347.07 ± 14.95
250	0	4.54 ± 0.54	3.30 ± 0.47	9.99 ± 0.92	9.10 ± 0.68	591.47 ± 47.18
250	125	19.49 ± 0.64	7.47 ± 0.40	22.49 ± 1.65	45.85 ± 1.04	1626.33 ± 73.72
250	250	23.03 ± 0.43	8.42 ± 0.19	26.66 ± 2.99	47.92 ± 1.009	1892.25 ± 29.66
250	500	27.77 ± 1.03	9.59 ± 1.17	32.17 ± 3.34	50.54 ± 0.52	2208.89 ± 49.85
F-ratio Treatment $_{(1,16)}$		24.46*	0.59	0.41	1.08	24.5*
F-ratio Dose $_{(3,16)}$		28.71*	9.01*	6.87*	366.005*	144.16*
F-ratio Treatment × Dose $_{(3,16)}$		250.58*	52.11*	54.57*	1549.31*	1424.96*
HSD		3.89	2.04	10.52	2.53	122.65

Parameter	MLR Equation	β-Regression Coefficient			Multiple correlation
		Cr(VI)	(MEL)	(Cr(VI) × MEL)	
Cystine	Y= 19.91 + −0.04 Cr(VI) + −0.01 (MEL) + 0.0002 (Cr(VI) × MEL)	−0.62	−0.4	1.23	0.74*
Ornithine	Y= 8.11 + −0.01 Cr(VI) + −0.003 (MEL) + 6 × 10^{-6} (Cr(VI) × MEL)	−0.84	−0.29	1.22	0.76*
Proline	Y= 31.96 + −0.07 Cr(VI) + −0.03 (MEL) + 0.0003 (Cr(VI) × MEL)	−0.87	−0.66	1.29	0.7*
Methionine	Y= 39.58 + −0.06 Cr(VI) + −0.004 (MEL) + 0.0003 (Cr(VI) × MEL)	−0.58	−0.05	0.89	0.65
Total Amino Acids	Y= 1835.4 + −3.57 Cr(VI) + −1.56 (MEL) + 0.01 (Cr(VI) × MEL)	−0.73	−0.47	1.25	0.71*

Two-way analysis of the variance of the amino acids revealed that glutamine, β-alanine and lysine, asparagine, serine, isoleucine, threonine, arginine, cystine, and the total amino acids were statistically significant for the Cr(VI), MEL, and Cr(VI) × MEL treatments, whereas the F ratios of aspartic acid, glutamic acid, glycine, citrulline, GABA, ornithine, proline, and methionine were statistically significant for MEL and Cr(VI) × MEL but not for theCr(VI) treatment. MLR analysis of aspartic acid resulted in positive values of the β-regression coefficient for Cr(VI) treatment and positive values of the β-regression coefficients for the combined Cr(VI) × MEL treatment, which reveals that Cr(VI) stress decreased, whereas, Cr(VI) × MEL increased the aspartic acid content in the *Vigna radiata* plants. The glutamine, β-alanine, lysine (Table 8), glutamic acid, asparagine, serine, isoleucine (Table 9), glycine, threonine, citrulline, arginine, GABA (Table 10), cystine, ornithine, proline, methionine, and total amino acids (Table 11) revealed negative values of the β-regression coefficients for Cr(VI) treatment, indicating its negative effect, and positive values of the β-regression coefficients for the interaction of Cr(VI) × MEL, indicating the increasing effect in the amino acids.

3.8. Organic Acids

Data obtained on the effect of Cr(VI) exposure and the combination of MEL with Cr(VI) treatment on the organic acids of the Krebs cycle (citrate, succinate, fumarate, and malate) in *Vigna radiata* plants are presented in Table 12. The fumaric acid content was elevated to 0.415 mg g^{-1} DW (dry weight) in Cr(VI)-treated plants in comparison to control plants (0.38 mg g^{-1} DW). The 500 ppm MEL application to the Cr(VI)-treated plants enhanced the fumaric acid content to 0.416 mg g^{-1} DW. The citric acid content also increased to 2.94 mg g^{-1} DW in Cr(VI)-exposed plants as compared to 2.32 mg g^{-1} DW in unexposed plants. Treatment with 500 ppm MEL further enhanced its content to 3.68 mg g^{-1} DW in Cr(VI)-treated plants as compared to non-treated plants. An enhancement was observed in the malic acid content of Cr(VI)-treated plants to 2.15 mg g^{-1} DW in comparison with control plants, which contained 1.47 mg malic acid content g^{-1} DW. Co-application of different MEL concentrations (125, 250, and 500 ppm) to Cr(VI)-stressed plants resulted in an increase of the malic acid content to 2.19,

2.48, and 2.50 mg g^{-1} DW, respectively. The succinic acid content followed the same trend. The succinic acid content in control plants was observed to be 0.870 mg g^{-1} DW, which increased to 0.876 mg g^{-1} DW with Cr(VI) application. Supplementation with 125, 250, and 500 ppm MEL to Cr(VI)-exposed plants increased the succinic acid content to 0.867, 0.884, and 0.899 mg g^{-1} DW, respectively.

Two-way ANOVA revealed that the F ratios of all the four organic acids were statistically significant under Cr(VI) treatment, whereas the F ratio of the MEL dose was only significant in case of the citric acid content. F ratios for the interaction of Cr(VI) × MEL were statistically significant for the fumaric acid, citric acid, and succinic acid but not significant for the malic acid content. HSD values for fumaric acid, malic acid, citric acid, and succinic acid contents were noted to be 0.02, 0.61, 0.4, and 0.09, respectively. MLR analysis revealed the positive impact of Cr(VI) treatment (positive β values) and the negative impact of the interaction of Cr(VI) × MEL (positive β values) on thefumaric acid and succinate contents, whereas Cr(VI) and Cr(VI) × MEL had a positive effect (positive β values) on the malic acid and citric acid contents.

Table 12. Effect of Cr(VI), different concentrations of methanol extract of *Rhododendron arboreum* leaves (MEL), and their combinations on fumaric acid, malic acid, citric acid, and succinic acid content in 7-day-old *Vigna radiata* seedlings. The values are the means of three replicates ± SD. (standard deviation). Tukeys test performed and significance checked at $^*p \leq 0.05$. The F ratio is the term in which ANOVA is represented; Cr(VI) = Dose, MEL treatments = Treatment and combination of Cr(VI) and MEL = Dose × treatment.

Concentrations		FumaricAcid Content (µg g^{-1} DW)	Malic Acid Content (µg g^{-1} DW)	Citric Acid Content (µg g^{-1} DW)	Succinic Acid Content (µg g^{-1} DW)
Cr(VI)(µM)	MEL (ppm)				
0	0	0.38 ± 0.009	1.47 ± 0.15	2.32 ± 0.02	0.870 ± 0.05
0	125	0.393 ± 0.005	1.45 ± 0.41	2.32 ± 0.15	0.783 ± 0.01
0	250	0.395 ± 0.005	1.46 ± 0.19	2.40 ± 0.01	0.824 ± 0.01
0	500	0.410 ± 0.008	1.47 ± 0.11	2.40 ± 0.01	0.866 ± 0.08
250	0	0.415 ± 0.010	2.15 ± 0.21	2.94 ± 0.34	0.876 ± 0.03
250	125	0.394 ± 0.008	2.19 ± 0.23	3.64 ± 0.10	0.867 ± 0.04
250	250	0.409 ± 0.007	2.48 ± 0.02	3.67 ± 0.05	0.884 ± 0.04
250	500	0.416 ± 0.008	2.50 ± 0.15	3.68 ± 0.08	0.899 ± 0.01
F-ratio Treatment $_{(1,16)}$		10.75*	95.55*	370.78*	11.30*
F-ratio Dose $_{(3,16)}$		1.97	1.11	11.18*	1.07
F-ratio Treatment × Dose $_{(3,16)}$		10.73*	1.05	8.20*	3.71*
HSD		0.02	0.61	0.4	0.09

Parameter	MLR Equation	β-Regression Coefficient			Multiple Correlation
		Cr(VI)	(MEL)	(Cr(VI) × MEL)	
Fumaric acid	Y= 0.38 + 0.0001 Cr(VI) + 5 × 10^{-6} (MEL) + 3 × 10^{-8} (Cr(VI) × MEL)	1.21	0.78	−1.16	0.8*
Malic acid	Y= 1.46 + 0.002 Cr(VI) + 1 × 10^{-6} 0.000001 (MEL) + 3 × 10^{-7} (Cr(VI) × MEL)	0.78	0.006	0.28	0.99*
Citric acid	Y= 2.23 + 0.003 Cr(VI) + 0.0002 (MEL) + 4 × 10^{-7} (Cr(VI) × MEL)	0.74	0.05	0.28	0.96*
Succinic acid	Y= 0.82 + 0.0002 Cr(VI) + 4 × 10^{-6} 0.000004 (MEL) + 3 × 10^{-8} (Cr(VI) × MEL)	0.88	0.21	−0.34	0.68

4. Discussion

Heavy metal contamination in soil is one of the foremost problems accountable for a reduction in agricultural yield. The most crucial segments of plant life are seed germination and growth of seedlings and these are adversely affected by metal stress [36]. Plants exposed to metal stress endure plentiful physiological and biochemical alterations because metal exposure involves oxidative stress [36]. The embarkment of stress endurance takes place when a stress causative agent approaches the cell surface or penetrates into the cytoplasm and injures the cell. The successive events intended to reinstate cell function are regarded as the stress response. The stress response is activated by a signal from a suitable receptor instantaneously after the commencement of the prevalence of the stress factor. At this stage, cell organization and utility is hindered. Oxidative damage generated by stress in the plant tissue is controlled by a combined operation of both antioxidant enzymes as well as non-enzymatic antioxidant systems. Apart from those, pigments, polyphenols, organic acids, and amino acids also play a key role in the plant defense against abiotic stress.

In the present study, the effect of Cr stress and MEL application on anthocyanin and xanthophyll pigments was observed. The results revealed that Cr treatment significantly enhanced the anthocyanin and xanthophyll content in comparison to untreated plants. Our results coincide with the findings of Kohli et al. [37], who observed an enhancement of anthocyanins and xanthophylls along with Pb metal stress in mustard, as well as Posmyk et al. [38], who reported an improvement in the anthocyanin content in red cabbage seedlings under Cu stress. The supplementation of MEL at 125, 250, and 500 ppm concentrations along with Cr(VI) metal stress further enhanced the pigment contents. Anthocyanins are involved in plants responses to various abiotic stresses and are reported to enhance stress tolerance by scavenging ROS or playing a role in stress signals. Kovinich et al. [39] reported that in response to various abiotic stresses in *Arabidopsis*, anthocyanins had diverse localizations at the tissue and organ levels. Anthocyanins amend the ROS level and the sensitivity to ROS-generating stresses in sustaining photosynthetic capacity [40]. Xanthophylls are considered as key antioxidants that defend plants subjected to heavy metal stress [41]. Xanthophyll is also vital for photo-protection; hence, its accumulation may downregulate Photosystem II actions to diminish oxidative damage [42]. Though reports explaining the approach to recover the pigment contents under stress are exclusively meager, the existing studies demonstrate that exogenous application of growth regulators improve the drought tolerance with increased activities of SOD, CAT, APX, ABA, and total improved pigment contents in maize [43]. The protein content in 7-day-old *Vigna radiata* Cr(VI)-stressed plants was observed to decrease as compared to the untreated control plants, whereas all the enzymes (SOD, POD, CAT, APOX, GR, DHAR, PPO, GST, and GPOX) in 7-day-old *Vigna radiata* plants showed enhanced activities under Cr(VI) stress. These results are in accordance with Rai et al. [44], who reported a decrease in protein and increase in enzymes with Cr stress. MEL supplementation showed an improvement in the protein content in Cr(VI)-treated plants. The combination of Cr(VI) and MEL further increased enzymatic activity at all concentrations when compared to Cr(VI) alone. SOD catalyzes the dismutation of the superoxide anion to di-oxygen and H_2O_2. Non-specific peroxidases are responsible for scavenging H_2O_2. Several reductases, like DHAR and GR, are responsible for keeping ascorbate and glutathione in the reduced form. SOD and catalase hold metal ions on their active sites as a fundamental fraction to combat the toxic effects of metal-stimulated ROS. The increase in both the protein content as well as enzymatic antioxidants supports the protective role of MEL towards the *Vigna radiata* plants to overcome the oxidative stress caused by Cr(VI). The non-enzymatic antioxidants, ascorbic acid, tocopherol, and glutathione, increased with Cr(VI) in comparison to untreated plants. The combination of Cr(VI) and MEL further enhanced the ascorbic acid, tocopherol, and glutathione contents and the maximum respective increase was observed in the case of Cr(VI) × 500 ppm MEL. Ascorbic acid is a well-known antioxidant involved in various processes, for example, cell wall expansion and cell division [45]. It scavenges free radicals directly in the aqueous phases of cells and guards the membrane and other hydrophobic sections from oxidative injury by redeveloping the antioxidant form of vitamin E [46]. It is also renowned for promoting photosynthetic pigments and improving the tolerance of plants in opposition to diverse stresses by scavenging ROS [47,48]. Therefore, the enhancement of anthocyanin and xanthophyll pigments in the present study might be attributed to the increased ascorbic acid content due to MEL application. Tocopherol is a water-soluble antioxidant found in chloroplasts that enhances pigment synthesis and modifies the biosynthesis pathways of pigments under stress conditions. Under salinity stress, it has been reported to play a role in several physiological functions, including growth regulation and the differentiation of plants [49,50]. Further, tocopherol has also been reported to protect cells from hydrogen peroxide and other free radicals generated in salinity stress by direct scavenging of ROS as well as collaborative action with antioxidant enzymes and other antioxidants [51–53]. An increase in the tocopherol content in *Arabidopsis thaliana* in response to Cd and Cu stress was observed [54]. The reason for this enhancement was the increase in transcripts encoding enzymes of the tocopherol biosynthetic pathway in response to metal exposure. The vitamin E-deficient (vte1) mutant was observed to be more prone to metal-induced stress compared to the wild-type (WT) control.

It was concluded that tocopherol plays a vital function in the tolerance of *Arabidopsis* to oxidative stress induced by heavy metals, such as Cu and Cd [54]. Glutathione plays a central role in the scavenging of ROS as well as in the chelation of heavy metals. GSH protects the plants from heavy metal-induced oxidative damage by chelation, detoxification, and compartmentalization of heavy metals. Additionally, it acts through its metabolizing enzymes, particularly glutathione peroxidase, glutathione-*S*-transferase, glutathione reductase, and dehydroascorbate reductase, for efficient protection against ROS [55].

Polyphenols are proven stress busters for their ROS scavenging potential. Five polyphenols, namely gallic acid, chlorogenic acid, caffeic acid, catechin, and coumaric acid, were found to be present in the plants and a considerable increase in the levels of all polyphenolic compounds with Cr was observed in the present study. Kohli et al. [37] also established the increase in polyphenol contents under metal stress. This could be due to the increased activity of a variety of enzymes, such as chalcone synthase and cinnamate 4-hydroxylase, which are responsible for polyphenol synthesis [56,57]. In the present study, the combination of Cr(VI) and MEL further enhanced the polyphenol contents. It has been previously reported that antioxidants produced within plants reduce the effects of stresses in plants [58,59]. Those polyphenols possess the affinity to bind with metal ions and thereby inhibit the production of several ROS and some of them sequester the heavy metals entering the cell [60].

Amino acids are the building blocks of proteins. Different amino acids, namely aspartic acid, glutamine, β-alanine, lysine, glutamic acid, asparagine, serine, isoleucine, glycine, threonine, citrulline, arginine, GABA, cystine, ornithine, proline, and methionine, decreased with Cr(VI) stress when compared to control plants. This trend correlates with the decreased amount of protein in the present study. MEL recovered the amino acid content in the Cr(VI)-stressed plants. As compared to control plants, the contents of four organic acids (fumarate, citrate, malate, and succinate) were increased with Cr(VI) stress. The present study is in agreement with the work of Kohli et al. [61] and Ma [62], who reported that organic acids increase in abiotic stress conditions. Co-application of MEL along with Cr(VI) further increased the organic acid contents as compared to only Cr(VI)-treated plants, indicating the stress protective role of organic acids. Amino acids and organic acids are metabolites that have been reported to impart tolerance against heavy metal stress [63–65]. The utility of some of them is still unknown due to the convolution in plant responses to these stresses. In case of heavy metal stress, two mechanisms of detoxification and tolerance take place. Detoxification is external whereas tolerance is an internal process. In detoxification, plant roots secrete organic acids, which bind with metal ions to alter their movement as well as bioavailability, leading to the prevention of heavy metal uptake by plants. Meanwhile, in internal heavy metal tolerance, organic acids might chelate the metal ions inside the cytosol, converting them to less toxic or totally nontoxic byproducts [66,67]. Different plants have been reported to create a variety of ligands for aluminum, cadmium, copper, nickel, cobalt, and zinc. Amino acids and organic acids, such as citrate and malate, are reported to be probable ligands for heavy metals and are established to be engaged in their tolerance and detoxification [66–69]. Oven et al. [70] observed that cobalt metal treatment to a Co-hyperaccumulator, namely *Crotalaria cobalticola,* and non-accumulators, *Raufolia serpentina* and *Silene cucubalus*, resulted in an enhancement of citrate, indicating its participation in the heavy metal ion complex formation. Malate is reported to chelate zinc in the cytosols of Zn hyperaccumulators [71]. Zhang et al. [72] observed that taking away the aluminum from the roots leads to a quick decrease in malate release to the non-aluminum point, which demonstrated a receptive aluminum and malate-secreting mechanism. Over 24 h of exposure to 50 μM aluminum, 10-fold elevated malate and three- to five-fold elevated succinate secretion was observed in aluminum-tolerant genotypes as compared to aluminum-sensitive seedlings [73].

Ellagic acid has been reported by Ascacio-Valdés et al. [74] to have the ability to protect plants against stresses because of its antioxidant activity. Moreover, Khan et al. [75] reported that ellagic acid is one of the best antioxidants to shield *Brassica napus* L. plants against salinity stress, and due to its antioxidant properties, ellagic acid can enhance the growth and yield of the crop. *Rhododendron* leaf extract (MEL) is enriched with ellagic acid and several other antioxidant polyphenolic compounds.

As a natural source of ellagic acid, it is as beneficial as pure ellagic acid for improving stress tolerance by regulating different physiological processes under Cr metal stress. Consequently, *Rhododendron* leaf extract can be used as an economical source of polyphenols, especially ellagic acid, for safeguarding plants from toxic effects of Cr stress.

5. Conclusions

In the present study, Cr stress brought about physiological as well as metabolic alterations in *Vigna radiata* plants even at small concentrations. The *Rhododendron* leaf extract supplementation to Cr-stressed *Vigna radiata* plants helps in tolerating Cr toxicity by modulating the contents of pigments and activation of the enzymatic as well as non-enzymatic antioxidativedefense system. It also restored polyphenols, organic acids, and amino acids, which also provides extra protection to the plants from Cr stress. We thus conclude that exogenous application of *Rhododendron* leaf extract (rich in ellagic acid) reduced the effect of chromium metal stress in *Vigna radiata* plants. Further prospects of the work will be intended towards fortification of the idea of the stress ameliorative activity of *Rhododendron* extract by investigating the mechanism attributed to the increased defense of plants against heavy metal stress.

Author Contributions: Conceptualization, V.G., R.B. and P.A.; methodology, V.G. and P.S.; software, B.A.P. and M.N.A.; validation, P.B., S.A. and P.A.; formal analysis, R.B. and M.N.A.; investigation, V.G. and P.S.; resources, R.B. and P.A.; data curation, P.B. and B.A.P.; writing—Original draft preparation, V.G.; P.S., P.B. and S.A.; writing—Review and editing, R.B., P.A. and M.N.A.; visualization, V.G.; P.A. and R.B.; supervision, R.B.; project administration, R.B.; funding acquisition, B.A.P.; M.N.A. and P.A. All authors have read and agreed to the published version of the manuscript.

Funding: RSP-2019/116.

Acknowledgments: The authors would like to extend their sincere appreciation to the Researchers Supporting Project Number (RSP-2019/116), King Saud University, Riyadh, Saudi Arabia.

Conflicts of Interest: The authors declare no conflict of interest.

References

1. Hasanuzzaman, M.; Fujita, M. Heavy metals in the environment: Current status, toxic effects on plants and possible phytoremediation. In *Phytotechnologies: Remediation of Environmental Contaminants*; Anjum, N.A., Pereira, M.A., Ahmad, I., Duarte, A.C., Umar, S., Khan, N.A., Eds.; Taylor and Francis/CRC Press: Boca Raton, FL, USA, 2012; pp. 7–73.
2. Jomovaa, K.; Valkob, M. Advances in metal-induced oxidative stress and human disease. *Toxicology* **2011**, *283*, 65–87. [CrossRef] [PubMed]
3. Dheeba, B.; Sampathkumar, P. Ameliorating effect of fertilizers on biochemical characteristics of *Vigna radiata* treated with hexavalent chromium. *Biosci. Biotechnol. Res. Asia* **2014**, *11*, 301–307. [CrossRef]
4. Oliveira, H. Chromium as an environmental pollutant: Insights on induced plant toxicity. *J. Bot.* **2012**. [CrossRef]
5. Panda, S.K. Chromium-mediated oxidative stress and ultrastructural changes in root cells of developing rice seedlings. *J. Plant Physiol.* **2007**, *164*, 1419–1428. [CrossRef]
6. Ahmad, P.; Jaleel, C.A.; Salem, M.A.; Nabi, G.; Sharma, S. Roles of enzymatic and nonenzymatic antioxidants in plants during abiotic stress. *Crit. Rev. Biotechnol.* **2010**, *30*, 161–175. [CrossRef]
7. Ighodaro, O.M.; Akinloye, O.A. First line defence antioxidants-superoxide dismutase (SOD), catalase (CAT) and glutathione peroxidase (GPX): Their fundamental role in the entire antioxidant defence grid. *Alex. J. Med.* **2018**, *54*, 287–293. [CrossRef]
8. Niwa, T.; Doi, U.; Kato, Y.; Osawa, T. Antioxidative Properties of Phenolic Antioxidants Isolated from Corn Steep Liquor. *J. Agric. Food Chem.* **2001**, *49*, 177–182. [CrossRef]
9. Singh, R.; Kaushik, S.; Wang, Y.; Xiang, Y.; Novak, I.; Komatsu, M.; Tanaka, K.; Cuervo, A.M.; Czaja, M.J. Autophagy regulates lipid metabolism. *Nature* **2009**, *458*, 1131–1135. [CrossRef]
10. Andresen, M.; Cedergreen, N. Plant growth is stimulated by tea-seed extract: A new natural growth regulator? *Hortscience* **2010**, *45*, 1848–1853. [CrossRef]

11. Gautam, V.; Kohli, S.; Arora, S.; Bhardwaj, R.; Kazi, M.; Ahmad, A.; Raish, M.; Ganaie, M.; Ahmad, P. Antioxidant and Antimutagenic Activities of Different Fractions from the Leaves of *Rhododendron arboreum* Sm. and Their GC-MS Profiling. *Molecules* **2018**, *23*, 2239. [CrossRef]
12. Painuli, S.; Rai, N.; Kumar, N. Gas chromatography and mass spectrometry analysis of methanolic extract of leaves of *Rhododendron arboreum*. *Asian J. Pharm. Clin. Res.* **2016**, *9*, 66–69.
13. Roy, J.D.; Handique, A.K.; Barua, C.C.; Talukdar, A.; Ahmed, F.A.; Barua, I.C. Evaluation of phytoconstituents and assessment of adaptogenic activity in vivo in various extracts of *Rhododendron arboreum* (leaves). *J. Pharm. Biol. Res.* **2014**, *2*, 49–56. [CrossRef]
14. Kiruba, S.; Mahesh, M.; Nisha, S.R.; Paul, Z.M.; Jeeva, S. Phytochemical analysis of the flower extracts of *Rhododendron arboreum* Sm. ssp. nilagiricum (Zenker) Tagg. *Asian Pac. J. Trop. Biomed.* **2011**, *1*, S284–S286. [CrossRef]
15. Rath, B.P.; Hota, S.; Subhadarshini, S.; Dash, D.; Das, P.K. Consequence of chromium-tainted soil on physical and biochemical responses of *Vigna radiata* L. *J. Appl. Biol. Biotechnol.* **2019**, *7*, 35–41.
16. Cobbett, C.S. Heavy metals and plants-model systems and hyper accumulators. *New Phytol.* **2003**, *159*, 289–293. [CrossRef]
17. Lawrence, J. Determination of total xanthophyll and marigold oleoresin. *J. Ass. Off. Anal. Chem.* **1990**, *2*, 970–975.
18. Mancinelli, A.L. Photoregulation of Anthocyanin Synthesis. *Plant Physiol.* **1984**, *75*, 447–453. [CrossRef]
19. Lowry, O.H.; Rosebrough, N.J.; Farr, A.L.; Randall, R.J. Protein measurement with the Folin phenol reagent. *J. Biol. Chem.* **1951**, *193*, 265–275.
20. Pütter, J. Peroxidases. In *Methods of Enzymatic Analysis*; Elsevier: Amsterdam, The Netherlands, 1974; pp. 685–690. [CrossRef]
21. Aebi, H. Catalase in vitro. In *Methods in Enzymology*; Elsevier: Amsterdam, The Netherlands, 1984; pp. 121–126. [CrossRef]
22. Kono, Y. Generation of superoxide radical during autoxidation of hydroxylamine and an assay for superoxide dismutase. *Arch. Biochem. Biophys.* **1978**, *186*, 189–195. [CrossRef]
23. Nakano, Y.; Asada, K. Hydrogen-Peroxide Is Scavenged by Ascorbate-Specific Peroxidase in Spinach-Chloroplasts. *Plant Cell Physiol.* **1981**, *22*, 867–880. [CrossRef]
24. Carlberg, I.; Mannervik, B. Purification and characterization of the flavoenzyme glutathione reductase from rat liver. *J. Biol. Chem.* **1975**, *250*, 5475–5480. [PubMed]
25. Dalton, D.A.; Russell, S.A.; Hanus, F.J.; Pascoe, G.A.; Evans, H.J. Enzymatic reactions of ascorbate and glutathione that prevent peroxide damage in soybean root nodules. *Proc. Natl. Acad. Sci. USA* **1986**, *83*, 3811–3815. [CrossRef] [PubMed]
26. Kumar, K.; Khan, P. Age-related changes in catalase and peroxidase activities in the excised leaves of *Eleusine coracana* Gaertn. cv PR 202 during senescence. *Exp. Gerontol.* **1982**, *18*, 409–417. [CrossRef]
27. Habig, W.H.; Jakoby, W.B. Assays for differentiation of glutathione S-Transferases. In *Methods in Enzymology*; Elsevier: Amsterdam, The Netherlands, 1981; pp. 398–405. [CrossRef]
28. Flohé, L.; Günzler, W.A. Assays of glutathione peroxidase. In *Methods in enzymology*; Elsevier: Amsterdam, The Netherlands, 1984; Volume 105, pp. 114–120.
29. Roe, J.H.; Kuether, C.A. The determination of ascorbic acid in whole blood and urine through the 2, 4-dinitrophenylhydrazine derivavative of dehydroascorbic acid. *J. Biol. Chem.* **1943**, *147*, 399–407.
30. Martinek, R.G. Method for the determination of vitamin E (total tocopherols) in serum. *Clin. Chem.* **1964**, *10*, 1078–1086. [CrossRef] [PubMed]
31. Sedlak, J.; Lindsay, R.H. Estimation of total, protein-bound, and nonprotein sulfhydryl groups in tissue with Ellman's reagent. *Anal. Biochem.* **1968**, *25*, 192–205. [CrossRef]
32. Hartmann, T.N.; Fricker, M.D.; Rennenberg, H.; Meyer, A.J. Cell-specific measurement of cytosolic glutathione in poplar leaves. *Plant Cell Environ.* **2003**, *26*, 965–975. [CrossRef]
33. Iriti, M.; Rossoni, M.; Borgo, M.; Ferrara, L.; Faoro, F. Induction of Resistance to Gray Mold with Benzothiadiazole Modifies Amino Acid Profile and Increases Proanthocyanidins in Grape: Primary versus Secondary Metabolism. *J. Agric. Food Chem.* **2005**, *53*, 9133–9139. [CrossRef]
34. Chen, M.-C.; Wang, M.-K.; Chiu, C.-Y.; Huang, P.-M.; King, H.-B. Determination of low molecular weight dicarboxylic acids and organic functional groups in rhizosphere and bulk soils of Tsuga and Yushania in a temperate rain forest. *Plant Soil* **2001**, *231*, 37–44. [CrossRef]

35. Bailey, N.T. *Statistical Methods in Biology*; Cambridge University Press: Cambridge, UK, 1995.
36. El-Soud, W.A.; Hegab, M.M.; AbdElgawad, H.; Zinta, G.; Asard, H. Ability of ellagic acid to alleviate osmotic stress on chickpea seedlings. *Plant Physiol. Biochem.* **2013**, *71*, 173–183. [CrossRef]
37. Kohli, S.K.; Handa, N.; Sharma, A.; Gautam, V.; Arora, S.; Bhardwaj, R.; Wijaya, L.; Alyemeni, M.N.; Ahmad, P. Interaction of 24-epibrassinolide and salicylic acid regulates pigment contents, antioxidative defense responses, and gene expression in *Brassica juncea* L. seedlings under Pb stress. *Environ. Sci. Pollut. Res.* **2018**, *25*, 15159–15173. [CrossRef] [PubMed]
38. Posmyk, M.M.; Kontek, R.; Janas, K.M. Antioxidant enzymes activity and phenolic compounds content in red cabbage seedlings exposed to copper stress. *Ecotoxicol. Environ. Saf.* **2009**, *72*, 596–602. [CrossRef] [PubMed]
39. Kovinich, N.; Kayanja, G.; Chanoca, A.; Otegui, M.S.; Grotewold, E. Abiotic stresses induce different localizations of anthocyanins in Arabidopsis. *Plant Signal Behav.* **2015**, *10*, e1027850. [CrossRef] [PubMed]
40. Xu, Z.; Mahmood, K.; Rothstein, S.J. ROS Induces Anthocyanin Production Via Late Biosynthetic Genes and Anthocyanin Deficiency Confers the Hypersensitivity to ROS-Generating Stresses in Arabidopsis. *Plant Cell Physiol.* **2017**, *58*, 1364–1377. [CrossRef]
41. Baek, S.-A.; Han, T.; Ahn, S.-K.; Kang, H.; Cho, M.R.; Lee, S.-C.; Im, K.-H. Effects of Heavy Metals on Plant Growths and Pigment Contents in *Arabidopsis thaliana*. *Plant Pathol. J.* **2012**, *28*, 446–452. [CrossRef]
42. Ji, Y.; Xie, X.; Wang, G. Effects of the heavy metal cadmium on photosynthetic activity and the xanthophyll cycle in *Phaeodactylum tricornutum*. *J. Oceanol. Limnol.* **2018**, *36*, 2194–2201. [CrossRef]
43. Ahmad, S.; Kamran, M.; Ding, R.; Meng, X.; Wang, H.; Ahmad, I.; Fahad, S.; Han, Q. Exogenous melatonin confers drought stress by promoting plant growth, photosynthetic capacity and antioxidant defense system of maize seedlings. *PeerJ* **2019**, *7*, e7793. [CrossRef]
44. Rai, V.; Vajpayee, P.; Singh, S.N.; Mehrotra, S. Effect of chromium accumulation on photosynthetic pigments, oxidative stress defense system, nitrate reduction, proline level and eugenol content of *Ocimum tenuiflorum* L. *Plant Sci.* **2004**, *167*, 1159–1169. [CrossRef]
45. Hameed, A.; Gulzar, S.; Aziz, I.; Hussain, T.; Gul, B.; Khan, M.A. Effects of salinity and ascorbic acid on growth, water status and antioxidant system in a perennial halophyte. *AoB PLANTS* **2015**, *7*. [CrossRef]
46. Hasanuzzaman, M.; Bhuyan, M.H.M.B.; Anee, T.I.; Parvin, K.; Nahar, K.; Mahmud, J.A.; Fujita, M. Regulation of ascorbate-glutathione pathway in mitigating oxidative damage in plants under abiotic stress. *Antioxidants* **2019**, *8*, 384. [CrossRef]
47. Dey, S.; Sidor, A.; O'Rourke, B. Compartment-specific Control of Reactive Oxygen Species Scavenging by Antioxidant Pathway Enzymes. *J. Biol. Chem.* **2016**, *291*, 11185–11197. [CrossRef] [PubMed]
48. Collin, V.C.; Eymery, F.; Genty, B.; Rey, P.; Havaux, M. Vitamin E is essential for the tolerance of *Arabidopsis thaliana* to metal-induced oxidative stress. *Plant Cell Environ.* **2008**, *31*, 244–257. [CrossRef] [PubMed]
49. Hossain, M.A.; Piyatida, P.; Teixeira-da-Silva, J.A.; Fujita, M. Molecular mechanism of heavy metal toxicity and tolerance in plants: Central role of glutathione in detoxification of reactive oxygen species and methylglyoxal and in heavy metal chelation. *J. Bot.* **2012**, *2012*, 872875. [CrossRef]
50. Gallie, D.R. L-Ascorbic Acid: A Multifunctional Molecule Supporting Plant Growth and Development. *Scientifica* **2013**, *2013*, 795964. [CrossRef] [PubMed]
51. Azooz, M.M.; Hassanein, A.M.; Faheed, F.A. Riboflavin (vitamin B2) treatments counteract the adverse effects of salinity on growth and some relevant physiological responses of Hibiscus sabdariffa L. seedlings. *Bull. Fac. Sci. Assuit Univ.* **2002**, *31*, 395–403.
52. Barakat, H. Interactive effects of salinity and certain vitamin on gene expression and cell division. *Int. J. Agric. Biol.* **2003**, *3*, 219–225.
53. Cvelkorska, M.; Rampitsch, C.; Bykova, N.; Xing, T. Genomic analysis of MAP kinase cascades in Arabidopsis defense responses *Plant Mol. Biol. Report.* **2005**, *23*, 331–343.
54. Pourcel, L.; Routaboul, J.M.; Cheynier, V. Flavonoid oxidation in plants: From biochemical properties to physiological functions. *Trends Plant Sci.* **2007**, *12*, 29–36. [CrossRef]
55. Hasanuzzaman, M.; Nahar, K.; Rahman, A.; Mahmud, J.A.; Alharby, H.F.; Fujita, M. Exogenous glutathione attenuates lead-induced oxidative stress in wheat by improving antioxidant defense and physiological mechanisms. *J. Plant Interact.* **2018**, *13*, 203–212. [CrossRef]
56. Dai, J.; Mumper, R.J. Plant Phenolics: Extraction, Analysis and Their Antioxidant and Anticancer Properties. *Molecules* **2010**, *15*, 7313–7352. [CrossRef]

57. Xi, Z.-m.; Zhang, Z.-w.; Huo, S.-s.; Luan, L.-y.; Gao, X.; Ma, L.-n.; Fang, Y.-l. Regulating the secondary metabolism in grape berry using exogenous 24-epibrassinolide for enhanced phenolics content and antioxidant capacity. *Food Chem.* **2013**, *141*, 3056–3065. [CrossRef] [PubMed]
58. Agada, O.O. Abiotic Stress, Antioxidants and Crop Productivity: The Mitigating Role of Exogenous Substances. *Greener J. Agric. Sci.* **2016**, *6*, 79–86. [CrossRef]
59. Alhasnawi, A.N.; Kadhimi, A.A.; Isahak, A.; Mohamad, A.; Yusoff, W.M.W.; Zain, C.R.C.M. Exogenous Application of Ascorbic Acid Ameliorates Detrimental Effects of Salt Stress in Rice (MRQ74 and MR269) Seedlings. *Asian J. Crop Sci.* **2015**, *7*, 186–196. [CrossRef]
60. Jung, C.; Maeder, V.; Funk, F.; Frey, B.; Sticher, H.; Frossard, E. Release of phenols from *Lupinus albus* L. roots exposed to Cu and their possible role in Cu detoxification. *Plant Soil* **2003**, *252*, 301–312. [CrossRef]
61. Kohli, S.K.; Handa, N.; Sharma, A.; Gautam, V.; Arora, S.; Bhardwaj, R.; Alyemeni, M.N.; Wijaya, L.; Ahmad, P. Combined effect of 24-epibrassinolide and salicylic acid mitigates lead (Pb) toxicity by modulating various metabolites in *Brassica juncea* L. seedlings. *Protoplasma* **2018**, *255*, 11–24. [CrossRef]
62. Ma, J.F. Role of Organic Acids in Detoxification of Aluminum in Higher Plants. *Plant Cell Physiol.* **2000**, *41*, 383–390. [CrossRef]
63. Callahan, D.L.; Baker, A.J.M.; Kolev, S.D.; Wedd, A.G. Metal ion ligands in hyperaccumulating plants. *J. Biol. Inorg. Chem.* **2006**, *11*, 2–12. [CrossRef]
64. Yusuf, M.; Fariduddin, Q.; Ahmad, A. 24-Epibrassinolide modulates growth, nodulation, antioxidant system, and osmolyte in tolerant and sensitive varieties of *Vigna radiata* under different levels of nickel: A shotgun approach. *Plant Physiol. Biochem.* **2012**, *57*, 143–153. [CrossRef]
65. Singh, S.; Parihar, P.; Singh, R.; Singh, V.P.; Prasad, S.M. Heavy metal tolerance in plants: Role of transcriptomics, proteomics, metabolomics, and ionomics. *Front. Plant Sci.* **2015**, *6*, 1143. [CrossRef]
66. Hall, J.L. Cellular mechanisms for heavy metal detoxification and tolerance. *J. Exp. Bot.* **2002**, *53*, 1–11. [CrossRef]
67. Yang, X.E.; Jin, X.F.; Feng, Y.; Islam, E. Molecular mechanisms and genetic basis of heavy metal tolerance/hyperaccumulation in plants. *J. Integr. Plant Biol.* **2005**, *47*, 1025–1035. [CrossRef]
68. Sharma, S.S.; Dietz, K.J. The relationship between metal toxicity and cellular redox imbalance. *Trends Plant Sci.* **2009**, *14*, 43–50. [CrossRef] [PubMed]
69. Clemens, S. Molecular mechanisms of plant metal tolerance and homeostasis. *Planta* **2001**, *212*, 475–486. [CrossRef] [PubMed]
70. Oven, M.; Grill, E.; Golan-Goldhirsh, A.; Kutchan, T.M.; Zenk, M.H. Increase of free cysteine and citric acid in plant cells exposed to cobalt ions. *Phytochemistry* **2002**, *60*, 467–474. [CrossRef]
71. Sinclair, S.A.; Krämer, U. The zinc homeostasis network of land plants. *Biochim. Et Biophys. Acta Mol. Cell Res.* **2012**, *1823*, 1553–1567. [CrossRef]
72. Zhang, W.H.; Ryan, P.R.; Tyerman, S.D. Malate permeable channels and cation channels activated by aluminum in the apical cells of wheat roots. *Plant Physiol.* **2001**, *125*, 1459–1472. [CrossRef]
73. Salazar, F.S.; Pandey, S.; Narro, L. Diallel analysis of acid-soil tolerant and intolerant tropical maize populations. *Crop Sci.* **1997**, *37*, 1457–1462. [CrossRef]
74. Ascacio-Valdés, J.A.; Buenrostro-Figueroa, J.J.; Aguilera-Carbo, A.; Prado-Barragán, A.; Rodríguez-Herrera, R.; Aguilar, C.N. Ellagitannins: Biosynthesis, biodegradation and biological properties. *J. Med. Plants Res.* **2011**, *5*, 4696–4703.
75. Khan, A.; Nazar, S.; Lang, I.; Nawaz, H.; Hussain, M.A. Effect of ellagic acid on growth and physiology of canola (*Brassica napus* L.) under saline conditions. *J. Plant Interact.* **2017**, *12*, 520–525. [CrossRef]

plants

MDPI

Review

Chromium Bioaccumulation and Its Impacts on Plants: An Overview

Anket Sharma [1,*,†], Dhriti Kapoor [2,†], Junfeng Wang [1,†], Babar Shahzad [3], Vinod Kumar [4], Aditi Shreeya Bali [5], Shivam Jasrotia [6], Bingsong Zheng [1], Huwei Yuan [1,*] and Daoliang Yan [1,*]

1 State Key Laboratory of Subtropical Silviculture, Zhejiang A&F University, Hangzhou 311300, China; 2017702431011@stu.zafu.edu.cn (J.W.); bszheng@zafu.edu.cn (B.Z.)
2 School of Bioengineering & Biosciences, Lovely Professional University, Punjab 144411, India; dhriti405@gmail.com
3 School of Land and Food, University of Tasmania, Hobart, Tasmania 7005, Australia; Babar.shahzad@utas.edu.au
4 State Higher Education Department, Jammu and Kashmir 180001, India; vinodverma507@gmail.com
5 Mehr Chand Mahajan D.A.V. College for Women, Chandigarh 160036, India; shreeyaaditi02@gmail.com
6 Department of Zoology, Guru Nanak Dev University, Amritsar 143005, India; shivbaba1990@gmail.com
* Correspondence: anketsharma@gmail.com (A.S.); hwyuan@zafu.edu.cn (H.Y.); liangsie@zafu.edu.cn (D.Y.); Tel.: +86-571-63732738 (H.Y.); +86-571-63732761 (D.Y.)
† Authors contributed equal.

Received: 29 October 2019; Accepted: 9 January 2020; Published: 13 January 2020

Abstract: Chromium (Cr) is an element naturally occurring in rocky soils and volcanic dust. It has been classified as a carcinogen agent according to the International Agency for Research on Cancer. Therefore, this metal needs an accurate understanding and thorough investigation in soil–plant systems. Due to its high solubility, Cr (VI) is regarded as a hazardous ion, which contaminates groundwater and can be transferred through the food chain. Cr also negatively impacts the growth of plants by impairing their essential metabolic processes. The toxic effects of Cr are correlated with the generation of reactive oxygen species (ROS), which cause oxidative stress in plants. The current review summarizes the understanding of Cr toxicity in plants via discussing the possible mechanisms involved in its uptake, translocation and sub-cellular distribution, along with its interference with the other plant metabolic processes such as chlorophyll biosynthesis, photosynthesis and plant defensive system.

Keywords: heavy metal; reactive oxygen species; oxidative burst; antioxidants

1. Introduction

Chromium (Cr) is a silver-colored hard metal naturally occurring in rocky soils, and volcanic dust. Chromium is the 24th element having a molecular weight of 51.1 a.m.u. and a density of 7.19 g/cm^3 [1]. The Agency for Toxic Substances and Disease Registry [2] has ranked Cr the 17th among the most hazardous substances. It has been classified as the number one carcinogen according to the International Agency for Research on Cancer [3]. Therefore, this metal needs an accurate understanding of its uptake, transport and bioaccumulation within plants and a thorough investigation in soil–plant systems.

Chromium can easily convert from one oxidation state to another due to its high redox potential and intricate electronic and valence shell chemistry [4,5]. It exists in a wide range of oxidation states, but the most common and stable states are Cr (VI) "hexavalent" and Cr (III) "trivalent" [6]. Both forms of Cr vary significantly with respect to their bioavailability in soil, translocation and toxicity within plants [4]. The Cr (III) occurs in the form of chromite ($FeOCr_2O_3$), whereas Cr (VI) occurs in association with oxygen to form chromate (CrO_4^{2-}) or dichromate ($Cr_2O_7^{2-}$), that are highly toxic to living organisms [7]. Based on its activity, former is the most stable form of Cr, whereas the highest noxious one for plants is the Cr (VI). Under physiological conditions, Cr (VI) enters the cells and may get reduced to Cr (V), Cr (IV), thiylradicals, hydroxyl radicals and finally Cr (III). All these oxidation states disrupt the cellular integrity of cells by attacking proteins, DNA and membrane lipids [8,9].

Hexavalent Cr (VI) is used in several industrial applications such as electroplating, dyeing of textiles, leather processing, steel production and tanning industry, resulting in discharge of chromium-containing effluents. This ultimately causes significant elevation in Cr contents in the environment [10]. Due to its high solubility in water and soil, Cr (VI) is regarded as a hazardous ion that contaminates groundwater and can be transferred through the food chain [10–12]. It also occurs in air, water and soil at different concentrations. Its contents in freshwater and seawater vary from 0.1 to 117 μg L^{-1} and 0.5 to 50 μg L^{-1} respectively. The weathering of Cr containing rocks and leaching of soils discharge significant Cr contents into the aquatic environment [7]. Cr contents in soil ranges between 10 and 50 mg kg^{-1} under natural conditions, however, its concentration in agricultural soils can reach up to 350 mg kg^{-1} of the soil [7]. The United States Environmental Protection Agency (USEPA) [13] has listed Cr among the 14 most dangerous substances that can cause serious health issues in living organisms. Cr can have both beneficial and harmful effects on human health depending on its uptake, exposure time and oxidation state. The trivalent form of Cr (III) is an important nutrient for humans and according to the World Health Organization [14], its daily ideal intake is between 50 and 200 μg day^{-1} for the metabolism of carbohydrates, proteins and fatty acids. However, its excess in the body poses serious health concerns. Moreover, hexavalent Cr (VI) is 10–100 folds more harmful than Cr (III), which can cause allergies and skin problems.

To date, Cr does not have any known biological role in plant physiology [15]. It is generally perceived that excessive Cr levels in plant tissues may provoke several morpho-physiological and biochemical processes in plants [16,17]. Any metal toxicity is attributed to a complex series of metal interactions with the genetic processes, signal transduction and pathways and cellular macromolecules [18–20]. Hence, Cr toxicity is reported to affect plant growth and impedes their essential metabolic processes [21]. Typically, Cr toxicity reduces plant growth by inducing ultrastructural modifications of the cell membrane and chloroplast, persuading chlorosis in the leaves, damaging root cells, reducing pigment content, disturbing water relations and mineral nutrition, affecting transpiration and nitrogen assimilation and by altering different enzymatic activities [15,22–25]. All these toxic effects of Cr might be due to the over production of reactive oxygen species (ROS), which ultimately disrupt the redox balance in plants [25]. Taking all into consideration, we review the literature that addresses Cr uptake, translocation and sub-cellular distribution in plants. We also discuss different effects of Cr on plant pigments, photosynthetic parameters, enzymatic and non-enzymatic antioxidative system and various endogenous levels of plant hormones (Table 1).

Table 1. Effects of chromium metal on different physiological processes in plants.

Plant Species	Physiological Response	Reference
Camellia sinensis	Increased SOD and CAT activities	Tang et al. [26]
Capsicum annuum	Increased carotenoid content	Oliveira [27]
Chamomilla recutita	Increased MDA level	Kováčik et al. [28]
Echinochloa colona	Increased CAT and POD activities	Samantaray et al. [29]
Kandelia candel	Increased MDA content, and activities of CAT and SOD	Rahman et al. [30]
Ocimum tenuiflorum	Increased proline level	Rai et al. [31]
Oryza sativa	Increased POD activity	Ma et al. [32]

Table 1. *Cont.*

Plant Species	Physiological Response	Reference
Oryza sativa	Increased ethylene synthesis	Trinh et al. [33]
Oryza sativa	Increased CAT and SOD activities	Zhang et al. [34]
Oryza sativa	Increased POD activity	Xu et al. [35]
Phaseolus vulgaris	Decreased carotenoids	Aldoobie and Beltagi [36]
Pisum sativum	Decreased APX activity	Duhan [37]
Pterogyne nitens	Increased spermidine level	Paiva et al. [38]
Raphanus sativus	Increased glycine-betaine content	Choudhary et al. [39]
Triticum aestivum	Increased MDA contents	Ali et al. [22]
Triticum aestivum	Increased lipid peroxidation	Zhang et al. [34]
Vigna radiata	Decreased glutathione level	Shanker et al. [40]
Zea mays	Increased SOD and GPX activities	Maiti et al. [41]
Zea mays	Increased lipid peroxidation and H_2O_2 content	Maiti et al. [41]

Abbreviations: Malondialdehyde—MDA; Superoxide dismutase—SOD; Catalase—CAT; Ascorbate peroxidase—APX; Peroxidase—POD; Guaiacol peroxidase—GPX; Hydrogen peroxide—H_2O_2.

2. Chromium Uptake, Translocation and Sub-Cellular Distribution

Plant roots secrete various organic acids such as citrate and malate that modify the solubility of metals present in insoluble form in the soil by acting as ligands [42–45]. Srivastava et al. [46] had shown an increased accumulation of Cr in tomato plants due to the presence of citrate, aspartate and oxalate, which converted inorganic Cr into organic complexes, which are readily available for the plant to uptake. Chromium appears to have no essential role in plant metabolism, hence, there is no specific mechanism for its uptake in plants [27]. Skeffington et al. [47] had proposed a mechanism for uptake of both Cr (III) and Cr (VI) in barley (*Hordeum vulgare*) plants. Nonetheless, specific carriers responsible for the absorption of essential ions also aid in the uptake of Cr [4]. The uptake of Cr (III) in plants undergoes through passive mechanism [48], however, Cr (VI) is uptaken through the plasma membrane, which is an active process involving carriers of essential anions such as sulfate [49,50]. Further, due to the structural similarity of Cr (VI) with phosphate and sulfate, its uptake by root cells involves phosphate or sulfate transporters [51,52].

The distribution and translocation of Cr within plants depend upon the plant species, the oxidation state of the Cr ions, and also its concentration in the growth medium [4]. Compared to other heavy metals, the mobility of Cr in the plant roots is low. Therefore, the concentration of Cr in the roots is sometimes 100 times higher than in the shoots [48,53]. For instance, Cr concentration was observed to be highest in the cytoplasm and intercellular spaces of rhizome and root cell wall of *Iris pseudacorus* [54]. The higher accumulation of Cr in roots might be attributed to the sequestration of Cr in the vacuoles of root cells as a protective mechanism [55]. Thus, this mechanism provides some natural tolerance to plants towards Cr toxicity [40]. Furthermore, the translocation of Cr from the roots to the aerial shoots is very limited and it depends on the chemical form of Cr inside the tissue [4]. In plant tissues, the Cr (VI) is converted to Cr (III) that has the tendency to bind to the cell walls, which hinders the further transport of Cr within plant tissues [56].

Numerous metal transporter gene families including CDF (cation diffusion facilitator), HMA (heavy metal ATPase), ATP binding cassette (ABC) superfamily and ZIP (ZRT, IRT-like protein) have been identified for different metals like Pb, Cd, Zn and As [4,57–61]. However, the role of transporter families in the translocation of Cr in plants is still unclear. The translocation of Cr (VI) to shoots is an active process that involves phosphate and sulfate transporters [4]. Hence, the translocation of Cr might be mediated by iron (Fe) and sulfur (S) channels in the roots that lead to the competition between metals e.g., Fe and Cr [4,62]. Cary et al. [63] had reported Cr uptake and translocation to the aerial shoots in Fe hyperaccumulators *Brassica rapa* and *Spinacia oleracea* signifying that Cr may be transported through Fe channels. However, the presence of Fe in the growth media reduced Cr translocation to the shoots [64], which could be due to the competition of carrier channels or due to the precipitation of Fe with Cr.

3. Effect of Cr on Nutrient Uptake

Heavy metal stress affects nutrient uptake in plants by interacting with other essential minerals. Chromium restricts the uptake of nutrients in soil by forming insoluble compounds [65]. Nutrient uptake is thereby inhibited by the metal toxicity especially when the concentration of the metal exceeds its permissible limits [66]. For instance, excessive Cr had been observed to reduce the uptake of essential minerals like iron (Fe), magnesium (Mg), phosphorus (P) and calcium (Ca) by masking the sorption sites and forming insoluble complexes [56,66]. However, Cr transport to different parts of *Citrullus* plants had increased leading to enhancement in the concentrations of manganese (Mn) and P, and reduction in sulphur (S), copper (Cu), zinc (Zn) and iron (Fe) contents in the leaves, suggesting that Cr disturbs the nutrient balance [67]. Turner and Rust [68] also suggested the similar effects of Cr on the uptake of various nutrients under Cr toxicity. A gradual decrease in the uptake of micronutrients like Zn, Cu, Fe, Mn and macronutrients like potassium (K), P and nitrogen (N) had been noticed in the paddy plants (*Oryza sativa* L.) under excessive Cr exposure [69]. This reduced nutrient uptake may occur due to decline in the root growth and impairment of the root penetration under Cr toxicity, or may be due to the decrease in essential element translocation because of the displacement of nutrients from the physiologically important binding sites [70,71].

4. Effect of Cr on Chlorophyll Molecules and Photosynthetic Performance

Foliar content of chlorophyll pigments including total chlorophyll, chlorophyll a (Chl a) and chlorophyll b (Chl b) were assayed under Cr treatment, which showed significant decrease in pigment accumulation of *Catharanthus roseus* plants [72]. This could be due to the inhibition of chlorophyll biosynthesis under Cr stress [73,74]. Increased concentration of Cr may lead to the deterioration of the chlorophyll content in many plants [75]. Plants exposed to Cr stress showed depleted chlorophyll contents that might be due to the disrupted chlorophyll biosynthesis [76]. Interestingly, an enzyme involved in chlorophyll biosynthesis, i.e., δ-aminolevulinic acid dehydratase (ALAD) is being inhibited by Cr due to the impairment in utilizing the δ-aminolevulinic acid [77].

Effect of Cr was also depicted on pigment contents viz. chlorophyll of vetiver [78] where alterations in the photosynthetic pigments were observed. Cr induced toxicity had been reported to decrease the chlorophyll contents in different plants species such as *Pistia stratiotes* [79], *Citrus limonia* and *Citrus reshni* [80], *Zea mays* [81,82], *Hibiscus esculantus* [83], *Camellia sinensis* [84], *Glycine max* [85] and *Ocimum tenuiflorum* [31]. Decrease in chlorophyll contents under Cr toxicity could be due to the impairment of chlorophyll biosynthesis enzymes, which are compromised under Cr toxicity [86,87]. Degradation of ALAD could occur under Cr toxicity leading to a decrease in chlorophyll level [88]. Hence, photosynthetic capacity of plants is compromised under Cr stress due to interaction with biosynthesis of chlorophyll molecules by inhibiting vital enzymes contributing in photosynthesis. Excessive Cr affects photosynthetic system by targeting the Calvin cycle enzymes, photosynthetic electron transport and thylakoid membrane [89]. Therefore, gradual decrease in the net photosynthetic rate can be observed in the plants treated with higher concentration of Cr [90].

Changes in the level of photosynthetic pigments give an important information regarding the toxic effects of heavy metals, e.g., Cr, Ni, Pb and Cd [70,74,91–94]. However, reduced chlorophyll contents may be observed due to the increased activity of enzymes like chlorophyllase and deficiency of nutrients, i.e., because of the translocation of the metals to shoots in higher concentration [91,95]. Moreover, a significant decrease in transpiration rate, net photosynthetic rate, intercellular CO_2 concentration and stomatal conductance in the leaves were observed where Cr toxicity reduced these parameters by 71%,

36%, 25% and 57% respectively [96]. Cr also poses hazardous effects on gas exchange as shown by the multiple linear regression (MLR) analysis that expressed negative β-regression coefficients for all the parameters of gas exchange [97]. Davies Jr et al. [96] had noticed that Cr inhibited photosynthetic process by targeting photosystem II (PSII). Hence, chlorophyll fluorescence seems quite useful tool to study photosynthetic apparatus and action of PSII under heavy metal stress.

5. Reactive Oxygen Species (ROS) and Oxidative Stress

A rather common and frequent effect of heavy metal stress is the overproduction of ROS including hydroxyl radicals (OH^-), hydroperoxyl radicals (HOO), superoxide (O_2^-), the peroxinitrite ($OONO^-$) ion, the paramagnetic singlet oxygen ($^1O_2$), nitrogen oxide radical (NO), hydrogen peroxide (H_2O_2), ozone (O_3) and hypochlorous acid (HOCl) molecules [98–101]. This process is considered as one of the primary cause for the alterations in plant biology at biochemical level under heavy metal toxicity [70,73,74,102–106]. Plants may suffer through various drastic physiological changes, which are mainly due to the imbalance in the generation and scavenging of ROS, termed as an oxidative burst [70,74,94,102,104,107–112]. Heavy metals like copper (Cu), nickel (Ni), cadmium (Cd), Cr and arsenic (As) have the tendency to generate ROS if they exceed permissible limits [70,113–115].

On the basis of physical and biochemical characteristics of bioactive-metals, these metals can be classified into two groups; redox metals like Cr, Cu, Fe and non-redox metals like Cd, Hg, Ni, Zn, etc. Redox active metals have the capacity to produce oxidative injuries in plants via Haber–Weiss and Fenton reactions, that consequently generate ROS and leading to disturb the balance between prooxidant and antioxidant level [116]. However, redox-inactive metals form covalent bonds with the protein sulfhydryl groups as these metals have the tendency of sharing the electrons.

When Cr metal interacts with the proteins at its catalytic site or any other site, it deactivates the active sites of enzymes by binding specific functional groups of proteins, thus leading to the alteration of enzymatic activities [117,118]. Furthermore, dislocation of critical cations from the specific enzyme binding sites disturbs the equilibrium of ROS in cells, and as a consequence ROS are generated in drastic amount [119]. Chromium metal has also tendency to bind and utilize the reduced form of glutathione (GSH) and its derivatives, which plays a significant role in ameliorating these ROS [120]. Besides, NADPH oxidase (present on plasma membrane) also leads to oxidative stress as they are linked with the Cr [121–123]. In the presence of Cr metal, the NADPH oxidases may consume cytosolic NADPH and produce free radical O_2^-, which is rapidly converted to H_2O_2 by superoxide dismutase enzyme (SOD) [124]. Free radicals generated by Cr in association with NADPH oxidase remains outside the plasma membrane, where pH remains usually low in comparison to inner side of the cell [125]. Reports suggest that the enhanced generation of ROS in plants under Cr toxicity leads to oxidative burst by causing damage to DNA, lipids, pigments, proteins and stimulates the process of lipid peroxidation (Figure 1) [126,127]. Carrier membrane stimulates the absorption of Cr and over productions of ROS further influences the plasma membrane [41].

There are several reports documented where drastic increase in ROS was observed [22,74,104,128] with increased malondialdehyde (MDA) contents under Cr toxicity [129]. Alterations in different physiological and biochemical activities have been observed in *Triticum aestivum*, *Vallisneria spiralis* and *Ocimum tenuiflorum* [22,31,88] where Cr metal stimulated the deterioration of membrane permeability by generating MDA. Similarly, increased levels of MDA were found in both roots and leaves of *Kandelia candel* (L.) plants in a dose dependent manner suggesting its gradual uptake under a timely manner [30].

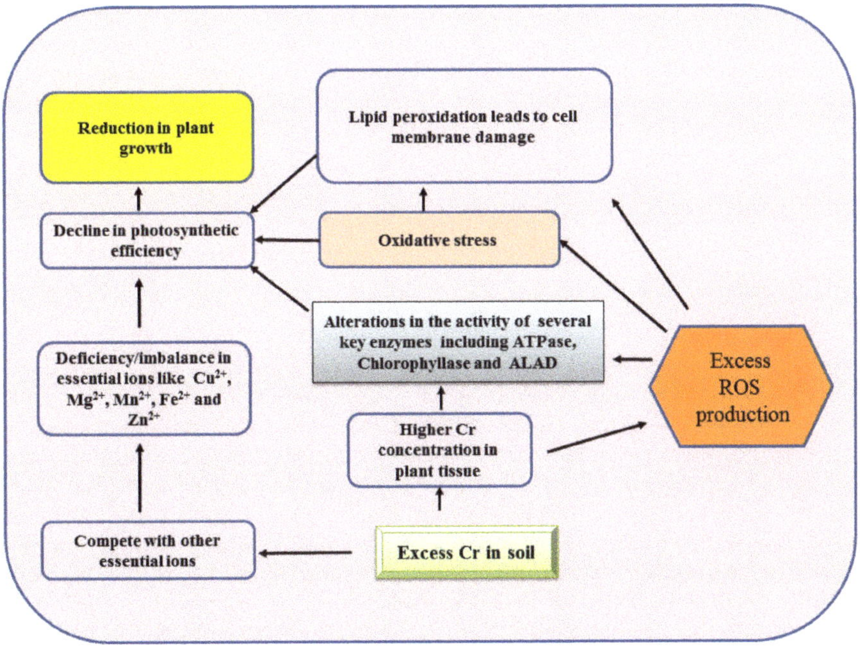

Figure 1. Consequences of oxidative stress generated under chromium toxicity.

6. Effect of Cr on Enzymatic Antioxidative System

Activities of various antioxidative enzymes drastically change in plants when subject to Cr toxicity. ROS leads to oxidative stress, which may affect different subcellular compartments sensitive to ROS. Superoxide dismutase (SOD) is considered as a first line of defense against various stresses in almost all the aerobic organisms [70,130]. Dismutation of superoxide ion is catalyzed by the SOD enzyme, which is localized in almost all the cellular compartments, leading to the production of hydrogen peroxide and release of oxygen [130]. SOD is involved in the Asada–Halliwell cycle in chloroplasts and also present in cytosol, apoplasts, mitochondria and peroxisomes [131]. For the removal of ROS, catalase (CAT) enzyme also plays crucial role, hence considered as important antioxidant enzyme [132]. Dismutation of H_2O_2 into O_2 and H_2O is undergone by this enzyme [133].

When exposed to Cr (III) stress for seven days, chamomile plants showed increased accumulation of Cr mainly in the roots of the plants, which contained high concentrations of ROS, nitric oxide and thiols. At higher concentration of Cr (III), SOD activity specifically was increased in the roots, while level of H_2O_2 showed irregular trend under different concentrations of Cr due to the altered activities of various peroxidases [28]. Different concentrations of Cr (VI) (50, 100, 200 and 300 μM L^{-1}) escalated the production of H_2O_2 leading to the lipid peroxidation and triggered the activities of antioxidative enzymes like SOD and guaiacol peroxidase (GPX) in comparison to control plants [41]. In contrast, activities of peroxidase (POD), SOD and CAT were decreased when subjected to Cr (III) in a dose dependent manner [26].

Maintaining metabolic functions under stress conditions is crucial for plants to survive. Therefore, a balance between generation and scavenging of ROS is required, which is achieved by regulating the production of enzymatic and non-enzymatic antioxidants [134]. Hence, the ability of plants to cope with oxidative stress is characterized by the degree of antioxidant activities [102,104]. However, activities of these antioxidants may vary with the duration, crop species and tissues under any stress condition [135]. For instance, *Echinochloa colona* plants showed increased activities of POD and CAT in tolerant calluses in comparison to non-tolerant ones [29]. At 0.5 mg L^{-1} concentration of Cr (VI), CAT activity was increased, however decreased at higher concentrations (1.0–2.0 mg L^{-1}). Activity of CAT was measured highest at 2.0 mg L^{-1} concentration of Cr in the roots of *K. candel* but decreased at higher concentration [30]. As CAT is an iron-porphyrine biomolecule, reduction in CAT activity indicates that Cr has the potential to interact with iron in metabolic pool or it may influence the presence of active form of iron [136]. Cr toxicity has detrimental effects on antioxidant enzymes such as POD, GPX, glutathione reductase (GR) and ascorbate peroxidase (APX), hence resulting in the inhibition of enzyme activities [129,137]. Plants equipped with an efficient antioxidant system are more capable to withstand and tolerate higher Cr concentrations. Failure to do so results in the breakdown of the plant defense system hence activities of antioxidants are jeopardized leading to reduced plant growth or even leading to plant death.

7. Effect of Cr on Non-Enzymatic Antioxidative System

Apart from the enzymatic antioxidants, plants are also comprised of a complex non-enzymatic antioxidant defense system to avoid the toxic effects of ROS. These non-enzymatic antioxidants consist of low molecular weight molecules such as ascorbic acid, glutathione (GSH), phenolic acids, carotenoids, flavonoids, etc. [94,138] and some high molecular weight secondary metabolites such as tannins [139]. Biosynthesis and accumulation of these non-enzymatic antioxidants by plants could be due to two main reasons. First, plants have an innate ability to synthesize a variety of phytochemicals to carry out their normal physiological functioning or to protect them from any pathogenic or herbivores. Second, plants also synthesize phytochemicals to respond to the environmental factors which could be due to their natural tendency of defense against any biotic and abiotic stress [140]. Therefore, these lower molecular weight antioxidants are synthesized and act as a redox buffer to interact with cellular components and directly influence plant growth and development by modulating different processes from mitosis to cell elongation and to senescence. Hence, it is crucial for plants to synthesize these antioxidants under stressed conditions. Glutathione is a redoxactive molecule that can be present in a reduced form GSH or an oxidized form GSSG. It plays important roles in the plant defensive system including biosynthetic pathways, detoxification, antioxidant biochemistry and redox homeostasis [141,142]. GSH acts as an antioxidant by quenching ROS and is involved in the ascorbate-glutathione cycle, which eliminates damaging peroxides [143]. In poplar trees, glutathione (GSH) biosynthesis was stimulated under Cr toxicity [144]. In the leaf extracts of tomato (*Lycopersicon esculentum*), maize (*Zea mays*) and cauliflower (*Brassica oleracea*) plants, GSH level increased subjected to Cr toxicity [145]. Alterations were observed in the glutathione pool dynamics where individual level of GSSG and GSH and GSH/GSSG ratio was affected, however sorghum (*Sorghum biclor*) plants showed potential to scavenge the free radicals generated under Cr toxicity [146].

Under sub-optimal conditions, level of antioxidants may decrease or increase depending on the severity of stress. For instance, a sharp decline in the GSH pool was observed under Cr stress, which severely affected the roots of the plants. Many reports suggested the oxidation of various cellular thiols like GSH and cysteine in plants subjected to Cr (VI) stress in in vitro conditions [147]. Therefore, to maintain the redox homeostasis of the cell and for scavenging of free radicals, the interconversion of reduced and oxidized forms of glutathione (GSH and GSSG respectively) is required.

A non-enzymatic antioxidant "carotenoid" contributes in providing protection to the chlorophylls against stress conditions by replacing peroxides and scavenging of photodynamic reactions [148]. For instance, carotenoid level increased in *Capsicum annuum* plants subjected to industrial effluents specifically containing Cr [27], acting as a defensive mechanism for capsicum plants to scavenge the free radicals. Other non-enzymatic antioxidants such as cysteine, proline, nonprotein thiol, etc. may also contribute in modulating resistance against Cr toxicity and protect the macromolecules from the free radicals generated during the oxidative burst [77]. In one of the reports, carotenoids, non-protein thiol (NP-SH) and cysteine level increased in the plants subjected to Cr [72].

For providing protection against different types of stresses, accumulation of compatible osmolytes like proline (Pro) occurs in most of the plants for providing membrane stability and osmotic adjustment [94,149]. Proline contents can increase against different types of biotic and abiotic stresses such as salinity, drought, temperature, heavy metal and pathogen attack [77]. For instance, contents of Pro increased in *Ocimum tenuiflorum* L. under Cr stress, which acted as an antioxidant by providing protection against the hazardous effects of metal [31]. Reports suggested that proline is the only amino acid that accumulates in the leaves of plants under stress conditions [150]. It starts accumulating even at low doses of stress and increases in a dose dependent manner. Hence, accumulation of proline contributes to osmotic adjustment when it gets accumulated in tissues and acts as a dependent marker for genotypes for the stress tolerance [150]. Polyamines are other non-enzymatic osmoprotectants, which increase under different abiotic stress conditions [151], and also associated with the boosting up of plant defensive mechanisms [92,152]. Polyamines including putrescine, spermine and spermidine have been investigated to increase Cr tolerance in plants under Cr toxicity [38].

8. Effect of Cr on the Endogenous Levels of Plant Hormones

Plant hormones control and regulate plant growth and development through different biochemical and physiological process. These hormones may act either close to or remote from their synthesis site to regulate responses to environmental stimuli or genetically programmed developmental changes [153]. Hormones thus have a vital role in plant adaptation to abiotic stress, from which the plant may attempt to escape or survive under stressful conditions [154]. Thus, abiotic stresses often alter the production, distribution or signal transduction of growth as well as stress hormones, which may promote specific protective mechanisms. The phytohormone abscisic acid (ABA) plays a vital role against abiotic stresses, thus it is considered as a stress hormone. Under different types of abiotic stress conditions, endogenous level of ABA are increased drastically in plants, which boosts up the signaling pathways and activates expression of ABA-responsive genes [155]. For instance, strong expression of ABA biosynthesis genes (*OsNCED2* and *OsNCED3*) and four ABA signaling genes were upregulated in response to heavy metal stress [156]. Salicylic acid (SA) also provides protection under Cr stress. SA plays a significant positive role in growth and development of plants as well as ripening and abiotic stress tolerance [157]. Interaction of ABA and SA also contributes in boosting the defensive strategies of plants against Cr toxicity [158]. Apart from this, SA also plays essential role in combination with jasmonic acid (JA) and ethylene, where the interaction of these hormones provides tolerance against Cr stress [159,160]. The biosynthesis, transportation and accumulation of these plant hormones boost the signaling pathways, activating certain antioxidant gene expressions and stimulating the production of osmoprotectants such as proline, soluble sugars, amino acids, etc. [161].

Indole acetic acid (IAA), a plant hormone of auxins contributes to the growth and development of plants under ideal as well as stressed conditions [162]. IAA plays a key role in plant adaptation to heavy metal stress by either increasing the membrane permeability or by increasing the concentration of osmotically active solutes [16,163]. The level of IAA usually increases in the plants under metal toxicity by affecting different metabolic activities of plants including growth and hormonal balance [164]. Gibberellins (GAs) are considered as comprehensive class of tetracyclic diterpenoid carboxylic acid compounds. Different forms of GAs have potential to play the role of growth hormone in higher plants such as GA1 and GA3 [165]. Generally, at low concentration of metal like Cr leads to the elevated level of GA3, although its high concentration reduces the GA3 content [166]. Apart from these, plant hormone like cytokinins (CKs), contributes in the regulation of plant development by stimulating cell division and elongation. Cr stress alters endogenous level of CKs suggesting that CKs are also involved in tolerating the stress [155]. A few reports also indicated the reduction in the synthesis of CKs and their transport from roots to other aerial parts of plants during Cr toxicity and they are also found to interact with other plant hormones [167].

9. Conclusions and Prospects

This review illustrates an overview of Cr metal effects on plant growth and development. Plants uptake Cr via roots, which causes nutrient imbalance, root injury as well as leaf chlorosis. Cr toxicity also targets chlorophyll biosynthesis by inhibiting the activity of vital enzymes. Additionally, it also results in oxidative stress by targeting cellular membranes and biomolecules resulting in retarded plant growth, induction of chlorosis and wilting of leaves. Although a handful of data is available that provides useful information to understand chromium interaction with other essential metal ions. The mechanisms generating Cr-induced toxicity at the protein and molecular level still need to be explored in detail. Furthermore, exploration of Cr tolerance mechanism and homeostasis are essential for sustainable crop production, which is poorly understood in many ecosystems. Therefore, it is essential to understand the possible means to reduce Cr uptake and its negative impacts on environment especially in plants.

Author Contributions: A.S., design of outline, wrote abstract, Sections 1 and 5 and revised initial draft; D.K., J.W., wrote Sections 2, 4 and 6; B.S., V.K., A.S.B. and S.J. wrote Sections 3, 7 and 8; B.Z., D.Y. and H.Y., were involved in designing outline, revision of initial draft and arranging resources. All authors have read and agreed to the published version of the manuscript.

Funding: We acknowledge the different funding agencies: This study was supported by National Key Research and Development Program of China (2018YFD1000600, 2018YFD1000604); Key Project of Zhejiang Provincial Natural Science Foundation (LZ18C160001); National Natural Science Foundation of China (31,901,346, 31,971,695, 31,470,683, 31,270,716 and 31,070,604); Independent Research Project of State Key Laboratory of Subtropical Silviculture, Zhejiang A&F University (ZY20180208, ZY20180308); Open Foundation of State Key Laboratory of Subtropical Silviculture, Zhejiang A&F University (KF201708); Overseas Expertise Introduction Project for Discipline Innovation (111 Project D18008); Key Research and Development Program of Zhejiang Province (2018C02004); National High Technology Research and Development Program of China (863 Program) (2013AA102605); Fruit Innovation Team Project of Zhejiang Province (2016C02052-12); Key Agricultural New Varieties Breeding Projects founded by Zhejiang Province Science and Technology Department (2016C02052-13); Zhejiang Provincial Natural Science Foundation for Distinguished Young Scholar (LR13C160001); Open Foundation of First-class Discipline of Forestry, Zhejiang Province (201703); The First-class General Financial Grant from the China Postdoctoral Science Foundation (2017M610377).

Conflicts of Interest: The authors declare no conflict of interest.

References

1. Nriagu, J.O.; Pacyna, J.; Milford, J.B.; Davidson, C.I. Distribution and characteristic features of chromium in the atmosphere. In *Chromium in Natural and Human Environments*; Wiley Interscience: New York, NY, USA, 1988; pp. 125–173.
2. CERCLA Priority List of Hazardous Substances. Agency for Toxic Substances and Disease Registry, USA. 2017. Available online: https://www.atsdr.cdc.gov/spl/ (accessed on 20 September 2019).
3. International Agency for Research on Cancer. Overall evaluations of carcinogenicity: And updating of IARC monographs, vol. 1 to 42. IARC monographs on the evaluation of the carcinogenic risk of chemicals to humans: Suppl 7. *IARC* **1987**, *7*, 1–440.
4. Shahid, M.; Shamshad, S.; Rafiq, M.; Khalid, S.; Bibi, I.; Niazi, N.K.; Dumat, C.; Rashid, M.I. Chromium speciation, bioavailability, uptake, toxicity and detoxification in soil-plant system: A review. *Chemosphere* **2017**, *178*, 513–533. [CrossRef] [PubMed]
5. Prado, C.; Chocobar Ponce, S.; Pagano, E.; Prado, F.E.; Rosa, M. Differential physiological responses of two Salvinia species to hexavalent chromium at a glance. *Aquat. Toxicol.* **2016**, *175*, 213–221. [CrossRef] [PubMed]
6. Ashraf, A.; Bibi, I.; Niazi, N.K.; Ok, Y.S.; Murtaza, G.; Shahid, M.; Kunhikrishnan, A.; Li, D.; Mahmood, T. Chromium (VI) sorption efficiency of acid-activated banana peel over organo-montmorillonite in aqueous solutions. *Int. J. Phytoremediation* **2017**, *19*, 605–613. [CrossRef] [PubMed]
7. Ertani, A.; Mietto, A.; Borin, M.; Nardi, S. Chromium in agricultural soils and crops: A review. *Water Air Soil Pollut.* **2017**, *228*, 190. [CrossRef]
8. Tchounwou, P.B.; Yedjou, C.G.; Patlolla, A.K.; Sutton, D.J. Heavy metal toxicity and the environment. In *Molecular, Clinical and Environmental Toxicology*; Springer: Berlin/Heidelberg, Germany, 2012; pp. 133–164.
9. Stambulska, U.Y.; Bayliak, M.M.; Lushchak, V.I. Chromium (VI) toxicity in legume plants: Modulation effects of rhizobial symbiosis. *BioMed Res. Int.* **2018**, *2018*, 1–13. [CrossRef]
10. Joutey, N.T.; Sayel, H.; Bahafid, W.; El Ghachtouli, N. Mechanisms of hexavalent chromium resistance and removal by microorganisms. In *Reviews of Environmental Contamination and Toxicology*; Springer: Berlin/Heidelberg, Germany, 2015; Volume 233, pp. 45–69.

11. Kumar, V.; Parihar, R.D.; Sharma, A.; Bakshi, P.; Sidhu, G.P.S.; Bali, A.S.; Karaouzas, I.; Bhardwaj, R.; Thukral, A.K.; Gyasi-Agyei, Y. Global evaluation of heavy metal content in surface water bodies: A meta-analysis using heavy metal pollution indices and multivariate statistical analyses. *Chemosphere* **2019**, *236*, 124364. [CrossRef]
12. Kumar, V.; Sharma, A.; Kaur, P.; Sidhu, G.P.S.; Bali, A.S.; Bhardwaj, R.; Thukral, A.K.; Cerda, A. Pollution assessment of heavy metals in soils of India and ecological risk assessment: A state-of-the-art. *Chemosphere* **2019**, *216*, 449–462. [CrossRef]
13. Environmental Protection Agency. Effluent Limitations Guidelines, Pretreatment Standards, and New Source Performance Standards for the Commercial Hazardous Waste Combustor Subcategory of the Waste Combustors Point Source Category. *Fed. Regist.* **2000**, *00-2019*, 4360–4385.
14. World Health Organization. *Air Quality Guidelines for Europe, 2000*; WHO Regional Office for Europe: Copenhagen, Denmark, 2000.
15. Reale, L.; Ferranti, F.; Mantilacci, S.; Corboli, M.; Aversa, S.; Landucci, F.; Baldisserotto, C.; Ferroni, L.; Pancaldi, S.; Venanzoni, R. Cyto-histological and morpho-physiological responses of common duckweed (*Lemna minor* L.) to chromium. *Chemosphere* **2016**, *145*, 98–105. [CrossRef]
16. UdDin, I.; Bano, A.; Masood, S. Chromium toxicity tolerance of *Solanum nigrum* L. and *Parthenium hysterophorus* L. plants with reference to ion pattern, antioxidation activity and root exudation. *Ecotoxicol. Environ. Saf.* **2015**, *113*, 271–278. [CrossRef] [PubMed]
17. Kamran, M.; Eqani, S.; Katsoyiannis, A.; Xu, R.; Bibi, S.; Benizri, E.; Chaudhary, H. Phytoextraction of chromium (Cr) and influence of *Pseudomonas putida* on *Eruca sativa* growth. *J. Geochem. Explor.* **2017**, *182*, 269–274. [CrossRef]
18. Santos, C.; Rodriguez, E. Review on some emerging endpoints of chromium (VI) and lead phytotoxicity. In *Botany*; InTech: Rijeka, Croatia, 2012; pp. 61–82.
19. Eleftheriou, E.; Adamakis, I.-D.; Panteris, E.; Fatsiou, M. Chromium-induced ultrastructural changes and oxidative stress in roots of *Arabidopsis thaliana*. *Int. J. Mol. Sci.* **2015**, *16*, 15852–15871. [CrossRef] [PubMed]
20. Kumari, V.; Yadav, A.; Haq, I.; Kumar, S.; Bharagava, R.N.; Singh, S.K.; Raj, A. Genotoxicity evaluation of tannery effluent treated with newly isolated hexavalent chromium reducing *Bacillus cereus*. *J. Environ. Manag.* **2016**, *183*, 204–211. [CrossRef]
21. Shanker, A.K.; Djanaguiraman, M.; Venkateswarlu, B. Chromium interactions in plants: Current status and future strategies. *Metallomics* **2009**, *1*, 375–383. [CrossRef] [PubMed]
22. Ali, S.; Bharwana, S.A.; Rizwan, M.; Farid, M.; Kanwal, S.; Ali, Q.; Ibrahim, M.; Gill, R.A.; Khan, M.D. Fulvic acid mediates chromium (Cr) tolerance in wheat (*Triticum aestivum* L.) through lowering of Cr uptake and improved antioxidant defense system. *Environ. Sci. Pollut. Res.* **2015**, *22*, 10601–10609. [CrossRef]
23. Farooq, M.; Ali, S.; Hameed, A.; Bharwana, S.; Rizwan, M.; Ishaque, W.; Farid, M.; Mahmood, K.; Iqbal, Z. Cadmium stress in cotton seedlings: Physiological, photosynthesis and oxidative damages alleviated by glycinebetaine. *S. Afr. J. Bot.* **2016**, *104*, 61–68. [CrossRef]
24. Cervantes, C.; Campos-García, J. Reduction and efflux of chromate by bacteria. In *Molecular Microbiology of Heavy Metals*; Springer: Berlin/Heidelberg, Germany, 2007; pp. 407–419.
25. Anjum, S.A.; Ashraf, U.; Khan, I.; Tanveer, M.; Shahid, M.; Shakoor, A.; Wang, L. Phyto-toxicity of chromium in maize: Oxidative damage, osmolyte accumulation, anti-oxidative defense and chromium uptake. *Pedosphere* **2017**, *27*, 262–273. [CrossRef]
26. Tang, M.; Mao, D.; Xu, L.; Li, D.; Song, S.; Chen, C. Integrated analysis of miRNA and mRNA expression profiles in response to Cd exposure in rice seedlings. *BMC Genom.* **2014**, *15*, 835. [CrossRef]
27. Oliveira, H. Chromium as an environmental pollutant: Insights on induced plant toxicity. *J. Bot.* **2012**, *2012*, 1–8. [CrossRef]
28. Kováčik, J.; Babula, P.; Klejdus, B.I.; Hedbavny, J. Chromium uptake and consequences for metabolism and oxidative stress in chamomile plants. *J. Agric. Food Chem.* **2013**, *61*, 7864–7873. [CrossRef] [PubMed]
29. Samantaray, S.; Rout, G.R.; Das, P. Induction, selection and characterization of Cr and Ni-tolerant cell lines of *Echinochloa colona* (L.) Link in vitro. *J. Plant Physiol.* **2001**, *158*, 1281–1290. [CrossRef]
30. Rahman, M.M.; Rahman, M.M.; Islam, K.S.; Chongling, Y. Effect of Chromium Stress on Antioxidative Enzymes and Malondialdehyde Content Activities in Leaves and Roots of Mangrove Seedlings *Kandelia candel* (L.) Druce. *J. For. Environ. Sci.* **2010**, *26*, 171–179.

31. Rai, V.; Vajpayee, P.; Singh, S.N.; Mehrotra, S. Effect of chromium accumulation on photosynthetic pigments, oxidative stress defense system, nitrate reduction, proline level and eugenol content of *Ocimum tenuiflorum* L. *Plant Sci.* **2004**, *167*, 1159–1169. [CrossRef]
32. Ma, J.; Lv, C.; Xu, M.; Chen, G.; Lv, C.; Gao, Z. Photosynthesis performance, antioxidant enzymes, and ultrastructural analyses of rice seedlings under chromium stress. *Environ. Sci. Pollut. Res.* **2016**, *23*, 1768–1778. [CrossRef] [PubMed]
33. Trinh, N.N.; Huang, T.L.; Chi, W.C.; Fu, S.F.; Chen, C.C.; Huang, H.J. Chromium stress response effect on signal transduction and expression of signaling genes in rice. *Physiol. Plant.* **2014**, *150*, 205–224. [CrossRef]
34. Zhang, X.; Zhang, S.; Xu, X.; Li, T.; Gong, G.; Jia, Y.; Li, Y.; Deng, L. Tolerance and accumulation characteristics of cadmium in *Amaranthus hybridus* L. *J. Hazard. Mater.* **2010**, *180*, 303–308. [CrossRef]
35. Xu, L.; Han, L.; Huang, B. Membrane fatty acid composition and saturation levels associated with leaf dehydration tolerance and post-drought rehydration in Kentucky bluegrass. *Crop Sci.* **2011**, *51*, 273–281. [CrossRef]
36. Aldoobie, N.; Beltagi, M. Physiological, biochemical and molecular responses of common bean (*Phaseolus vulgaris* L.) plants to heavy metals stress. *Afr. J. Biotechnol.* **2013**, *12*, 4614–4622. [CrossRef]
37. Duhan, J.S. Chromium stress on peroxidase, ascorbate peroxidase and acid invertase in pea (*Pisum sativum* L.) seedling. *Int. J. Biotechnol. Mol. Biol. Res.* **2012**, *3*, 15–21.
38. Paiva, L.B.; Correa, S.F.; Santa Catarina, C.; Floh, E.I.S.; Silva, M.G.D.; Vitória, A.P. Ecophysiological and biochemical parameters for assessing Cr^{+6} stress conditions in *Pterogyne nitens* Tul.: New and usual methods for the management and restoration of degraded areas. *Environ. Eng. Manag. J.* **2014**, *13*, 3073–3081. [CrossRef]
39. Choudhary, S.P.; Kanwar, M.; Bhardwaj, R.; Yu, J.-Q.; Tran, L.-S.P. Chromium stress mitigation by polyamine-brassinosteroid application involves phytohormonal and physiological strategies in *Raphanus sativus* L. *PLoS ONE* **2012**, *7*, e33210. [CrossRef] [PubMed]
40. Shanker, A.K.; Djanaguiraman, M.; Sudhagar, R.; Chandrashekar, C.; Pathmanabhan, G. Differential antioxidative response of ascorbate glutathione pathway enzymes and metabolites to chromium speciation stress in green gram (*Vigna radiata* (L.) R. Wilczek. cv CO 4) roots. *Plant Sci.* **2004**, *166*, 1035–1043. [CrossRef]
41. Maiti, S.; Ghosh, N.; Mandal, C.; Das, K.; Dey, N.; Adak, M.K. Responses of the maize plant to chromium stress with reference to antioxidation activity. *Braz. J. Plant Physiol.* **2012**, *24*, 203–212. [CrossRef]
42. Khanna, K.; Jamwal, V.L.; Sharma, A.; Gandhi, S.G.; Ohri, P.; Bhardwaj, R.; Al-Huqail, A.A.; Siddiqui, M.H.; Ali, H.M.; Ahmad, P. Supplementation with plant growth promoting rhizobacteria (PGPR) alleviates cadmium toxicity in Solanum lycopersicum by modulating the expression of secondary metabolites. *Chemosphere* **2019**, *230*, 628–639. [CrossRef] [PubMed]
43. Kaur, R.; Yadav, P.; Sharma, A.; Thukral, A.K.; Kumar, V.; Kohli, S.K.; Bhardwaj, R. Castasterone and citric acid treatment restores photosynthetic attributes in *Brassica juncea* L. under Cd (II) toxicity. *Ecotoxicol. Environ. Saf.* **2017**, *145*, 466–475. [CrossRef]
44. Kaur, R.; Yadav, P.; Thukral, A.K.; Sharma, A.; Bhardwaj, R.; Alyemeni, M.N.; Wijaya, L.; Ahmad, P. Castasterone and citric acid supplementation alleviates cadmium toxicity by modifying antioxidants and organic acids in *Brassica juncea*. *J. Plant Growth Regul.* **2018**, *37*, 286–299. [CrossRef]
45. Kohli, S.K.; Handa, N.; Sharma, A.; Gautam, V.; Arora, S.; Bhardwaj, R.; Alyemeni, M.N.; Wijaya, L.; Ahmad, P. Combined effect of 24-epibrassinolide and salicylic acid mitigates lead (Pb) toxicity by modulating various metabolites in *Brassica juncea* L. seedlings. *Protoplasma* **2018**, *255*, 11–24. [CrossRef]
46. Srivastava, S.; Srivastava, S.; Prakash, S.; Srivastava, M. Fate of trivalent chromium in presence of organic acids: A hydroponic study on the tomato plant. *Chem. Speciat. Bioavailab.* **1998**, *10*, 147–150. [CrossRef]
47. Skeffington, R.; Shewry, P.; Peterson, P. Chromium uptake and transport in barley seedlings (*Hordeum vulgare* L.). *Planta* **1976**, *132*, 209–214. [CrossRef]
48. Shanker, A.K.; Cervantes, C.; Loza-Tavera, H.; Avudainayagam, S. Chromium toxicity in plants. *Environ. Int.* **2005**, *31*, 739–753. [CrossRef] [PubMed]
49. Cervantes, C.; Campos-García, J.; Devars, S.; Gutiérrez-Corona, F.; Loza-Tavera, H.; Torres-Guzmán, J.C.; Moreno-Sánchez, R. Interactions of chromium with microorganisms and plants. *FEMS Microbiol. Rev.* **2001**, *25*, 335–347. [CrossRef] [PubMed]
50. Appenroth, K.J.; Luther, A.; Jetschke, G.; Gabrys, H. Modification of chromate toxicity by sulphate in duckweeds (Lemnaceae). *Aquat. Toxicol.* **2008**, *89*, 167–171. [CrossRef] [PubMed]

51. De Oliveira, L.M.; Ma, L.Q.; Santos, J.A.; Guilherme, L.R.; Lessl, J.T. Effects of arsenate, chromate, and sulfate on arsenic and chromium uptake and translocation by arsenic hyperaccumulator *Pteris vittata* L. *Environ. Pollut.* **2014**, *184*, 187–192. [CrossRef] [PubMed]
52. De Oliveira, L.M.; Gress, J.; De, J.; Rathinasabapathi, B.; Marchi, G.; Chen, Y.; Ma, L.Q. Sulfate and chromate increased each other's uptake and translocation in As-hyperaccumulator *Pteris vittata*. *Chemosphere* **2016**, *147*, 36–43. [CrossRef] [PubMed]
53. Gupta, A.K.; Sinha, S. Chemical fractionation and heavy metal accumulation in the plant of *Sesamum indicum* (L.) var. T55 grown on soil amended with tannery sludge: Selection of single extractants. *Chemosphere* **2006**, *64*, 161–173. [CrossRef]
54. Caldelas, C.; Bort, J.; Febrero, A. Ultrastructure and subcellular distribution of Cr in *Iris pseudacorus* L. using TEM and X-ray microanalysis. *Cell Biol. Toxicol.* **2012**, *28*, 57–68. [CrossRef]
55. Mangabeira, P.A.; Ferreira, A.S.; de Almeida, A.-A.F.; Fernandes, V.F.; Lucena, E.; Souza, V.L.; dos Santos Júnior, A.J.; Oliveira, A.H.; Grenier-Loustalot, M.F.; Barbier, F. Compartmentalization and ultrastructural alterations induced by chromium in aquatic macrophytes. *Biometals* **2011**, *24*, 1017–1026. [CrossRef]
56. Kabata-Pendias, A.; Szteke, B. *Trace Elements in Abiotic and Biotic Environments*; CRC Press: Boca Raton, FL, USA, 2015.
57. Assuncao, A.G.; Herrero, E.; Lin, Y.F.; Huettel, B.; Talukdar, S.; Smaczniak, C.; Immink, R.G.; van Eldik, M.; Fiers, M.; Schat, H.; et al. Arabidopsis thaliana transcription factors bZIP19 and bZIP23 regulate the adaptation to zinc deficiency. *Proc. Natl. Acad. Sci. USA* **2010**, *107*, 10296–10301. [CrossRef]
58. Nishida, S.; Mizuno, T.; Obata, H. Involvement of histidine-rich domain of ZIP family transporter TjZNT1 in metal ion specificity. *Plant Physiol. Biochem.* **2008**, *46*, 601–606. [CrossRef]
59. Gustin, J.L.; Loureiro, M.E.; Kim, D.; Na, G.; Tikhonova, M.; Salt, D.E. MTP1-dependent Zn sequestration into shoot vacuoles suggests dual roles in Zn tolerance and accumulation in Zn-hyperaccumulating plants. *Plant J. Cell Mol. Biol.* **2009**, *57*, 1116–1127. [CrossRef] [PubMed]
60. Kobae, Y.; Sekino, T.; Yoshioka, H.; Nakagawa, T.; Martinoia, E.; Maeshima, M. Loss of AtPDR8, a plasma membrane ABC transporter of Arabidopsis thaliana, causes hypersensitive cell death upon pathogen infection. *Plant Cell Physiol.* **2006**, *47*, 309–318. [CrossRef] [PubMed]
61. Kim, D.Y.; Bovet, L.; Maeshima, M.; Martinoia, E.; Lee, Y. The ABC transporter AtPDR8 is a cadmium extrusion pump conferring heavy metal resistance. *Plant J. Cell Mol. Biol.* **2007**, *50*, 207–218. [CrossRef]
62. Zayed, A.; Lytle, C.M.; Qian, J.-H.; Terry, N. Chromium accumulation, translocation and chemical speciation in vegetable crops. *Planta* **1998**, *206*, 293–299. [CrossRef]
63. Cary, E.E.; Allaway, W.H.; Olson, O.E. Control of chromium concentrations in food plants. 1. Absorption and translocation of chromium by plants. *J. Agric. Food Chem.* **1977**, *25*, 300–304. [CrossRef] [PubMed]
64. Mallick, S.; Sinam, G.; Kumar Mishra, R.; Sinha, S. Interactive effects of Cr and Fe treatments on plants growth, nutrition and oxidative status in *Zea mays* L. *Ecotoxicol. Environ. Saf.* **2010**, *73*, 987–995. [CrossRef] [PubMed]
65. Chigonum, W.J.; Ikenna, O.C.; John, C.U. Dynamic Impact of Chromium on Nutrient Uptake from Soil by Fluted Pumpkin (*Telfairia occidentalis*). *Am. J. Biosci. Bioeng.* **2019**, *7*, 1–9.
66. Osu Charles, I.; Onyema, M.O. Vanadium inhibition capacity on nutrients and heavy metal uptake by *Cucumis Sativus*. *J. Am. Sci.* **2016**, *12*. [CrossRef]
67. Dube, B.; Tewari, K.; Chatterjee, J.; Chatterjee, C. Excess chromium alters uptake and translocation of certain nutrients in citrullus. *Chemosphere* **2003**, *53*, 1147–1153. [CrossRef]
68. Turner, M.; Rust, R. Effects of Chromium on Growth and Mineral Nutrition of Soybeans 1. *Soil Sci. Soc. Am. J.* **1971**, *35*, 755–758. [CrossRef]
69. Sundaramoorthy, P.; Chidambaram, A.; Ganesh, K.S.; Unnikannan, P.; Baskaran, L. Chromium stress in paddy: (i) nutrient status of paddy under chromium stress; (ii) phytoremediation of chromium by aquatic and terrestrial weeds. *C. R. Biol.* **2010**, *333*, 597–607. [CrossRef] [PubMed]
70. Shahzad, B.; Tanveer, M.; Rehman, A.; Cheema, S.A.; Fahad, S.; Rehman, S.; Sharma, A. Nickel; whether toxic or essential for plants and environment-A review. *Plant Physiol. Biochem.* **2018**, *132*, 641–651. [CrossRef] [PubMed]
71. Mengel, K.; Kirkby, E. *Principles of Plant Nutrition*; International Potash Institute: Worblaufen, Switzerland, 1987; pp. 687–695.

72. Rai, V.; Tandon, P.K.; Khatoon, S. Effect of chromium on antioxidant potential of *Catharanthus roseus* varieties and production of their anticancer alkaloids: Vincristine and vinblastine. *BioMed Res. Int.* **2014**, *2014*, 1–10. [CrossRef] [PubMed]
73. Lushchak, V.I. Adaptive response to oxidative stress: Bacteria, fungi, plants and animals. *Comp. Biochem. Physiol. Part C Toxicol. Pharmacol.* **2011**, *153*, 175–190. [CrossRef]
74. Sharma, A.; Kumar, V.; Shahzad, B.; Ramakrishnan, M.; Sidhu, G.P.S.; Bali, A.S.; Handa, N.; Kapoor, D.; Yadav, P.; Khanna, K. Photosynthetic Response of Plants Under Different Abiotic Stresses: A Review. *J. Plant Growth Regul.* **2019**, *38*, 1–23. [CrossRef]
75. Valko, M.; Rhodes, C.; Moncol, J.; Izakovic, M.; Mazur, M. Free radicals, metals and antioxidants in oxidative stress-induced cancer. *Chem.-Biol. Interact.* **2006**, *160*, 1–40. [CrossRef]
76. Chandra, P.; Kulshreshtha, K. Chromium accumulation and toxicity in aquatic vascular plants. *Bot. Rev.* **2004**, *70*, 313–327. [CrossRef]
77. Hayat, S.; Khalique, G.; Irfan, M.; Wani, A.S.; Tripathi, B.N.; Ahmad, A. Physiological changes induced by chromium stress in plants: An overview. *Protoplasma* **2012**, *249*, 599–611. [CrossRef]
78. Rajendran, M.; An, W.-H.; Li, W.-C.; Perumal, V.; Wu, C.; Sahi, S.V.; Sarkar, S.K. Chromium detoxification mechanism induced growth and antioxidant responses in vetiver (*Chrysopogon zizanioides* (L.) Roberty). *J. Cent. South Univ.* **2019**, *26*, 489–500. [CrossRef]
79. Sinha, S.; Saxena, R.; Singh, S. Chromium induced lipid peroxidation in the plants of *Pistia stratiotes* L.: Role of antioxidants and antioxidant enzymes. *Chemosphere* **2005**, *58*, 595–604. [CrossRef]
80. Balal, R.M.; Shahid, M.A.; Vincent, C.; Zotarelli, L.; Liu, G.; Mattson, N.S.; Rathinasabapathi, B.; Martínez-Nicolas, J.J.; Garcia-Sanchez, F. Kinnow mandarin plants grafted on tetraploid rootstocks are more tolerant to Cr-toxicity than those grafted on its diploids one. *Environ. Exp. Bot.* **2017**, *140*, 8–18. [CrossRef]
81. Islam, F.; Yasmeen, T.; Arif, M.S.; Riaz, M.; Shahzad, S.M.; Imran, Q.; Ali, I. Combined ability of chromium (Cr) tolerant plant growth promoting bacteria (PGPB) and salicylic acid (SA) in attenuation of chromium stress in maize plants. *Plant Physiol. Biochem.* **2016**, *108*, 456–467. [CrossRef] [PubMed]
82. Zou, J.; Yu, K.; Zhang, Z.; Jiang, W.; Liu, D. Antioxidant response system and chlorophyll fluorescence in chromium (VI)-treated *Zea mays* L. seedlings. *Acta Biol. Crac. Ser. Bot.* **2009**, *51*, 23–33.
83. Amin, H.; Arain, B.A.; Amin, F.; Surhio, M.A. Phytotoxicity of chromium on germination, growth and biochemical attributes of *Hibiscus esculentus* L. *Am. J. Plant Sci.* **2013**, *4*, 720–726. [CrossRef]
84. Tang, J.; Xu, J.; Wu, Y.; Li, Y.; Tang, Q. Effects of high concentration of chromium stress on physiological and biochemical characters and accumulation of chromium in tea plant (*Camellia sinensis*. L.). *Afr. J. Biotechnol.* **2012**, *11*, 2248–2255.
85. Amin, H.; Arain, B.A.; Amin, F.; Surhio, M.A. Analysis of growth response and tolerance index of *Glycine max* (L.) Merr. under hexavalent chromium stress. *Adv. Life Sci.* **2014**, *1*, 231–241.
86. Muslu, A.; Ergün, N. Effects of copper and chromium and high temperature on growth, proline and protein content in wheat seedlings. *Bangladesh J. Bot.* **2013**, *42*, 105–112. [CrossRef]
87. Van Assche, F.; Clijsters, H. Effects of metals on enzyme activity in plants. *Plant Cell Environ.* **1990**, *13*, 195–206. [CrossRef]
88. Vajpayee, P.; Rai, U.; Ali, M.; Tripathi, R.; Yadav, V.; Sinha, S.; Singh, S. Chromium-induced physiologic changes in *Vallisneria spiralis* L. and its role in phytoremediation of tannery effluent. *Bull. Environ. Contam. Toxicol.* **2001**, *67*, 246–256. [CrossRef]
89. Zlobin, I.; Kholodova, V.; Rakhmankulova, Z.; Kuznetsov, V.V. Brassica napus responses to short-term excessive copper treatment with decrease of photosynthetic pigments, differential expression of heavy metal homeostasis genes including activation of gene NRAMP4 involved in photosystem II stabilization. *Photosynth. Res.* **2015**, *125*, 141–150. [CrossRef]
90. Vernay, P.; Gauthier-Moussard, C.; Hitmi, A. Interaction of bioaccumulation of heavy metal chromium with water relation, mineral nutrition and photosynthesis in developed leaves of *Lolium perenne* L. *Chemosphere* **2007**, *68*, 1563–1575. [CrossRef] [PubMed]
91. Khan, I.; Iqbal, M.; Ashraf, M.Y.; Ashraf, M.A.; Ali, S. Organic chelants-mediated enhanced lead (Pb) uptake and accumulation is associated with higher activity of enzymatic antioxidants in spinach (*Spinacea oleracea* L.). *J. Hazard. Mater.* **2016**, *317*, 352–361. [CrossRef] [PubMed]
92. Sharma, S.S.; Dietz, K.-J. The significance of amino acids and amino acid-derived molecules in plant responses and adaptation to heavy metal stress. *J. Exp. Bot.* **2006**, *57*, 711–726. [CrossRef] [PubMed]

93. Sharma, D.; Sharma, C. Chromium uptake and toxicity effects on growth and metabolic activities in wheat, *Triticum aestivum* L. cv. UP 2003. *Indian J. Exp. Biol.* **1996**, *34*, 689–691. [PubMed]
94. Sharma, A.; Shahzad, B.; Rehman, A.; Bhardwaj, R.; Landi, M.; Zheng, B. Response of phenylpropanoid pathway and the role of polyphenols in plants under abiotic stress. *Molecules* **2019**, *24*, 2452. [CrossRef]
95. Shakoor, M.B.; Ali, S.; Hameed, A.; Farid, M.; Hussain, S.; Yasmeen, T.; Najeeb, U.; Bharwana, S.A.; Abbasi, G.H. Citric acid improves lead (Pb) phytoextraction in *Brassica napus* L. by mitigating Pb-induced morphological and biochemical damages. *Ecotoxicol. Environ. Saf.* **2014**, *109*, 38–47. [CrossRef]
96. Davies, F., Jr.; Puryear, J.; Newton, R.; Egilla, J.; Saraiva Grossi, J. Mycorrhizal fungi increase chromium uptake by sunflower plants: Influence on tissue mineral concentration, growth, and gas exchange. *J. Plant Nutr.* **2002**, *25*, 2389–2407. [CrossRef]
97. Handa, N.; Kohli, S.K.; Sharma, A.; Thukral, A.K.; Bhardwaj, R.; Abd_Allah, E.F.; Alqarawi, A.A.; Ahmad, P. Selenium modulates dynamics of antioxidative defence expression, photosynthetic attributes and secondary metabolites to mitigate chromium toxicity in *Brassica juncea* L. plants. *Environ. Exp. Bot.* **2019**, *161*, 180–192. [CrossRef]
98. Radi, R. Oxygen radicals, nitric oxide, and peroxynitrite: Redox pathways in molecular medicine. *Proc. Natl. Acad. Sci. USA* **2018**, *115*, 5839–5848. [CrossRef]
99. Pisoschi, A.M.; Pop, A. The role of antioxidants in the chemistry of oxidative stress: A review. *Eur. J. Med. Chem.* **2015**, *97*, 55–74. [CrossRef]
100. Turkan, I.; Uzilday, B.; Dietz, K.J.; Brautigam, A.; Ozgur, R. Reactive oxygen species and redox regulation in mesophyll and bundle sheath cells of C4 plants. *J. Exp. Bot.* **2018**, *69*, 3321–3331. [CrossRef] [PubMed]
101. Valko, M.; Leibfritz, D.; Moncol, J.; Cronin, M.T.; Mazur, M.; Telser, J. Free radicals and antioxidants in normal physiological functions and human disease. *Int. J. Biochem. Cell Biol.* **2007**, *39*, 44–84. [CrossRef] [PubMed]
102. Shahzad, B.; Tanveer, M.; Che, Z.; Rehman, A.; Cheema, S.A.; Sharma, A.; Song, H.; Ur Rehman, S.; Zhaorong, D. Role of 24-epibrassinolide (EBL) in mediating heavy metal and pesticide induced oxidative stress in plants: A review. *Ecotoxicol. Environ. Saf.* **2018**, *147*, 935–944. [CrossRef] [PubMed]
103. Anjum, S.A.; Tanveer, M.; Hussain, S.; Ullah, E.; Wang, L.; Khan, I.; Samad, R.A.; Tung, S.A.; Anam, M.; Shahzad, B. Morpho-physiological growth and yield responses of two contrasting maize cultivars to cadmium exposure. *Clean–Soil Air Water* **2016**, *44*, 29–36. [CrossRef]
104. Anjum, S.A.; Tanveer, M.; Hussain, S.; Bao, M.; Wang, L.; Khan, I.; Ullah, E.; Tung, S.A.; Samad, R.A.; Shahzad, B. Cadmium toxicity in Maize (*Zea mays* L.): Consequences on antioxidative systems, reactive oxygen species and cadmium accumulation. *Environ. Sci. Pollut. Res.* **2015**, *22*, 17022–17030. [CrossRef]
105. Grover, P.; Rekhadevi, P.; Danadevi, K.; Vuyyuri, S.; Mahboob, M.; Rahman, M. Genotoxicity evaluation in workers occupationally exposed to lead. *Int. J. Hyg. Environ. Health* **2010**, *213*, 99–106. [CrossRef]
106. Opdenakker, K.; Remans, T.; Keunen, E.; Vangronsveld, J.; Cuypers, A. Exposure of Arabidopsis thaliana to Cd or Cu excess leads to oxidative stress mediated alterations in MAPKinase transcript levels. *Environ. Exp. Bot.* **2012**, *83*, 53–61. [CrossRef]
107. Tanveer, M.; Shahzad, B.; Sharma, A.; Khan, E.A. 24-Epibrassinolide application in plants: An implication for improving drought stress tolerance in plants. *Plant Physiol. Biochem.* **2018**, *135*, 295–303. [CrossRef]
108. Tanveer, M.; Shahzad, B.; Sharma, A.; Biju, S.; Bhardwaj, R. 24-Epibrassinolide; an active brassinolide and its role in salt stress tolerance in plants: A review. *Plant Physiol. Biochem.* **2018**, *130*, 69–79. [CrossRef]
109. Shahzad, B.; Mughal, M.N.; Tanveer, M.; Gupta, D.; Abbas, G. Is lithium biologically an important or toxic element to living organisms? An overview. *Environ. Sci. Pollut. Res.* **2017**, *24*, 103–115. [CrossRef]
110. Anjum, S.A.; Tanveer, M.; Hussain, S.; Shahzad, B.; Ashraf, U.; Fahad, S.; Hassan, W.; Jan, S.; Khan, I.; Saleem, M.F. Osmoregulation and antioxidant production in maize under combined cadmium and arsenic stress. *Environ. Sci. Pollut. Res.* **2016**, *23*, 11864–11875. [CrossRef] [PubMed]
111. Morina, F.; Jovanovic, L.; Mojovic, M.; Vidovic, M.; Pankovic, D.; Veljovic Jovanovic, S. Zinc-induced oxidative stress in *Verbascum thapsus* is caused by an accumulation of reactive oxygen species and quinhydrone in the cell wall. *Physiol. Plant.* **2010**, *140*, 209–224. [CrossRef] [PubMed]
112. Kováčik, J.; Klejdus, B.; Hedbavny, J.; Bačkor, M. Effect of copper and salicylic acid on phenolic metabolites and free amino acids in *Scenedesmus quadricauda* (Chlorophyceae). *Plant Sci.* **2010**, *178*, 307–311. [CrossRef]
113. Duquesnoy, I.; Champeau, G.M.; Evray, G.; Ledoigt, G.; Piquet-Pissaloux, A. Enzymatic adaptations to arsenic-induced oxidative stress in *Zea mays* and genotoxic effect of arsenic in root tips of *Vicia faba* and *Zea mays*. *C. R. Biol.* **2010**, *333*, 814–824. [CrossRef]

114. Márquez-García, B.; Pérez-López, R.; Ruíz-Chancho, M.J.; López-Sánchez, J.F.; Rubio, R.; Abreu, M.M.; Nieto, J.M.; Córdoba, F. Arsenic speciation in soils and Erica andevalensis Cabezudo & Rivera and Erica australis L. from São Domingos Mine area, Portugal. *J. Geochem. Explor.* **2012**, *119*, 51–59.
115. Körpe, D.A.; Aras, S. Evaluation of copper-induced stress on eggplant (*Solanum melongena* L.) seedlings at the molecular and population levels by use of various biomarkers. *Mutat. Res./Genet. Toxicol. Environ. Mutagenesis* **2011**, *719*, 29–34. [CrossRef]
116. Jozefczak, M.; Remans, T.; Vangronsveld, J.; Cuypers, A. Glutathione is a key player in metal-induced oxidative stress defenses. *Int. J. Mol. Sci.* **2012**, *13*, 3145–3175. [CrossRef]
117. Gupta, D.; Nicoloso, F.; Schetinger, M.; Rossato, L.; Pereira, L.; Castro, G.; Srivastava, S.; Tripathi, R. Antioxidant defense mechanism in hydroponically grown *Zea mays* seedlings under moderate lead stress. *J. Hazard. Mater.* **2009**, *172*, 479–484. [CrossRef]
118. Gupta, D.; Huang, H.; Yang, X.; Razafindrabe, B.; Inouhe, M. The detoxification of lead in *Sedum alfredii* H. is not related to phytochelatins but the glutathione. *J. Hazard. Mater.* **2010**, *177*, 437–444. [CrossRef]
119. Shahzad, B.; Tanveer, M.; Hassan, W.; Shah, A.N.; Anjum, S.A.; Cheema, S.A.; Ali, I. Lithium toxicity in plants: Reasons, mechanisms and remediation possibilities—A review. *Plant Physiol. Biochem.* **2016**, *107*, 104–115. [CrossRef]
120. Lee, S.; Moon, J.S.; Ko, T.-S.; Petros, D.; Goldsbrough, P.B.; Korban, S.S. Overexpression of Arabidopsis phytochelatin synthase paradoxically leads to hypersensitivity to cadmium stress. *Plant Physiol.* **2003**, *131*, 656–663. [CrossRef] [PubMed]
121. Weyemi, U.; Dupuy, C. The emerging role of ROS-generating NADPH oxidase NOX4 in DNA-damage responses. *Mutat. Res./Rev. Mutat. Res.* **2012**, *751*, 77–81. [CrossRef]
122. Potocký, M.; Pejchar, P.; Gutkowska, M.; Jiménez-Quesada, M.J.; Potocká, A.; de Dios Alché, J.; Kost, B.; Žárský, V. NADPH oxidase activity in pollen tubes is affected by calcium ions, signaling phospholipids and Rac/Rop GTPases. *J. Plant Physiol.* **2012**, *169*, 1654–1663. [CrossRef] [PubMed]
123. Pourrut, B.; Shahid, M.; Douay, F.; Dumat, C.; Pinelli, E. Molecular mechanisms involved in lead uptake, toxicity and detoxification in higher plants. In *Heavy Metal Stress in Plants*; Springer: Berlin/Heidelberg, Germany, 2013; pp. 121–147.
124. Pourrut, B.; Perchet, G.; Silvestre, J.; Cecchi, M.; Guiresse, M.; Pinelli, E. Potential role of NADPH-oxidase in early steps of lead-induced oxidative burst in *Vicia faba* roots. *J. Plant Physiol.* **2008**, *165*, 571–579. [CrossRef] [PubMed]
125. Sagi, M.; Fluhr, R. Production of reactive oxygen species by plant NADPH oxidases. *Plant Physiol.* **2006**, *141*, 336–340. [CrossRef]
126. Choudhury, S.; Panda, S.K. Toxic effects, oxidative stress and ultrastructural changes in moss *Taxithelium nepalense* (Schwaegr.) Broth. under chromium and lead phytotoxicity. *Water Air Soil Pollut.* **2005**, *167*, 73–90. [CrossRef]
127. Ullah, A.; Shahzad, B.; Tanveer, M.; Nadeem, F.; Sharma, A.; Lee, D.J.; Rehman, A. Abiotic Stress Tolerance in Plants Through Pre-sowing Seed Treatments with Mineral Elements and Growth Regulators. In *Priming and Pretreatment of Seeds and Seedlings: Implication in Plant Stress Tolerance and Enhancing Productivity in Crop Plants*; Hasanuzzaman, M., Fotopoulos, V., Eds.; Springer: Singapore, 2019; pp. 427–445. [CrossRef]
128. Fahad, S.; Rehman, A.; Shahzad, B.; Tanveer, M.; Saud, S.; Kamran, M.; Ihtisham, M.; Khan, S.U.; Turan, V.; Ur Rahman, M.H. Chapter 14—Rice Responses and Tolerance to Metal/Metalloid Toxicity. *Advances in Rice Research for Abiotic Stress Tolerance* **2019**, 299–312. [CrossRef]
129. Adrees, M.; Ali, S.; Iqbal, M.; Bharwana, S.A.; Siddiqi, Z.; Farid, M.; Ali, Q.; Saeed, R.; Rizwan, M. Mannitol alleviates chromium toxicity in wheat plants in relation to growth, yield, stimulation of anti-oxidative enzymes, oxidative stress and Cr uptake in sand and soil media. *Ecotoxicol. Environ. Saf.* **2015**, *122*, 1–8. [CrossRef]
130. Gill, S.S.; Tuteja, N. Reactive oxygen species and antioxidant machinery in abiotic stress tolerance in crop plants. *Plant Physiol. Biochem.* **2010**, *48*, 909–930. [CrossRef]
131. Bhaduri, A.M.; Fulekar, M. Antioxidant enzyme responses of plants to heavy metal stress. *Rev. Environ. Sci. Bio/Technol.* **2012**, *11*, 55–69. [CrossRef]
132. Panda, S.; Choudhury, S. Chromium stress in plants. *Braz. J. Plant Physiol.* **2005**, *17*, 95–102. [CrossRef]

133. Mhamdi, A.; Hager, J.; Chaouch, S.; Queval, G.; Han, Y.; Taconnat, L.; Saindrenan, P.; Gouia, H.; Issakidis-Bourguet, E.; Renou, J.-P. Arabidopsis GLUTATHIONE REDUCTASE1 plays a crucial role in leaf responses to intracellular hydrogen peroxide and in ensuring appropriate gene expression through both salicylic acid and jasmonic acid signaling pathways. *Plant Physiol.* **2010**, *153*, 1144–1160. [CrossRef] [PubMed]
134. Wu, H.; Tito, N.; Giraldo, J.P. Anionic Cerium Oxide Nanoparticles Protect Plant Photosynthesis from Abiotic Stress by Scavenging Reactive Oxygen Species. *ACS Nano* **2017**, *11*, 11283–11297. [CrossRef] [PubMed]
135. Anjum, S.A.; Ashraf, U.; Tanveer, M.; Khan, I.; Hussain, S.; Shahzad, B.; Zohaib, A.; Abbas, F.; Saleem, M.F.; Ali, I.; et al. Drought Induced Changes in Growth, Osmolyte Accumulation and Antioxidant Metabolism of Three Maize Hybrids. *Front. Plant Sci.* **2017**, *8*, 69. [CrossRef]
136. Sharma, D.; Sharma, C.; Tripathi, R. Phytotoxic lesions of chromium in maize. *Chemosphere* **2003**, *51*, 63–68. [CrossRef]
137. Ali, S.; Chaudhary, A.; Rizwan, M.; Anwar, H.T.; Adrees, M.; Farid, M.; Irshad, M.K.; Hayat, T.; Anjum, S.A. Alleviation of chromium toxicity by glycinebetaine is related to elevated antioxidant enzymes and suppressed chromium uptake and oxidative stress in wheat (*Triticum aestivum* L.). *Environ. Sci. Pollut. Res.* **2015**, *22*, 10669–10678. [CrossRef]
138. Alscher, R.G.; Donahue, J.L.; Cramer, C.L. Reactive oxygen species and antioxidants: Relationships in green cells. *Physiol. Plant.* **1997**, *100*, 224–233. [CrossRef]
139. Blokhina, O.; Virolainen, E.; Fagerstedt, K.V. Antioxidants, oxidative damage and oxygen deprivation stress: A review. *Ann. Bot.* **2003**, *91*, 179–194. [CrossRef]
140. Kasote, D.M.; Katyare, S.S.; Hegde, M.V.; Bae, H. Significance of antioxidant potential of plants and its relevance to therapeutic applications. *Int. J. Biol. Sci.* **2015**, *11*, 982–991. [CrossRef]
141. Meyer, A.J.; Hell, R. Glutathione homeostasis and redox-regulation by sulfhydryl groups. *Photosynth. Res.* **2005**, *86*, 435–457. [CrossRef] [PubMed]
142. Noctor, G.; Mhamdi, A.; Chaouch, S.; Han, Y.; Neukermans, J.; Marquez-Garcia, B.; Queval, G.; Foyer, C.H. Glutathione in plants: An integrated overview. *Plant Cell Environ.* **2012**, *35*, 454–484. [CrossRef] [PubMed]
143. Galant, A.; Preuss, M.L.; Cameron, J.C.; Jez, J.M. Plant glutathione biosynthesis: Diversity in biochemical regulation and reaction products. *Front. Plant Sci.* **2011**, *2*, 45. [CrossRef] [PubMed]
144. Noctor, G.; Foyer, C.H. Simultaneous measurement of foliar glutathione, γ-glutamylcysteine, and amino acids by high-performance liquid chromatography: Comparison with two other assay methods for glutathione. *Anal. Biochem.* **1998**, *264*, 98–110. [CrossRef] [PubMed]
145. Di Toppi, L.S.; Fossati, F.; Musetti, R.; Mikerezi, I.; Favali, M.A. Effects of Hexavalent Chromium on Maize, Tomato, and Cauliflower Plants. *J. Plant Nutr.* **2002**, *25*, 701–717. [CrossRef]
146. Shanker, A.K.; Pathmanabhan, G. Speciation dependant antioxidative response in roots and leaves of sorghum (*Sorghum bicolor* (L.) Moench cv CO 27) under Cr (III) and Cr (VI) stress. *Plant Soil* **2004**, *265*, 141–151. [CrossRef]
147. McAuley, A.; Olatunji, M.A. Metal-ion oxidations in solution. Part XIX. Redox pathways in the oxidation of penicillamine and glutathione by chromium (VI). *Can. J. Chem.* **1977**, *55*, 3335–3340. [CrossRef]
148. Marwa, E.M.; Meharg, A.A.; Rice, C.M. Risk assessment of potentially toxic elements in agricultural soils and maize tissues from selected districts in Tanzania. *Sci. Total. Environ.* **2012**, *416*, 180–186. [CrossRef]
149. Kumar, V.; Yadav, S.K. Proline and betaine provide protection to antioxidant and methylglyoxal detoxification systems during cold stress in *Camellia sinensis* (L.) O. Kuntze. *Acta Physiol. Plant.* **2009**, *31*, 261–269. [CrossRef]
150. Ganesh, K.S.; Baskaran, L.; Chidambaram, A.; Sundaramoorthy, P. Influence of chromium stress on proline accumulation in soybean (*Glycine max* L. Merr.) genotypes. *Glob. J. Environ. Res.* **2009**, *3*, 106–108.
151. Bitrián, M.; Zarza, X.; Altabella, T.; Tiburcio, A.F.; Alcázar, R. Polyamines under abiotic stress: Metabolic crossroads and hormonal crosstalks in plants. *Metabolites* **2012**, *2*, 516–528. [CrossRef]
152. Hussain, M.I.; González, L.; Souto, C.; Reigosa, M. Ecophysiological responses of three native herbs to phytotoxic potential of invasive Acacia melanoxylon R. *Br. Agrofor. Syst.* **2011**, *83*, 149–166. [CrossRef]
153. Davies, P.J. *Plant Hormones: Biosynthesis, Signal Transduction, Action*; Springer: Berlin/Heidelberg, Germany, 2004.
154. Skirycz, A.; Inzé, D. More from less: Plant growth under limited water. *Curr. Opin. Biotechnol.* **2010**, *21*, 197–203. [CrossRef] [PubMed]

155. O'Brien, J.A.; Benková, E. Cytokinin cross-talking during biotic and abiotic stress responses. *Front. Plant Sci.* **2013**, *4*, 451. [CrossRef] [PubMed]
156. Huang, T.L.; Nguyen, Q.T.; Fu, S.F.; Lin, C.Y.; Chen, Y.C.; Huang, H.J. Transcriptomic changes and signalling pathways induced by arsenic stress in rice roots. *Plant Mol. Biol.* **2012**, *80*, 587–608. [CrossRef]
157. Rivas-San Vicente, M.; Plasencia, J. Salicylic acid beyond defence: Its role in plant growth and development. *J. Exp. Bot.* **2011**, *62*, 3321–3338. [CrossRef]
158. Miura, K.; Tada, Y. Regulation of water, salinity, and cold stress responses by salicylic acid. *Front. Plant Sci.* **2014**, *5*, 4. [CrossRef]
159. Jia, C.; Zhang, L.; Liu, L.; Wang, J.; Li, C.; Wang, Q. Multiple phytohormone signalling pathways modulate susceptibility of tomato plants to *Alternaria alternata* f. sp. lycopersici. *J. Exp. Bot.* **2013**, *64*, 637–650. [CrossRef]
160. Khanna, P.; Kaur, K.; Gupta, A.K. Salicylic acid induces differential antioxidant response in spring maize under high temperature stress. *Indian J. Exp. Biol.* **2016**, *54*, 386–393.
161. Matilla-Vazquez, M.; Matilla, A. Ethylene: Role in plants under environmental stress. In *Physiological Mechanisms and Adaptation Strategies in Plants under Changing Environment*; Springer: Berlin/Heidelberg, Germany, 2014; pp. 189–222.
162. Kazan, K. Auxin and the integration of environmental signals into plant root development. *Ann. Bot.* **2013**, *112*, 1655–1665. [CrossRef]
163. Agami, R.A.; Mohamed, G.F. Exogenous treatment with indole-3-acetic acid and salicylic acid alleviates cadmium toxicity in wheat seedlings. *Ecotoxicol. Environ. Saf.* **2013**, *94*, 164–171. [CrossRef]
164. Fahad, S.; Hussain, S.; Bano, A.; Saud, S.; Hassan, S.; Shan, D.; Khan, F.A.; Khan, F.; Chen, Y.; Wu, C. Potential role of phytohormones and plant growth-promoting rhizobacteria in abiotic stresses: Consequences for changing environment. *Environ. Sci. Pollut. Res.* **2015**, *22*, 4907–4921. [CrossRef] [PubMed]
165. Sponsel, V.; Hedden, P. Gibberellin biosynthesis and inactivation. In *Plant Hormones*; Davis, P.J., Ed.; Springer: Dordrecht, The Netherlands, 2010; pp. 63–94. [CrossRef]
166. Atici, Ö.; Ağar, G.; Battal, P. Changes in phytohormone contents in chickpea seeds germinating under lead or zinc stress. *Biol. Plant.* **2005**, *49*, 215–222. [CrossRef]
167. Ha, S.; Vankova, R.; Yamaguchi-Shinozaki, K.; Shinozaki, K.; Tran, L.-S.P. Cytokinins: Metabolism and function in plant adaptation to environmental stresses. *Trends Plant Sci.* **2012**, *17*, 172–179. [CrossRef] [PubMed]

plants

MDPI

Review

Role of Silicon in Mitigation of Heavy Metal Stresses in Crop Plants

Javaid Akhter Bhat [1,†], S. M. Shivaraj [2,†], Pritam Singh [3], Devanna B. Navadagi [4], Durgesh Kumar Tripathi [5], Prasanta K. Dash [4], Amolkumar U. Solanke [4], Humira Sonah [3,*] and Rupesh Deshmukh [3,*]

1 Department of Genetics and Plant Breeding, Indian Agricultural Research Institute, New Delhi 110012, India; javid.akhter69@gmail.com
2 Département de Phytologie, Université Laval, Québec City, QC G1V 0A6, Canada; sraj100@gmail.com
3 National Agri-Food Biotechnology Institute, Mohali 140306, India; raghav.pritam26dec@gmail.com
4 National Research Centre on Plant Biotechnology, New Delhi 110012, India; devnova2460@gmail.com (D.B.N.); prasanta01@yahoo.com (P.K.D.); amolsgene@gmail.com (A.U.S.)
5 Amity Institute of Organic Agriculture, Amity University, Uttar Pradesh, Noida 201313, India; dktripathiau@gmail.com
* Correspondence: biohuma@gmail.com (H.S.); rupesh0deshmukh@gmail.com (R.D.); Tel.: +91-6239715281 (H.S.); +91-9650792638 (R.D.)
† Authors contributed equally.

Received: 9 February 2019; Accepted: 19 March 2019; Published: 21 March 2019

Abstract: Over the past few decades, heavy metal contamination in soil and water has increased due to anthropogenic activities. The higher exposure of crop plants to heavy metal stress reduces growth and yield, and affect the sustainability of agricultural production. In this regard, the use of silicon (Si) supplementation offers a promising prospect since numerous studies have reported the beneficial role of Si in mitigating stresses imposed by biotic as well as abiotic factors including heavy metal stress. The fundamental mechanisms involved in the Si-mediated heavy metal stress tolerance include reduction of metal ions in soil substrate, co-precipitation of toxic metals, metal-transport related gene regulation, chelation, stimulation of antioxidants, compartmentation of metal ions, and structural alterations in plants. Exogenous application of Si has been well documented to increase heavy metal tolerance in numerous plant species. The beneficial effects of Si are particularly evident in plants able to accumulate high levels of Si. Consequently, to enhance metal tolerance in plants, the inherent genetic potential for Si uptake should be improved. In the present review, we have discussed the potential role and mechanisms involved in the Si-mediated alleviation of metal toxicity as well as different approaches for enhancing Si-derived benefits in crop plants.

Keywords: metal stress; toxicity; silicon; Si-fertilization; genomics; transporter genes

1. Introduction

Plants, being sessile, are continuously exposed to many biotic and abiotic stresses affecting their growth and development. Among the abiotic factors affecting plants, heavy metal stresses have received increasing attention over the last several decades. The term heavy metal refers to any metallic element with relatively high density that is toxic even at low concentration. In general, heavy metals relate to a group of metals and metalloids with greater than 4 $g \cdot cm^{-3}$ atomic density [1]. Among the naturally occurring elements, 53 are categorized as heavy metals. The heavy metals include cadmium (Cd), nickel (Ni), lead (Pb), iron (Fe), zinc (Zn), cobalt (Co), arsenic (As), chromium (Cr), silver (Ag) and platinum (Pt), and the majority of them do not play an essential role in plants. Although naturally present in the soil, concentration of these heavy metals increases as a result of geologic and anthropogenic activities causing a harmful/toxic effect on both plants and animals [2]. Heavy metals retard plant growth by marginalizing the cellular functions of proteins, lipids, and elemental components of thylakoid membranes [3]. Moreover, heavy metals can be transported through the food chain into animals and humans, so their presence will cause a significant threat to human health [4]. Among means to reduce heavy metal toxicity in agricultural production, silicon (Si) is often reported for its potential to mitigate their adverse effects [3,5].

Silicon, a second most abundant element after oxygen in the earth crust, is considered as a quasi-essential element because of the numerous benefits it confers to plants, specifically under biotic and abiotic stress [6,7]. For instance, it is reported to alleviate a number of abiotic factors in plants including drought, salinity stress, lodging and heavy metal toxicity [8,9]. A possible role of Si in metal detoxification is attributed to alteration of plant cellular mechanisms and biochemical interactions with the external growth medium [10]. The positive effects of Si vary with the crop species and are usually more pronounced in plants that accumulate high concentrations of Si in their tissues [11,12]. The beneficial effects of Si are predominantly if not exclusively manifest when plants are subjected to stress [13]. Despite the abundant availability of Si in soils, the plant-available form is often limited in most soil types. Plant roots uptake Si in the form of silicic acid (H_4SiO_4) where the concentration ranges from 0.1 to 0.6 $mmol{\cdot}L^{-1}$ in the soil solution [14]. The Si concentrations in plants will vary according to the plant species, ranging from 1 to 100 $g{\cdot}kg^{-1}$, which represents the largest range of mineral elements [6]. This phenomenon is related to the different ability of plant species to uptake and transport Si, through a dedicated system of Si transporters [14,15]. In the present review, we discuss the role of Si in enhancing heavy metal tolerance in crop plants and the possible mechanisms by which Si achieves this feat.

2. Heavy Metal: From Essentiality to Toxicity

Heavy metals are classified into non-essential elements (Cd, Pd, Hg, Cr, As and Ag) being potentially toxic to plants and essential micronutrients (Cu, Zn, Fe, Mn, Mo, Ni, and Co) which are important for healthy growth and development of plants [16]. The essential heavy metals are involved in many important biochemical and physiological processes of plants. The principal functions of essential heavy metals include participation in the redox reaction of cellular processes and other molecular activities by being an integral part of several enzymes. In general, a plant grows normally as long as the supply of a given nutrient matches the plant's requirement. Deficiency of nutrients will result in symptoms leading as far as mortality under extreme conditions. Typical deficiency symptoms arising from different metals in plants are presented in Supplementary Text 1. The presence of both essential and non-essential heavy metals in excess can lead to the reduction and inhibition of growth in plants, caused by biochemical, structural and physiological changes [17]. Higher concentrations of heavy metals also alter the uptake, accumulation, and translocation of the essential elements in plants [18]. Common toxic effects of heavy metals include inhibition of growth and photosynthesis, chlorosis, low biomass accumulation, altered nutrient assimilation, and water balance, and senescence, which ultimately can cause plant death [17].

3. Heavy Metal Toxicity and Crop Plants

Heavy metal contamination of agricultural soils has emerged as a critical and significant concern because of unfavorable ecological effects. The land used for crop production has been reported to be polluted with the excess of heavy metals, especially by Cd, Pb, and Zn in many parts of the world [19]. The agricultural land that has been polluted by the elevated concentration of heavy metals was shown to have highly adverse effects on plant metabolism and growth, soil biological activity, fertility, biodiversity, and health of humans and animals [18,19]. Several studies carried out to understand the toxic impact of heavy metals on plants highlighted many direct and indirect toxic effects of heavy metals in different crop species (Figure 1; Table 1; Supplementary Text 1). Several different approaches including the application of phyto-extracts, mobilizers and, more sustainably, the use of Si have contributed to the mitigation of heavy metal stress in plants [20,21].

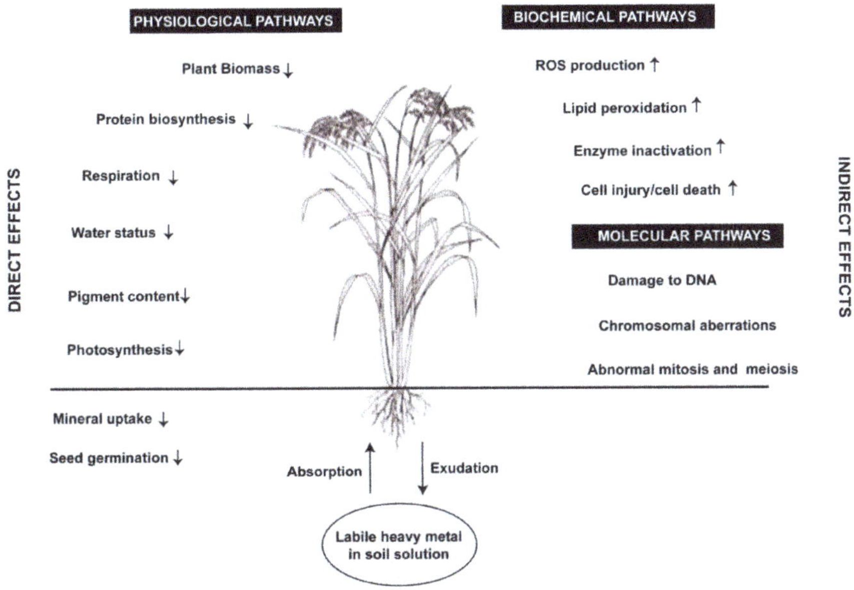

Figure 1. Heavy metal uptake by plant roots as well as their possible direct and indirect negative effects on crop productivity. The sign ↓ indicates decrease and the sign ↑ indicates increase.

Table 1. Phytotoxicity of heavy metals on growth, development, and metabolism of different crop species.

Crop Species	Heavy Metal	Phytotoxicity	References #
Rice	Hg	Reduction in grain yield; reduced tiller and panicle formation; decrease in plant height	S1
	As	Reduction in seed germination and seedling height; reduced leaf area and dry matter production	S2
	Cd	Inhibition of root growth; The appearance of black spots in the cortex and pericycle	S3
Wheat	Pb	Reduced shoot/root length, shoot fresh/dry weights, number of tillers; decreased photosynthetic pigments such as chla and chlb	S4
	Pb	The decrease in the contents of Chl*a*, Chl*b* and proline content; growth inhibition and a decrease in dry weights of plant parts	S5
	Ni	Reduction in the total shoot and root lengths; decrease in Chl*a* and Chl*b* contents and enhancement of chlorophyll a/b ratio	S6
	Cd	Reduction in seed germination; decrease in plant nutrient content; reduced shoot and root length	S7
	Cr	Inhibition of shoot and root growth	S8
	Ni	Reduction in plant nutrient acquisition	S9
Maize	Cd & Zn	Reduction in shoot/root biomass; decrease in length as well as dry weights of shoot/root	S10
	Cd & Ni	Inhibition of root growth; reduction in the length of primary roots	S11
	Cd	Inhibition of shoot and root growth	S12
Barley	Cd	Reduction of Chl*a*, Chl*b*, and carotenoids; decreased plant growth and biomass	S13
	Cu & Cd	Reduction of plant dry weights, root length, and shoot height; alteration of photosynthetic pigments and lipid peroxidation	S14
	Cd & Zn	Reduction of total biomass; decrease in the length of roots and leaves	S15
Sorghum	Cu & Cd	The decrease in root diameter, width, and thickness of leaf midrib and diameter of xylem vessels; reduction of yield and yield contributing traits	S16
	Cd	The decrease in chlorophyll pigments, plant growth, and root characteristics	S17
Oat	Pd	Inhibition of enzyme activity which affected CO_2 fixation	S18
Soybean	Hg	Reduction of oil content; changes in major and minor fatty acid concentration of soybean seed	S19
	Cd	The decrease in the accumulated amount of dry matter as well as the content of mineral elements; inhibition of chlorophyll biosynthesis; decrease in the Mg uptake	S20
	Co & Pb	Reduced seed germination, seedling growth, leaf area, root development, and biomass production	S21
	Cd	Reduction of nodulation and leghaemoglobin content; a decrease in crop productivity	S22
Bean	Cu	Root malformation and accumulation of Cu in plant roots; reduction of root growth	S23
	Zn	The decrease in plant growth, development, and metabolism; induction of oxidative damage	S24
	Zn	Reduction of photosynthetic pigments including Chl*a* and Chl*b*; disruption of absorption and translocation of Fe and Mg into the chloroplast	S25
Chickpea	Cd	Inhibition of seed germination and root growth; disruption of vascular tissues as well as associated tissues	S26
	Cu	Reduction of root/shoot length and RWC; lipid peroxidation	S27
	Co	Inhibition of photosynthetic process and activity of antioxidative enzymes; The increase in proline content and lipid peroxidation	S28
Pigeonpea	Ni	The decrease in stomatal conductance and chlorophyll content; decreased enzyme activity which affected the Calvin cycle and CO_2 fixation	S29
	Pd & Cd	Reduction in photosynthetic activity; decrease in chlorophyll content and stomatal conductance	S30
	Hg & Cd	Reduction of germination percentage, root and shoot length, fresh and dry weight of seedlings; inhibition of root elongation	S31

Table 1. *Cont.*

Crop Species	Heavy Metal	Phytotoxicity	References #
Faba bean	Mn	Mn accumulation in shoot and root; reduction of the shoot and root length; chlorosis	S32
	Cd	Cd accumulation in shoot and root; decrease in photosynthetic pigments as well as root fresh and dry weights	S33
Pea	Mn	The decrease in Chl*a* and Chl*b* content; reduction in relative growth rate; reduced photosynthetic O_2 evolution activity and photosystem II activity	S34
	Zn	Reduction in chlorophyll content; alteration in the structure of chloroplast; reduction in photosystem II activity; reduced plant growth	S35
Mung bean	Co	Reduction in antioxidant enzyme activities; decrease in plant sugar, starch, amino acids, and protein content	S36
Groundnut	Cd, Pb & As	Reduction in germination rate, root length, shoot length, biomass and seedling vigor index	S37
	Cr	Increase in lipid peroxidation; decrease in photosynthetic pigments; induced oxidative stress	S38
	Ni	Reduction of seed germination, root/shoot growth and fresh weight of seedlings	S39
Cotton	Cd	The decrease in plant height, biomass and leaf area; reduction of Chl*a*, Chl*b*, photosynthetic rate, stomatal conductance, and transpiration rate	S40
	Pb	The decrease in net photosynthetic rate, stomatal conductance, transpiration rate, water use efficiency, chlorophyll, carotenoids, and the Soil Plant Analysis Development (SPAD) chlorophyll meter value	S41
Tomato	As	The decrease in leaf fresh weight; Reduced fruit yield	S42
	Co	Reduction in plant nutrient content	S43
	Cr	Decrease in plant nutrient acquisition	S44
	Hg	Reduction in germination percentage; reduced plant height; reduction in flowering and fruit weight; chlorosis	S45
Canola	As	Wilting, chlorosis and stunted growth	S46
Brassica juncea	Cd & Pb	Reduction of growth and biomass yield; decrease of chlorophyll and carotenoid pigments	S47
Garlic	Cd	Reduced shoot growth; Cd accumulation	S48
Onion	Cr	Inhibition of germination process; reduction of plant biomass	S49
Radish	Co	Decrease in shoot and root length as well as total leaf area; reduction in chlorophyll content, plant nutrient content and antioxidant enzyme activity	S50

Detailed list of references are provided in the Supplementary Text 2.

4. Silicon: A Multifaceted Element for Alleviating Heavy Metal Toxicity in Crops

Silicon derived enhancement in plant tolerance to heavy metal toxicity is well documented, and the beneficial role of Si in detoxification can be ascribed to both external (growth media) and internal plant mechanisms [21,22]. The external mechanism of elevating heavy metal tolerance is mainly due to the increased pH by silicate application resulting in metal silicate precipitates that decrease the metal phyto-availability [22]. In plants, Si affects the translocation and distribution of metals in various plant parts and allows them to survive under higher metal stress [23]. Given that plants vary in their ability to accumulate Si, higher accumulators such as monocots will usually obtain greater benefits, even though metal toxicity in both monocots and dicots can be alleviated by Si [24,25]. In rice, Si-rich amendments showed a reduction of heavy metal accumulation, as well as increased growth in multi-metal (Cd, Zn, Cu, and Pb), contaminated acidic soil [26]. Silicon is also reported to increase seedling biomass and decrease Zn concentration in both roots and shoots of rice seedlings, and the xylem sap flow. In addition, Zn accumulation was significantly suppressed by Si supply in different plant parts such as roots and leaves of cotton and maize [27,28]. Shi et al. [29] reported that Si minimizes Cd metal toxicity by reducing ion absorption and translocation from root to shoot in rice seedlings. Silicon application was also found to reduce lipid peroxidation and fatty acid desaturation in plant tissues and improve the growth and biomass of plants under heavy metal stress [4].

Silicon can also be effective in alleviating Al toxicity in barley plants [30]. Similarly, decreased Al content with Si application was observed in the stem, roots, and leaves of peanut and rice seedlings [31,32]. Silicon amendments as an *alternative* detoxification method for Al toxicity have been reported in sorghum, tomato, soybean and maize [21]. Considerable reduction of Si pools in agricultural soils results from regular removal of Si-rich litter during the crop harvest [33]. Hence, external application of Si in agriculture will become a trend in the near future to compensate its depletion in soils, simultaneously reaping its benefits of improving plant growth and alleviating heavy metal toxicity.

5. Silicon-Mediated Mechanisms of Metal Detoxification

Several mechanisms explaining Si-derived benefits towards metal detoxification have been proposed. The widely accepted mechanisms include toxic metal immobilization in the soil (at soil level), and stimulation of enzymatic and non-enzymatic antioxidants, co-precipitation of metals, metal ions chelation, compartmentation, and structural alterations of plant tissues and alteration in molecular responses (at plant level) (Figure 2; Table 2).

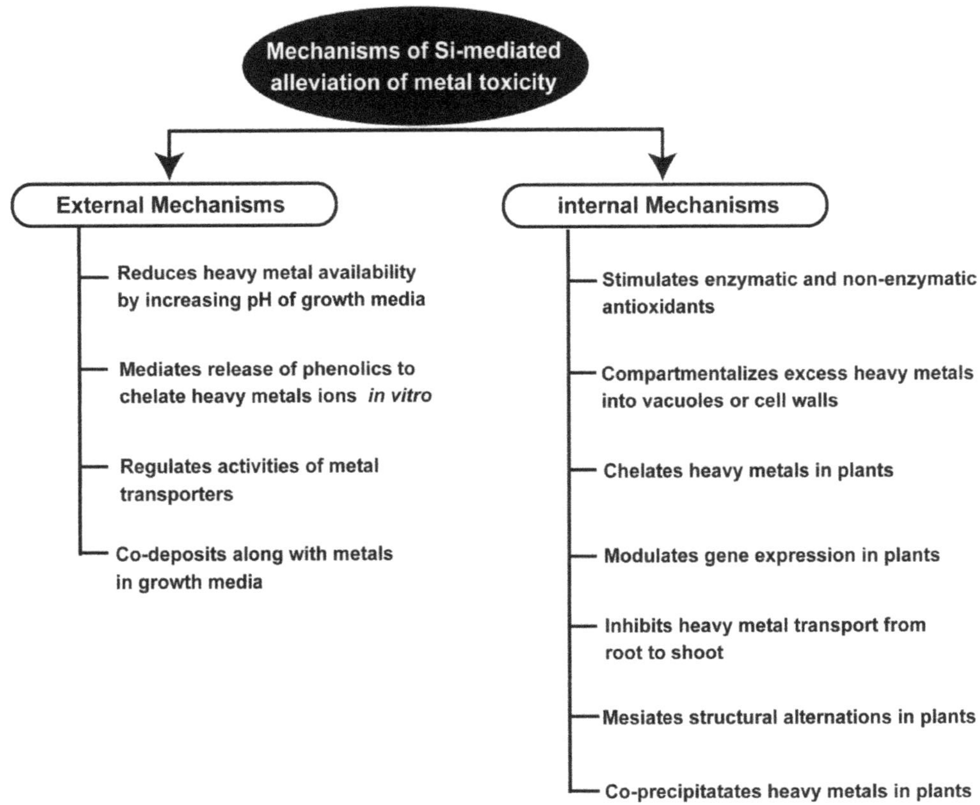

Figure 2. Different external and internal mechanisms used by silicon to mitigate the toxic effects of heavy metals.

Table 2. External and internal silicon-mediated mechanisms for enhancing tolerance of plant's against heavy metal toxicity.

Crop	Heavy Metal	Mechanisms	References [#]
Rice	Cd, Zn, Cu & Pb	Immobilization of heavy metals in culture media and decrease of phytoavailability which further suppressed metal uptake	S51
	As	Overcomes heavy metal uptake by competes with arsenate ions for root entry point	S52
	Zn	Strong binding of Zn in the cell wall of less bioactive tissues, especially in sclerenchyma of root	S51
	Pb	Preventing Pb transfer from rice roots to aboveground parts, and blocking Pb accumulation in rice grains	S53
	Cd	Si bound to cell wall inhibits apoplastic Cd uptake by covalently bonding with Cd and trapping Cd as it diffuses through the cell wall and intracellular spaces.	S54
	Cd	Increased the activities of antioxidant enzymes and preventing membrane oxidative damage of plant tissue	S55
	Cd	Decreased Cd accumulation in rice shoots by compartmentalization of Cd in the root cell walls	S56
Maize	Cd & Zn	Increased diameter of xylem, thickness of leaf mesophyll and epidermis, and transversal area occupied by collenchyma and mid vein	S57
	Al	Formation of hydroxyaluminosilicates in the apoplast of the root apex reducing the mobility of apoplastic Al	S58
	Cd	Formation of colloidal silicon in cell walls which has high specific adsorption property to Cd preventing Cd uptake into the cell	S59
	Cd	Cd co-precipitation with silicates, resulting in strong binding of Cd to cell walls, thereby reducing the concentration of Cd in the symplast	S60
	Zn	Formation of less soluble zinc-silicates in the cytoplasm	S61
Wheat	Cd	Enhanced antioxidant enzymes activities, and preventing lipid peroxidation as well as membrane oxidative damage of plant tissue	S62
	Cd	Decreased Cd uptake as well as translocation of Cd to shoots and grains	S63
	Cu	Cu form complex with organic acids and reduced the Cu translocation to shoots	S64
	Cd	Formation of apoplasmic barriers in endodermis closer to the wheat root apex	S65
Barley	Al	Exclusion of Al from the subtending tissue as a result of silicon deposition at the epidermis, restricting total overall Al uptake into the root	S66
	Cr	Increased plant height, number of tillers, root length and leaf size of barley plants	S67
Cucumber	Mn	Reduction of lipid peroxidation, and increase of enzymatic and non-enzymatic antioxidants levels	S68
	Mn	Si increased Mn fraction in the cell wall of shoots, thereby reduced Mn concentration of symplast	S69

Table 2. *Cont.*

Crop	Heavy Metal	Mechanisms	References #
Cowpea	Mn	Si reduced the apoplastic Mn concentration and modify the cation binding capacity of the cell wall	S70
	Mn	Enhanced adsorption of Mn on cell walls reducing the amount of soluble apoplastic Mn	S71
	Mn	Co-precipitation of Si and Mn in leaf apoplast of cowpea plants, and increases Mn fraction in the cell wall of shoots	S71
Peanut	Cd	Increased activities of antioxidant enzymes; inhibition of Cd transport from roots to shoots	S72
Arabidopsis	Cu	Stimulated the genes responsible for the production of metallothioneins (MTs) that can chelate toxic metals	S73
Rapeseed	Cd	Mediated extensive development of suberin lamellae in endoderm closer to the root tips	S74
Minuartia verna	Zn	Co-precipitation as Zn silicates in the cell walls of leaf epidermis	S75
Cardaminopsis halleri	Zn	Formation of Si–Zn complexes in leaves surface of Cardaminopsishalleri grown on a Zn contaminated soil amended with Si.	S76
Pakchoi	Cd	Increased activities of enzymatic and non-enzymatic antioxidants levels, protein thiols (NPT) and ascorbic acid	S77
Cotton	Cd	Enhanced activities of antioxidant enzymes as well as reduced electrolytic leakage, malondialdehyde and hydrogen peroxide contents, thereby preventing plant tissue from oxidative damage	S78
	Pb	Increased the activities of antioxidant enzymes and preventing membrane oxidative damage of plant tissue	S41

Detailed list of references are provided in the Supplementary Text 2.

5.1. Silicon Mediated Immobilization of Toxic Metal in the Soil

Immobilization of toxic metals is the much simpler mechanism to explain Si-derived benefits. The toxic metal immobilization in the soil through Si application has been reported in several studies [21]. The immobilization took place either due to the increased soil pH or changing metal speciation in soil solution through the formation of silicate complexes. In rice, application of Si-rich amendments (fly ash and steel slag) was found to increase soil pH from 4.0 to 5.0–6.4, and decreased the phyto-availability of heavy metals by at least 60%, which further suppressed metal uptake [26]. Similarly, in banana, reduced uptake of Pb has been reported with Si application in Pb contaminated soil [34]. The decreased bio-availability of Pb in banana was found to be associated with significantly increased soil pH and decreased proportion of exchangeable Pb in soil [34].

In addition, Si helps by changing the speciation of metals from toxic to nontoxic form by the formation of silicate complexes in the soil solution. In Si-amended soil, mostly Cd was found in the form of oxides or adsorbed by Fe-Mn oxides [35]. It has been proposed that the bio-availability of Al to plants in the presence of Si can be limited by forming Al-Si complexes like hydroxyaluminosilicate (HAS) [36]. In aqueous solution, soluble silicate hydrolyzes and produces gelatinous metasilicic acid (H_2SiO_3) which can absorb heavy metals, or lead to deposition of heavy metals into their silicates in Si-rich soil [26], both of which reduce the metal concentration available for uptake. Zhang et al. [37] have shown that the Si application in Cr-contaminated soil can markedly decrease the amount of exchangeable Cr by accelerating the precipitation of organic matter bound Cr fraction. Similarly, Shim et al. [38] have observed the reduced metal mobility in Pb contaminated soil by the Si application. Subsequent investigation with X-ray diffraction analysis revealed the formation of insoluble Pb-silicate in the soil. Similarly, effects have been observed in the case of Cd and Zn contaminated soil where Si application accelerated the formation of more stable fractions of Cd and Zn [28].

5.2. Stimulation of Antioxidant Defense System

Heavy metal stress induces an excess formation of reactive oxygen species (ROS), which results in several metabolic disorders in crop plants [39,40]. In this context, the enzymatic and non-enzymatic antioxidant system stimulated by Si helps to lower oxidative stress by reducing the production of ROS. Alleviation of Mn toxicity by Si in cucumber was attributed to a significant reduction in lipid peroxidation (LPO) intensity caused by excess Mn, and to a significant increase in enzymatic (*Superoxide dismutase*, *ascorbate peroxidase*, and *glutathione reductase*), and non-enzymatic (ascorbate and glutathione) antioxidants [41]. Similarly, under Cd stress, Si application reduced the H_2O_2 and electrolytic leakage (EL) in *Solanum nigrum* [42]. Decreased contents of an oxidative stress related compound like malondialdehyde (MDA), H_2O_2 and EL were observed with Si application in plants under Cd [43], Zn [27], and Pb stress [44]. Thiobarbituric acid reactive substances (TBARS), a widely used marker for reactive oxygen contents, were found to be reduced significantly with Si supplementation in rice and maize plants grown under Cd stress condition [39]. Similarly, under Cd stress, the effect of Si supplementation on antioxidants has also been observed in many crop plants including maize [45], wheat [43], rice [23,46] and peanut [47]. Enhanced antioxidant enzyme activities as well as activities of non-enzymatic antioxidants like glutathione, non-protein thiols, and ascorbic acid have been reported in several plant species under heavy metal stress [48,49]. Si-mediated detoxification through stimulating enzymatic and non-enzymatic antioxidants has also been observed under Pb, Mn, Zn and Cu stress [20]. In light of the previous studies, application of Si appears to induce the antioxidant system in plants, thereby improving stress resistance, but it is still unclear if this results from a direct or indirect action of Si [50].

5.3. Compartmentation within Plants

Enhanced compartmentation of metal elements in plant tissues has been observed with Si supplementation in several studies. Williams and Vlamis [51] observed for the first time that the effect of Si in alleviating Mn toxicity in barley was not the result of a reduction in Mn concentration, but rather of improved compartmentation within the leaf tissues. Another level of compartmentation, mostly regulated through the translocation activity leading to the increased metal concentration in plant roots compared to shoots, has been widely reported with Si supplementation [52,53]. The Si treatment was observed to reduce transport of Zn from roots to shoots and increase the binding of Zn to the cell wall, thus decreasing Zn concentration in the rice shoots [53]. In wheat, Si application found to reduce Cd translocation from root to shoots and grains [54]. In rice, Shi et al. [29] observed a Si-decreased root to shoot translocation of Cd by 33%. Microscopic analysis performed by Shi et al. [29] clearly showed deposition of Cd near the endodermis and epidermis; however, a high amount of Si was found to be deposited near the endodermis than in the epidermis. The Si deposition at endodermis seems to play a crucial role in decreasing Cd translocation from the epidermis to the endodermis. In addition, with Si application in rice, decreased Cd accumulation in shoots was found to be associated with the increased compartmentalization of Cd in the root cell walls [23]. Considerable reduction in the symplasmic concentration of Cd and increase in the apoplastic concentration in maize grown with Si supplementation have been reported by Vaculik et al. [55]. Silicon was also observed to increase the Mn localization in the cell wall in cucumber [56] and cowpea [57] under Mn stress and in rice, under Zn stress [26]. Furthermore, in Si-treated plants, less Mn was located in the symplast (<10%) and more Mn was bound to the cell wall (>90%) compared to control plants (about 50% in each compartment) [58]. Hence, Si-mediated detoxification of heavy metals through their compartmentation into different plant tissues might be a key mechanism supporting the beneficial role of Si.

5.4. Chelation Mediated Metal Toxicity Reduction with Silicon Application

The Si-mediated heavy metal detoxification predominantly includes the chelation of metal by flavonoid-phenolics or organic acids. Increased phenol exudation up to 15 times has been observed in maize upon Si supplementation. The phenolic compounds like catechin and quercetin have high Al-chelating activity, which can alleviate Al toxicity in the root tip apoplast [10,59]. Barceló et al. [60] revealed a considerable increase of malic acid concentration upon Si application in maize under Al stress. The reduced Al toxicity was found to be correlated with the chelating of Al with malic acid. In bamboo, Si has also been observed to increase the concentration of Cu(I) S-ligands that chelate Cu as well as increase the Cu sequestration in a less toxic form [61]. Keller et al. [52] have observed reduction in translocation of Cu from root to shoot in wheat grown with Si supplementation. The altered translocation of Cu may be because of the increased proportion of citrate, malate, and aconitate in roots of wheat seedlings. Taken together, these studies suggest that Si may indirectly promote the chelation of heavy metals in plants, thereby reducing their phytotoxicity.

Schaller et al. [62] recently reported that Si availability was significantly correlated to phosphorus (P) mobilization in Arctic soils. The results confirmed that the addition of Si significantly increases P mobilization by mobilizing Fe(II)-P phases from mineral surfaces as well as increases soil respiration in P deficient soils. The study highlights the Si as an important component regulating P mobilization in Arctic soils. Besides an important nutrient element, P supplementation also plays an important role in biochemical activities in the soil. In this regard, significant efforts have been performed to understand the effects of P mobilization on bioavailability of heavy metals [63]. A considerably high number of studies have shown the importance of P compounds to reduce bioavailability of heavy metals by immobilizing the metal ions in the soil [64]. Therefore, Si mediated P mobilization seems a valuable option for the sustainable management of P availability as well as for the minimizing losses that occurred due to heavy metals in agro-ecosystems.

5.5. Regulation of Gene Expression

Phytochelatin synthesis represents one of the major heavy metal detoxification mechanisms in plants [65]. The Si-mediated mitigation of metal toxicity is also attributed to its role in altering gene expression. In Arabidopsis grown under Cu stress, Si supplementation was observed to stimulate the genes governing the production of metallothioneins (MTs), a well-known chelating agent [66]. Similarly, significantly downregulated expression of the genes encoding heavy metal transporters (*OsHMA2* and *OsHMA3*), and upregulation of the genes responsible for Si transport (*OsLSi1* and *OsLSi2*) has been observed with Si supplementation in rice [3]. Similarly, the enhanced expression of *phytochelatin synthase* 1 (*PCS1*) and decreased expression of the *metallothionein* gene (*MT1a*) was associated with Si supplementation in Arabidopsis under Cu stress [67]. Recently, Ma et al. [68] have shown that Si supply under Cd stress in rice upregulated the expression of *OsLsi1* (encoding for Si transport NIP-III (Nodulin 26-like intrinsic proteins-III) Aquaporin) and downregulated the expression of *Nramp5*, a gene involved in the Cd transport. However, many plant species particularly belonging to family *brassicaceae* [69], *solanaceae* [70], and *Linaceae* [71] does not carry any Si-transporter (NIP-III Aquaporin) and are known as poor Si accumulators. However, several reports suggesting Si derived benefits in such species make it more difficult to understand the molecular consequences [72,73]. So far, the Si-mediated mechanisms for the reduction of metal toxicity are less understood at the genetic and molecular levels. A better understanding of the gene expression dynamics involved in Si-driven alleviation of metal toxicity is necessary to properly decipher the molecular mechanism underlying this phenomenon. Still, limited efforts have been directed toward explaining Si-mediated transcriptomic changes in plants, and none of those were related to metal toxicity.

5.6. Structural Alterations Related to Metal Stress Tolerance in Plants

Morphological and anatomical features of crop plants get improved with Si supplementation that helps to overcome the adverse effects of heavy metal stress. Notable examples where increased plant height, root length, number of leaves and leaf size have been observed with Si application to plants under Cd, Zn and Pb stress [74,75]. Ali et al. [76] observed that Si + Cr treatments increased plant height, the number of tillers, root length, and leaf size of barley plants compared to plants treated with Cr only. Similarly, root length and shoot size were significantly increased with Si compared to without Si treatment [74]. Doncheva et al. [77] reported that the Si application increased the leaf-epidermal-layer thickness in maize plant under Mn stress. Silicon was also reported to increase xylem diameter, epidermis, mesophyll and the transverse area of collenchymas and mid-vein under Cd and Zn stress [28,78]. The accelerated development of the Casparian bands, suberin lamellae, and root vascular tissues was observed in maize with Cd + Si treatments [45,55]. Similarly, higher growth of suberin lamellae in the endodermis particularly near the root tips has been observed with Si application in rapeseed and Indian mustard grown under Cd stress [79]. In wheat, Greger et al. [80] reported the formation of apoplastic barriers in the endodermis closer to the root apex in the presence of Si in Cd-treated plants. Thus, structural alterations induced by Si under metal stress may explain the alleviation of heavy metal toxicity.

5.7. Co-Precipitation of Metals by Silicon Application in Soil

Many studies suggest that the co-precipitation of metals by Si leads to the alleviation of heavy metal stress in plants. For example, Si treatment in plants under Al stress was suggested to form aluminosilicates or hydroxyaluminosilicates (or both) in the apoplast of the plant root apex leading to Al detoxification [22,60]. In *Minuartia verna* (Si-accumulating dicot), Si was reported to co-precipitate Zn as their silicates in the leaf epidermis cell wall [81]. Gu et al. [26] observed the co-precipitation of Si with Cd in the stem of rice that lowered heavy metal concentration in leaves. Similarly, Si was proposed to form a complex with Zn at the leaf surface of *Cardaminopsis halleri* [82]. Zhang et al. [23] observed synchronous accumulation of Si and Cd in the middle and border of phytoliths in rice shoots. However, there are some contrasting reports, such as ones by Rizwan et al. [83] and Keller et al. [52], where Cu and Cd were not found in the leaf phytoliths of wheat. These reports suggest the possibility of a mechanism other than the co-precipitation involved in Cu detoxification in wheat plants. Similarly, Dresler et al. [84] report also support the notion since Si–Cd complexes were not observed in maize plants treated with Si under Cd stress.

6. Approaches for Improving Silicon Accumulation in Crop Plants

The beneficial effect of Si for mitigating the toxic effects of heavy metals is usually apparent in plants that accumulate high levels of Si in their shoots [53,85]. Silicon accumulation in plants depends upon the availability of silicic acid (H_4SiO_4) in the soil as well as the inherent capacity of the plant for Si uptake. Silicon availability in a soil can be improved by applying Si-rich fertilizers or by modifying soil properties, whereas genetic modification or selection can improve the inherent capacity a species to accumulate Si. The different approaches used for increasing Si accumulation in crop plants are presented in Figure 3 and are discussed in the following sections.

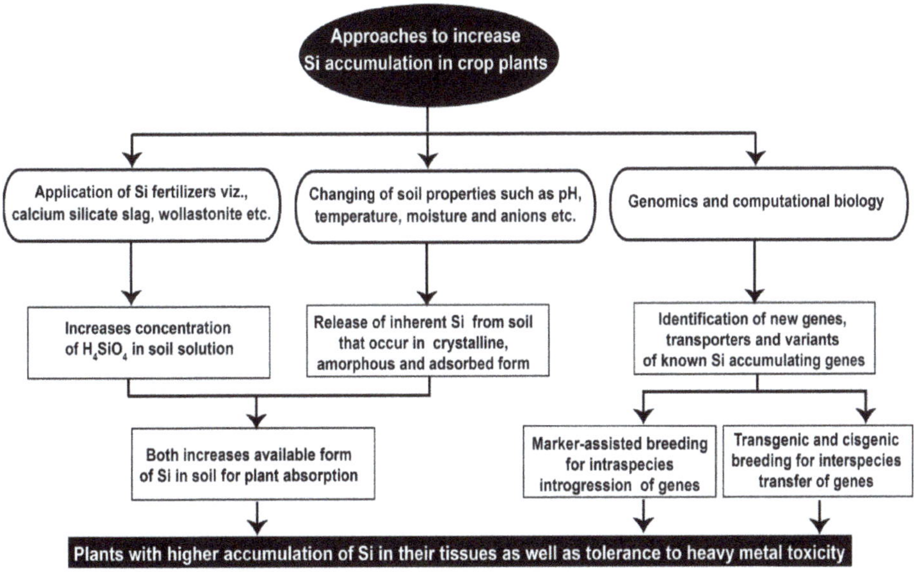

Figure 3. The different approaches that can be utilized for improving silicon accumulation to enhance metal stress tolerance in crop plants.

6.1. Silicon Fertilization

The estimated amount of Si removed annually by different crops on a global scale is between 210 and 224 million tons [86]. Continuous and intensive cropping of Si-accumulator crops results in a significant reduction of plant-available Si in the soil [87]. Desplanques et al. [88] noted that five years of continuous cultivation of rice could exhaust most of the available Si from the soil. In addition, certain soils contain low levels of Si, particularly the plant-available form, and these soils include Oxisols, Ultisols, and Histosols as well as soils that are composed of a large fraction of quartz [89]. Silicon is absorbed by plants in the form of monosilicic acid (H_4SiO_4) and its concentration in the soil solution will determine the amount a plant accumulates [90]. Fertilization can rapidly increase the concentration of H_4SiO_4 in the soil and has become a standard practice in areas with intensive cropping systems, particularly for those soils that are inherently low in soluble silicon [91,92]. Several studies have shown a beneficial effect of an adequate supply of Si in growth medium not only for high Si-accumulating plants such as rice [46] and sugarcane but also for poor accumulators like tomato, cucumber, strawberry and orange [93,94] (Table 2). On the other hand, some studies have also shown that the fertilization regime can influence the amount of Si that will be available to the plant [94]. With respect to the alleviation of heavy metal tolerance, Ning et al. [95] and Jarosz et al. [93] observed less concentration of Zn and Cu in fruits of plants grown with Si-based fertilizer. However, most of the studies to date have been carried out in greenhouses, and there is a lack of data supporting the performance of Si-based fertilizers in large-scale field experiments.

6.2. Modification of Soil Properties

The abundance of Si in soils does not necessarily relate to the concentration of plant-available Si [91]. The concentration of H_4SiO_4 in the soil solution is influenced by many soil factors such as pH, temperature, weathering of soil, moisture, redox potential, amounts of clay, mineral, organic matter and Fe/Al oxides/hydroxides [86]. The solubility of both the crystalline and the amorphous silica is approximately constant between pH values 2 and 8.5 and increases rapidly at pH ~9. The soil pH also affects the formation of Si-complexes with other elements, for instance, the amount of monosilicic acids that is absorbed by Fe/Al oxides increases from pH 4 to pH 10. Kaczorek and Sommer [96] have revealed that, under the conditions of soil acidification, the number of free silica increases. Similarly, Höhn et al. [97] reported an increase of Si available in the soil with a decrease in pH. In this regard, the application of acid-producing fertilizer increases the concentration of H_4SiO_4 in the soil solution, whereas liming and high organic matter content results in a reduction in the concentration and mobility of the H_4SiO_4.

6.3. Approaches for Genetic Gain

Identification of two Si-transporters genes using low silicon (*Lsi*) rice mutants by Ma et al. [98,99] are the milestone discoveries that accelerated Si research many folds. One of the genes is a passive influx transporter (*OsLsi1*) belonging to an NIP group of the aquaporin family, which is responsible for the uptake of Si from soil into the root cells [15,100]. Several homologs of the *Lsi1* have been identified and functionally validated in different plant species (Table 3, Supplementary Text 1: Table S1). Another gene, *Lsi2*, encodes an active efflux transporter and belongs to the cation transporter family (Table 3; Figure S1). The information about the *Lsi1* and *Lsi2* was helpful to understand the uptake of Si from the root and subsequent trans-location from roots to aerial parts of the plant [101]. It also provided an opportunity to explore transgenic approaches for the enhancement of Si-uptake, particularly in poor accumulator plant species. In Arabidopsis, a well-known poor accumulator, heterologous expression of wheat Si-transporter (*TaLsi1*) showed increased Si absorption by several folds [102]. Similarly, functionally characterized Si-transporter (mostly with *Xenopus oocyte* assays) from different species could be utilized efficiently for the development of transgenic crop plants (Table S1). Recently, Deshmukh et al. [103] showed categorization of plant species as poor or high Si-accumulator based on the presence of *Lsi1* homolog. The interspecies variation can be sufficiently explained by the characterization of *Lsi1* homolog. Considerable interspecies variation for Si concentration has been reported, ranging from 0.1 to 10%, although these variations appear to be more limited at the intraspecies level [14]. For example, in sugarcane (*Saccharum officinarum*), the Si concentration in the shoots of different genotypes has been observed to range from 6.4 to 10.2 mg^{-1} [104]. Similarly, in a survey of about 400 cultivars of barley (*Hordeum vulgare*), the Si concentration in barley grain revealed a variation, ranging from 1.24 to 3.80 $mg \cdot g^{-1}$ in barley grains [105]. In rice, japonica rice cultivars usually accumulate more Si than indica rice cultivars [106]. Quantitative trait loci (QTLs) governing the intra-species variation have been identified in rice. However, these QTLs do not collocate with the major genes *Lsi1* and *Lsi2*, and the discovery of *Lsi1* and *Lsi2* enhanced molecular understanding of Si-uptake has limited use for crop improvement. The gene(s) responsible for the genotypic differences have yet to be identified. Efforts towards the identification of such genes will help to develop cultivars with better Si uptake through breeding approaches.

Table 3. Influx/efflux Si transporters as well as their ortholog's identified in different crop species.

Crop Species	Transporter	Type	Expression Site	References #
Maize	ZmLsi1	Influx	Root	S79
	ZmLsi6	Influx	Leaf Sheaths and blades	S80
	ZmLsi2	Efflux	Roots	S80
Rice	OsLsi1	Influx	Roots	S81
	OsLsi2	Efflux	Roots	S82
	OsLsi6	Influx	Leaf	S83
	OsLsi3	Influx	Panicles	S84
Barley	HvLsi2	Efflux	Root	S80
	HvLsi1	Influx	Root	S85
	HvLsi6	Influx	Leaf Sheaths and blades	S86
Soybean	GmNIP2-1	Influx	Root and shoot	S87
	GmNIP2-2	Influx	Root and shoot	S87
Wheat	TaLsi1	Influx	Root	S88
Pumpkin	CmLsi1	Influx	Root and shoot	S89
	CmLsi2	Efflux	Root and shoot	S89
Horsetail	EaNIP3;1	Influx	Root and shoot	S90
	EaNIP3;3,	Influx	Root and shoot	S90
	EaLsi2-1	Efflux	Root and shoot	S91
	EaLsi2-2	Efflux	Root and shoot	S91
Potato	StLsi1	Influx	Root and leaves	S92
Tomato	SlNIP2-1 (V140del) *	Influx	Root and leaves	S93

* mutated version where removal of Valine at potion 140 in non-functional wildtype turn it into a functional Si-transporter; # Detailed list of references are provided in the Supplementary Text 2.

7. Conclusions

Natural and anthropogenic activities have resulted in a higher concentration of toxic metals in the agricultural farmlands leading to the severe adverse effects on crop production and productivity as well as human health. In this regard, Si has emerged as a practical option to reduce phytotoxicity and accumulation of toxic metal in plants. Beneficial effects derived from Si have been reported in several studies and have been explained through different possible mechanisms, although more experiments are needed to validate them. Many studies argue that the use of Si-based fertilizers constitutes a promising approach to alleviate heavy metal toxicity and large-scale field studies should be undertaken to define with precision the extent of Si-derived benefits. At the same time, the enhancement of the genetic potential of plants to uptake Si would represent an interesting avenue to optimize their responses to Si. The recent advances in the field of genomics, computational biology, and high-throughput phenotyping will facilitate the deciphering of the genetic basis of Si accumulation in crop plants, which in turn should accelerate the development of varieties accumulating higher levels of Si. This strategy would fit very well in a program of sustainable agriculture to improve heavy metal tolerance in plants.

Supplementary Materials: The following are available online at http://www.mdpi.com/2223-7747/8/3/71/s1, **Figure S1:** Diagram showing the absorption of Si in rice plants. Silicon enters the exodermis in the form of silicic acid through specific influx transporter (Lsi1) and leaves into the cortex through active transporters (Lsi2). In the cortex, silicic acid moves apoplastically until it reaches the endodermis, where the silicic acid is loaded into the stele by Lsi1 and Lsi2 transporters. The solid green line shows the path of Si transport. **Table S1:** Details of silicon transporter genes from different plant species were validated using oocyte assay or transgenic approaches. **Supplementary Text 1:** Detailed discussion about the deficiency symptoms caused by different metals in plants, toxic effects of heavy metals on growth, development, and metabolism of different crop species. **Supplementary Text 2**: Details of references provided in Tables 1–3.

Author Contributions: J.A.B. and S.M.S. wrote the initial draft, P.S., D.B.N., D.K.T., P.K.D., A.U.S., H.S. and R.D. provided input and were involved in the compilation of the final draft.

Funding: This research received no external funding.

Acknowledgments: The authors are thankful to the Department of Biotechnology, Government of India for the financial support in the form of the Ramalingaswami Fellowship to H.S. and R.D.

Conflicts of Interest: The authors declare no conflict of interest.

References

1. Hawkes, S.J. What Is a" Heavy Metal"? *J. Chem. Educ.* **1997**, *74*, 1374. [CrossRef]
2. Chibuike, G.U.; Obiora, S.C. Heavy metal polluted soils: Effect on plants and bioremediation methods. *Appl. Environ. Soil Sci.* **2014**, *2014*. [CrossRef]
3. Kim, Y.-H.; Khan, A.L.; Kim, D.-H.; Lee, S.-Y.; Kim, K.-M.; Waqas, M.; Jung, H.-Y.; Shin, J.-H.; Kim, J.-G.; Lee, I.-J. Silicon mitigates heavy metal stress by regulating P-type heavy metal ATPases, Oryza sativa low silicon genes, and endogenous phytohormones. *BMC Plant Biol.* **2014**, *14*, 13. [CrossRef] [PubMed]
4. Nagajyoti, P.C.; Lee, K.D.; Sreekanth, T. Heavy metals, occurrence and toxicity for plants: A review. *Environ. Chem. Lett.* **2010**, *8*, 199–216. [CrossRef]
5. Liang, Y.; Sun, W.; Zhu, Y.-G.; Christie, P. Mechanisms of silicon-mediated alleviation of abiotic stresses in higher plants: A review. *Environ. Pollut.* **2007**, *147*, 422–428. [CrossRef]
6. Deshmukh, R.K.; Ma, J.F.; Bélanger, R.R. Role of silicon in plants. *Front. Plant Sci.* **2017**, *8*, 1858. [CrossRef]
7. Rasoolizadeh, A.; Labbé, C.; Sonah, H.; Deshmukh, R.K.; Belzile, F.; Menzies, J.G.; Bélanger, R.R. Silicon protects soybean plants against Phytophthora sojae by interfering with effector-receptor expression. *BMC Plant Biol.* **2018**, *18*, 97. [CrossRef]
8. Liang, Y.; Nikolic, M.; Bélanger, R.; Gong, H.; Song, A. *Silicon in Agriculture: From Theory to Practice*; Springer: Berlin, Germany, 2015.
9. Chérif, M.; Asselin, A.; Bélanger, R. Defense responses induced by soluble silicon in cucumber roots infected by *Pythium* spp. *Phytopathology* **1994**, *84*, 236–242. [CrossRef]
10. Wang, Y.; Stass, A.; Horst, W.J. Apoplastic binding of aluminum is involved in silicon-induced amelioration of aluminum toxicity in maize. *Plant Physiol.* **2004**, *136*, 3762–3770. [CrossRef] [PubMed]
11. Ma, J.F.; Yamaji, N. Silicon uptake and accumulation in higher plants. *Trends Plant Sci.* **2006**, *11*, 392–397. [CrossRef]
12. Hosseini, S.A.; Naseri Rad, S.; Ali, N.; Yvin, J.-C. The Ameliorative Effect of Silicon on Maize Plants Grown in Mg-Deficient Conditions. *Int. J. Mol. Sci.* **2019**, *20*, 969. [CrossRef] [PubMed]
13. Fauteux, F.; Chain, F.; Belzile, F.; Menzies, J.G.; Bélanger, R.R. The protective role of silicon in the Arabidopsis–powdery mildew pathosystem. *Proc. Natl. Acad. Sci. USA* **2006**, *103*, 17554–17559. [CrossRef]
14. Hodson, M.; White, P.; Mead, A.; Broadley, M. Phylogenetic variation in the silicon composition of plants. *Ann. Bot.* **2005**, *96*, 1027–1046. [CrossRef] [PubMed]
15. Deshmukh, R.; Bélanger, R.R. Molecular evolution of aquaporins and silicon influx in plants. *Funct. Ecol.* **2016**, *30*, 1277–1285. [CrossRef]
16. Kalaivanan, D.; Ganeshamurthy, A.N. Mechanisms of heavy metal toxicity in plants. In *Abiotic Stress Physiology of Horticultural Crops*; Springer: Berlin, Germany, 2016; pp. 85–102.
17. Ali, S.; Bai, P.; Zeng, F.; Cai, S.; Shamsi, I.H.; Qiu, B.; Wu, F.; Zhang, G. The ecotoxicological and interactive effects of chromium and aluminum on growth, oxidative damage and antioxidant enzymes on two barley genotypes differing in Al tolerance. *Environ. Exp. Bot.* **2011**, *70*, 185–191. [CrossRef]
18. Zia-ur-Rehman, M.; Sabir, M.; Nadeem, M. Remediating cadmium-contaminated soils by growing grain crops using inorganic amendments. In *Soil Remediation and Plants: Prospects and Challenges*; Elsevier Inc., Academic Press: Amsterdam, The Netherlands, 2015; pp. 367–396.
19. Foucault, Y.; Lévèque, T.; Xiong, T.; Schreck, E.; Austruy, A.; Shahid, M.; Dumat, C. Green manure plants for remediation of soils polluted by metals and metalloids: Ecotoxicity and human bioavailability assessment. *Chemosphere* **2013**, *93*, 1430–1435. [CrossRef]
20. Wu, J.; Shi, Y.; Zhu, Y.; Wang, Yi.; Gong, H. Mechanisms of enhanced heavy metal tolerance in plants by silicon: A review. *Pedosphere* **2013**, *23*, 815–825. [CrossRef]

21. Sahebi, M.; Hanafi, M.M.; Siti Nor Akmar, A.; Rafii, M.Y.; Azizi, P.; Tengoua, F.; Nurul Mayzaitul Azwa, J.; Shabanimofrad, M. Importance of silicon and mechanisms of biosilica formation in plants. *BioMed Res. Int.* **2015**, *2015*. [CrossRef] [PubMed]
22. Cocker, K.M.; Evans, D.E.; Hodson, M.J. The amelioration of aluminium toxicity by silicon in higher plants: Solution chemistry or an in planta mechanism? *Physiol. Plant.* **1998**, *104*, 608–614. [CrossRef]
23. Zhang, C.; Wang, L.; Nie, Q.; Zhang, W.; Zhang, F. Long-term effects of exogenous silicon on cadmium translocation and toxicity in rice (*Oryza sativa* L.). *Environ. Exp. Bot.* **2008**, *62*, 300–307. [CrossRef]
24. Shi, Z.; Yang, S.; Han, D.; Zhou, Z.; Li, X.; Liu, Y.; Zhang, B. Silicon alleviates cadmium toxicity in wheat seedlings (*Triticum aestivum* L.) by reducing cadmium ion uptake and enhancing antioxidative capacity. *Environ. Sci. Pollut. Res.* **2018**, *25*, 7638–7646. [CrossRef] [PubMed]
25. Greger, M.; Landberg, T.; Vaculík, M. Silicon Influences Soil Availability and Accumulation of Mineral Nutrients in Various Plant Species. *Plants* **2018**, *7*, 41. [CrossRef] [PubMed]
26. Gu, H.-H.; Qiu, H.; Tian, T.; Zhan, S.-S.; Chaney, R.L.; Wang, S.-Z.; Tang, Y.-T.; Morel, J.-L.; Qiu, R.-L. Mitigation effects of silicon rich amendments on heavy metal accumulation in rice (*Oryza sativa* L.) planted on multi-metal contaminated acidic soil. *Chemosphere* **2011**, *83*, 1234–1240. [CrossRef] [PubMed]
27. Anwaar, S.A.; Ali, S.; Ali, S.; Ishaque, W.; Farid, M.; Farooq, M.A.; Najeeb, U.; Abbas, F.; Sharif, M. Silicon (Si) alleviates cotton (*Gossypium hirsutum* L.) from zinc (Zn) toxicity stress by limiting Zn uptake and oxidative damage. *Environ. Sci. Pollut. Res.* **2015**, *22*, 3441–3450. [CrossRef]
28. Patrícia Vieira da Cunha, K.; Williams Araújo do Nascimento, C.; José da Silva, A. Silicon alleviates the toxicity of cadmium and zinc for maize (*Zea mays* L.) grown on a contaminated soil. *J. Plant Nutr. Soil Sci.* **2008**, *171*, 849–853. [CrossRef]
29. Shi, X.; Zhang, C.; Wang, H.; Zhang, F. Effect of Si on the distribution of Cd in rice seedlings. *Plant Soil* **2005**, *272*, 53–60. [CrossRef]
30. Hammond, K.E.; Evans, D.E.; Hodson, M.J. Aluminium/silicon interactions in barley (*Hordeum vulgare* L.) seedlings. *Plant Soil* **1995**, *173*, 89–95. [CrossRef]
31. Singh, V.P.; Tripathi, D.K.; Kumar, D.; Chauhan, D.K. Influence of exogenous silicon addition on aluminium tolerance in rice seedlings. *Biol. Trace Elem. Res.* **2011**, *144*, 1260–1274. [CrossRef] [PubMed]
32. Shen, X.; Xiao, X.; Dong, Z.; Chen, Y. Silicon effects on antioxidative enzymes and lipid peroxidation in leaves and roots of peanut under aluminum stress. *Acta Physiol. Plant.* **2014**, *36*, 3063–3069. [CrossRef]
33. Haynes, R.J. A contemporary overview of silicon availability in agricultural soils. *J. Plant Nutr. Soil Sci.* **2014**, *177*, 831–844. [CrossRef]
34. Li, L.; Zheng, C.; Fu, Y.; Wu, D.; Yang, X.; Shen, H. Silicate-mediated alleviation of Pb toxicity in banana grown in Pb-contaminated soil. *Biol. Trace Elem. Res.* **2012**, *145*, 101–108. [CrossRef]
35. Liang, Y.; Wong, J.; Wei, L. Silicon-mediated enhancement of cadmium tolerance in maize (*Zea mays* L.) grown in cadmium contaminated soil. *Chemosphere* **2005**, *58*, 475–483. [CrossRef] [PubMed]
36. Hodson, M.J.; Evans, D.E. Aluminium/silicon interactions in higher plants. *J. Exp. Bot.* **1995**, *46*, 161–171. [CrossRef]
37. Ding, X.; Zhang, S.; Li, S.; Liao, X.; Wang, R. Silicon Mediated the Detoxification of Cr on Pakchoi (*Brassica Chinensis* L.) in Cr-contaminated Soil. *Procedia Environ. Sci.* **2013**, *18*, 58–67. [CrossRef]
38. Shim, J.; Shea, P.J.; Oh, B.-T. Stabilization of heavy metals in mining site soil with silica extracted from corn cob. *WaterAir Soil Pollut.* **2014**, *225*, 2152. [CrossRef]
39. Adrees, M.; Ali, S.; Rizwan, M.; Zia-ur-Rehman, M.; Ibrahim, M.; Abbas, F.; Farid, M.; Qayyum, M.F.; Irshad, M.K. Mechanisms of silicon-mediated alleviation of heavy metal toxicity in plants: A review. *Ecotoxicol. Environ. Saf.* **2015**, *119*, 186–197. [CrossRef] [PubMed]
40. Ahmad, P.; Tripathi, D.K.; Deshmukh, R.; Singh, V.P.; Corpas, F.J. Revisiting the role of ROS and RNS in plants under changing environment. *Environ. Exp. Bot.* **2019**, in press. [CrossRef]
41. Shi, Q.; Bao, Z.; Zhu, Z.; He, Y.; Qian, Q.; Yu, J. Silicon-mediated alleviation of Mn toxicity in Cucumis sativus in relation to activities of superoxide dismutase and ascorbate peroxidase. *Phytochemistry* **2005**, *66*, 1551–1559. [CrossRef] [PubMed]
42. Liu, J.; Zhang, H.; Zhang, Y.; Chai, T. Silicon attenuates cadmium toxicity in Solanum nigrum L. by reducing cadmium uptake and oxidative stress. *Plant Physiol. Biochem.* **2013**, *68*, 1–7. [CrossRef]

43. Hussain, I.; Ashraf, M.A.; Rasheed, R.; Asghar, A.; Sajid, M.A.; Iqbal, M. Exogenous application of silicon at the boot stage decreases accumulation of cadmium in wheat (*Triticum aestivum* L.) grains. *Braz. J. Bot.* **2015**, *38*, 223–234. [CrossRef]
44. Bhatti, K.; Anwar, S.; Nawaz, K.; Hussain, K.; Siddiqi, E.; Sharif, R.; Talat, A.; Khalid, A. Effect of heavy metal lead (Pb) stress of different concentration on wheat (*Triticum aestivum* L.). *Middle-East J. Sci. Res.* **2013**, *14*, 148–154.
45. Lukačová, Z.; Švubová, R.; Kohanová, J.; Lux, A. Silicon mitigates the Cd toxicity in maize in relation to cadmium translocation, cell distribution, antioxidant enzymes stimulation and enhanced endodermal apoplasmic barrier development. *Plant Growth Regul.* **2013**, *70*, 89–103. [CrossRef]
46. Tripathi, D.K.; Singh, V.P.; Kumar, D.; Chauhan, D.K. Impact of exogenous silicon addition on chromium uptake, growth, mineral elements, oxidative stress, antioxidant capacity, and leaf and root structures in rice seedlings exposed to hexavalent chromium. *Acta Physiol. Plant.* **2012**, *34*, 279–289. [CrossRef]
47. Shi, G.; Cai, Q.; Liu, C.; Wu, L. Silicon alleviates cadmium toxicity in peanut plants in relation to cadmium distribution and stimulation of antioxidative enzymes. *Plant Growth Regul.* **2010**, *61*, 45–52. [CrossRef]
48. Li, P.; Song, A.; Li, Z.; Fan, F.; Liang, Y. Silicon ameliorates manganese toxicity by regulating manganese transport and antioxidant reactions in rice (*Oryza sativa* L.). *Plant Soil* **2012**, *354*, 407–419. [CrossRef]
49. Song, A.; Li, Z.; Zhang, J.; Xue, G.; Fan, F.; Liang, Y. Silicon-enhanced resistance to cadmium toxicity in *Brassica chinensis* L. is attributed to Si-suppressed cadmium uptake and transport and Si-enhanced antioxidant defense capacity. *J. Hazard. Mater.* **2009**, *172*, 74–83. [CrossRef]
50. Coskun, D.; Deshmukh, R.; Sonah, H.; Menzies, J.G.; Reynolds, O.; Ma, J.F.; Kronzucker, H.J.; Bélanger, R.R. The controversies of silicon's role in plant biology. *New Phytol.* **2019**, *221*, 67–85. [CrossRef] [PubMed]
51. Williams, D.E.; Vlamis, J. The effect of silicon on yield and manganese-54 uptake and distribution in the leaves of barley plants grown in culture solutions. *Plant Physiol.* **1957**, *32*, 404–409. [CrossRef]
52. Keller, C.; Rizwan, M.; Davidian, J.-C.; Pokrovsky, O.; Bovet, N.; Chaurand, P.; Meunier, J.-D. Effect of silicon on wheat seedlings (*Triticum turgidum* L.) grown in hydroponics and exposed to 0 to 30 µM Cu. *Planta* **2015**, *241*, 847–860. [CrossRef]
53. Yamaji, N.; Mitatni, N.; Ma, J.F. A transporter regulating silicon distribution in rice shoots. *Plant Cell* **2008**, *20*, 1381–1389. [CrossRef]
54. Naeem, A.; Ghafoor, A.; Farooq, M. Suppression of cadmium concentration in wheat grains by silicon is related to its application rate and cadmium accumulating abilities of cultivars. *J. Sci. Food Agric.* **2015**, *95*, 2467–2472. [CrossRef]
55. Vaculík, M.; Landberg, T.; Greger, M.; Luxová, M.; Stoláriková, M.; Lux, A. Silicon modifies root anatomy, and uptake and subcellular distribution of cadmium in young maize plants. *Ann. Bot.* **2012**, *110*, 433–443. [CrossRef]
56. Dragišić Maksimović, J.; Mojović, M.; Maksimović, V.; Römheld, V.; Nikolic, M. Silicon ameliorates manganese toxicity in cucumber by decreasing hydroxyl radical accumulation in the leaf apoplast. *J. Exp. Bot.* **2012**, *63*, 2411–2420. [CrossRef] [PubMed]
57. Iwasaki, K.; Maier, P.; Fecht, M.; Horst, W.J. Leaf apoplastic silicon enhances manganese tolerance of cowpea (*Vigna unguiculata*). *J. Plant Physiol.* **2002**, *159*, 167–173. [CrossRef]
58. Rogalla, H.; Römheld, V. Role of leaf apoplast in silicon-mediated manganese tolerance of Cucumis sativus L. *Plant Cell Environ.* **2002**, *25*, 549–555. [CrossRef]
59. Kidd, P.; Llugany, M.; Poschenrieder, C.; Gunse, B.; Barcelo, J. The role of root exudates in aluminium resistance and silicon-induced amelioration of aluminium toxicity in three varieties of maize (*Zea mays* L.). *J. Exp. Bot.* **2001**, *52*, 1339–1352. [PubMed]
60. Barcelo, J.; Guevara, P.; Poschenrieder, C. Silicon amelioration of aluminium toxicity in teosinte (*Zea mays* L. ssp. *mexicana*). *Plant Soil* **1993**, *154*, 249–255. [CrossRef]
61. Collin, B.; Doelsch, E.; Keller, C.; Cazevieille, P.; Tella, M.; Chaurand, P.; Panfili, F.; Hazemann, J.-L.; Meunier, J.-D. Evidence of sulfur-bound reduced copper in bamboo exposed to high silicon and copper concentrations. *Environ. Pollut.* **2014**, *187*, 22–30. [CrossRef] [PubMed]
62. Schaller, J.; Faucherre, S.; Joss, H.; Obst, M.; Goeckede, M.; Planer-Friedrich, B.; Peiffer, S.; Gilfedder, B.; Elberling, B. Silicon increases the phosphorus availability of Arctic soils. *Sci. Rep.* **2019**, *9*, 449. [CrossRef]

63. Bolan, N.S.; Adriano, D.C.; Naidu, R. Role of phosphorus in (im) mobilization and bioavailability of heavy metals in the soil-plant system. In *Reviews of Environmental Contamination and Toxicology*; Springer: Berlin, Germany, 2003; pp. 1–44.
64. Gupta, D.; Chatterjee, S.; Datta, S.; Veer, V.; Walther, C. Role of phosphate fertilizers in heavy metal uptake and detoxification of toxic metals. *Chemosphere* **2014**, *108*, 134–144. [CrossRef]
65. Rea, P.A. Phytochelatin synthase: Of a protease a peptide polymerase made. *Physiol. Plant.* **2012**, *145*, 154–164. [CrossRef]
66. Li, J.; Leisner, S.M.; Frantz, J. Alleviation of copper toxicity in Arabidopsis thaliana by silicon addition to hydroponic solutions. *J. Am. Soc. Hortic. Sci.* **2008**, *133*, 670–677. [CrossRef]
67. Khandekar, S.; Leisner, S. Soluble silicon modulates expression of Arabidopsis thaliana genes involved in copper stress. *J. Plant Physiol.* **2011**, *168*, 699–705. [CrossRef]
68. Ma, J.; Cai, H.; He, C.; Zhang, W.; Wang, L. A hemicellulose-bound form of silicon inhibits cadmium ion uptake in rice (*Oryza sativa*) cells. *New Phytol.* **2015**, *206*, 1063–1074. [CrossRef] [PubMed]
69. Sonah, H.; Deshmukh, R.K.; Labbé, C.; Bélanger, R.R. Analysis of aquaporins in Brassicaceae species reveals high-level of conservation and dynamic role against biotic and abiotic stress in canola. *Sci. Rep.* **2017**, *7*, 2771. [CrossRef] [PubMed]
70. Shivaraj, S.; Deshmukh, R.; Bhat, J.A.; Sonah, H.; Bélanger, R.R. Understanding Aquaporin Transport System in Eelgrass (*Zostera marina* L.), an Aquatic Plant Species. *Front. Plant Sci.* **2017**, *8*, 1334. [CrossRef] [PubMed]
71. Shivaraj, S.; Deshmukh, R.K.; Rai, R.; Bélanger, R.; Agrawal, P.K.; Dash, P.K. Genome-wide identification, characterization, and expression profile of aquaporin gene family in flax (*Linum usitatissimum*). *Sci. Rep.* **2017**, *7*, 46137. [CrossRef] [PubMed]
72. Zargar, S.M.; Mahajan, R.; Bhat, J.A.; Nazir, M.; Deshmukh, R. Role of silicon in plant stress tolerance: Opportunities to achieve a sustainable cropping system. *3 Biotech* **2019**, *9*, 73. [CrossRef]
73. Jiang, N.; Fan, X.; Lin, W.; Wang, G.; Cai, K. Transcriptome Analysis Reveals New Insights into the Bacterial Wilt Resistance Mechanism Mediated by Silicon in Tomato. *Int. J. Mol. Sci.* **2019**, *20*, 761. [CrossRef]
74. Farooq, M.A.; Ali, S.; Hameed, A.; Ishaque, W.; Mahmood, K.; Iqbal, Z. Alleviation of cadmium toxicity by silicon is related to elevated photosynthesis, antioxidant enzymes; suppressed cadmium uptake and oxidative stress in cotton. *Ecotoxicol. Environ. Saf.* **2013**, *96*, 242–249. [CrossRef]
75. Bharwana, S.; Ali, S.; Farooq, M.; Iqbal, N.; Abbas, F.; Ahmad, M. Alleviation of lead toxicity by silicon is related to elevated photosynthesis, antioxidant enzymes suppressed lead uptake and oxidative stress in cotton. *J. Bioremed. Biodeg* **2013**, *4*, 10.4172.
76. Ali, S.; Farooq, M.A.; Yasmeen, T.; Hussain, S.; Arif, M.S.; Abbas, F.; Bharwana, S.A.; Zhang, G. The influence of silicon on barley growth, photosynthesis and ultra-structure under chromium stress. *Ecotoxicol. Environ. Saf.* **2013**, *89*, 66–72. [CrossRef]
77. Doncheva, S.; Poschenrieder, C.; Stoyanova, Z.; Georgieva, K.; Velichkova, M.; Barceló, J. Silicon amelioration of manganese toxicity in Mn-sensitive and Mn-tolerant maize varieties. *Environ. Exp. Bot.* **2009**, *65*, 189–197. [CrossRef]
78. da Cunha, K.P.V.; do Nascimento, C.W.A. Silicon effects on metal tolerance and structural changes in maize (*Zea mays* L.) grown on a cadmium and zinc enriched soil. *Water Air Soil Pollut.* **2009**, *197*, 323. [CrossRef]
79. Vatehová, Z.; Kollárová, K.; Zelko, I.; Richterová-Kučerová, D.; Bujdoš, M.; Lišková, D. Interaction of silicon and cadmium in *Brassica juncea* and *Brassica napus*. *Biologia* **2012**, *67*, 498–504. [CrossRef]
80. Greger, M.; Landberg, T.; Vaculik, M.; Lux, A. Silicon influences nutrient status in plants. In Proceedings of the 5th International Conference on Silicon in Agriculture, Beijing, China, 13–18 September 2011; The Organizing Committee of the 5th Silicon in Agriculture Conference: Beijing, China, 2011.
81. Neumann, D.; zur Nieden, U.; Schwieger, W.; Leopold, I.; Lichtenberger, O. Heavy metal tolerance of *Minuartia verna*. *J. Plant Physiol.* **1997**, *151*, 101–108. [CrossRef]
82. Neumann, D.; Zur Nieden, U. Silicon and heavy metal tolerance of higher plants. *Phytochemistry* **2001**, *56*, 685–692. [CrossRef]
83. Rizwan, M.; Meunier, J.-D.; Miche, H.; Keller, C. Effect of silicon on reducing cadmium toxicity in durum wheat (*Triticum turgidum* L. cv. Claudio W.) grown in a soil with aged contamination. *J. Hazard. Mater.* **2012**, *209*, 326–334. [CrossRef]
84. Dresler, S.; Wójcik, M.; Bednarek, W.; Hanaka, A.; Tukiendorf, A. The effect of silicon on maize growth under cadmium stress. *Russ. J. Plant Physiol.* **2015**, *62*, 86–92. [CrossRef]

85. Ma, J.F.; Goto, S.; Tamai, K.; Ichii, M. Role of root hairs and lateral roots in silicon uptake by rice. *Plant Physiol.* **2001**, *127*, 1773–1780. [CrossRef] [PubMed]
86. Savant, N.K.; Datnoff, L.E.; Snyder, G.H. Depletion of plant-available silicon in soils: A possible cause of declining rice yields. *Commun. Soil Sci. Plant Anal.* **1997**, *28*, 1245–1252. [CrossRef]
87. Meunier, J.; Guntzer, F.; Kirman, S.; Keller, C. Terrestrial plant-Si and environmental changes. *Mineral. Mag.* **2008**, *72*, 263–267. [CrossRef]
88. Desplanques, V.; Cary, L.; Mouret, J.-C.; Trolard, F.; Bourrié, G.; Grauby, O.; Meunier, J.-D. Silicon transfers in a rice field in Camargue (France). *J. Geochem. Explor.* **2006**, *88*, 190–193. [CrossRef]
89. Datnoff, L.; Deren, C.; Snyder, G. Silicon fertilization for disease management of rice in Florida. *Crop Prot.* **1997**, *16*, 525–531. [CrossRef]
90. Henriet, C.; Draye, X.; Oppitz, I.; Swennen, R.; Delvaux, B. Effects, distribution and uptake of silicon in banana (*Musa* spp.) under controlled conditions. *Plant Soil* **2006**, *287*, 359–374. [CrossRef]
91. Tubaña, B.S.; Heckman, J.R. Silicon in soils and plants. In *Silicon and Plant Diseases*; Springer: Berlin, Germany, 2015; pp. 7–51.
92. Guntzer, F.; Keller, C.; Meunier, J.-D. Benefits of plant silicon for crops: A review. *Agron. Sustain. Dev.* **2012**, *32*, 201–213. [CrossRef]
93. Jarosz, Z. The effect of silicon application and type of substrate on yield and chemical composition of leaves and fruit of cucumber. *J. Elem.* **2013**, *18*, 403–414. [CrossRef]
94. Ouellette, S.; Goyette, M.-H.; Labbé, C.; Laur, J.; Gaudreau, L.; Gosselin, A.; Dorais, M.; Deshmukh, R.K.; Bélanger, R.R. Silicon transporters and effects of silicon amendments in strawberry under high tunnel and field conditions. *Front. Plant Sci.* **2017**, *8*, 949. [CrossRef] [PubMed]
95. Ning, D.; Song, A.; Fan, F.; Li, Z.; Liang, Y. Effects of slag-based silicon fertilizer on rice growth and brown-spot resistance. *PLoS ONE* **2014**, *9*, e102681. [CrossRef] [PubMed]
96. Kaczorek, D.; Sommer, M. Silikon cycle in terrestrial biogeosystems of temperate climate. *Soil Sci. Annu.* **2004**, *55*, 221–230.
97. Höhn, A.; Sommer, M.; Kaczorek, D.; Schalitz, G.; Breuer, J. Silicon fractions in histosols and gleysols of a temperate grassland site. *J. Plant Nutr. Soil Sci.* **2008**, *171*, 409–418. [CrossRef]
98. Ma, J.F.; Tamai, K.; Yamaji, N.; Mitani, N.; Konishi, S.; Katsuhara, M.; Ishiguro, M.; Murata, Y.; Yano, M. A silicon transporter in rice. *Nature* **2006**, *440*, 688–691. [CrossRef]
99. Ma, J.F.; Yamaji, N.; Mitani, N.; Tamai, K.; Konishi, S.; Fujiwara, T.; Katsuhara, M.; Yano, M. An efflux transporter of silicon in rice. *Nature* **2007**, *448*, 209–212. [CrossRef]
100. Mitani, N.; Chiba, Y.; Yamaji, N.; Ma, J.F. Identification and characterization of maize and barley Lsi2-like silicon efflux transporters reveals a distinct silicon uptake system from that in rice. *Plant Cell* **2009**, *21*, 2133–2142. [CrossRef] [PubMed]
101. Pandey, A.K.; Gautam, A.; Dubey, R.S. Transport and detoxification of metalloids in plants in relation to plant-metalloids tolerance. *Plant Gene* **2019**, *17*, 1–7. [CrossRef]
102. Deshmukh, R.K.; Vivancos, J.; Guérin, V.; Sonah, H.; Labbé, C.; Belzile, F.; Bélanger, R.R. Identification and functional characterization of silicon transporters in soybean using comparative genomics of major intrinsic proteins in Arabidopsis and rice. *Plant Mol. Biol.* **2013**, *83*, 303–315. [CrossRef]
103. Deshmukh, R.K.; Vivancos, J.; Ramakrishnan, G.; Guérin, V.; Carpentier, G.; Sonah, H.; Labbé, C.; Isenring, P.; Belzile, F.J.; Bélanger, R.R. A precise spacing between the NPA domains of aquaporins is essential for silicon permeability in plants. *Plant J.* **2015**, *83*, 489–500. [CrossRef] [PubMed]
104. Deren, C. Plant genotype, silicon concentration, and silicon-related responses. In *Studies in Plant Science*; Elsevier: Amsterdam, The Netherlands, 2001; Volume 8, pp. 149–158.
105. Ma, J.F.; Higashitani, A.; Sato, K.; Takeda, K. Genotypic variation in silicon concentration of barley grain. *Plant Soil* **2003**, *249*, 383–387.
106. Ma, J.F.; Yamaji, N.; Tamai, K.; Mitani, N. Genotypic difference in silicon uptake and expression of silicon transporter genes in rice. *Plant Physiol.* **2007**, *145*, 919–924. [CrossRef]

plants MDPI

Review

Assisting Phytoremediation of Heavy Metals Using Chemical Amendments

Md. Mahadi Hasan [1], Md. Nashir Uddin [2], Iffat Ara-Sharmeen [3], Hesham F. Alharby [4], Yahya Alzahrani [4], Khalid Rehman Hakeem [4] and Li Zhang [5,*]

1 State Key Laboratory of Grassland Agro-Ecosystems, School of Life Sciences, Lanzhou University, Lanzhou 730000, China
2 Department of Biochemistry and Microbiology, School of Health and Life Sciences, North South University, Dhaka 1229, Bangladesh
3 Department of Biochemistry, School of Life Science, Independent University, Dhaka 1229, Bangladesh
4 Department of Biological Sciences, Faculty of Science, King Abdulaziz University, Jeddah 21577, Saudi Arabia
5 Institute of Cell Biology, School of Life Sciences, Lanzhou University, No.222 South TianShui Road, Lanzhou 730000, China
* Correspondence: zllily@lzu.edu.cn; Tel.: +86-182-9841-7508

Received: 29 July 2019; Accepted: 16 August 2019; Published: 21 August 2019

Abstract: Phytoremediation is one of the safer, economical, and environment-friendly techniques in which plants are used to recover polluted soils, particularly those containing toxic organic substances and heavy metals. However, it is considered as a slow form of remediation, as plants take time to grow and flourish. Various amendments, including the augmentation of certain chemical substances i.e., ethylenediamine tetraacetic acid (EDTA), ethylene glycol tetra acetic acid (EGTA), and sodium dodecyl sulfate (SDS) have been used to induce and enhance the phytoextraction capacity in plants. Several reports show that chemical amendments can improve the metal accumulation in different plant parts without actually affecting the growth of the plant. This raises a question about the amount and mechanisms of chemical amendments that may be needed for potentially good plant growth and metal phytoremediation. This review provides a detailed discussion on the mechanisms undertaken by three important chemical amendments that are widely used in enhancing phytoremediation (i.e., EDTA, EGTA, and SDS) to support plant growth as well as soil phytoremediation. A core part of this review focuses on the recent advances that have been made using chemical amendments in assisting metal phytoremediation.

Keywords: environmental pollution; phytoextraction; cadmium; biostimulation; oxidative damage

1. Introduction

Heavy metal pollution is regarded as a serious problem for crop production [1]. In soil ecosystems, the gradual increase in heavy metal levels is a major concern throughout the world [2]. Heavy metal pollution is a vital issue for environmental management due to the rapid increase in anthropogenic activities, including industrialization, transportation, and urbanization. Several sources of heavy metals include medical waste, the combustion of coal, petrol, mining, fertilizers, smelting, and pesticides, which are adding heavy metals to the environment [3,4]. Among these heavy metals, cadmium (Cd), lead (Pb), aluminum (Al), zinc (Zn), manganese (Mn), chromium (Cr), and copper (Cu) are considered as common toxic heavy metals [1].

Toxic heavy metals considered as phytotoxic agents to plants that affect the plant morphological and physiological processes such as lower growth rate, stomatal movement and nutrient imbalance, and photosynthetic processes result in the oxidative damage [5]. Due to increased contamination of soil by toxic metal components, it is important to use necessary techniques for cleaning up heavy metals from contaminated soil, which requires an effective and reasonable solution. Some modern techniques that have been used for the remediation of heavy metals are based on the biological, chemical, and physical approaches [6]. Phytoremediation is one of the biological approaches used for the remediation of soils. Phytoremediation is preferable due to its safety and lower cost as compared to physical and chemical remediation [6]. The technologies of metal phytoremediation include phytoextraction, phytostabilization, and phytovolatilization [7]. Among these, phytoextraction is one of the promising techniques being used for reclaiming the metal polluted soils [1].

However, the heavy metal extraction by plants is usually limited by the availability of heavy metals in soils. The application of chemical enhanced technology is one of the prevalent approaches, which helps the uptake of heavy metals and their translocation in the aboveground parts of plants [4,8].

Several chemical amendments have been used for the enhancement of phytoextraction processes. In recent years, chemical amendments including ethylenediamine tetraacetic acid (EDTA), ethylene glycol tetraacetic acid (EGTA), and sodium dodecyl sulfate (SDS) are widely used for soil remediation applications, which can bring potential metal leaching risk. EDTA is considered as having the most potential and most studied chemical amendment, and is used to increase the phytoextraction of metals from contaminated soils [9].

Therefore, it is important to increase phytoremediation efficiency through the phytoextraction processes of heavy metals. The phytoremediation of heavy metals in plants may be enhanced through an emerging chemical amendments technology. Our present review attempts to describe the potential benefits of chemical amendments in phytoremediation research.

2. Source of Heavy Metal Pollution, Ecotoxicity, and Approaches for Remediation

In general, heavy metals originate from anthropogenic and natural sources. Different sources of heavy metals were identified such as (a) industrial sources, (b) domestic effluent, (c) agricultural sources, and (d) natural sources such as the atmosphere (Figure 1). It has been reported that most of the areas in the world such as China, Japan, and Indonesia have been contaminated by Cd, Cu, and Zn due to mining and agricultural operations [6].

In terrestrial ecosystems, the heavy metal contaminations are increasing due to anthropogenic activities, which are known to be liable for ecotoxicity. The largest availability of heavy metals toxicity occurs in soil and aquatic ecosystems, whereas the smaller portion of metals available in the atmosphere form of vapors or particulate. In soil, the major heavy metal ranges vary from 0.01 to 0.7 ppm dw of Cd, 2 to 200 ppm dw of Pb, 10 to 300 ppm dw of Zn, 5 to 3000 ppm dw of Cr, and 7000 to 55,000 ppm dw of Fe etc. [10]. Heavy metal pollution in soils is becoming increasingly common in the agricultural sector, and affects food safety and crop growth. The metal toxicity in plants is not the same; it varies with plant species, metal concentration, soil pH, soil composition, specific metal and chemical form, etc. According to Misra and Mani [11], the range of vital heavy metal in plants is 0.1 to 2.4 μg^{-1} dw for Cd, 1 to 13 μg^{-1} dw for Pb, 0.02 to 7 μg^{-1} dw for As, 8 to 100 μg^{-1} dw for Zn, 0.2 to 1 μg^{-1} dw for Cr, 140 μg^{-1} dw for Fe, etc.

Toxic heavy metal is hazardous to the environment. Therefore, heavy metal remediation in the soil is very important [12]. Over the past few decades, various techniques were employed for the remediation of toxic heavy metals. The preferred strategies were selected for the remediation of heavy metal from the environment based on their technical complexity and cost [13]. The techniques include physical, chemical, and biological methods [14].

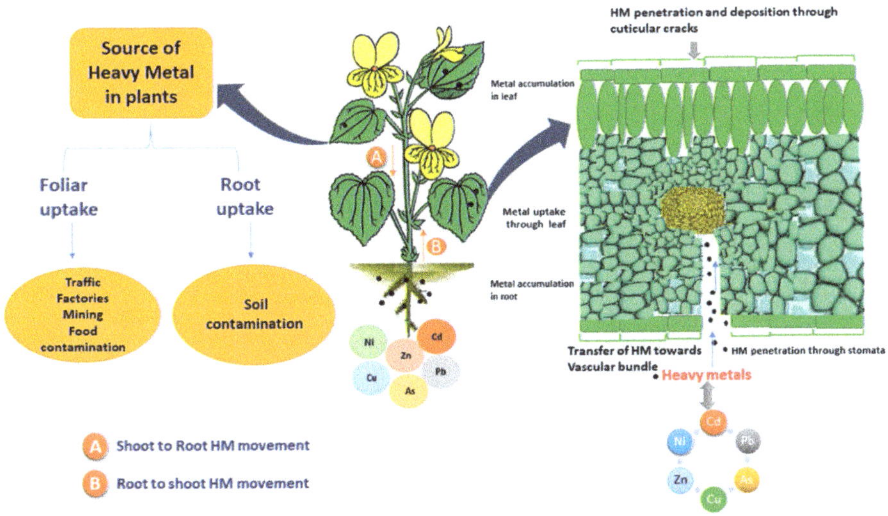

Figure 1. Sources of heavy metals, and foliar, root uptake of heavy metals in plants.

Traditionally, the contaminated soils are cleaned up by excavation or removing the soils from the land sites. The toxic metals problems in the soil may also arise due to the transportation of contaminants that are closely adjacent to the soils [15]. Soil washing is another strategy to eliminate soil metal contaminants, but it has been reported that this is unsuitable for plant growth and development due to the hindrance of biological and chemical activities [16]. Chemical methods are not preferable due to alterations in the soil texture and structure, expense, and the generation of high quantities of sludge [16,17].

2.1. *Physiochemical Techniques*

The physiochemical technique includes excavation, leaching, landfill, and thermal treatment (bioreactor) approaches. Nevertheless, these processes are fast but costly, and have detrimental effects on the soil properties [13,18]. These techniques are not completely suitable for heavy metal remediation, and only change the form of the problem without remediating the pollutants thoroughly [19].

2.2. *Biological Approaches*

Heavy metals are removed from the environment through natural remediation coordinated by microorganisms and plants [14]. Biological remediation is a preferable method as it is natural, cost-effective, environmentally friendly, and has wider public support [20]. There are several approaches, such as bioaugmentation, biostimulation, bioleaching, composting, bioreactors, bioremediation, and phytoremediation [21]. Biological approaches are considered superior to physiological approaches, because these processes use solar energy and ensure that the soil properties are conserved [20]. Bioremediation is a technique by which heavy metals are removed from the environment [19,22]. Bacterial strains such as *P. aeruginosa* and *Bacillus* spp. can remediate metals such as zinc and copper in this context [23]. Bioremediation can be done using biostimulation, biofilters, treated and pumped methods, bioventing, bioreactors, composting, land forming, bioaugmentation, and intrinsic bioremediation [24]. The efficiency of phytoremediation can be increased if microbes are used [25].

2.2.1. Phytoremediation of Heavy Metals

Phytoremediation is a part of the emerging green technology being used for the uptake of various heavy metals in different amounts from the soil and storage of them in parts of the plant that can be harvested [26]. With changes in soil properties, plants can tolerate pollutants [27]. Soils containing heavy metals can be easily treated through phytoremediation, and the biomass that is formed during the process can be further applied in biodiesel production. Thus, bioenergy crops such as *Brassica* species, which are known to accumulate toxic metals, are increasingly suitable for this purpose. Some plants can accumulate pollutants in tissues [28]. The Jerusalem artichoke (*Helianthus tuberosus* L.) is known to be an energy crop that is used for the phytoremediation of soils contaminated by heavy metals. It is more suitable for the phytoextraction of heavy metals among the energy crops. The highest heavy metal uptake was observed at a dose of 60 Mg DMha^{-1} in the Jerusalem artichoke [29].

Hyperaccumulating plants are fit to grow on soils contaminated with heavy metals and can be used to remove pollutants [19]. Plants that contain greater than 10,000 mg/kg dry weight of Zn or Mn or more than 1000 mg/kg dry weight of Ni, Cu, or Pb or greater than 100 mg/kg dry weight of Cd in contaminated areas are considered as hyperaccumulating plants. There are several families of plants that are known to hyperaccumulate toxic heavy metals. Some such important plant families are *Lamiaceae*, Fabaceae, Scrophulariaceae, Asteraceae, Euphorbiaceae, and Brassicaceae, which are usually used in the phytoremediation processes. Other than that, there have been reports of heavy metal hyperaccumulation in about 500 plant species [30]. Plants with greater hyperaccumulating abilities include *Alyssum bertolonii*, *Thlaspi caerulescens*, *Calendula officinalis*, and *Tagetes erecta* [31]. Higher concentrations of Ni, Zn, and Cd are best gathered by *Thlaspi caerulescens* [32]. This plant can accumulate 500–52,000 mg kg^{-1} of Zn and 0.3–1020 mg kg^{-1} of Cd. Trees are more suitable for phytoremediation because of their greater root systems and biomass [19], although trees take more time in accumulating metals, even in low quantities. This issue could be solved by using fast-growing plants instead of trees [33]. For an example, the *Poplus alba* is a deciduous tree, which can accumulate zinc (Zn) in different plant parts such as leaves, stems, and roots. An increasing trend of Zn accumulation was observed in the leaves of *Poplus alba* with the application of SDS [34]. Phytoremediation depends on the ability of the plants to gather increased levels of toxic metals within their tissues [35]. Some plants contain enzymes that can break down a number of organic compounds. However, pollutants that are inorganic cannot be degraded with these enzymes. Thus, there is a need to ensure that inorganic pollutants be less available in soils or extracted and accumulated in different parts of the plants and also reduce volatile versions of inorganic pollutants [36]. Energy recovery strategies can be used to produce bioenergy from plant biomass, such as to form biodiesel. Fuel gas, char, and bio-oil can be produced via pyrolysis, during which the biomass undergoes thermal degradation without oxygen [37]. Soils greatly polluted with toxic heavy metals can be remediated easily by growing plants that are tolerant to more than one heavy metal, can produce a good amount of harvestable biomass with enhanced growth rates, and are highly competitive [30].

2.2.2. Problems with Heavy Metals of Remediation with Phytoremediation

Phytoremediation performance could be affected by several processes such as the rate of contaminant uptake by plant roots, the availability of toxic metal ions in the soil, and the root-to-shoot movement of the metal ions [38]. Plants store the heavy metals in the different parts of the plant such as in the leaf, stem, and root [26]. Although phytoremediation is considered as a green technology, it has some problems or limitations in the case of remediation of soils. According to Koptsik [39], there are some problems or limitations of phytoremediation as follows:

- It depends on the local climate, weather, and seasonal conditions.
- It requires more time to remediate pollutants from the soil.
- It is suitable only for low-polluted territories.
- It depends on the depth of the root system and solubility of the pollutants.
- Pollutants may enter the trophic chains and adjacent media.

3. Assistance of Chemical Amendments to Increase the Efficiency of Phytoremediation

High heavy metal concentrations in plant tissues and biomass are considered as a key factor for the successful phytoremediation of heavy metal polluted soils [40]. The natural levels of heavy metals are relatively high in the Earth's crust, and most of these are phytoavailable and low soluble [41]. Chemical amendments have a key role to compensate for relatively low heavy metal availability in soil, and it helps the plants uptake and translocate metals toward the shoot [42]. Different kinds of chemical amendments or chelating agents have been used and tested to increase the bioavailability of metals in plants and facilitate the transport of metals between the roots and shoots. The most important of the three chemical amendments are EDTA, EGTA, and SDS, which have been widely used in recent years (Figure 2).

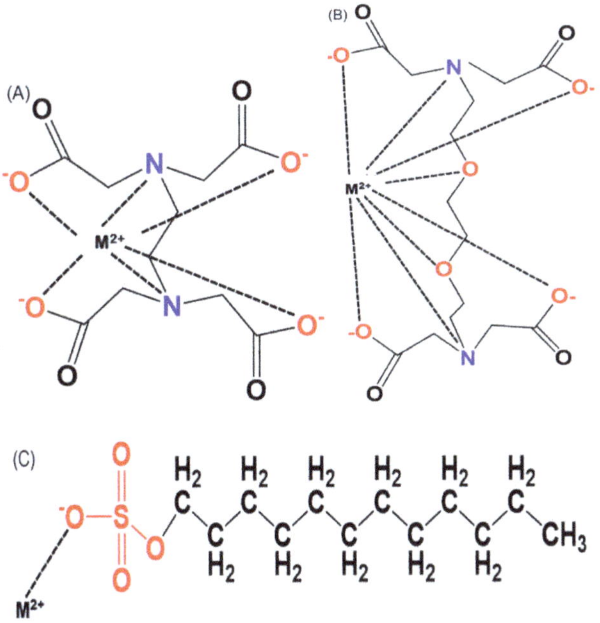

Figure 2. Chemical structure of (**A**) ethylene diamine tetraacetic acid (EDTA), (**B**) ethylene glycol tetraacetic acid (EGTA), (**C**) sodium dodecyl sulfate (SDS), and binding with metals, M^{2+}(Cd^{2+}, Pb^{2+} etc.).

3.1. Assisting of Phytoremediation by EDTA

Ethylenediamine tetraacetic acid, also known as EDTA, is used frequently in the agricultural sector due to its ability to mobilize heavy metals. EDTA enhances metal uptake through the roots and also supports metal xylem loading (Figure 3).

This has also been reported in previous studies [44]. The increased uptake occurs due to the production of soluble metal–EDTA complexes [45]. Most plants are able to take up metal–EDTA complexes [46], especially hyperaccumulator species [47]. The effect of EDTA depends on the plant species, heavy metal, and type of soil, ranging from 0 to 200-fold higher accumulations [42]. Cu, Zn, Pb, Ni, and Cd uptake was enhanced by EDTA in *Zea mays* and *Lolium perenne* [48]. Lead accumulation increased fourfold in the *Sedum alfredii* roots [49] and twofold in *Vicia faba* seedlings in a 24-h experiment [50].

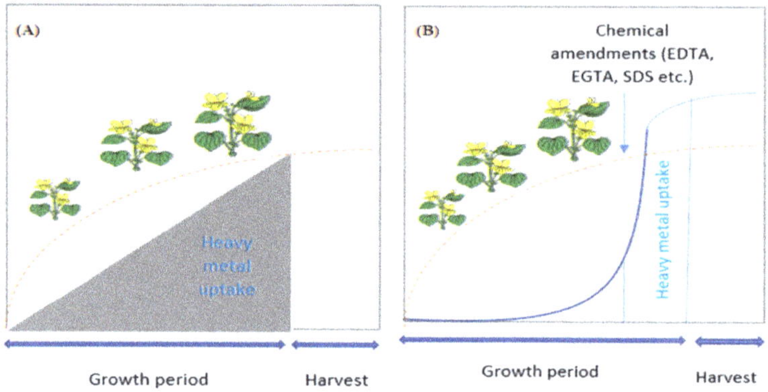

Figure 3. Chemical amendments assisting the heavy metal uptake in the plants, adapted from Souza et al. [43]. Figure (**A**) depicts that a hyperaccumulator plant accumulates the heavy metal gradually and constantly during the entire life cycle of the plant, whereas Figure (**B**) shows that the heavy metal behavior of a non-hyperaccumulator plant is relatively slow, but the metal uptake increases quickly after the application of chemical amendments along with metals.

The metal uptake rate or concentration in plants depends on the chemical amendments dose during the application of chemical amendments. The uptake of toxic metals rate may reach a maximum in plants at a certain amendments dose level. Before treatment, the lead (Pb) concentration was 0.025 mg/g in *Typha* sp. The concentration was increased to 0.846 mg/g, while 1 mL of EDTA was added along with 1 ppm of lead. Similarly, the copper (Cu) was increased 0.030 mg/g to 0.522 mg/g, when 1 mL of EDTA was added in combination with 1 ppm of Cu. Cd concentration was increased by 51.98% in the shoot during the application of EDTA with Cd [45].

In another study, EDTA caused a 15-fold increased uptake of lead in the roots of *Vetiveria zizanioides*. The concentrations of EDTA solution (0–10 mmol·kg^{-1} soil) were added under the exposure of 1000 mg·kg^{-1} of Pb in the form of $Pb(NO_3)_2$ for 14 days. They concluded that 10 mmol·kg^{-1} of EDTA treatment was best regarding the lead (Pb) uptake in *Vetiveria zizanioides* [51]. The maximum Pb concentrations in the shoot were found in *Canavalia ensiformis* L. when 0.5 g·kg^{-1} of EDTA was used for 40 days with the application of 1800 mg·kg^{-1} of Pb as $Pb(NO_3)_2$ [52]. A study showed a higher uptake of lead, but no other heavy metals from contaminated soils [53]. Under the exposure of Cd as $CdCl_2$ (50 mg·kg^{-1}), the shoot concentrations of Cd significantly increased at a rate of 0.5 g·kg^{-1} EDTA in *Helianthus annuus* [54] (Table 1).

Phytoextraction using EDTA can be made more efficient with the proper mix of organic chelators, metals, and appropriate plant species [70].

The mechanism of EDTA increasing metal uptake is not fully understood yet. There are various steps involved in the entrance of metals from the soil to the roots, which determines the rate of uptake and also the capability of a plant to take in heavy metals. The uptake of metals into the roots involves: (i) the movement of soluble metals to plant roots through mass flow or diffusion [71], (ii) adsorption on roots, and (iii) attachment to functional groups within the rhizoderm cell surface [72]. The adsorption of metals into the plant root surface has been observed in various studies [73]. The metal–EDTA complexes form affects almost all of the steps previously mentioned of metals uptaking through plant roots. Initially, EDTA allows the diffusion of metals through the roots by (i) increasing their concentration in soil by desorbing metals and (ii) lowering the apparent diffusion coefficient of metals in metal–EDTA complex forms [71]. Since metal–EDTA complexes carry a neutral charge, they are not attached or blocked by polysaccharides or carboxyl groups in the rhizoderm cell surface. In this way, EDTA allows the movement of metals directly into the roots. However, there have been various hypotheses about whether metal–EDTA complexes dissociate just before entering the plant roots or enter as they are [74]. In some studies, it was shown that EDTA form complexes in solution, then enter plants [61]. A study involving 14C-labeling showed that indeed, the full metal–EDTA complex is absorbed, with particular selectivity toward lower charged complexes in Swiss chard. Sarret et al. [75] mentioned that these metal–EDTA complexes are nontoxic and break down after entering the roots, forming free heavy metal ions that could induce phytotoxicity. Inductively coupled plasma mass spectrometry (ICP-MS) analysis of xylem sap showed the presence of metal–EDTA complexes and the absence of EDTA individually in *Hordeum vulgare* grown in contaminated soil amended with EDTA. Schaider et al. [76] showed the presence of complexes in xylem sap such as Cd–EDTA, Pb–EDTA, and Fe–EDTA.

Table 1. Effects of heavy metals on the plants with different growing conditions along with chemical amendments.

Scientific Name	Heavy Metals	Growing Conditions	Chemical Amendments	References
Calendula officinalis	Cd	Pot	EDTA, EGTA, SDS	[55]
Tagetes erecta	Pb	Pot	EDTA	[56]
Impatiens walleriana	Cd	Pot	EDTA	[57]
Medicago sativa	Cr	Pot	EDTA	[58]
Tribulus terrestris	Cd, Pb	Pot	EDTA	[59]
Helianthus annuus	Cd, Ni	Pot	EDTA	[54]
Dianthus chinensis	Cd, Zn, Pb	Pot	EDTA	[60]
Vetiver zizanioides	Cd, Zn, Pb	Pot	EDTA	[60]
Canavalia ensiformis L.	Pb	Pot	EDTA	[52]
Brassica carinata	Cd, Cr, Pb	Pot	EDTA	[44]
Brassica juncea	Cd, Cr, Pb	Pot	EDTA	[44]
Brassica juncea	Pb	Pot	EDTA	[61]
Phaseolus vulgaris L.	Pb, Zn, Cu	Pot	EDTA	[62]
Zea mays L. cv. Nongda 108	Pb, Zn, Cu	Pot	EDTA	[62]
Brassica juncea L. Czern.	Pb, Zn	Pot	EDTA	[63]
Triticum aestivum	Pb, Zn	Pot	EDTA	[63]
Mirabilis jalapa L.	Cd	Pot	EDTA, EGTA	[64]
Althaea rosea	Cd	Pot	EDTA, EGTA, SDS	[65]
Mirabilis jalapa	Cd	Pot	EDTA, EGTA	[64]
Cicer arietinum	Cd	Pot	EGTA	[66]
Sesbania exaltata	Pb	Pot	EDTA, EGTA	[67]
Calendula officinalis	Cd	Pot	EGTA, SDS	[68]
Althaea rosea	Cd	Pot	EGTA, SDS	[68]
Halimione portulacoides	Cu	Field	SDS	[69]
Populus alba	Zn	Pot	SDS	[34]

3.2. Assisting of Phytoremediation by EGTA

EGTA, or ethylene glycol tetraacetic acid, is a widely used chelating agent. Similar to EDTA, the four carboxyl groups dissociate and produce four protons (Figure 2). The two N atoms of the two amino groups have unshared pairs of electrons each. EGTA has been shown to have more affinity for Ca ions, but less affinity toward Mg ions. Such synthetic chelators have been successfully used for phytoextraction [77]. The factor to consider for using EGTA is that it increases the uptake of heavy metals by plants more efficiently. Sakouhi et al. [66] reported that applying EGTA increased Pb accumulation by more than 80% in parts of *Cicer arietinum* plants above ground. After the application of 1 mmol kg^{-1} of EGTA in *A. rosea*, the maximum total Cd content was observed, which was increased by 72% [68]. In *Mirabilis Jalapa*, Cd concentration was increased by 43.27% in the shoot under EGTA treatment along with Cd (25 mg·kg^{-1}) [64]. The Cd concentration was at a maximum in the shoots when 1.0 mmol·kg^{-1} was used. In *Calendula officinalis*, 30–100 mg·kg^{-1} of Cd as $CdCl_2$ was added to the soil, and it was observed that the total Cd increases up to 217% with the application of EGTA alone. They concluded that the use of 1.0 mmol kg^{-1} EGTA showed the greatest effect among the treatments [55] (Table 1).

3.3. *Assisting of Phytoremediation by SDS*

SDS is a surfactant that is most commonly used in detergents, but it can also be used for heavy metal and organic contaminant remediation from soil [34]. SDS can ameliorate solubilities of various hydrocarbons and heavy metals such as zinc, cadmium, lead, and copper, making their removal easier, both in phytoremediation trails involving herbaceous species [78] and soil flushing [79]. Surfactants contain a hydrophobic portion that has less affinity for aqueous solutions, and the hydrophilic polar portion, which has a higher affinity for aqueous solutions. Thus, surfactants are amphiphilic. Anionic surfactants such as SDS are amphipathic, as they can interact with both non-polar and polar macromolecules, causing membrane damage and even oxidative stress [80]. However, the direct involvement of SDS in plant remediation trials has not yet been widely investigated, because some strains of *Pseudomonas* can degrade SDS by using it as a carbon source [81,82] and photoelectrochemical reactions [83]. It has been observed that SDS can increase the dry biomass of plants such as *Althaea rosea*, and also promote Cd accumulation in roots and shoots [65]. *Calendula* and *Althaea rosea* provide some evidence that Zn accumulation changes with the presence of SDS [34]. When 1 mM of Zn was added in combination with 0.5 mM of SDS, the Zn translocation was increased toward basal leaves in *Poplus alba* [34]. The maximum Cd concentration was observed in the shoots and roots in *Althaea rosea*, when 1.0 mmol kg^{-1} was added among the single SDS treatments (0.5 mmol·kg^{-1} to 2 mmol·kg^{-1}) [68]. In *Calendula officinalis*, when the applied soil Cd contamination was 30 mgkg^{-1}, the Cd concentrations in the shoot increased significantly under 0.5–2.0 mmolkg^{-1} SDS treatments. For the higher concentration of 100 mg·kg^{-1} Cd as $CdCl_2$, the application of SDS (0.5–2.0 mmol·kg^{-1}) was observed to increase the Cd concentrations in the shoot. They concluded that the maximal shoot Cd concentrations were observed when 2.0 mmol·kg^{-1} SDS were used [55] (Table 1).

4. Challenges with Chemical Amendments

Chemical amendments are useful in phytoremediation methods used to phytoextract heavy metals from contaminated soil; however, adding EDTA, EGTA, and SDS also have a few limitations [84], such as toxic effects toward soil microbes [85], soil enzyme activities, and on cultivated plant species [86]. Due to metal mobilization during extended periods, the chemical-assisted phytoextraction increases the risk of adverse environmental effects. Chemical amendments can disrupt chemical properties and the physical structures of soil by dissolving minerals.

EDTA and its metal complexes are not just highly toxic, but also non-biodegradable and could persist for many weeks [87]. EDTA is an exogenous substance that has adverse environmental effects on soils. EDTA is said to persist within the soil for six months or more [86]. EDTA-facilitated metal movement through soil could occur post-growing season. A monitoring study conducted by [88] mentioned a similar movement of metals such as Zn, Cu, Pb, and Cd using EDTA post-treatment while observing the behavior of heavy metals. Grcman et al. [89] observed that EDTA could leave phytotoxic effects on plants such as red clover. Luo et al. [90] reported similar phytotoxic effects by EDTA even after six months since addition to the soil.

A recent study by Krujatz [91] found that EDTA reduced the toxicity of Ni^{2+} and Cd^{2+} within the stoichiometric ratio, but still inhibited the growth of *Pseudomonas brassicacearum* above the ratio. EDTA also has potentially toxic effects on fungi and bacteria living in soil [88] and plants [84]. Ruley et al. [92] observed lower photosynthetic activity in seedlings of *Sesbania drummondii* that were exposed to EDTA solutions. The toxicity of metals induced by EDTA is due to increased metal uptake [93]. Other phytotoxic effects caused by EDTA may occur due to enhanced metal uptake by plants [94]. Other authors reported results of lower plant biomass when EDTA is present [95]. The toxicity symptom seen in *Brassica juncea* and *Lolium perenne* showed a significant decrease in their biomass [87]. In *Typha angustifolia*, there was a significant decrease in biomass and plant height, resulting in stunted growth [96].

Metal leaching enhanced by chemical amendments depends on various factors, such as:

- Plant-related: root type, species, type and amount of root exudates, age [52,97];
- Soil related: soil texture, pH, organic contents, competing ions, carbonates, biological and microbial conditions, soil water holding capacity, cation exchange capacity (CEC), soil redox potential, soil-buffering capacity [98];
- Metal related: concentration and type of metal, EDTA, EGTA, and SDS-binding capacity to a specific metal, whether single or more than one metal contamination [61,99].

5. Conclusions and Future Direction

Heavy metals can cause serious environmental pollution, as they can accumulate in soils and persist for a long time, and even possibly enter the food chain. The chemical amendment-induced remediation of heavy metals by plant species is an effective technique, because it has high binding capacities toward the majority of the heavy metals. Among the three chemical amendments, EDTA is known to be the most efficient organic ligand that can increase metal uptake, solubilization, and translocation, as it can form highly stable and soluble metal–EDTA complexes. Most probably, iron (Fe) is the best metal to bind with EDTA due to its high affinity to ferric ions. Metal phytoremediation enhanced by EDTA, EGTA, and SDS can be affected depending on various biogeochemical processes found in plants, metal, and the soil. Chemical amendment capacity is an important aspect, in order to reduce the time and cost needed for heavy metal contaminated soil. This can be done by increasing the heavy metal bioaccumulation index in plants.

Author Contributions: M.M.H. wrote the initial draft of the manuscript, M.N.U. modified and edited the figure and table, I.A.-S., H.F.A., K.R.H. and Y.A. were involved in the compilation of the final draft, and L.Z. revised the final manuscript.

Funding: This research did not receive any external funding.

Acknowledgments: The authors are thankful to the School of Life Sciences, Lanzhou University, China, for the financial support in the form of a postdoctorate fellowship to M.M.H. We thank Mirza Hasanuzzaman, Department of Agronomy, Sher-e-Bangla Agricultural University, and Katherine Li, Director of the office of external affairs, North South University for the critical reading and formatting of the final manuscript.

Conflicts of Interest: The authors declare no conflict of interest.

References

1. Bhat, J.A.; Shivaraj, S.M.; Singh, P.; Navadagi, D.B.; Tripathi, D.K.; Dash, P.K.; Solanke, A.U.; Sonah, H.; Deshmukh, R. Role of Silicon in Mitigation of Heavy Metal Stresses in Crop Plants. *Plants* **2019**, *8*, 71. [CrossRef]
2. Fuentes, A.; Almonacid, L.; Ocampo, J.A.; Arriagada, C. Synergistic interactions between a saprophytic fungal consortium and *Rhizophagus irregularis* alleviate oxidative stress in plants grown in heavy metal contaminated soil. *Plant Soil* **2016**, *407*, 355–366. [CrossRef]
3. Li, Y.; Luo, J.; Yu, J.; Xia, L.; Zhou, C.; Cai, L.; Ma, X. Improvement of the phytoremediation efficiency of *Neyraudia reynaudiana* for lead-zinc mine-contaminated soil under the interactive effect of earthworms and EDTA. *Sci. Rep.* **2018**, *8*, 6417. [CrossRef] [PubMed]
4. Jiang, M.; Liu, S.; Li, Y.; Li, X.; Luo, Z.; Song, H.; Chen, Q. EDTA-facilitated toxic tolerance, absorption and translocation and phytoremediation of lead by dwarf bamboos. *Ecotoxicol. Environ. Saf.* **2019**, *170*, 502–512. [CrossRef] [PubMed]
5. Nakbanpote, W.; Meesungnoen, O.; Prasad, M.N.V. Potential of Ornamental Plants for Phytoremediation of Heavy Metals and Income Generation. In *Bioremediation and Bioeconomy*; Elsevier: Amsterdam, The Netherlands, 2016; pp. 179–217.

6. Ashraf, M.A.; Hussain, I.; Rasheed, R.; Iqbal, M.; Riaz, M.; Arif, M.S. Advances in microbe-assisted reclamation of heavy metal contaminated soils over the last decade: A review. *J. Environ. Manag.* **2017**, *198*, 132–143. [CrossRef]
7. Ernst, W.H.O. Phytoextraction of mine wastes—Options and impossibilities. *Chem. Erde-Geochem.* **2005**, *65*, 29–42. [CrossRef]
8. Yan, D.Y.S.; Lo, I.M.C. Phyrophosphate coupling with chelant-enhanced soil flushing of field contaminated soils for heavy metal extraction. *J. Hazard. Mater.* **2012**, *199*, 51–57. [CrossRef] [PubMed]
9. Chen, Y.; Li, X.D.; Shen, Z.G. Leaching and uptake of heavy metals by ten different species of plants during an EDTA-assisted phytoextraction process. *Chemosphere* **2004**, *57*, 187–196. [CrossRef] [PubMed]
10. Nagajyoti, P.C.; Lee, K.D.; Sreekanth, T.V.M. Heavy metals, occurrence and toxicity for plants: A review. *Environ. Chem. Lett.* **2010**, *8*, 199–216. [CrossRef]
11. Misra, S.G.; Mani, D. *Soil Pollution*; Ashish Publishing House: Punjabi Bagh, India, 1991.
12. Taj, Z.Z.; Rajkumar, M. Perspectives of plant growth-promoting actinomycetes in heavy metal phytoremediation. In *Plant Growth Promoting Actinobacteria*; Springer: Singapore, 2016; pp. 213–231.
13. Mahar, A.; Wang, P.; Ali, A.; Awasthi, M.K.; Lahori, A.H.; Wang, Q.; Li, R.; Zhang, Z. Challenges and opportunities in the phytoremediation of heavy metals contaminated soils: A review. *Ecotoxicol. Environ. Saf.* **2016**, *126*, 111–121. [CrossRef]
14. Alamgir, M. The effect of soil properties to the extent of soil contamination with metals. In *Environmental Remediation Technologies for Metal-Contaminated Soils*; Springer: Cham, Switzerland, 2016; pp. 1–19.
15. Thakur, S.; Singh, L.; Wahid, Z.A.; Siddiqui, M.F.; Atnaw, S.M.; Din, M.F.M. Plant-driven removal of heavy metals from soil: Uptake, translocation, tolerance mechanism, challenges, and future perspectives. *Environ. Monit. Assess.* **2016**, *188*, 206. [CrossRef] [PubMed]
16. Tangahu, B.V.; Sheikh, A.S.R.; Basri, H.; Idris, M.; Anuar, N.; Mukhlisin, M. A review on heavy metals (As, Pb, and Hg) uptake by plants through phytoremediation. *Int. J. Chem. Eng.* **2011**. [CrossRef]
17. Zubair, M.; Shakir, M.; Ali, Q.; Rani, N.; Fatima, N.; Farooq, S.; Shafiq, S.; Kanwal, N.; Ali, F.; Nasir, I.A. Rhizobacteria and phytoremediation of heavy metals. *Environ. Technol. Rev.* **2016**, *5*, 112–119. [CrossRef]
18. Jegatheesan, V.; Ravishankar, H.; Shu, L.; Wang, J. Application of Green and Physico-Chemical Technologies in Treating Water Polluted by Heavy Metals. In *Green Technologies for Sustainable Water Management*; American Society of Civil Engineers: Reston, VA, USA, 2016; pp. 579–614.
19. Da Conceição Gomes, M.A.; Hauser-Davis, R.A.; de Souza, A.N.; Vitoria, A.P. Metal phytoremediation: General strategies, genetically modified plants and applications in metal nanoparticle contamination. *Ecotoxicol. Environ. Saf.* **2016**, *134*, 133–147. [CrossRef] [PubMed]
20. Kang, C.H.; Kwon, Y.J.; So, J.S. Bioremediation of heavy metals by using bacterial mixtures. *Ecol. Eng.* **2016**, *89*, 64–69. [CrossRef]
21. Mani, D.; Kumar, C. Biotechnological advances in bioremediation of heavy metals contaminated ecosystems: An overview with special reference to phytoremediation. *Int. J. Environ. Sci. Technol.* **2014**, *11*, 843–872. [CrossRef]
22. Chibuike, G.U.; Obiora, S.C. Heavy metal polluted soils: Effect on plants and bioremediation methods. *Appl. Environ. Soil Sci.* **2014**. [CrossRef]
23. Kumar, A.; Bisht, B.S.; Joshi, V.D.; Dhewa, T. Review on bioremediation of polluted environment: A management tool. *Int. J. Environ. Sci.* **2011**, *1*, 1079–1093.
24. Ullah, A.; Heng, S.; Munis, M.F.H.; Fahad, S.; Yang, X. Phytoremediation of heavy metals assisted by plant growth promoting (PGP) bacteria: A review. *Environ. Exp. Bot.* **2015**, *117*, 28–40. [CrossRef]
25. Chang, P.; Gerhardt, K.E.; Huang, X.D.; Yu, X.M.; Glick, B.R.; Gerwing, P.D.; Greenberg, B.M. Plant growth-promoting bacteria facilitate the growth of barley and oats in salt-impacted soil: Implications for phytoremediation of saline soils. *Int. J. Phytoremediat.* **2014**, *16*, 1133–1147. [CrossRef]
26. Chaturvedi, R.; Varun, M.; Paul, M.S. Phytoremediation: Uptake and role of metal transporters in some members of Brassicaceae. In *Phytoremediation*; Springer International Publishing: Cham, Switzerland, 2016; pp. 453–468.
27. Mench, M.; Schwitzguebel, J.P.; Schroeder, P.; Bert, V.; Gawronski, S.; Gupta, S. Assessment of successful experiments and limitations of phyto technologies: Contaminant uptake, detoxification and sequestration, and consequences for food safety. *Environ. Sci. Pollut. Res.* **2009**, *16*, 876–900. [CrossRef] [PubMed]

28. Munir, I.; Faisal, M. Plant growth-promoting bacteria: A good source for phytoremediation of metal-contaminated soil. In *Phytoremediation*; Springer: Cham, Switzerland, 2016; pp. 119–129.
29. Antonkiewicz, J.; Kołodziej, B.; Bielińska, E.J.; Witkowicz, R.; Tabor, S. Using Jerusalem Artichoke to Extract Heavy Metals from Municipal Sewage Sludge Amended Soil. *Pol. J. Environ. Stud.* **2018**, *27*, 513–527. [CrossRef]
30. Jaffre, T.; Pillon, Y.; Thomine, S.; Merlot, S. The metal hyperaccumulators from New Caledonia can broaden our understanding of nickel accumulation in plants. *Front. Plant Sci.* **2013**, *4*, 279. [CrossRef]
31. Glick, B. Plant Growth-promoting Bacteria: Mechanisms and Applications. *Scientifica* **2012**. [CrossRef]
32. Assunçao, A.; Schat, H.; Aarts, M. *Thlaspi caerulescens*, an attractive model species to study heavy metal hyperaccumulation in plants. *New Phytol.* **2003**, *159*, 351–360. [CrossRef]
33. He, C.; Tan, G.; Liang, X.; Du, W.; Chen, Y.; Zhi, G. Effect of Zn-tolerant bacterial strains on growth and Zn accumulation in Orychophragmus violaceus. *Appl. Soil. Ecol.* **2010**, *44*, 1–5. [CrossRef]
34. Pierattini, E.C.; Francini, A.; Raffaelli, A.; Sebastiani, L. Surfactant and heavy metal interaction in poplar: A focus on SDS and Zn uptake. *Tree Physiol.* **2017**, *38*, 109–118. [CrossRef]
35. Ma, Y.; Rajkumar, M.; Zhang, C.; Freitas, H. Beneficial role of bacterial endophytes in heavy metal phytoremediation. *J. Environ. Manag.* **2016**, *174*, 14–25. [CrossRef] [PubMed]
36. Ribiero de Souza, S.C.; Andrade, S.A.L.; Souza, L.A.; Schiavinato, M.A. Lead tolerance and phytoremediation potential of Brazilian leguminous tree species at the seedling stage. *J. Environ. Manag.* **2012**, *110*, 299–307. [CrossRef] [PubMed]
37. Lievens, C.; Yperman, J.; Vangronsveld, J.; Carleer, R. Study of the potential valorisation of heavy metal contaminated biomass via phytoremediation by fast pyrolysis: Part I. Influence of temperature, biomass species and solid heat carrier on the behaviour of heavy metals. *Fuel* **2008**, *87*, 1894–1905. [CrossRef]
38. Prasad, M.N.V.; Freitas, H.M.O. Metal hyper accumulation in plants—Biodiversity prospecting for phytoremediation technology. *Electron. J. Biotechnol.* **2003**, *6*, 285–321. [CrossRef]
39. Koptsik, G.N. Problems and Prospects Concerning the Phytoremediation of Heavy Metal Polluted Soils: A Review. *Eurasian Soil Sci.* **2014**, *47*, 923–939. [CrossRef]
40. Mignardi, S.; Corami, A.; Ferrini, V. Evaluation of the effectiveness of phosphate treatment for the remediation of mine waste soils contaminated with Cd, Cu, Pb, and Zn. *Chemosphere* **2012**, *86*, 354–360. [CrossRef]
41. Arshad, M.; Silvestre, J.; Merlina, G.; Dumat, C.; Pinelli, E.; Kallerhoff, J. Thidiazuron induced shoot organogenesis from mature leaf explants of scented *Pelargonium capitatum* cultivars. *Plant Cell Tissue Organ Cult.* **2012**, *108*, 315–322. [CrossRef]
42. Evangelou, M.W.H.; Ebel, M.; Schaeffer, A. Chelate assisted phytoextraction of heavy metals from soil: Effect, mechanism, toxicity, and fate of chelating agents. *Chemosphere* **2007**, *68*, 989–1003. [CrossRef]
43. Souza, L.A.; Piotto, F.A.; Nogueiro, R.C.; Azevedo, R.A. Use of Non-Hyperaccumulator Plant Species for the Phytoextraction of Heavy Metals Using Chelating Agents. *Sci. Agric.* **2013**, *70*, 290–295. [CrossRef]
44. Iqbal, M.; Bakht, J.; Shafi, M.; Ullah, R. Effect of heavy metal and EDTA application on heavy metal uptake and gene expression in different Brassica species. *Afr. J. Biotechnol.* **2012**, *11*, 7649–7658.
45. Dipu, S.; Kumar, A.A.; Thanga, S.G. Effect of chelating agents in phytoremediation of heavy metals. *Remediat. J.* **2012**, *22*, 133–146. [CrossRef]
46. Shahid, M.; Dumat, C.; Pourrut, B.; Silvestre, J.; Laplanche, C.; Pinelli, E. Influence of EDTA and citric acid on lead-induced oxidative stress to *Vicia faba* roots. *J. Soils Sediments* **2014**, *14*, 835–843. [CrossRef]
47. Awokunmi, E.E.; Asaolu, S.S.; Ajayi, O.O.; Adebayo, O.A. The role of EDTA on heavy metals phytoextraction by *Jatropha gossypifolia* grown on soil collected from dumpsites in Ekiti state Nigeria. *Br. J. Environ. Clim. Chang.* **2012**, *2*, 153–162. [CrossRef]
48. Zhao, Z.; Xi, M.; Jiang, G.; Liu, X.; Bai, Z.; Huang, Y. Effects of IDSA, EDDS and EDTA on heavy metals accumulation in hydroponically grown maize (*Zea mays*, L.). *J. Hazard. Mater.* **2010**, *181*, 455–459. [CrossRef] [PubMed]
49. Sun, Y.; Zhou, Q.; Wang, L.; Liu, W. The influence of different growth stages and dosage of EDTA on Cd uptake and accumulation in Cd-hyperaccumulator (*Solanum nigrum* L.). *Bull. Environ. Contam. Toxicol.* **2009**, *82*, 348–353. [CrossRef] [PubMed]
50. Shahid, A.; Austruy, G.; Echevarria, M.; Arshad, M.; Sanaullah, M.; Aslam, M.; Nadeem, M.; Nasim, W.; Dumat, C. EDTA-Enhanced Phytoremediation of Heavy Metals: A Review. *Soil Sediment Contam. Int. J.* **2014**, *23*, 389–416. [CrossRef]

51. Andra, S.S.; Datta, R.; Sarkar, D.; Saminathan, S.K.M.; Mullens, C.P.; Bach, S.B.H. Analysis of phytochelatin complexes in the lead tolerant vetiver grass *Vetiveria zizanioides* (L.) using liquid chromatography and mass spectrometry. *Environ. Pollut.* **2009**, *157*, 2173–2183. [CrossRef]
52. Gabos, M.B.; Abreu, C.A.D.; Coscione, A.R. EDTA assisted phytoremediation of a pb contaminated soil: Metal leaching and uptake by jack beans. *Sci. Agric.* **2009**, *66*, 506–514. [CrossRef]
53. Khan, S.; Hesham, A.E.L.; Qiao, M.; Rehman, S.; He, J.Z. Effects of Cd and Pb on soil microbial community structure and activities. *Environ. Sci. Pollut. Res.* **2010**, *17*, 288–296. [CrossRef]
54. Chen, Y.; Cutright, T. EDTA and HEDTA Effects on Cd, Cr, and Ni Uptake by Helianthus annuus. *Chemosphere* **2001**, *45*, 21–28. [CrossRef]
55. Jianv, L.; Qixing, Z.; Song, W. Evaluation of Chemical Enhancement on Phytoremediation Effect of Cd-Contaminated Soils with *Calendula officinalis* L. *Int. J. Phytoremediat.* **2010**, *12*, 503–515.
56. Malarkodi, M.; Krishnasamy, R.; Chitdeshwari, T. Phytoextraction of Nickel Contaminated Soil Using Castor Phytoextractor. *J. Plant Nutr.* **2008**, *31*, 219–229. [CrossRef]
57. Wei, J.; Lai, H.; Chen, Z. Chelator effects on bioconcentration and translocation of cadmium by hyperaccumulators, *Tagetes patula* and *Impatiens walleriana*. *Ecotoxicol. Environ. Saf.* **2012**, *84*, 173–178. [CrossRef] [PubMed]
58. Chigbo, C.; Batty, L. Effect of EDTA and citric acid on phytoremediation of Cr-B[a]P-co-contaminated soil. *Environ. Sci. Pollut. Res.* **2013**, *20*, 8955–8963. [CrossRef] [PubMed]
59. Markovskaa, Y.; Genevab, M.; Petrovb, P.; Boychinovab, M.; Lazarovac, I.; Todorovb, I.; Stancheva, I. EDTA Reduces Heavy Metal Impacts on *Tribulus terrestris* Photosynthesis and Antioxidants. *Russ. J. Plant Physiol.* **2013**, *60*, 623–632. [CrossRef]
60. Lai, H.-Y.; Chen, Z.S. Effects of EDTA on solubility of cadmium, zinc, and lead and their uptake by rainbow pink and vetiver grass. *Chemosphere* **2004**, *55*, 421–430. [CrossRef] [PubMed]
61. Vassil, A.D.; Kapulnik, Y.; Raskin, I.; Salt, D.E. The Role of EDTA in Lead Transport and Accumulation by Indian Mustard. *Plant Physiol.* **1998**, *117*, 447–453. [CrossRef] [PubMed]
62. Luo, C.; Shen, Z.; Li, X. Enhanced phytoextraction of Cu, Pb, Zn, and Cd with EDTA and EDDS. *Chemosphere* **2005**, *59*, 1–11. [CrossRef]
63. Cui, Y.; Wang, Q.; Dong, Y.; Li, H.; Christie, P. Enhanced uptake of soil Pb and Zn by Indian mustard and winter wheat following combined soil application of elemental sulphur and EDTA. *Plant Soil* **2004**, *261*, 181–188. [CrossRef]
64. Wang, S.; Liu, J. The effectiveness and risk comparison of EDTA with EGTA in enhancing Cd phytoextraction by *Mirabilis jalapa* L. *Environ. Monit. Assess.* **2014**, *186*, 751–759. [CrossRef] [PubMed]
65. Liu, J.N.; Zhou, Q.X.; Wang, S.; Sun, T. Cadmium tolerance and accumulation of *Althaea rosea* Cav. And its potential as a hyperaccumulation under chemical enhancement. *Environ. Monit. Assess.* **2009**, *149*, 419–427. [CrossRef]
66. Sakouhi, L.; Rahoui, S.; Massoud, M.B.; Munemasa, S.; Ferjani, E.F.; Murata, Y.; Chaou, A. Calcium and EGTA Alleviate Cadmium Toxicity in Germinating Chickpea Seeds. *J. Plant Growth Regul.* **2016**, *35*, 1064–1073. [CrossRef]
67. Miller, G.; Begonia, G.; Begonia, M.; Ntoni, J.; Hundley, O. Assessment of the Efficacy of Chelate-Assisted Phytoextraction of Lead by Coffeeweed (*Sesbania exaltata* Raf.). *Int. J. Environ. Res. Public Health* **2008**, *5*, 428–435. [CrossRef] [PubMed]
68. Liu, J.-N.; Zhou, Q.; Sun, T.; Ma, L.Q.; Wang, S. Identification of and Chemical Enhancement of Two Ornamental Plants for Phytoremediation. *Bull. Environ. Contam. Toxicol.* **2008**, *80*, 260–265. [CrossRef] [PubMed]
69. Almeida, C.M.R.; Mucha, A.P.; Delgado, M.F.C.; Caçador, M.I.; Bordalo, A.A.; Vasconcelos, M.T.S.D. Can PAHs influence Cu accumulation by salt marsh plants. *Mar. Environ. Res.* **2008**, *66*, 311–318. [CrossRef]
70. Saifullah, M.E.; Qadir, M.; de Caritat, P.; Tack, F.M.G.; Du Laing, G.; Zia, M.H. EDTA-assisted Pbphytoextraction. *Chemosphere* **2009**, *74*, 1279–1291. [CrossRef]
71. Degryse, F.; Smolders, E.; Merckx, R. Labile Cd complexes increase Cd availability to plants. *Environ. Sci. Technol.* **2006**, *40*, 830–836. [CrossRef]
72. Seregin, I.V.; Ivanov, V.B. Physiological aspects of cadmium and lead toxic effects on higher plants. *Russ. J. Plant Physiol.* **2001**, *48*, 523–544. [CrossRef]

73. Pourrut, B.; Shahid, M.; Dumat, C.; Winterton, P.; Pinelli, E. Lead uptake, toxicity and detoxification in plants. *Rev. Environ. Contam. Toxicol.* **2011**, *213*, 113–131. [PubMed]
74. Wu, L.; Li, Z.; Akahane, I.; Liu, L.; Han, C.; Makino, T.; Luo, Y.; Christie, P. Effects of organic amendments on Cd, Zn and Cu bioavailability in soil with repeated phytoremediation by *Sedum plumbizincicola*. *Int. J. Phytoremediat.* **2012**, *14*, 1024–1038. [CrossRef]
75. Sarret, G.; Vangronsveld, J.; Manceau, A.; Musso, M.; D'Haen, J.; Menthonnex, J.J.; Hazemann, J.L. Accumulation forms of Zn and Pb in Phaseolus vulgaris in the presence and absence of EDTA. *Environ. Sci. Technol.* **2001**, *35*, 2854–2859. [CrossRef]
76. Schaider, L.; Parker, D.; Sedlak, D. Uptake of EDTA-complexed Pb, Cd and Fe by solution-and sand-cultured Brassica juncea. *Plant Soil* **2006**, *286*, 377–391. [CrossRef]
77. Pereira, B.F.F.; de Abreu, C.A.; Herpin, U.; de Abreu, M.F.; Berton, R.S. Phytoremediation of lead by jack beans on a RhodicHapludox amended with EDTA. *Sci. Agric.* **2010**, *67*, 308–318. [CrossRef]
78. Messina, C.M.; Faggio, C.; Laudicella, V.A.; Sanfilippo, M.; Trischitta, F.; Santulli, A. Effect of sodium dodecyl sulfate (SDS) on stress response in the Mediterranean mussel (*Mytilus galloprovincialis*): Regulatory volume decrease (Rvd) and modulation of biochemical markers related to oxidative stress. *Aquat. Toxicol.* **2014**, *157*, 94–100. [CrossRef] [PubMed]
79. Ramamurthy, A.; Schalchian, H. Surfactant assisted removal of Cu(II), Cd(II) and Pb(II) from contaminated soils. *Environ. Prot. Eng.* **2013**, *39*, 87–99.
80. Wang, J.; Wang, J.; Xu, C.; Liu, R.; Chen, Y. Molecular mechanism of catalase activity change under sodium dodecyl sulfate-induced oxidative stress in the mouse primary hepatocytes. *J. Hazard. Mater.* **2016**, *307*, 173–183. [CrossRef] [PubMed]
81. Chaturvedi, V.; Kumar, A. Isolation of a strain of *Pseudomonas putida* capable of metabolizing anionic detergent sodium dodecyl sulfate (SDS). *Iran. J. Microbiol.* **2011**, *3*, 47–53.
82. Paulo, A.M.S.; Plugge, C.M.; García-Encina, P.A.; Stams, A.J.M. Anaerobic degradation of sodium dodecyl sulfate (SDS) by denitrifying bacteria. *Int. Biodeterior. Biodegrad.* **2013**, *84*, 14–20. [CrossRef]
83. Nguyen, H.M.; Phan, C.M.; Sen, T. Degradation of sodium dodecyl sulfate by photoelectrochemical and electrochemical processes. *Chem. Eng. J.* **2016**, *287*, 633–639. [CrossRef]
84. Lambrechts, T.; Gustot, Q.; Couder, E.; Houben, D.; Iserentant, A.; Lutts, S. Comparison of EDTA-enhanced phytoextraction and phytostabilisation strategies with Loliumperenne on a heavy metal contaminated soil. *Chemosphere* **2011**, *85*, 1290–1298. [CrossRef] [PubMed]
85. Muhlbachov, A.G. Soil microbial activities and heavy metal mobility in long-term contaminated soils after addition of EDTA and EDDS. *Ecol. Eng.* **2011**, *37*, 1064–1071. [CrossRef]
86. Neugschwandtner, R.W.; Tlustos, P.; Komarek, M.; Szakov, J.; Jakoubkova, L. Chemically enhanced phytoextraction of risk elements from a contaminated agricultural soil using Zea mays and Triticumaestivum: Performance and metal mobilization over a three year period. *Int. J. Phytoremediat.* **2012**, *14*, 754–771. [CrossRef]
87. Goel, S.; Gautam, A. Effect of chelating agents on mobilization of metal from waste catalyst. *Hydrometallurgy* **2010**, *101*, 120–125. [CrossRef]
88. Meers, E.; Ruttens, A.; Hopgood, M.; Lesage, E.; Tack, F.M.G. Potential of Brassicrapa, Cannabis sativa, Helianthus annuus and Zea mays for phytoextraction of heavy metals from calcareous dredged sediment derived soils. *Chemosphere* **2005**, *61*, 561–572. [CrossRef] [PubMed]
89. Grcman, H.; Velikonja-Bolta, S.; Vodnik, D.; Kos, B.; Lestan, D. EDTA enhanced heavy metal phytoextraction: Metal accumulation, leaching and toxicity. *Plant Soil* **2001**, *235*, 105–114. [CrossRef]
90. Luo, C.; Shen, Z.; Li, X.; Baker, A.J.M. The role of root damage in the chelate-enhanced accumulation of lead by Indian mustard plants. *Int. J. Phytoremdiat.* **2006**, *8*, 323–337. [CrossRef]
91. Krujatz, F. Assessing the toxic effects of nickel, cadmium and EDTA on growth of the plant growth-promoting rhizobacterium Pseudomonas brassicacearum. *Water Air Soil Pollut.* **2012**, *223*, 1281–1293. [CrossRef]
92. Ruley, A.T.; Sharma, N.C.; Sahi, S.V. Antioxidant defense in a lead accumulating plant, Sesbaniadrummondii. *Plant Physiol. Biochem.* **2004**, *42*, 899–906. [CrossRef]
93. Hadi, F.; Bano, A.; Fuller, M.P. The improved phytoextraction of lead (Pb) and the growth of maize (*Zea mays* L.): The role of plant growth regulators (GA3 and IAA) and EDTA alone and in combinations. *Chemosphere* **2010**, *80*, 457–462. [CrossRef]

94. Ruley, A.T.; Sharma, N.C.; Sahi, S.V.; Singh, S.R.; Sajwan, K.S. Effects of lead and chelators on growth, photosynthetic activity and Pb uptake in Sesbaniadrummondii grown in soil. *Environ. Pollut.* **2001**, *144*, 11–18. [CrossRef] [PubMed]
95. Xu, W.; Li, W.; He, J.; Balwant, S.; Xiong, Z. Effects of insoluble Zn, Cd, and EDTA on the growth, activities of antioxidant enzymes and uptake of Zn and Cd in Vetiveriazizanioides. *J. Environ. Sci.* **2010**, *21*, 186–192. [CrossRef]
96. Muhammad, D.; Chen, F.; Zhao, J.; Zhang, G.; Wu, F. Comparison of EDTA- and citric acid-enhanced phytoextraction of heavy metals in artificially metal contaminated soil by Typhaangustifolia. *Int. J. Phytoremediat.* **2009**, *11*, 558–574. [CrossRef]
97. Sanaullah, M.; Blagodatskaya, E.; Chabbi, A.; Rumpel, C.; Kuzyakov, Y. Drought effects on microbial biomass and enzyme activities in the rhizosphere of grasses depend on plant community composition. *Appl. Soil Ecol.* **2011**, *48*, 38–44. [CrossRef]
98. Saifullah, G.A.; Zia, M.H.; Murtaza, G.; Waraich, E.A.; Bibi, S.; Srivastava, P. Comparison of organic and inorganic amendments for enhancing soil lead phytoextraction by wheat (*Triticum aestivum* L.). *Int. J. Phytoremediat.* **2010**, *12*, 633–649. [CrossRef] [PubMed]
99. Jean-Soro, L.; Bordas, F.; Bollinger, J.C. Column leaching of chromium and nickel from a contaminated soil using EDTA and citric acid. *Environ. Pollut.* **2012**, *164*, 175–181. [CrossRef] [PubMed]

plants

Review

Jute: A Potential Candidate for Phytoremediation of Metals—A Review

Muhammad Hamzah Saleem [1], Shafaqat Ali [2,3,*], Muzammal Rehman [4], Mirza Hasanuzzaman [5], Muhammad Rizwan [2], Sana Irshad [6], Fahad Shafiq [7,8], Muhammad Iqbal [7], Basmah M. Alharbi [9], Taghreed S. Alnusaire [10] and Sameer H. Qari [11]

1 MOA Key Laboratory of Crop Ecophysiology and Farming System in the Middle Reaches of the Yangtze River, College of Plant Science and Technology, Huazhong Agricultural University, Wuhan 430070, China; saleemhamza312@webmail.hzau.edu.cn
2 Department of Environmental Sciences and Engineering, Government College University Allama Iqbal Road, 38000 Faisalabad, Pakistan; mrazi1532@yahoo.com
3 Department of Biological Sciences and Technology, China Medical University (CMU), Taichung City 40402, Taiwan
4 School of Agriculture, Yunnan University, Kunming 650504, China; muzammal@ynu.edu.cn
5 Department of Agronomy Sher-e-Bangla Agricultural University Dhaka 1207, Bangladesh; mhzsauag@yahoo.com
6 School of Environmental Studies, China university of Geosciences, Wuhan 430070, China; sanairshad55@gmail.com
7 Department of Botany, Government College University Allama Iqbal Road, 38000 Faisalabad, Pakistan; fahadsheikh1800@gmail.com (F.S.); iqbaluaf@yahoo.com (M.I.)
8 Institute of Molecular Biology and Biotechnology (IMBB), University of Lahore, Lahore 54590, Pakistan
9 Department of Biology, Faculty of Science, University of Tabuk, Tabuk-71491, Saudi Arabia; b.alharbi@ut.edu.sa
10 Biology Department, College of Science, Jouf University, Sakaka, 2014, Saudi Arabia; taghreed0804@hotmail.com
11 Biology Department, Aljumum, University College, Umm Al-Qura University, Mecca 21955, Saudi Arabia; sqarinet@gmail.com
* Correspondence: shafaqataligill@yahoo.com or shafaqataligill@gcuf.edu.pk

Received: 11 January 2020; Accepted: 6 February 2020; Published: 17 February 2020

Abstract: Jute (*Corchorus capsularis*) is a widely cultivated fibrous species with important physiological characteristics including biomass, a deep rooting system, and tolerance to metal stress. Furthermore, *Corchorus* species are indigenous leafy vegetables and show phytoremediation potential for different heavy metals. This species has been used for the phytoremediation of different toxic pollutants such as copper (Cu), cadmium (Cd), zinc (Zn), mercury (Hg) and lead (Pb). The current literature highlights the physiological and morphological characteristics of jute that are useful to achieve successful phytoremediation of different pollutants. The accumulation of these toxic heavy metals in agricultural regions initiates concerns regarding food safety and reductions in plant productivity and crop yield. We discuss some innovative approaches to increase jute phytoremediation using different chelating agents. There is a need to remediate soils contaminated with toxic substances, and phytoremediation is a cheap, effective, and in situ alternative, and jute can be used for this purpose.

Keywords: fibrous crop; phytoextraction; environmental pollutants; morphological traits; soil remediation; chelating agents

1. Introduction

Heavy metals are elements with a high density, high relative atomic mass (their atomic mass number is greater than 20), and metallic properties such as conductivity and cation stability [1–5]. Examples of heavy metals are Cu, Ni, Zn, Cd, Pb, and many others which can cause toxicity in plants. Furthermore, some heavy metals such as Cu, Fe, Mo, and Zn are also micronutrients, i.e., required by plants in minute quantities, but they can also cause toxicity when their levels go beyond the permissible limits. In contrast, some other heavy metals such as Cd, Pb, and Hg are not essential for plants and cause toxicity even in low quantities [6–13]. Cultivating plant species in highly contaminated soil can cause morphological and physiological alterations in the plants which ultimately decrease crop yield and productivity. Furthermore, these elements also play an integral role in many physiological and biochemical processes such as the oxidation of many compounds, DNA synthesis, formation of many important carbohydrates and proteins, and cell wall metabolism [3,6,14–16]. In contrast, high contents of these elements in the soil are toxic for plants. Previously, in many studies, it was noticed that high concentrations of metals in the soil affected crop yield and plant productivity [2,4,14,17–20]. Studies on the phytotoxicity of heavy metals show that plant physiological process such as photosynthesis and respiration are also affected in higher plants. [2,21–24]. In many studies, heavy metals such as Cd and Pb are not essential for plant growth and are considered the "main threats" to the plants [25–27]. A report by Chen et al. [23] revealed that heavy metal pollution in the soil has become a critical issue worldwide, and about 16% of the soil in China is contaminated by different heavy metals. Moreover, heavy metals originate within the Earth's crust; hence the occurrence of heavy metals in soil is simply a product of weathering process [2,3,18]. Accumulation of these toxic pollutants in the soil is dangerous to plant tissues and can alter many important biological and physiological processes in plants. In addition, these toxic pollutants are non-biodegradable in the soil, which also makes them very difficult to remove [6,17,28]. Heavy metal toxicity in plant tissues depends upon the plant species. For instance, metal-tolerant species, also known as hyperaccumulators (able to accumulate a large amount of heavy metals in their aboveground tissues) [29], have barrier mechanisms against the toxicity caused by the heavy metal-initiated pressure, but the duration and magnitude of exposure and other natural conditions add to the effect of heavy metals [3,4]. Some of the heavy metals such as Cd, Ni, Zn, As, Se, and Cu are abundantly available in agricultural land, while Co, Mn, and Fe are moderately bio-available, and Pb, Cr, and Ur are not readily available in the soil [30–32]. Hence, it is necessary to measure, understand, and control toxic heavy metal contamination in the soil.

For this purpose, there are several techniques for removing heavy metals from metal-contaminated soil such as soil washing, thermal desorption, incineration, stabilization, and soil flushing [33,34]. However, these techniques have many disadvantages such as cost, they require 24 h monitoring, and they are not an efficient method to remove toxic heavy metals and other contaminants from agricultural land [31,35]. Phytoremediation is the direct use of living green plants and is an effective, cheap, non-invasive, and environmentally friendly technique used to transfer or stabilize all the toxic metals and environmental pollutants in polluted soil or ground water [33,36]. Phytoremediation (*phyto* = plant, *remediation* = correct evil) means re-vegetation of land which is spoiled by toxic substances and phytoremediation might be successful when plant using for phytoremediation material can accumulate high concentration of heavy metals in their shoots parts [2,37–39]. Furthermore, phytoremediation is concerned with the potential of a plant species to accumulate high concentrations of toxic pollutants in their tissues. A number of plant metabolic processes come into play to degrade various organic compounds. By contrast, inorganic pollutants cannot be degraded in these processes easily. Hence, these inorganic pollutants should be less bio-available for the plants and less easily transported in different parts of the plant tissues; there is also a need to reduce the volatile forms of inorganic pollutants [2,40–46]. Hence, phytoremediation is the best technique to deal with low amounts of organic or inorganic contaminants in the soil and can be applied with other traditional soil remediation methods for efficient removal of toxic pollutants from the soil [33,34,47]. There are many types of phytoremediation for agricultural land and water bodies, e.g., phytotransformation, rhizosphere bioremediation, phytostabilization, phytoextraction (phyto-accumulation), rhizofiltration, phytovolatization, phytodegradtion, and hydraulic control. The most common type of phytoremediation is phytoextraction in which green plants are cultivated in metal-contaminated soil and accumulate large amounts of metals in their above ground parts; these are hyperaccumulator species [31,48,49]. Previously, many plant species have been used as hyperaccumulators for a range of heavy metals in the soil [23,50–55]. A report presented by Muszynska and Hanus-Fajerska [31] mentioned that around 500 different plant species are known as hyperaccumulators and can accumulate a large amount of toxic pollutants in their aboveground parts without any toxic effects.

Recently, jute has been used in phytoextraction in different studies related to heavy metal toxicity [47,56–59]. Jute is more tolerant to heavy metal stress among different fibrous crop probably due to its distinct biological and physiological activities [27,56,60]. Jute is a vegetative fibrous crop (also called allyott and golden fibre) and is a herbaceous annual plant belonging to the family Malvaceae; there are two different types: *Corchorus capsularis* known as white jute (a variety that originates from poor villagers in India) and *Corchorus olitorius,* known as Tosaa jute (a variety that originates from native South Asia) [60–63]. Furthermore, it is a low-cost bast fibre crop, and its fibre is commonly used for bags, sacks, packs, and wrappings. Jute is a rain-fed crop similar to rice and needs a very small amount of fertilizer and due to its versatility; jute's fibre is the second most important fibre source after cotton [61,63,64]. Jute originates from the Mediterranean, but many countries in Asia such as India, China, and Bangladesh and even many countries of South America are exporting premium jute. These countries export 2300×10^3 to 2850×10^3 tons of jute fibre. In contrast to the fibre, jute is widely consumed as a vegetable in almost all countries of Africa, and its leaves are used as a soup and are also cultivated for fibre production [44,61,63]. China imports a large amount of jute fibre and its products from Bangladesh, India, and other countries [65]. Jute is the cheapest fibre for mass consumption, mainly glass fibre, while in terms of volume, it is the most important bast fibre after cotton. In the modern era, jute fibres have replaced synthetic fibre in the composition of different households such as carpets, ropes, sacks, etc. To ensure a sensible return to the farmers, a lot of new plant species have been explored, and one of such avenue is the area of fibre -reinforced composites [60–63,65].

Although several excellent investigations have been done on jute [47,55–57,60,66,67] cultivated in metal-contaminated soil, there is no comprehensive review on the phytoremediation potential of jute plants grown in metal-contaminated soil. In this review article, the role of different heavy metals in the morpho-physiological and phytoremediation potential of jute plant is discussed, and some practical options are presented on how jute plays crucial role in the phytoextraction of different heavy metals.

2. Habitat, Growth, and Morphology of Jute

Jute is cultivated in hot and moist tropical regions of almost all continents, especially Asia and Africa. Commercial jute cultivation is mainly restricted between 80°18′ E–92° E and 21°24′ N–26°30′ N on the Indo-Bangladesh sub-continent. Other major jute- and kenaf-growing countries are China, Thailand, Myanmar, Indonesia, Brazil, and Nepal. Africa and the Indo-Myanmar region are the primary and secondary centers of origin, respectively, for jute, while the Indo-Myanmar region including South China is considered the center of origin of jute [68–70]. Some major exporters of jute fibres along their annual yields per tons are presented in Figure 1. African jute is also cultivated as a green leafy vegetable for edible purposes [71]. Delta Ganga (India) is considered the best region for jute growth because this region has alluvial soil types and a sufficient amount of rainfall. India was the greatest exporter of jute worldwide but in 1947 India suffered a great setback because most of the area of jute went to Bangladesh (East Pakistan), but fortunately, most of jute industries remained in India. India has been the greatest producer of jute in the last 37 years; production increased from 4.1 million bales to 233 10 million bales by 1998. After 1998, there was an irregular increase in jute production. India is the leading producer of jute, with 1,968,000 tons /year followed by Bangladesh (1,349,000), and China (29,628 tons /year) [62,72].

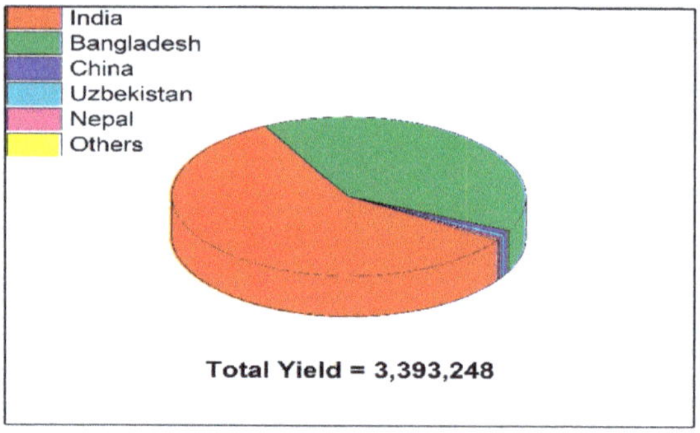

Figure 1. Some major exporters of jute fibres and their annual yields per ton.

Jute is a C3 plant and is a long, soft, shiny vegetable fibre. Jute is a Kharif crop, and the optimum seed rate for a fibre crop is 4–6 and 6–8 kg/ha, while for seed jute, the rate is 3–4 and 4–5 kg/ha when sown in mid-May to mid-June. Consistent water logging is very dangerous to jute, and the optimum temperature required for the growth of jute is from 24–30 oC with 70–80% humidity and rainfall of 160-200 cm annually. Moreover, a small amount of pre-monsoon rainfall (25–55 cm) is also very useful for jute as it helps in the proper growth of the plant before the arrival of the proper monsoon. However, in the month of sowing, rainfall between 2.5–7.5 cm is sufficient for germinating jute seedlings. Light sandy or clayey loams or new grey alluvial soil is considered best for jute growth. Jute is mostly sown in clay or sandy loams and in river basins. Jute can also be cultivated in red soil which requires a high amount of animal manure. Acidic soil pH is required for the best growth, ranging from 4.8–5.8, and plains or low land is ideal for jute cultivation. The seeds of jute are purgative [62,68]. Jute is a rainy seasonal crop and needs water amounts similar to rice, sown mostly in rainy seasons from March to April depending upon the water availability, with harvesting in August to September depending upon the sowing period. Nitrogen, phosphorus, and potassium are required as the fertilizers for jute growth. The most common sowing methods for jute are line sowing and broadcasting [60,69,73]. Jute should be harvested within 120–150 days immediately after the shedding of the flowers. Early harvest provides good and healthy fibres. Plants with a height of 8–12 ft. are harvested with sickles close to the ground level. In flooded areas, the plants are harvested without roots above the ground level. After harvesting, plants should be left in the ground until all flowers are shed [62,72].

3. Plant Selection Considerations

The adaptation of a plant to metal-contaminated soil for the purpose of phytoremediation has been widely accepted for the cleaning of metal-contaminated soil [3,23,74]. Furthermore, the phytoremediation potential of a plant depends upon fertilizer application and growth conditions [14,24,75,76]. For phytoremediation of heavy metals, a plant species should have fast growing vegetation, be easy to maintain, use a large amount of water through evaporation, and convert toxic pollutants into less toxic products [59,77]. Plants selected for heavy metal remediation should have high biomass (length and weight) with some economic importance and should have active defense systems which help to tolerate the metal stress environment [19,34,78]. Furthermore, phytoremediation plants are selected on the basis of root depth, the nature of pollutants, soil type, and regional climate [79,80]. In the hotter regions of the world, phreatophytes (such as willow and cottonwood) are mostly selected because of their unique properties such as fast growth, a deep root system, and a large number of stomata on their leaves (i.e., high transpiration rate); the plants should be native to the country. Hybrid poplar is selected as a terrestrial species while pondweed and coontail are selected as aquatic species [81,82]. The cleaning capacity of different types of producers, i.e., grass, shrubs, and trees is 3%, 10%, and 20%, respectively. The nature of pollutants is also a key factor in the selection of a plant for phytoremediation [34,80]. Grasses are mostly planted in metal-contaminated soil or soil with organic pollutants in tandem with trees as the primary remediation method. Moreover, grasses provide a tremendous amount of fine roots in the surface soil which is effective at binding and transforming hydrophobic contaminants such as total petroleum hydrocarbons and polynuclear aromatic hydrocarbons [34].

The selection of a plant for phytoremediation is based on plant species, soil type, and climatic conditions. Among different plant species, indigenous plant species are more favorable with greater tolerance to the stress environment, require less maintenance, and are less toxic for human health than non-native or genetically altered species [2,33,34]. Furthermore, there are some plant species with the ability to take up a large amount of metals in plant tissues without noxious effects (hyperaccumulators) [55,56,60]. Hyperaccumulators are plants with the ability to accumulate a large amount of heavy metals in their aboveground parts compared with their belowground parts without any toxic damage caused to their tissues [23,49]. Jute (a widely used fibrous specie) has the ability to accumulate a large amount of heavy metals in its harvestable parts when grown in metal-contaminated soil [57–60]. Among different fibrous crops, jute has higher potential to tolerate metal-contaminated soil possibly due to its specific physiological and biochemical activities [47,55].

4. Studies Related to Phytoremediation Using Jute

Despite the limited scientific knowledge on heavy metal behavior in jute based on the early nineties of the last century, on-site studies were initiated in different countries of Asia, Europe and North America—the reason was the long-lasting tradition in the breeding and growing of jute in these countries and also the search for new roles of this crop both in agriculture and industry at the end of the century. In addition, phytoremediation technology emerged at this time as a new approach to clean the environment. Later, jute plants were used in phytoremediation of different heavy metals [83–85].

Nizam et al. [57] and Uddin et al. [59] used natural contaminated soil from the Mymensingh district in Bangladesh (As-contaminated soil) and Bhalukaupazila, Bangladesh (Pb-contaminated soil), respectively, cultivated with different varieties of jute. Their results depicted that all fibrous crops accumulated considerable amounts of heavy metals, and high concentrations of these toxic pollutants were transported to the above ground parts of the plants compared with the below ground parts. Jute plants can germinate in metal-contaminated soil and are hence considered As and Pb accumulators, exhibiting remediation capability in contaminated soil. The author concluded that jute plant can be considered for phytoremediation technology in metal-contaminated soil.

In our previous study, [47], we also tried to investigate the potential and tolerance mechanism of jute under the controlled condition with different levels of Cu (0, 100, 200, 300, and 400 mg kg^{-1}) which was artificially contaminated $CuSO_4.5H_2O$. The objective of this experiment was to determine the tolerance mechanism of jute and the effect of Cu toxicity on the morphological and physiological behavior of the plant. The results illustrated that *C. capsularis* can tolerate Cu concentrations of up to 300 mg kg^{-1} without significant decreases in growth or biomass, but further increases in Cu concentration (i.e., 400 mg kg^{-1}) led to significant reductions in plant growth and biomass. Furthermore, increasing levels of Cu in the soil caused oxidative damage in the leaves of jute plants which was overcome by the action of antioxidative enzymes. They noticed that, the concentrations of Cu in different parts of the plant (the roots, leaves, stem core, and fibres) were also investigated at four different stages of the life cycle of jute, i.e., 30, 60, 90, and 120 days after sowing (DAS). Cu was highly accumulated in the roots in earlier stages of the growth while transported to the above ground parts in the lateral stage of the growth. To support our results, we also conducted a Petri dish experiment, providing short-term exposure of Cu to investigate Cu-sensitive and Cu-resistant varieties of jute under the same level of Cu (50 µmol L^{-1}) in the medium [55]. A similar pattern was observed: a high concentration of Cu in the medium caused a significant decrease in plant height, plant fresh and dry weights, total chlorophyll content, and seed germination. Similarly, Cu toxicity caused lipid peroxidation which might be a result of hydrogen peroxide and electrolyte leakage. However, reactive oxygen species are toxic for plants, and a plant has a strong antioxidative defense system to overcome the effect of ROS production. It was also noticed that jute has considerable potential to absorb a large amount of Cu from the soil. Hence, in both of our studies, we noticed that jute can tolerate Cu-stress due to a strong antioxidative defense system and can be used as a phytoremediation tool in Cu-contaminated soil.

Vegetative jute has also been examined under untreated industrial wastewater irrigation to assess the effect on the growth measurements as well as analyses of soils, irrigation waters, and plants for heavy metal and nutrient concentrations [60]. The authors noticed that in the harvestable parts of the plants the concentration of different heavy metals such as Pb, Cd, Cr, Cu, Fe, and Zn were higher than the roots of wastewater-irrigated plant. A remarkable reduction in the growth parameters of the plant were observed when irrigated with untreated industrial wastewater. Hence, jute is a hyperaccumulator species for different heavy metals such as Pb, Cr, Cu, Fe, and Zn.

Previously conducted studies also showed that phytoextraction of different heavy metals can be enhanced by using different organic chelators or organic acids. Usually, many of the heavy metals are adsorbed on soil particles, forming soil aggregates that are difficult to be integrated by plants. Thus, organic acids or chelators with a low molecular weight are crucial to alter the chemical activity/bioavailability of heavy metals and improve phytoextraction [19,52]. This method was successfully used in different plant species such as rapeseed [19], maize [86], and sunflower [87]. Limited literature is available on the enhancement of the phytoextraction of different heavy metals using jute plants. Mazen [88] conducted a hydroponic experiment with the exogenous application of salicylic acid (SA) with the same levels (5 $\mu g\ cm^{-3}$) of Cd, Pb, Al, and Cu and harvested all the plants after 6 days. The authors noticed that application of SA is the safer option which significantly increased the uptake of Cd and Pb and cysteine (cyst) and also increased plant growth and biomass of jute seedlings. In another study, jute was cultivated in Pb-contaminated soil with exogenous application of citric acid (CA), and it was illustrated that application of CA enhanced (Pb) uptake and minimized Pb stress in plants [27]. Furthermore, CA is the most common chelating agent exogenously applied in the nutrient solution of Cd-contaminated mixtures [89]. The authors also noticed the similar trend that exogenous application of CA significantly improved plant growth and biomass in jute seedlings and also increased phytoextraction of Cd when grown in Cd contaminated nutrient solution. Based on this information, exogenous application of organic acids or chelators is a useful strategy to reduce environmental risks associated with metal mobilization and an innovative approach for increasing metal accumulation by jute plants and biomass production.

The accumulation of different heavy metals in the roots and shoots of jute along with the dry weight of plants and removal of metals from the soil is shown in Table 1.

Table 1. Accumulation of different heavy metal in roots (mg/kg) and shoots (mg/kg) of jute along with the dry weight (g/plant) and metal removal from the contaminated soil.

Heavy Metal (Type)	Concentration in Roots	Concentration in Shoots	Dry Biomass	Experiment Type	Reference
As	2.1	4.6	28	Pot	[57]
Pb	1.97	16.3	42.5	Pot	[59]
Cu	60	61	0.1	Petri dish	[55]
As	1.2	2.4	-	Pot	[66]
Pb	722	624	8.9	Hydroponic	[27]
Cd	-	167	4.9	Hydroponic	[88]
Pb	-	143	4.1	Hydroponic	[88]
Cu	-	495	7	Hydroponic	[88]
Cd	261	41	1.28	Pot	[60]
Pb	367	370	1.28	Pot	[60]
Cr	563	631	1.28	Pot	[60]
Cu	169	179	1.28	Pot	[60]
Fe	679	682	1.28	Pot	[60]

Table 1. *Cont.*

Heavy Metal (Type)	Concentration in Roots	Concentration in Shoots	Dry Biomass	Experiment Type	Reference
Ni	83	64	1.28	Pot	[60]
Mn	120	102	1.28	Pot	[60]
Zn	148	152	1.28	Pot	[60]
Zn	-	22	-	Pot	[58]
Cu	-	16.3	-	Pot	[58]
Cd	-	1	-	Pot	[58]
Pb	-	5.2	-	Pot	[58]
Pb	17	22	-	Pot	[67]
Cu	20	23	-	Pot	[67]
Fe	292	305	-	Pot	[67]
Cr	42	53	-	Pot	[67]
Cu	260	534	43	Pot	[47]
Cd	163	48	-	Hydroponic	[89]
Mn	0.017	-	-	Field	[90]
Fe	0.218	-	-	Field	[90]
Cu	0.009	-	-	Field	[90]
Zn	0.025	-	-	Field	[90]
Ni	-	0.17	-	Pot	[91]
Zn	-	6.40	-	Pot	[91]
Pb	-	2.35	-	Pot	[91]
Cu	-	5.78	-	Pot	[91]
Cd	-	0.09	-	Pot	[91]
As	-	0.04	-	Pot	[91]

5. Advantages of Jute as a Phytoremediation Candidate

Jute is mostly cultivated for its fibre production in most parts of the world. In addition, in the rural areas of Africa and Asia, leaves of jute plants are also consumed as a green leafy vegetable [71]. These leafy vegetables are rich with nutritious proteins, carbohydrates, lipids, fats, minerals, and some hormone precursors. Moreover, it is an indigenous leafy vegetable which has many important nutrients in the human diet and has the ability to tolerate stress conditions [60,62]. Jute leaves are important sources of some essential nutrients such as K, Mg, Fe, Cu, and Mn as well as some important pigments that are essential for human and animal nutrition. Among these essential elements, jute is also a good source of vitamins (A, C, and E) [92]. Furthermore, indigenous leafy vegetables can produce a high yield even under stress conditions and are easy to cultivate and can be used to monitor pest and disease control [62].

Woven jute fibres (a type of natural fibre) are used as comparison of carbon and glass fibres in polymer matrix composites. Jute is also used as iron natural fibres in Mercedes-Benz E-class cars. Furthermore, this unique natural plant has been used in many different components of vehicles such as insulation parts, Cpillar trim, and seat cushion parts. At a commercial scale, jute has been used as a material in packaging, for roping, and many other purposes. Among these applications, upcoming applications of natural jute fibres include building/construction, home/garden furniture, and the toy industry [62,68,72]. Hence, these are some industrial applications of jute which may applicable when jute is grown in metal-contaminated soil. The flow chart diagram of jute and its uses is shown in Figure 2.

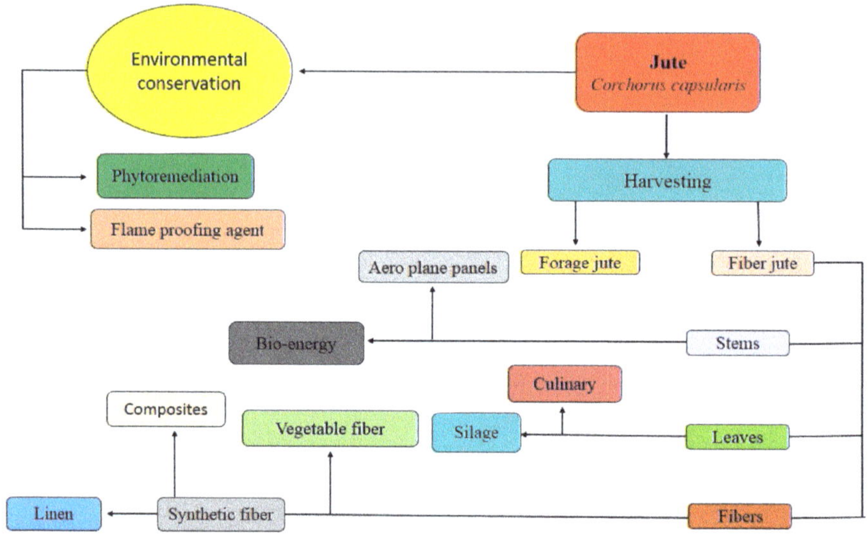

Figure 2. The flow chart diagram of jute and its uses.

6. Role of Chelating Agents in Assisting Phytoremediation of Heavy Metals

In some previous studies, chelating agents such as ethylenediaminetetraacetic acid (EDTA), Ethylenediamine-N,N'-disuccinic acid (EDDS), and citric acid (CA) have been used as soil extractants, and some chelates have also proved to be a source of micronutrient fertilizers that uphold the solubility of micronutrients in hydroponic solutions or contaminated soil [19,93–95]. In contrast, some plant species have the potential to generate organic acids in their rhizosphere, which may also act as a chelate, creating a complex known as an organic acid complex with metal interactions, and can take up a large amount of metal from the soil [52,53,96,97]. Moreover, chelates aid efficient metal phytoextraction but not their elimination, e.g. an increase in low molecular weight organic acids concentration in the rhizosphere provides carbon sources for soil microorganisms that facilitate metal mobilization from the soil to the plant by (a) replacing adsorbed metals at the surface of soil particles through ligand-exchange reactions, and (b) developing metal-organic complexes [97].

Hence using the chelators, we can increase the phytoextraction of metals in jute and also improved plant growth and biomass in jute plants. Although, a few literatures are available of phytoremediation potential of jute using the chelating agents but a lot of experiments has been conducted on different plant species under metal contaminated soil [26,51,86,98–101]. Niazy and Wahdan [27] studied jute in Pb-contaminated soil with the application of CA in the nutrient solution and found that application of CA is helpful jute by increasing plant growth and biomass and the phytoextraction of Pb when jute is grown in a Pb-contaminated nutrient mixture. Hence, application of different chelating agents may be helpful to jute plants to increase the phytoremediation potential when grown in metal-contaminated soil.

7. Conclusions and Future Prospects

Jute has considerable potential to tolerate metal-contaminated soil and accumulate a large amount of metals in its body parts. Moreover, it has the ability to survive under different environmental conditions with different growth types and can accumulate a large amount of metals in its harvestable parts. When using jute plants in metal-contaminated soil, full-scale investigations of the long-term phytoremediation of contaminated sediments is needed to evaluate the influence and the bioavailability of contaminates. Hence, the presented investigations on jute when grown in metal-contaminated soil can help researchers to select this plant as a phytoremediation tool. Besides phytoremediation, jute is a fibrous crop; when cultivated in metal-contaminated soil, jute can provide fibres which have many industrial and medical advantages. The soft, uniform, cheap, and lengthy fibres are the unique properties of jute to attract scientists to grow this species at a commercial level for the best economic status of a country. It seems that phytoremediation of heavy metals using jute is an efficient approach in modern agriculture all over the world. The utilization and importance of jute for its bast fibre have widely been documented, but several other aspects still require further study to decrease the fibre demand. For example, jute used as animal feed also needs to be studied because it might be a good prospective to decrease the demand for animal feed in urban areas. The utilization of jute as a culinary and medicinal herb also needs to be studied. Sonali bags (invented by Dr. Mubarak Ahmad Khan from Bangladesh) should be more common than regular polythene bags as these are biodegradable and eco-friendly. The extension of jute cultivation throughout the world also decreased when cultivated only for fibre purposes. The application of chelating agents is an innovative approach to increase the phytoremediation potential of a plant species. However, future research is needed to study the effects of chelating agents on the quality of the fibre from jute to assess the viability of the potential of jute application in the phytoremediation of metal contaminated soil. A lot of research is going on in the field on jute as a phytoremediation candidate, but a lot of effort is needed to explore the commercial utility of the natural fibre composites.

The following conclusions of the literature review are summarized:

- A few studies on the phytoremediation potential of jute have been published. These studies concluded that jute is a hyperaccumulator species of different heavy metals.
- The published articles also revealed that jute can absorb different heavy metals from every source, i.e., soil, water, and a field environment.
- Jute has a strong antioxidant defense system which can overcomes the effect of oxidative stress caused by metal toxicity.
- After phytoremediation, the biomass of jute can be used for the production of value-added by-products such as sheet, roof tiles, chairs, and tables.
- Application of chelating agents is effective. However, it not only increases phytoextraction of heavy metals but also improves plant growth and biomass, even under stress conditions.

Funding: This research was supported by Higher Education Commission, Pakistan.

Acknowledgments: This research was supported by Higher Education Commission, Pakistan.

Conflicts of Interest: The authors declare no conflict of interest.

References

1. Eissa, M.A.; Almaroai, Y.A. Phytoremediation Capacity of Some Forage Plants Grown on a Metals-Contaminated Soil. *Soil Sediment Contam. Int. J.* **2019**, *28*, 569–581. [CrossRef]
2. Mahar, A.; Wang, P.; Ali, A.; Awasthi, M.K.; Lahori, A.H.; Wang, Q.; Li, R.; Zhang, Z. Challenges and opportunities in the phytoremediation of heavy metals contaminated soils: A review. *Ecotoxicol. Environ. Saf.* **2016**, *126*, 111–121. [CrossRef]
3. Nagajyoti, P.C.; Lee, K.D.; Sreekanth, T. Heavy metals, occurrence and toxicity for plants: A review. *Environ. Chem. Lett.* **2010**, *8*, 199–216. [CrossRef]
4. Vardhan, K.H.; Kumar, P.S.; Panda, R.C. A review on heavy metal pollution, toxicity and remedial measures: Current trends and future perspectives. *J. Mol. Liq.* **2019**, *290*, 111197. [CrossRef]
5. Li, L.; Hou, M.; Cao, L.; Xia, Y.; Shen, Z.; Hu, Z. Glutathione S-transferases modulate Cu tolerance in Oryza sativa. *Environ. Exp. Bot.* **2018**, *155*, 313–320. [CrossRef]
6. Rehman, M.; Liu, L.; Wang, Q.; Saleem, M.H.; Bashir, S.; Ullah, S.; Peng, D. Copper environmental toxicology, recent advances, and future outlook: A review. *Environ. Sci. Pollut. Res.* **2019**, *26*, 18003–18016. [CrossRef]

7. Jiang, M.; Liu, S.; Li, Y.; Li, X.; Luo, Z.; Song, H.; Chen, Q. EDTA-facilitated toxic tolerance, absorption and translocation and phytoremediation of lead by dwarf bamboos. *Ecotoxicol. Environ. Saf.* **2019**, *170*, 502–512. [CrossRef]
8. Chen, L.; Wang, D.; Zeng, C.; Long, C. Improving Cobalt Phytoextraction by Astragalus Sinicus L. Grown in Co-Contaminated Soils Using Biodegradable Chelators. *Soil Sediment Contam. Int. J.* **2019**, *28*, 461–472. [CrossRef]
9. Li, L.; Zhang, K.; Gill, R.A.; Islam, F.; Farooq, M.A.; Wang, J.; Zhou, W. Ecotoxicological and Interactive Effects of Copper and Chromium on Physiochemical, Ultrastructural, and Molecular Profiling in Brassica napus L. *BioMed Res. Int.* **2018**, *2018*. [CrossRef]
10. Amna; Masood, S.; Syed, J.H.; Munis, M.F.H.; Chaudhary, H.J. Phyto-extraction of Nickel by Linum usitatissimum in Association with Glomus intraradices. *Int. J. Phytoremediation* **2015**, *17*, 981–987. [CrossRef]
11. ul Hassan, Z.; Ali, S.; Rizwan, M.; Hussain, A.; Akbar, Z.; Rasool, N.; Abbas, F. Role of zinc in alleviating heavy metal stress. In *Essential Plant Nutrients*; Springer: Berlin/Heidelberg, Germany, 2017; pp. 351–366.
12. Feller, U.; Anders, I.; Wei, S. Distribution and Redistribution of 109Cd and 65Zn in the Heavy Metal Hyperaccumulator Solanum nigrum L.: Influence of Cadmium and Zinc Concentrations in the Root Medium. *Plants* **2019**, *8*, 340. [CrossRef] [PubMed]
13. Fabbricino, M.; Ferraro, A.; Luongo, V.; Pontoni, L.; Race, M. Soil washing optimization, recycling of the solution, and ecotoxicity assessment for the remediation of Pb-contaminated sites using EDDS. *Sustainability* **2018**, *10*, 636. [CrossRef]
14. Adrees, M.; Ali, S.; Rizwan, M.; Zia-ur-Rehman, M.; Ibrahim, M.; Abbas, F.; Farid, M.; Qayyum, M.F.; Irshad, M.K. Mechanisms of silicon-mediated alleviation of heavy metal toxicity in plants: A review. *Ecotoxicol. Environ. Saf.* **2015**, *119*, 186–197. [CrossRef] [PubMed]
15. Tan, J.; He, S.; Yan, S.; Li, Y.; Li, H.; Zhang, H.; Zhao, L.; Li, L. Exogenous EDDS modifies copper-induced various toxic responses in rice. *Protoplasma* **2014**, *251*, 1213–1221. [CrossRef] [PubMed]
16. Xiao, T.; Mi, M.; Wang, C.; Qian, M.; Chen, Y.; Zheng, L.; Zhang, H.; Hu, Z.; Shen, Z.; Xia, Y. A methionine-R-sulfoxide reductase, OsMSRB5, is required for rice defense against copper toxicity. *Environ. Exp. Bot.* **2018**, *153*, 45–53. [CrossRef]
17. Adrees, M.; Ali, S.; Rizwan, M.; Ibrahim, M.; Abbas, F.; Farid, M.; Zia-ur-Rehman, M.; Irshad, M.K.; Bharwana, S.A. The effect of excess copper on growth and physiology of important food crops: A review. *Environ. Sci. Pollut. Res.* **2015**, *22*, 8148–8162. [CrossRef] [PubMed]
18. Al Naggar, Y.; Khalil, M.S.; Ghorab, M.A. Environmental pollution by heavy metals in the aquatic ecosystems of Egypt. *Open Acc. J. Toxicol.* **2018**, *3*, 555603.
19. Zaheer, I.E.; Ali, S.; Rizwan, M.; Farid, M.; Shakoor, M.B.; Gill, R.A.; Najeeb, U.; Iqbal, N.; Ahmad, R. Citric acid assisted phytoremediation of copper by Brassica napus L. *Ecotoxicol. Environ. Saf.* **2015**, *120*, 310–317. [CrossRef]
20. Wuana, R.A.; Okieimen, F.E. Heavy metals in contaminated soils: A review of sources, chemistry, risks and best available strategies for remediation. *Isrn Ecol.* **2011**, *2011*. [CrossRef]
21. Kamran, M.; Malik, Z.; Parveen, A.; Huang, L.; Riaz, M.; Bashir, S.; Mustafa, A.; Abbasi, G.H.; Xue, B.; Ali, U. Ameliorative Effects of Biochar on Rapeseed (Brassica napus L.) Growth and Heavy Metal Immobilization in Soil Irrigated with Untreated Wastewater. *J. Plant Growth Regul.* **2019**, 1–16. [CrossRef]
22. Chen, X.; Lu, X. Contamination characteristics and source apportionment of heavy metals in topsoil from an area in Xi'an city, China. *Ecotoxicol. Environ. Saf.* **2018**, *151*, 153–160. [CrossRef] [PubMed]
23. Chen, J.; Shafi, M.; Li, S.; Wang, Y.; Wu, J.; Ye, Z.; Peng, D.; Yan, W.; Liu, D. Copper induced oxidative stresses, antioxidant responses and phytoremediation potential of Moso bamboo (Phyllostachys pubescens). *Sci. Rep.* **2015**, *5*, 13554. [CrossRef] [PubMed]
24. Jan, R.; Khan, M.A.; Asaf, S.; Lee, I.-J.; Kim, K.M. Metal Resistant Endophytic Bacteria Reduces Cadmium, Nickel Toxicity, and Enhances Expression of Metal Stress Related Genes with Improved Growth of Oryza Sativa, via Regulating Its Antioxidant Machinery and Endogenous Hormones. *Plants* **2019**, *8*, 363. [CrossRef] [PubMed]
25. Merrington, G. The good, the bad and the ugly: Copper and arsenic in soils. In *Soil Health: the Foundation of Sustainable Agriculture*; Wollongbar Agricultural Institute: Wollongbar, Australia, 2018.

26. Saleem, M.H.; Fahad, S.; Rehman, M.; Saud, S.; Jamal, Y.; Khan, S.; Liu, L. Morpho-physiological traits, biochemical response and phytoextraction potential of short-term copper stress on kenaf (Hibiscus cannabinus L.) seedlings. *PeerJ* **2020**, *8*, e8321. [CrossRef]
27. Niazy Abdou, M.; Wahdan, M. *Citric Acid-Enhanced Phytoremediation of Lead Using Corchorus Capsularis, L, and Eucalyptus Camaldulensis*; ResearchGate: Berlin, Germany, 2017.
28. Xu, L.; Wang, W.; Guo, J.; Qin, J.; Shi, D.; Li, Y.; Xu, J. Zinc improves salt tolerance by increasing reactive oxygen species scavenging and reducing Na+ accumulation in wheat seedlings. *Biol. Plant.* **2014**, *58*, 751–757. [CrossRef]
29. Watanabe, M. Phytoremediation on the brink of commercialization. 1997. *Environ. Sci. Technol.* **1997**, *31*, 182–186. [CrossRef]
30. Vareda, J.P.; Durães, L. Efficient adsorption of multiple heavy metals with tailored silica aerogel-like materials. *Environ. Technol.* **2019**, *40*, 529–541. [CrossRef]
31. Muszynska, E.; Hanus-Fajerska, E. Why are heavy metal hyperaccumulating plants so amazing? *Biotechnol. J. Biotechnol. Comput. Biol. Bionanotechnol.* **2015**, *96*, 265–271. [CrossRef]
32. Zhang, X.; Zhao, X.; Li, B.; Xia, J.; Miao, Y. SRO1 regulates heavy metal mercury stress response in Arabidopsis thaliana. *Chin. Sci. Bull.* **2014**, *59*, 3134–3141. [CrossRef]
33. Ashraf, S.; Ali, Q.; Zahir, Z.A.; Ashraf, S.; Asghar, H.N. Phytoremediation: Environmentally sustainable way for reclamation of heavy metal polluted soils. *Ecotoxicol. Environ. Saf.* **2019**, *174*, 714–727. [CrossRef]
34. Ashraf, M.A.; Hussain, I.; Rasheed, R.; Iqbal, M.; Riaz, M.; Arif, M.S. Advances in microbe-assisted reclamation of heavy metal contaminated soils over the last decade: A review. *J. Environ. Manag.* **2017**, *198*, 132–143. [CrossRef] [PubMed]
35. Kołaciński, Z.; Rincón, J.; Szymański, T.; Sobiecka, E. Thermal plasma vitrification process as the effective technology for hospital incineration fly ash immobilization. In *Vitrification and Geopolimerization of Wastes for Immobilization or Recycling*; Universidad Miguel Hernandez: Alicante, Spain, 2017; Volume 51.
36. Saleem, M.; Ali, S.; Rehman, M.; Rana, M.; Rizwan, M.; Kamran, M.; Imran, M.; Riaz, M.; Hussein, M.; Elkelish, A.; et al. Influence of phosphorus on copper phytoextraction via modulating cellular organelles in two jute (Corchorus capsularis L.) varieties grown in a copper mining soil of Hubei Province, China. *Chemosphere* **2020**, *248*, 126032. [CrossRef]
37. Zhang, B.; Zheng, J.; Sharp, R. Phytoremediation in engineered wetlands: Mechanisms and applications. *Procedia Environ. Sci.* **2010**, *2*, 1315–1325. [CrossRef]
38. Yahaghi, Z.; Shirvani, M.; Nourbakhsh, F.; De La Pena, T.C.; Pueyo, J.J.; Talebi, M. Isolation and Characterization of Pb-Solubilizing Bacteria and Their Effects on Pb Uptake by Brassica juncea: Implications for Microbe-Assisted Phytoremediation. *J. Microbiol. Biotechnol.* **2018**, *28*, 1156–1167. [CrossRef] [PubMed]
39. Parmar, S.; Singh, V. Phytoremediation approaches for heavy metal pollution: A review. *J. Plant Sci. Res.* **2015**, *2*, 135.
40. Yadav, R.; Arora, P.; Kumar, S.; Chaudhury, A. Perspectives for genetic engineering of poplars for enhanced phytoremediation abilities. *Ecotoxicology* **2010**, *19*, 1574–1588. [CrossRef]
41. Sangeeta, M.; Maiti, S.K. Phytoremediation of metal enriched mine waste: A review. *Am. Eurasian J. Agric. Environ. Sci.* **2010**, *9*, 560–575.
42. Wiszniewska, A.; Hanus-Fajerska, E.; MUSZYŃSKA, E.; Ciarkowska, K. Natural organic amendments for improved phytoremediation of polluted soils: A review of recent progress. *Pedosphere* **2016**, *26*, 1–12. [CrossRef]
43. Akhtar, M.S.; Chali, B.; Azam, T. Bioremediation of arsenic and lead by plants and microbes from contaminated soil. *Res. Plant Sci.* **2013**, *1*, 68–73.
44. Saleem, M.H.; Fahad, S.; Khan, S.U.; Din, M.; Ullah, A.; Sabagh, A.E.; Hossain, A.; Llanes, A.; Liu, L. Copper-induced oxidative stress, initiation of antioxidants and phytoremediation potential of flax (Linum usitatissimum L.) seedlings grown under the mixing of two different soils of China. *Environ. Sci. Pollut. Res.* **2019**. [CrossRef]
45. Saleem, M.H.; Kamran, M.; Zhou, Y.; Parveen, A.; Rehman, M.; Ahmar, S.; Malik, Z.; Mustafa, A.; Anjum, R.M.A.; Wang, B. Appraising growth, oxidative stress and copper phytoextraction potential of flax (Linum usitatissimum L.) grown in soil differentially spiked with copper. *J. Environ. Manag.* **2020**, *257*, 109994. [CrossRef] [PubMed]

46. Imran, M.; Sun, X.; Hussain, S.; Ali, U.; Rana, M.S.; Rasul, F.; Saleem, M.H.; Moussa, M.G.; Bhantana, P.; Afzal, J. Molybdenum-Induced Effects on Nitrogen Metabolism Enzymes and Elemental Profile of Winter Wheat (Triticum aestivum L.) Under Different Nitrogen Sources. *Int. J. Mol. Sci.* **2019**, *20*, 3009. [CrossRef] [PubMed]
47. Saleem, M.H.; Fahad, S.; Khan, S.U.; Ahmar, S.; Khan, M.H.U.; Rehman, M.; Maqbool, Z.; Liu, L. Morpho-physiological traits, gaseous exchange attributes, and phytoremediation potential of jute (Corchorus capsularis L.) grown in different concentrations of copper-contaminated soil. *Ecotoxicol. Environ. Saf.* **2019**, *189*, 109915. [CrossRef] [PubMed]
48. Liu, X.; Fu, J.W.; Da Silva, E.; Shi, X.X.; Cao, Y.; Rathinasabapathi, B.; Chen, Y.; Ma, L.Q. Microbial siderophores and root exudates enhanced goethite dissolution and Fe/As uptake by As-hyperaccumulator Pteris vittata. *Environ. Pollut.* **2017**, *223*, 230–237. [CrossRef] [PubMed]
49. Usman, K.; Al-Ghouti, M.A.; Abu-Dieyeh, M.H. Phytoremediation: Halophytes as Promising Heavy Metal Hyperaccumulators. *Heavy Met.* **2018**. [CrossRef]
50. Rehman, M.; Liu, L.; Bashir, S.; Saleem, M.H.; Chen, C.; Peng, D.; Siddique, K.H. Influence of rice straw biochar on growth, antioxidant capacity and copper uptake in ramie (Boehmeria nivea L.) grown as forage in aged copper-contaminated soil. *Plant Physiol. Biochem.* **2019**, *138*, 121–129. [CrossRef]
51. Najeeb, U.; Xu, L.; Ali, S.; Jilani, G.; Gong, H.; Shen, W.; Zhou, W. Citric acid enhances the phytoextraction of manganese and plant growth by alleviating the ultrastructural damages in Juncus effusus L. *J. Hazard. Mater.* **2009**, *170*, 1156–1163. [CrossRef]
52. Kanwal, U.; Ali, S.; Shakoor, M.B.; Farid, M.; Hussain, S.; Yasmeen, T.; Adrees, M.; Bharwana, S.A.; Abbas, F. EDTA ameliorates phytoextraction of lead and plant growth by reducing morphological and biochemical injuries in Brassica napus L. under lead stress. *Environ. Sci. Pollut. Res.* **2014**, *21*, 9899–9910. [CrossRef]
53. Sidhu, G.P.S.; Bali, A.S.; Singh, H.P.; Batish, D.R.; Kohli, R.K. Ethylenediamine disuccinic acid enhanced phytoextraction of nickel from contaminated soils using Coronopus didymus (L.) Sm. *Chemosphere* **2018**, *205*, 234–243. [CrossRef]
54. Jiang, L.Y.; Yang, X.; He, Z. Growth response and phytoextraction of copper at different levels in soils by Elsholtzia splendens. *Chemosphere* **2004**, *55*, 1179–1187. [CrossRef]
55. Saleem, M.H.; Ali, S.; Seleiman, M.F.; Rizwan, M.; Rehman, M.; Akram, N.A.; Liu, L.; Alotaibi, M.; Al-Ashkar, I.; Mubushar, M. Assessing the Correlations between Different Traits in Copper-Sensitive and Copper-Resistant Varieties of Jute (Corchorus capsularis L.). *Plants* **2019**, *8*, 545. [CrossRef] [PubMed]
56. Ogunkunle, C.O.; Ziyath, A.M.; Adewumi, F.E.; Fatoba, P.O. Bioaccumulation and associated dietary risks of Pb, Cd, and Zn in amaranth (Amaranthus cruentus) and jute mallow (Corchorus olitorius) grown on soil irrigated using polluted water from Asa River, Nigeria. *Environ. Monit. Assess.* **2015**, *187*, 281. [CrossRef] [PubMed]
57. Uddin Nizam, M.; Mokhlesur Rahman, M.; Kim, J.E. Phytoremediation Potential of Kenaf (Hibiscus cannabinus L.), Mesta (Hibiscus sabdariffa L.), and Jute (Corchorus capsularis L.) in Arsenic-contaminated Soil. *Korean J. Environ. Agric.* **2016**, *35*, 111–120. [CrossRef]
58. Abubakari, M.; Moomin, A.; Nyarko, G.; Dawuda, M. Heavy metals concentrations and risk assessment of roselle and jute mallow cultivated with three compost types. *Ann. Agric. Sci.* **2017**, *62*, 145–150. [CrossRef]
59. Uddin, M.N.; Wahid-Uz-Zaman, M.; Rahman, M.M.; Islam, M.S.; Islam, M.S. Phytoremediation Potentiality of Lead from Contaminated Soils by Fibrous Crop Varieties. *Am. J. Appl. Sci. Res.* **2016**, *2*, 22.
60. Ahmed, D.A.; Slima, D.F. Heavy metal accumulation by Corchorus olitorius L. irrigated with wastewater. *Environ. Sci. Pollut. Res.* **2018**, *25*, 14996–15005. [CrossRef] [PubMed]
61. Faruk, O.; Bledzki, A.K.; Fink, H.-P.; Sain, M. Biocomposites reinforced with natural fibers: 2000–2010. *Prog. Polym. Sci.* **2012**, *37*, 1552–1596. [CrossRef]
62. Choudhary, S.B.; Sharma, H.K.; Karmakar, P.G.; Kumar, A.; Saha, A.R.; Hazra, P.; Mahapatra, B.S. Nutritional profile of cultivated and wild jute ('Corchorus') species. *Aust. J. Crop Sci.* **2013**, *7*, 1973.
63. Thiruchitrambalam, M.; Athijayamani, A.; Sathiyamurthy, S.; Thaheer, A.S.A. A review on the natural fiber-reinforced polymer composites for the development of roselle fiber-reinforced polyester composite. *J. Nat. Fibers* **2010**, *7*, 307–323. [CrossRef]
64. Saleem, M.H.; Rehman, M.; Zahid, M.; Imran, M.; Xiang, W.; Liu, L. Morphological changes and antioxidative capacity of jute (Corchorus capsularis, Malvaceae) under different color light-emitting diodes. *Braz. J. Bot.* **2019**. [CrossRef]

65. Kozlowski, R.; Mieleniak, B.; Helwig, M.; Przepiera, A. Flame resistant lignocellulosic-mineral composite particleboards. *Polym. Degrad. Stab.* **1999**, *64*, 523–528. [CrossRef]
66. Bhattacharya, S.; Guha, G.; Gupta, K.; Chattopadhyay, D.; Mukhopadhyay, A.; Ghosh, U.C. Trend of arsenic pollution and subsequent bioaccumulation in Oryza sativa and Corchorus capsularis in Bengal Delta. *Int. Lett. Nat. Sci.* **2014**, *16*. [CrossRef]
67. Osundiya, M.; Ayejuyo, O.; Olowu, R.; Bamgboye, O.; Ogunlola, A. Bioaccumulation of heavy metals in frequently consumed leafy vegetable grown along Nigeria-Benin Seme Border, West Africa. *Pelagia Res. Libr.* **2014**, *5*, 1–7.
68. Mahapatra, B.; Mitra, S.; Ramasubramanian, T.; Sinha, M. Research on jute (Corchorus olitorius and C. capsularis) and kenaf (Hibiscus cannabinus and H. sabdariffa): Present status and future perspective. *Indian J. Agric. Sci.* **2014**, *79*, 951–967.
69. Mohanty, A.; Misra, M. Studies on jute composites—A literature review. *Polym. Plast. Technol. Eng.* **1995**, *34*, 729–792. [CrossRef]
70. Haghdan, S.; Smith, G.D. Natural fiber reinforced polyester composites: A literature review. *J. Reinf. Plast. Compos.* **2015**, *34*, 1179–1190. [CrossRef]
71. Ndlovu, J.; Afolayan, A. Nutritional analysis of the South African wild vegetable Corchorus olitorius L. *Asian J. Plant Sci.* **2008**, *7*, 615–618.
72. Singh, H.; Singh, J.I.P.; Singh, S.; Dhawan, V.; Tiwari, S.K. A Brief Review of Jute Fibre and Its Composites. *Mater. Today Proc.* **2018**, *5*, 28427–28437. [CrossRef]
73. Islam, M.; Rahman, M. Hand book on agricultural technologies of Jute, Kenaf and Mesta crops. *Bangladesh Jute Res. Inst. Dhaka* **2008**, *2*.
74. Sarwar, N.; Imran, M.; Shaheen, M.R.; Ishaque, W.; Kamran, M.A.; Matloob, A.; Rehim, A.; Hussain, S. Phytoremediation strategies for soils contaminated with heavy metals: Modifications and future perspectives. *Chemosphere* **2017**, *171*, 710–721. [CrossRef]
75. Khan, M.N.; Zhang, J.; Luo, T.; Liu, J.; Ni, F.; Rizwan, M.; Fahad, S.; Hu, L. Morpho-physiological and biochemical responses of tolerant and sensitive rapeseed cultivars to drought stress during early seedling growth stage. *Acta Physiol. Plant.* **2019**, *41*, 25. [CrossRef]
76. Kamran, M.; Parveen, A.; Ahmar, S.; Malik, Z.; Hussain, S.; Chattha, M.S.; Saleem, M.H.; Adil, M.; Heidari, P.; Chen, J.T. An Overview of Hazardous Impacts of Soil Salinity in Crops, Tolerance Mechanisms, and Amelioration through Selenium Supplementation. *Int. J. Mol. Sci.* **2019**, *21*, 148. [CrossRef] [PubMed]
77. Usman, A.R.; Lee, S.S.; Awad, Y.M.; Lim, K.J.; Yang, J.E.; Ok, Y.S. Soil pollution assessment and identification of hyperaccumulating plants in chromated copper arsenate (CCA) contaminated sites, Korea. *Chemosphere* **2012**, *87*, 872–878. [CrossRef] [PubMed]
78. Qadir, M.; Oster, J. Vegetative bioremediation of calcareous sodic soils: History, mechanisms, and evaluation. *Irrig. Sci.* **2002**, *21*, 91–101.
79. Sharma, H.D.; Reddy, K.R. *Geoenvironmental Engineering: Site Remediation, Waste Containment, and Emerging Waste Management Technologies*; John Wiley & Sons: Hoboken, NJ, USA, 2004.
80. Laghlimi, M.; Baghdad, B.; El Hadi, H.; Bouabdli, A. Phytoremediation mechanisms of heavy metal contaminated soils: A review. *Open J. Ecol.* **2015**, *5*, 375. [CrossRef]
81. Malik, R.N.; Husain, S.Z.; Nazir, I. Heavy metal contamination and accumulation in soil and wild plant species from industrial area of Islamabad, Pakistan. *Pak. J. Bot.* **2010**, *42*, 291–301.
82. Sinha, S.; Mishra, R.; Sinam, G.; Mallick, S.; Gupta, A. Comparative evaluation of metal phytoremediation potential of trees, grasses, and flowering plants from tannery-wastewater-contaminated soil in relation with physicochemical properties. *Soil Sediment Contam. Int. J.* **2013**, *22*, 958–983. [CrossRef]
83. Haque, S.; Ferdous, A.S.; Sarker, S.; Islam, M.; Hossain, K.; Khan, H. Identification and expression profiling of microRNAs and their corresponding targets related to phytoremediation of heavy metals in jute (Corchorus olitorius var. O-9897). *Biores. Comm.* **2016**, *2*, 152–157.
84. Nwaichi, E.O.; Dhankher, O.P. Heavy metals contaminated environments and the road map with phytoremediation. *J. Environ. Prot.* **2016**, *7*, 41. [CrossRef]
85. Oyedele, D.; Asonugho, C.; Awotoye, O. Heavy metals in soil and accumulation by edible vegetables after phosphate fertilizer application. *Electron. J. Environ. Agric. Food Chem.* **2006**, *5*, 1446–1453.
86. Anwer, S.; Ashraf, M.Y.; Hussain, M.; Ashraf, M.; Jamil, A. Citric acid mediated phytoextraction of cadmium by maize (Zea mays L.). *Pak. J. Bot.* **2012**, *44*, 1831–1836.

87. Azhar, N.; Ashraf, M.Y.; Hussain, M.; Hussain, F. Phytoextraction of lead (Pb) by EDTA application through sunflower (Helianthus annuus L.) cultivation: Seedling growth studies. *Pak. J. Bot.* **2006**, *38*, 1551–1560.
88. Mazen, A. Accumulation of four metals in tissues of Corchorus olitorius and possible mechanisms of their tolerance. *Biol. Plant.* **2004**, *48*, 267–272. [CrossRef]
89. Hassan, M.; Dagari, M.; Babayo, A. Effect of citric acid on cadmium ion uptake and stress response of hydroponically grown jute mallow (Corchorus olitorius). *J. Environ. Anal. Toxicol.* **2016**, *6*, 375.
90. Oguntade, O.A.; Olagbenro, T.S.; Odusanya, O.A.; Olagunju, S.O.; Adewusi, K.M.; Adegoke, A.T. Assessment of composted kitchen waste and poultry manure amendments on growth, yield and heavy metal uptake by Jute mallow Corchorus olitorius Linn. *Int. J. Recycl. Org. Waste Agric.* **2019**, *8*, 187–195. [CrossRef]
91. Ogoko, E. Accumulation of heavy metal in soil and their transfer to leafy vegetables with phytoremediation potential. *Am. J. Chem.* **2015**, *5*, 125–131.
92. Idirs, S.; Yisa, J.; Ndamitso, M. Nutritional composition of Corchorus olitorius leaves. *Anim. Prod. Res. Adv.* **2009**, *5*. [CrossRef]
93. Nörtemann, B. Biodegradation of chelating agents: EDTA, DTPA, PDTA, NTA, and EDDS. *Biogeochem. Chelating Agents* **2005**, *910*, 150–170. [CrossRef]
94. Habiba, U.; Ali, S.; Farid, M.; Shakoor, M.B.; Rizwan, M.; Ibrahim, M.; Abbasi, G.H.; Hayat, T.; Ali, B. EDTA enhanced plant growth, antioxidant defense system, and phytoextraction of copper by Brassica napus L. *Environ. Sci. Pollut. Res.* **2015**, *22*, 1534–1544. [CrossRef]
95. Afshan, S.; Ali, S.; Bharwana, S.A.; Rizwan, M.; Farid, M.; Abbas, F.; Ibrahim, M.; Mehmood, M.A.; Abbasi, G.H. Citric acid enhances the phytoextraction of chromium, plant growth, and photosynthesis by alleviating the oxidative damages in Brassica napus L. *Environ. Sci. Pollut. Res.* **2015**, *22*, 11679–11689. [CrossRef]
96. Mohsin, M.; Kuittinen, S.; Salam, M.M.A.; Peräniemi, S.; Laine, S.; Pulkkinen, P.; Kaipiainen, E.; Vepsäläinen, J.; Pappinen, A. Chelate-assisted phytoextraction: Growth and ecophysiological responses by Salix schwerinii EL Wolf grown in artificially polluted soils. *J. Geochem. Explor.* **2019**, *205*, 106335. [CrossRef]
97. Kwak, J.h.; Park, K.; Chang, P.c.; Liu, W.; Kim, J.Y.; Kim, K.W. Influence of phosphate and citric acid on the phytoextraction of As from contaminated soils. *Int. J. Environ. Waste Manag.* **2013**, *11*, 1–12. [CrossRef]
98. Han, Y.; Zhang, L.; Gu, J.; Zhao, J.; Fu, J. Citric acid and EDTA on the growth, photosynthetic properties and heavy metal accumulation of Iris halophila Pall. cultivated in Pb mine tailings. *Int. Biodeterior. Biodegrad.* **2018**, *128*, 15–21. [CrossRef]
99. Farid, M.; Ali, S.; Shakoor, M.B.; Bharwana, S.A.; Rizvi, H.; Ehsan, S.; Tauqeer, H.M.; Iftikhar, U.; Hannan, F. EDTA assisted phytoremediation of cadmium, lead and zinc. *Int. J. Agron. Plant Prod.* **2013**, *4*, 2833–2846.
100. Ali, S.Y.; Chaudhury, S. EDTA-enhanced phytoextraction by Tagetes sp. and effect on bioconcentration and translocation of heavy metals. *Environ. Processes* **2016**, *3*, 735–746. [CrossRef]
101. Ehsan, S.; Ali, S.; Noureen, S.; Mahmood, K.; Farid, M.; Ishaque, W.; Shakoor, M.B.; Rizwan, M. Citric acid assisted phytoremediation of cadmium by Brassica napus L. *Ecotoxicol. Environ. Saf.* **2014**, *106*, 164–172. [CrossRef]

MDPI
St. Alban-Anlage 66
4052 Basel
Switzerland
Tel. +41 61 683 77 34
Fax +41 61 302 89 18
www.mdpi.com

Plants Editorial Office
E-mail: plants@mdpi.com
www.mdpi.com/journal/plants

www.ingramcontent.com/pod-product-compliance
Lightning Source LLC
Chambersburg PA
CBHW061757210926
51850CB00006B/577
9783039361984